Green Ethics and Philosophy

Green Ethics and Philosophy

An A-to-Z Guide

Julie Newman, General Editor
Yale University

Paul Robbins, Series Editor
University of Arizona

Los Angeles | London | New Delhi
Singapore | Washington DC

Los Angeles | London | New Delhi
Singapore | Washington DC

FOR INFORMATION:

SAGE Publications, Inc.
2455 Teller Road
Thousand Oaks, California 91320
E-mail: order@sagepub.com

SAGE Publications Ltd.
1 Oliver's Yard
55 City Road
London EC1Y 1SP
United Kingdom

SAGE Publications India Pvt. Ltd.
B 1/I 1 Mohan Cooperative Industrial Area
Mathura Road, New Delhi 110 044
India

SAGE Publications Asia-Pacific Pte. Ltd.
33 Pekin Street #02-01
Far East Square
Singapore 048763

Publisher: Rolf A. Janke
Assistant to the Publisher: Michele Thompson
Senior Editor: Jim Brace-Thompson
Production Editors: Kate Schroeder, Tracy Buyan
Reference Systems Manager: Leticia Gutierrez
Reference Systems Coordinator: Laura Notton
Typesetter: C&M Digitals (P) Ltd.
Proofreader: Rae-Ann Goodwin
Indexer: Jeanne Busemeyer
Cover Designer: Gail Buschman
Marketing Manager: Kristi Ward

Golson Media
President and Editor: J. Geoffrey Golson
Author Manager: Ellen Ingber
Editors: Mary Jo Scibetta, Kenneth Heller
Copy Editors: Tricia Lawrence, Barbara Paris

Printed in the United States of America

Library of Congress Cataloging-in-Publication Data

Green ethics and philosophy : an A-to-Z guide / Julie Newman, editor.

p. cm. — (The Sage reference series on green society: toward a sustainable future)
Includes bibliographical references and index.

ISBN 978-1-4129-9687-7 (cloth) — ISBN 978-1-4129-7460-8 (ebk)

1. Environmental ethics. 2. Human ecology—Philosophy. I. Newman, Julie.

GE42.G735 2011 178—dc22 2011006529

11 12 13 14 15 10 9 8 7 6 5 4 3 2 1

Contents

About the Editors

Green Series Editor: Paul Robbins

Paul Robbins is a professor and the director of the University of Arizona School of Geography and Development. He earned his Ph.D. in Geography in 1996 from Clark University. He is General Editor of the *Encyclopedia of Environment and Society* (2007) and author of several books including *Environment and Society: A Critical Introduction* (2010), *Lawn People: How Grasses, Weeds, and Chemicals Make Us Who We Are* (2007), and *Political Ecology: A Critical Introduction* (2004).

Robbins's research focuses on the relationships between individuals (homeowners, hunters, professional foresters), environmental actors (lawns, elk, mesquite trees), and the institutions that connect them. He and his students seek to explain human environmental practices and knowledge, the influence nonhumans have on human behavior and organization, and the implications these interactions hold for ecosystem health, the local community, and social justice. Past projects have examined chemical use in the suburban United States, elk management in Montana, forest product collection in New England, and wolf conservation in India.

Green Ethics and Philosophy General Editor: Julie Newman

Julie Newman, Ph.D., has worked in the field of sustainable development and campus sustainability since 1993. Her research has focused on the role of decision-making processes and organizational behavior in institutionalizing sustainability into higher education. In 2004, Newman was recruited to be the founding director of the Office of Sustainability for Yale University. At Yale, Newman also holds a lecturer appointment with the Yale School of Forestry and Environmental Studies, where she teaches an undergraduate course titled Sustainability: From Innovation to Transformation in Institutions. Prior to her work at Yale, Newman assisted in the establishment of the longest-standing sustainability office in the United States—the University of New Hampshire Office of Sustainability. Newman is a pioneer in the field of campus sustainability, beginning in 1995 when she worked for University Leaders for a Sustainable Future (ULSF) while a graduate student at Tufts University. In 2004, Newman cofounded the Northeast Campus Sustainability Consortium to advance education and action for sustainable development on university campuses in the northeast and maritime region. This has led to a 10-year regional commitment and a set of annual meetings for sustainability professionals in the northeast and maritime region. She also co-coordinates a sustainability working group for the International Alliance of

Research Universities and a sustainability working group for the Council of Ivy Presidents. In addition, Newman is a member of the editorial board of *Sustainability: Journal of Record*. Newman lectures at and consults for universities both nationally and internationally, participates on a variety of boards and advisory committees, and has contributed to a series of edited books and peer-reviewed journals. She holds a B.S. from the University of Michigan, an M.S. from Tufts University, and a Ph.D. from the University of New Hampshire.

Introduction

The discussion about the consequences of human impacts on the environment is as old as civilization. Obtaining clean water and keeping water sources clean for continued supply is a human concern as old and fundamental as obtaining food and shelter. Ancient cultures had to deal with the effect of the weather on their irrigation systems and aqueducts and the effects of those engineering projects on the water cycle and ecosystem. They had waste to dispose of, from human and animal waste to food waste to the by-products of early industries like tanning, mining, and soap making. By the Middle Ages, legal and scientific writers were dealing with the problems of air pollution, contaminated water, waste management, and the effects of soil contamination on agriculture and famine. As civilization and technology became more sophisticated, standards of cleanliness and responsibility became higher and more complicated: just as technology made visible the bacteria and harmful chemicals that might lie in otherwise clean-seeming water, our increased understanding of the world made visible the long-term consequences of our impacts on it, raising concerns about erosion and other local environment change, wildfire vulnerability, biodiversity, nonrenewable resources, carcinogenicity and genetic damage, and, of course, pollution in its myriad expressions.

Humanity's discussions of environmental concerns have for thousands of years debated the allocation of responsibility, the need for preventative measures, the role of the human species in the biosphere, and various issues of fairness and justice as they pertain to what economists call externalities. Externalities are the costs or benefits produced by a transaction but not accounted for in its prices, incurred by parties outside the transaction; pollution, for instance, results from the business of a factory but affects everyone regardless of their dealings with that factory. This becomes even more ethically complicated when the transactors can shield themselves from the externality, as when a company operates a factory that pollutes the soil and water of a region in which the soil and water resources don't matter to them (though they may matter very much to the factory's workers).

But until recently, the world's major schools of philosophical and ethical thought have primarily not been directly concerned with what we now call green issues. Rather, these matters came up in frameworks concerned with other ends. Utilitarianism, for instance, provides a framework in which green issues can be addressed, but is not specifically formulated as a green school of thought and is as often used to justify polluting behavior. Christian and Jewish schools of thought may make much of mankind's role as steward of the Earth but can come to a number of different conclusions about the behavior that role should or can involve.

Modern ecological thought begins perhaps with the scientific forest management pioneered in Europe in the 17th century and developed more fully in the 19th century by foresters in the United States and India. These early conservation efforts were aimed at "keeping nature's household," encouraging tree growth while protecting the forest from fires. The rise of progressivism in the United States at the end of the 19th century led to discussions about the proper relationships between government, business, and the natural world. While the pure capitalists held that property owners could do whatever they liked with that property, their opponents were divided between the environmentalists and the conservationists. Environmentalists, whose position was argued most prominently by Sierra Club founder John Muir (1838–1914), held nature as sacred unto itself and opposed development, treating it as a necessary evil at best and one to be avoided in public lands. Environmentalists particularly opposed the use of public lands (such as national parks) for logging and timber cutting and the building of dams, which destroy ecosystems. Both of these, on the other hand, were supported by the conservationists, whose strongest voice was that of President Theodore Roosevelt, who despite his disagreement with the environmentalists was perhaps the greatest advocate of green thinking in the history of the presidency. While he supported exploiting natural resources, he was one of the only presidents to make a high national priority the responsible use of those resources and established the first American national forests, the United States Forest Service, the first American bird reserve, and five new national parks. But equally as important, historically, as his actions were the attention he brought to the debate.

Of course, during this time, the nature of mankind's relationship with the environment changed. Wood, peat, and dung were replaced by fossil and fissile fuels, which may someday be replaced by solar, wind, or ocean power. The world's population expanded considerably, no longer fettered by rampant infant mortality or the diseases that in the past had killed so many children and young adults. Frontier zones filled up with cities, and previously unspoiled areas of wilderness were razed, developed, strip-mined, or turned into tourist attractions. There was no organized agenda to convert the raw material of the wilderness into the processed product of civilization, and had there been, there would have been more opportunity for a discussion of methods and means. The process was as natural as locusts stripping greenery from a landscape and moving on in search of more, except that Man is the reasoning locust, able to internalize, model, and debate the consequences of his appetites.

In the 1930s, the effects of mankind's exploitation of nature were made clear by the Dust Bowl, which resulted in one of the greatest population migrations in American history. Great Plains farms settled during a particularly wet period dried up and were abandoned when drought struck during the Great Depression. Agriculture having killed off the native prairie grasses, there was nothing to prevent soil loss once those farms' crops died off, and the winds of dust storms scooped up a foot or more of loose, light, dried-out topsoil, creating storm systems that blew across the middle of the country and churned enough particulate matter into the atmosphere that months later, the snow in the northeast was a dirty rusty brown. But it wasn't until the 1960s and 1970s that green concerns took center stage again, as concerns about pesticides and other chemicals, climate change, overlogging, acid rain, smog, carcinogens, and other effects of pollution and anthropogenic climate impact became part of the national conversation. In the political sphere, this led to the creation of the Environmental Protection Agency and various forms of environmentalist legislation, as well as debates about the relationship between the private and public sectors and their impact on the natural world.

Much of this discussion centered in its early moments around the 1962 publication of *Silent Spring* by marine biologist Rachel Carson, who soon became a household name, referenced everywhere from *Peanuts* to Johnny Carson's monologues. Already an established popular writer of natural history, with *Silent Spring* Carson became a critic and advocate, documenting the effects of pesticides on the environment and especially on birds. She challenged the chemicals industry head-on and found an audience in all political walks of life. Though she met with opposition—green issues had not yet been as highly politicized as they are today—within the first couple of years after publication, her views had been mainstreamed, drawing comparisons to Upton Sinclair's *The Jungle*.

Similar works followed from other authors, founding the body of modern environmental literature and green philosophy. A lecture by history professor Lynn White, Jr., "The Historical Roots of Our Ecological Crisis," delivered in 1966 and published in *Science* the following year, popularized the idea of the environmental crisis as an inevitable outcome of medieval Western Christianity, which provided a philosophical framework that redefined the relationship between mankind and the natural world. In White's view, Judeo-Christian theology, especially as it was formulated in the Middle Ages, supported anthropocentrism and the elevation of Man—God's favorite creation, gifted with a soul and reason—over the rest of creation, which serves at his pleasure. While many early Christian and Jewish movements emphasized harmony with one's natural surroundings, this medieval view led to an indifference to nature, one so deep that even many environmentalist arguments were phrased in terms of avoiding ill effects on mankind rather than considering the health of the natural world as a good unto itself. White's work did not have the public impact of Carson's, concerned as it was with general principles, trends, and intellectual history rather than specific contemporary issues. But the impact it had on green philosophy, environmental thought, and religious thought continues to be felt more than 40 years later. The perceived accusation that Christianity is complicit in the destruction of the environment strengthened the resolve of Christian environmentalists both in the political sphere (former vice president and senator Al Gore) and the academic/theological sphere, where "ecotheology" has become a school of religious thought both in Christianity and in Judaism. Ecotheology is an exploration of the relationship between religious principles and mankind's relationship with (and generally, stewardship of) nature, and religious groups with a strong ecotheological focus have founded a number of outreach programs in the developing world in which humanitarian aid is strongly informed by an environmental sensitivity and aims toward sustainability. There is in some cases an overlap between ecotheology and the Gaia theory, but it's important to note that ecotheology need not challenge mainstream Christian or Jewish beliefs, but rather explores what constitutes a responsible attitude toward the Earth, from a religious perspective. Various groups invested in ecotheology come down on different sides of the evolution/creation debate, which sometimes impacts the articulation of their ideas.

White's work also influenced the formulation of "Deep Ecology," a phrase coined in 1973 by Norwegian philosopher Arne Naess. Though metaphysical, deep ecology is not inherently theological or supernatural but is a response to the perceived need for "ecological wisdom" in guiding mankind's relationship with the natural world. Naess particularly rejected the use of the existence of a human soul and human reason to justify the elevation of the human species above other animals and argued that all forms of life have a right to live, a right that cannot be quantified and weighted in relative terms any more than citizens' rights can be expressed fractionally. Deep ecology has provided a philosophical basis for some environmental activists and ecologists, particularly advocating the ideas that ecosystems can only tolerate limited, gradual change, and that human activity is complicit in mass extinctions. For many, there is a spiritual element to deep ecology and a rejection of the Judeo-Christian notion of "stewardship" as, in Naess's words, "arrogance."

The journal *Environmental Ethics* (*EE*) was founded in 1979 by philosophy professor Eugene C. Hargrove, and for a long time, the name of the journal doubled as the name of the field this volume is devoted to. Published by the University of North Texas's Center for Environmental Philosophy and now made available online through the Philosophy Documentation Center, *EE*'s recent articles demonstrate the breadth of the field: Cecilia Wee on "Mencius and the Natural Environment," Aaron Simmons on "Do Animals Have an Interest in Continued Life?," Piers H. G. Stephens on "Towards a Jamesian Environmental Philosophy," Stephen Quilley on "The Land Ethic as an Ecological Civilizing Process," Matthew Hall on "Plant Autonomy and Human-Plant Ethics," Rita Turner on "The Discursive Construction of Anthropocentrism," and Ben A. Minteer on "Biocentric Farming."

When the Deepwater Horizon oil spill of 2010 began, the most remarkable thing about it, illustrated by the highly politicized discussions that followed, was how little could be done, how few immediate solutions suggested themselves. The seemingly simple act of plugging a man-made hole lay beyond the everyday reach of our technology, again raising the question of whether our ability to damage the world had outstripped our ability to repair it, as it had during the Dust Bowl—never mind our interest or desire in doing so, but simply our ability. It is the continued occurrence of events like this that provide a focus to the ongoing discussions of green ethics, the proving ground for principles of green philosophy, as covered in the material that follows.

The Editors

Reader's Guide

Consumption Ethics and Philosophy (Public/Private)

Carbon Offsets
Consumption, Business Ethics and
Consumption, Consumer Ethics and
Ecological Footprint
Ethical Vegetarianism
Local Food Movement
Marketing, Consumption Ethics and
Organic Trend
Sustainability, Business Ethics and
Sustainability, Consumer Ethics and
Western "Way of Life"

Ethics and Philosophy in Practice

Adaptive Management
Agriculture
Biodiversity
Bright Green Environmentalism
Business Ethics, Shades of Green
China
Conservation
Cost-Benefit Analysis
Democracy
Development, Ethical Sustainability and
Ecological Restoration
Ecology
Ecopedagogy and Ecodidactics
Ecopolitics
Environmental Justice
Environmental Policy
Ethics and Science
Genetic Engineering
Globalization
Green Liberalism
Green Party, German
Greenwashing
Precautionary Principle
Preservation
Sierra Club
Social Ecology
Sustainability and Spiritual Values
Technology
United Nations Environment Programme
Urbanization

Ethics and Philosophy in Theory

Animal Ethics
Anthropocentrism
Biocentric Egalitarianism
Biocentrism
Civic Environmentalism
Climate Ethics
Deep Ecology
Deep Green Theory
Ecocentrism
Ecofascism
Ecofeminism/Ecological Feminism
Economism
Ecophenomenology
Environmental Pluralism
Gaia Hypothesis
Green Party Ethical System, Four Pillars of the
Hannover Principles
Human Values and Sustainability
Instrumental Value

Green Ethicists and Philosophers

History of Green Ethics and Philosophy

Law and Green Ethics

List of Articles

List of Contributors

Ackom, Emmanuel Kofi
University of British Columbia

Ahteensuu, Marko
University of Turku

Ard, Kerry
University of Michigan

Arney, Jo
University of Wisconsin–La Crosse

Barry, John
Queen's University Belfast

Booth, Kelvin J.
Thompson Rivers University

Boslaugh, Sarah
Washington University in St. Louis

Boudes, Philippe
Paris 7 University, LADYSS-CNRS

Branch, Matthew
Pennsylvania State University

Bustos, Keith
University of St Andrews

Byrne, Jason
Griffith University

Certomà, Chiara
Sant'Anna School of Advanced Studies

Collier, William M.
Independent Scholar

Collins, Timothy
Western Illinois University

Cooley, Amanda Harmon
South Texas College of Law

Cruger, Katherine M.
University of Colorado, Boulder

Danielson, Stentor
Slippery Rock University

Davis, Edward F.
Knox College

de Souza, Lester
Independent Scholar

Dedekorkut, Aysin
Griffith University

Edelglass, William
Marlboro College

Erhard, Nancie
Saint Mary's University, Halifax

Fiskio, Janet
Oberlin College

Fletcher, Jonathan R.
Knox College

Forbes, William
Stephen F. Austin State University

Groves, Chris
Cardiff University

Gunn, Alastair S.
University of Waikato

Hayden, Anders
Dalhousie University

Helfer, Jason A.
Knox College

Heller, Matthew E.
University of Colorado, Boulder

Hicks, Stephen
Western Illinois University

Howes, Michael
Griffith University

Islam, Md Saidul
Nanyang Technological University

Jarvie, Michelle E.
Independent Scholar

Joldersma, Clarence W.
Calvin College

Kassiola, Joel Jay
San Francisco State University

Kte'pi, Bill
Independent Scholar

Lamb, Vanessa
York University, Toronto

Lankowski, Carl
U.S. Department of State

Lautensach, Alexander K.
University of Northern British Columbia

Leonard, Liam
Institute of Technology, Sligo

LeVasseur, Todd
University of Florida

Lewis, James G.
Forest History Society

May, Shannon
University of California, Berkeley

Minor, John Jesse
University of Arizona

Myers, Justin
The Graduate Center, City University of New York

Nascimento, Susana
ISCTE-IUL, Lisbon University Institute

O'Sullivan, John
Gainesville State College

Palmer, Daniel E.
Kent State University, Trumbull

Parker, Jonathan
University of North Texas

Plutynski, Anya
University of Utah

Roser, Dominic
University of Zurich

Roth-Johnson, Danielle
University of Nevada, Las Vegas

Rowley, Brandon B.
University of Idaho

Salsedo, Carl A.
University of Connecticut

Schroth, Stephen T.
Knox College

Scott, Austin Elizabeth
University of Florida

Shahar, Dan C.
University of Arizona

Shear, Boone
University of Massachusetts, Amherst

Shooter, Wynn
Monash University

Silva, Carlos Nunes
University of Lisbon

Thornton, Fanny
Australian National University

Tomozeiu, Daniel
University of Westminster

Turrell, Sophie
Independent Scholar

Uebel, Michael
University of Texas at Austin

van Bueren, Ellen
Delft University of Technology

Vanderheiden, Steve
University of Colorado, Boulder

Vaz, Sofia Azevedo Guedes
New University of Lisbon

Weiland, Sabine
Catholic University of Louvain

Witt, Joseph
University of Florida

Woods, Mark
University of San Diego

Yuhas, Stephanie
University of Denver

Green Ethics and Philosophy Chronology

1635: A law is passed in Ireland prohibiting the pulling of wool off live sheep and the attaching of a plow to a horse's tail.

1641: Massachusetts' "Body of Liberties" is passed, containing legislation that forbids "Tirranny or Crueltie" against domestic animals.

1864: George Perkins Marsh publishes *Man and Nature*, a book that argues that many civilizations have fallen because of environmental degradation.

1872: President Ulysses Grant signs into law a bill designating the area of Yellowstone as the world's first national park.

1873: The first federal law against animal cruelty is passed in the United States. Called the "Twenty-Eight Hour Law," it requires that "livestock transported across country be provided with water and rest at least once every 28 hours."

1892: Social reformer Henry Salt publishes *Animals' Rights: Considered in Relation to Social Progress.*

1892: The Sierra Club is founded in the city of San Francisco by preservationist John Muir.

1896: The National Forestry Commission is established.

1903: President Theodore Roosevelt signs an executive order selecting Pelican Island in Florida as the United States' first national wildlife refuge.

1905: The United States Forest Service is established.

1911: Congress passes the Weeks Act, authorizing the Secretary of Agriculture to "examine, locate and recommend for purchase . . . such lands within the watersheds of navigable streams as . . . may be necessary to the regulation of flow of navigable streams."

1916: The National Park Organic Act is signed into law, establishing the National Park Service.

1935: The Wilderness Society is founded. Over time, the organization's membership grows to more than 300,000 people.

1946: The Bureau of Land Management is formed when the Grazing Service and the General Land Office are merged.

1949: Aldo Leopold publishes *The Sand County Almanac* in which he states that humanity should adopt an "ethic dealing with man's relation to land and to the animals and plants which grow upon it."

1962: Rachel Carson's *Silent Spring* is released and is subsequently credited with spurring the creation of the modern environmental movement.

1962: Murray Bookchin's *Our Synthetic Environment* is published.

1964: The Wilderness Act is signed into law by President Lyndon Johnson. The act sets aside 9 million acres for environmental protection.

1967: Roderick Nash's *Wilderness and the American Mind,* a book espousing the theory of environmental preservation, is published.

1968: The Club of Rome is founded with the intent of convincing industrialists that unlimited economic growth and preserving the environment are two incompatible ideas.

1970: Norman Borlaug, known as "the father of the Green Revolution," is awarded the Nobel Peace Prize for his work in developing high-yielding strains of wheat, which leads to a significant worldwide reduction in food scarcity.

1970: The National Environmental Policy Act (NEPA) is passed, requiring all federal agencies to compile and submit an Environmental Impact Statement.

April 22, 1970: The world's first Earth Day is celebrated.

1973: Philosopher Arne Naess coins the term *deep ecology* to describe an environmental movement that focuses not only on reducing pollution but also on recognizing that nature has significant intrinsic value.

1973: Congress passes the Endangered Species Act.

1974: French feminist Françoise d'Eaubonne coins the term *ecofeminism* to describe a set of beliefs that lay blame for environmental problems on the worldwide inequality between men and women.

1974: Australian philosopher John Passmore publishes *Man's Responsibility for Nature.*

1976: Bioethics professor Peter Singer publishes *Animal Liberation: A New Ethics for Our Treatment of Animals.*

1976: The United Nations Convention on the Prohibition of Military or Any Other Hostile Use of Environmental Modification Techniques (ENMOD) is held.

1977: Wendell Berry publishes *The Unsettling of America*, a critique on the mechanization of agriculture.

1978: Dr. Eugene C. Hargrove founds *Environmental Ethics*, the first journal to focus entirely on environmental concerns.

1978: A tiny neighborhood in Niagara Falls, New York, named Love Canal makes national news when it is revealed that tons of toxic waste have inundated the neighborhood's landscape. Later, U.S. President Jimmy Carter declares the site a federal disaster area.

1979: James Lovelock introduces the Gaia hypothesis, a controversial theory stating that the Earth is a constantly evolving organism that must maintain a "preferred homeostasis."

1983: Philosophy professor Robin Attfield publishes *The Ethics of Environmental Concern*, exploring the history of human-centered thinking and its effects on the environment.

1986: American philosopher Paul Taylor publishes *Respect for Nature: A Theory of Environmental Ethics.*

1986: Environmentalist Jay Westerveld coins the term *greenwashing* to describe how companies often use deceptive marketing to make it seem as though their products are more environmentally friendly than they actually are. Examples include British Petroleum rebranding itself as "Beyond Petroleum."

1987: The World Commission on Environment and Development (WCED) convenes and releases its report titled *Our Common Future.*

1989: Dr. Eugene C. Hargrove publishes *Foundations of Environmental Ethics.*

1990: The International Society for Environmental Ethics (ISEE) is founded.

1991: Freya Mathews publishes *The Ecological Self*, a book arguing that modern environmental philosophy is not adequate enough to solve most environmental problems.

1992: Representatives from 172 governments around the world gather in Rio de Janeiro for the United Nations Conference on Environment and Development (UNCED).

1992: Professor William Rees coins the term *ecological footprint*, meaning the effect humans have on the Earth's ecosystems.

1992: California genetic engineering company Calgene develops the Flavr Savr tomato, the first genetically engineered food to be introduced to the world market.

1993: The Environmental Law and Policy Center (ELPC) is founded.

1994: U.S. President Bill Clinton signs Executive Order 12898, requiring federal agencies to determine the impact that environmental degradation has on low-income communities.

1994: The Science and Environmental Health Network, a consortium of environmental organizations, is founded.

1996: Researchers at the Roslin Institute in Scotland successfully clone the first mammal in history, a sheep named Dolly.

1997: The Kyoto Protocol, an international document whose signatories pledge to reduce greenhouse gases, is officially adopted.

1997: The International Association of Environmental Philosophers (IAEP) is founded by professors Bruce Foltz and Robert Freedman.

2000: Edward Freeman, Jessica Pierce, and Richard Dodd publish *Environmentalism and the New Logic of Business: How Firms Can Be Profitable and Leave Our Children a Living Planet.*

2000: The Earth Charter, a 2,400-word document that outlines global environmental sustainability efforts, is unveiled to the public.

2000: The United Nations Millennium Summit is held and eight Millennium Development Goals (MDGs) are adopted by various countries.

2002: The World Summit on Sustainable Development is held in Johannesburg, South Africa.

2002: The United Nations designates 2002 as the International Year of Ecotourism.

2003: Journalist Alex Steffen coins the term *bright green environmentalism* to describe a new wing of environmentalism that espouses the effectiveness of technological innovation.

2006: The United Nations releases a study showing that the U.S. meat industry has the largest impact on global warming than any other industry.

2006: The documentary film *An Inconvenient Truth*, featuring a slide-show-like presentation about the environment by former vice president Al Gore, is released. The film goes on to gross approximately $25 million in the United States.

2006: Studies are released showing that only 0.65 percent of the total agricultural lands worldwide are "organic" farms.

2009: The Barack Obama administration announces it will increase corporate average fuel economy standards.

2010: The Obama administration proposes to open certain offshore areas for oil and natural gas drilling.

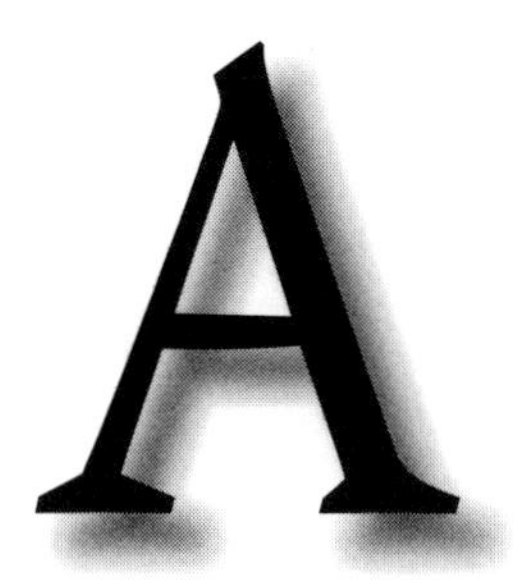

Adaptive Management

Because of the complexity of the systems undergoing manipulation, efforts in ecological restoration have frequently been met with ambiguous outcomes and results. Biophysical complexity in terms of the structure and function of ecosystems is not the only confounding factor, however. Often, multiple competing social factors influence the outcomes and interpretations of ecological restoration work. In addition, scientific uncertainty and ambiguity can cloud the analysis of restoration success. Finally, environmental changes contribute the possibility of a moving baseline by which to judge the success of ecological restoration. In light of these challenges, a novel approach to management and scientific study of the process and outcomes of management has begun to be widely adopted: adaptive management.

Adaptive management, in its simplest formulation, is a technique by which resource managers work toward a restoration goal, while simultaneously monitoring and studying the effects and impacts of the management techniques. Typically, adaptive management allows for shifting management goals, as the results of scientific monitoring and study can alter subsequent management effort. Equally important, projects utilizing adaptive management techniques can seek to harness scientific uncertainty via an iterative process of hypothesis testing, the outcomes of which are used to resolve decisions over how to implement the next stage of management effort. Adaptive management is particularly useful for creating management strategies or plans under the following situations, each of which is discussed below. When optimal or sustainable reference conditions for ecological restoration or resource management are unclear, adaptive management may provide useful approaches for planning. Adaptive management also provides possibilities for management action in the face of moving targets and baselines, such as in the case of resources potentially affected by climate change. Adaptive management plans can provide for scientific uncertainty through active hypothesis testing and through further study of systems about which substantial knowledge gaps remain. Perhaps most importantly, adaptive management can be used to enlist broad public support, because adaptive management plans can be open to diverse societal values.

A U.S. Fish & Wildlife Service official checks on a prescribed burn to improve wildlife habitat in the Lower Klamath National Wildlife Refuge in Oregon in April 2008. The emphasis on continued monitoring and research in adaptive management projects provides vital data for future management efforts.

Source: U.S. Fish & Wildlife Service

Uncertain Reference Conditions for Restoration or Management

When planning an ecological restoration project, it is important to consider what "original" conditions the restoration project is aimed at recreating. In practice, these reference conditions are often chosen to reflect environmental states that existed prior to European colonization and settlement or prior to changing land management and resource-use patterns altered ecosystem structure and function. In many cases, though, it can be difficult or impossible to say with absolute certainty what the actual state of the environment was prior to alteration. Various ecological proxies can afford insight into historical and prerecorded environmental conditions. Each proxy has its limitations in terms of spatial and temporal resolution. Archival sources and oral histories can yield explicit insights into previous environmental states and resource-use patterns, but may be incomplete or reflect the cultural biases of colonizing groups. Dendrochronology can be used to reconstruct historical age structure, growth rates, fire histories, and species composition, but this technique can only yield inferences back to the oldest samples still in existence. Sedimentary charcoal can be utilized to date not only periods of burning, but also land cover conversion and rates of soil loss, but typically with low spatial resolution. Palynology provides fairly high spatial and temporal resolution in terms of which plants, including agricultural plants, were present in a landscape.

In addition to potential inadequacies in terms of spatial and temporal resolution, each proxy method used for environmental reconstruction has unique issues in terms of its historical reach and the proxy's ability to shed light on the actual resources or values of concern for managers. In light of the difficulty of making concrete decisions about a stable prerecord reference condition, adaptive management offers several distinct benefits. Adaptive management can set the stage for informed management activity to commence, with additional ongoing research offering insight into original reference conditions as knowledge continues to accrue. This flexible facet of adaptive management allows decisions and understandings regarding reference conditions to be arrived at after initiation of management activity and restoration planning. Further, adaptive management allows the restoration site itself to be used for research into reference conditions, which is ideal because the planning scale can be aligned with the scale of research used to determine optimal reference conditions.

Climate Change

Even in cases where reference conditions can be identified and agreed upon, the ultimate goal of management or restoration may be uncertain. While unique local factors can contribute to this situation, the most important cause of variable management goals is the predicted outcomes of climate change on diverse systems.

The mechanisms of global climate change are well understood, but the local or regional effects of climate change are often difficult to model or predict. In light of this fact, resource managers are often faced with difficult decisions in terms of how to allocate restoration and management assets. For example, endangered species endemic to isolated high-elevation forests in the southwestern United States are further threatened by diverse climate change effects, including amplified disturbance regimes, competition from invasive species, forest conversion to savanna and open grassland, and concurrent habitat loss. Managers who have the responsibility of assuring the survival of endangered species in these tenuous habitats face difficult choices in terms of how to meet their objectives. In the case of the Mount Graham Red Squirrel, which is endemic to the Pinaleño Mountains of southeastern Arizona, forest thinning would reduce the risk of damaging wildfires that might destroy the squirrel's last habitat refuge. But forest canopy thinning might also lead to higher squirrel predation by hawks, and also might prove disruptive to breeding and foraging.

Adaptive management could provide a template for action in cases where management activity has the potential for threatening the values of concern. The results of smaller case studies can be used to guide later management action, once more is known about the outcomes of management activity on the resource in question. In addition, the flexibility of adaptive management techniques might allow for essentially planning for the new regime that will be ushered in by climate change.

Human Values and Perceptions

Public land managers often face the difficult task of balancing competing and, in some cases, contradictory goals and visions. Many public agencies must weigh extractive economic activities such as mining, timbering, and grazing against competing activities such as tourism and sightseeing, while also managing listed species and sensitive natural and cultural values. Establishing management plans in such an environment can be a contentious process, particularly when diverse constituencies are active in the management planning process. If, however, adaptive management projects are built into the planning process, public agencies have the opportunity to attempt to balance competing interests in as objective a way as possible.

Nearly a century of fire suppression has led to the buildup of dangerous levels of fuels in North American forests. The ballooning costs of fire suppression, coupled with the risk that heavy fuel loads pose for human settlements and ecosystems as a whole, have led public land management agencies to consider landscape-scale forest restoration projects as a cost-effective and realistic way of coping with the scale of the management dilemma.

A prime example of this trend is FireScape, a series of cooperative efforts between agencies such as the National Forest Service, National Park Service, Bureau of Land Management (BLM), U.S. Fish and Wildlife Service, the state of Arizona, and multi-stakeholders, including the Nature Conservancy and private landowners and permit holders. FireScape projects seek to increase ecosystem resilience through fuel treatments and the return of high-frequency yet low-intensity fire to forests and grasslands in southern Arizona. Managing a project that includes lands administered by multiple federal and state agencies is quite challenging, and the addition of stakeholders such as ranchers, mountain-cabin owners, environmental groups, and various Forest Service permit holders only adds complexity to management decisions. A multi-shareholder approach to synthesizing management activities across diverse land units under assorted agency control is bound to give rise to debates that center on the role of different fuels treatments in various ecosystems, on the rightful place and scale of fire on the land, and on the role of anthropogenic and natural disturbance of these ecosystems. Ultimately, these debates come down to competing environmental visions and beliefs about how humans should manage western lands.

Adaptive management tools and techniques can assist in such a potentially volatile scenario as a FireScape restoration effort. Because adaptive management projects often reincorporate the lessons learned from earlier management activities into ongoing or future work, stakeholders have the opportunity to have their opinions heard at multiple stages of the project. In some cases, opposition to certain management activities will fall away as the results of the projects become apparent. Furthermore, the emphasis on continued monitoring and research in many adaptive management projects will satisfy many stakeholders and agency personnel, for whom robust scientific findings are critical to justify management effort and expenditures.

Scientific Uncertainty

At times, management action should or must be taken in the absence of broad scientific consensus on an issue. At the most basic level, predictions can be made about the outcomes of management activity—this fundamental step involves some, but not necessarily complete, knowledge of the system under consideration. Adaptive management can allow for an iterative process of hypothesis testing, whereby research and monitoring of ongoing management activities provide the basis by which to refine future management of a site. In this way, science-based resource management can effectively reduce uncertainty over the best course of action; over time, managers and scientists more fully comprehend the system.

Shifting Targets or Metrics for Success

In many resource management applications, the targets by which management successes are determined move over time. Shifting metrics can be a result of political processes or changes to the system under management such as those brought about by unprecedented disturbance, adjacent human development, or climate change. For example, if an endangered plant species is discovered growing in or near a day-use area in a national forest, two competing and incompatible land uses are suddenly at stake: recreation access and endangered species protection. Likewise, if burgeoning rural development begins to encroach on land classified as open range, conflicts over grazing, livelihoods, and land use are likely to erupt. Whatever the cause, as competing values and visions affect the rationales behind resource management, the techniques offered by adaptive management can help adjudicate competing claims and can assist in decisions over how resource management should proceed.

Brief History of Adaptive Management

Early efforts at adaptive management of renewable resources grew out of a long history of management failures, particularly in the arenas of fisheries and timber. As scientific knowledge about natural systems increased, so did the belief that scientifically informed groups of people could devise sustainable systems of harvest and of resource extraction. Despite burgeoning scientific understandings of complex systems, mismanagement of resources continued, with fisheries collapsing and forests experiencing overharvest. This ongoing mismatch between science-based research-management decisions and ecological effects led to the conclusion that researchers had a poorly developed understanding of the social and economic systems that influence resource depletion, and that the science itself was not practiced in such a way as to lead to robust management decisions.

The development and diffusion of powerful and less-expensive computing systems enabled the expansion of mathematical models of sociobiophysical systems. Faster model development meant that at the same time that deductive and predictive power was increasing, a wider suite of actors and experts could be brought into resource management projects, such as those focusing on Pacific salmon or continent-wide insect outbreaks.

As the social, economic, and research pressures on natural systems continued to increase, managers and researchers began to conclude that adaptive management techniques could produce more rapid gains in knowledge than traditional monitoring or small-scale research could independently. Scientific uncertainty could be decreased at the same time that management efficiencies and successes could be increased, although full scientific knowledge and paramount management effectiveness will probably never be achieved.

See Also: Ecological Restoration; Ecology; Environmental Policy.

Further Readings

Allen, Catherine and George Henry Stankey, eds. *Adaptive Environmental Management: A Practitioner's Guide.* New York: Springer, 2009.

Holling, C. S., ed. *Adaptive Environmental Assessment and Management.* Toronto, Canada: Wiley-Interscience Publication, 1978.

Meffe, Gary K., et al., eds. *Ecosystem Management: Adaptive, Community-Based Conservation.* Washington, DC: Island Press, 2002.

Walters, Carl J. *Adaptive Management of Renewable Resources.* New York: Macmillan, 1986.

Williams, Byron K., et al. *Adaptive Management: The U.S. Department of the Interior Technical Guide.* Washington, DC: U.S. Department of the Interior, 2007.

John Jesse Minor
University of Arizona

Agriculture

Agriculture is one of the primary ways that humans interact with, modify, and deplete or enrich the biodiversity and resilience of their environments. Since the Neolithic revolution, the shift from primarily hunter-gatherer lifestyle in the Paleolithic to settled farming communities circa 10,000 B.C.E., agriculture has been not only a means of survival but also a complex cultural institution that converges with political and economic institutions and

Sustainable farming and animal rights movements have protested the increased industrialization of agriculture, especially the consolidation of farms, the introduction of chemical pesticides and genetically modified crops, and the spread of confined animal feeding operations such as this one used for cattle in Yuma, Arizona, in 2002.

Source: U.S. Department of Agriculture, Natural Resources Conservation Service

with ecological events. In spite of its long history, agriculture has been somewhat marginalized within philosophical discourse.

Nevertheless, philosophers have emphasized the crucial philosophical and ethical issues raised by contemporary agriculture, including the decline of the family farm, labor, sustainable development, biotechnology, and urban farming. The history of agriculture in the 20th and 21st centuries brings each of these questions into focus, replete with meanings that go beyond subsistence to questions of democracy and justice.

Ethical and philosophical questions about agriculture have a long history, extending from the classical period with works such as the Greek poet Hesiod's *Works and Days* (8th century B.C.E.) and the Roman poet Virgil's *Georgics* (37–30 B.C.E.). The Georgic ideal of agriculture, named for Virgil's poem, is related to the pastoral ideal of rural life and harmony with nature, but emphasizes the value and dignity of labor.

Agrarianism, a variant of the Georgic ideal, was championed by Thomas Jefferson, who envisioned the United States as a nation of independent farmers. For Jefferson, the health of the nation depended on the moral strength of the farmer, whose character was shaped by close work with the land. In *Notes of the State of Virginia,* Jefferson writes, "Those who labour in the earth are the chosen people of God, if ever he had a chosen people, whose breasts he has made his peculiar deposit for substantial and genuine virtue." This valorization of the farmer contrasted sharply with Jefferson's depiction of "the mobs of great cities" as "sores" on the body politic. This dichotomy between farm and industry, rural and urban, virtue and vice continues to influence environmental and popular thinking about farming today.

Agrarian philosophy in the United States has been developed by a series of thinkers, including Liberty Hyde Bailey and Wendell Berry. Berry—philosopher, farmer, and writer—is perhaps the best-known contemporary agrarian theorist. Throughout the course of his long career he has advocated for the value of farming for the formation of culture and character. Berry's work includes incisive critiques of industrial agriculture as both ecologically and morally unsound in its exploitation of land and people. This critique links to Berry's ongoing defense of the family farm in response to the depopulation of rural communities. The U.S. Environmental Protection Agency estimates that less than 2 percent of the U.S. population lives on farms, down from a peak of 6.8 million farms in 1935.

Although the popular image of agriculture focuses on the virtues of small farms and rural life, both the concept and the practice of agriculture in the United States has a more complex history. David B. Danbom describes the ways that agriculture has been both a form of and impetus for imperialism, including acquisition of land through war, purchase, and dispossession of indigenous inhabitants, and a reliance on slave labor up to the Civil War. The 20th century has witnessed the convergence of agricultural practices with political and economic events in sometimes disastrous ways, such as the Great Depression and the Dust Bowl in the 1930s. It is during the 20th century that we see the increasing industrialization of agriculture, including consolidation of farms, heavy machinery, monocultures (first of hybrid seeds and later of genetically modified seeds), and the introduction of inorganic fertilizers and of chemical pesticides and herbicides. This movement toward larger farms and reduced diversity was accompanied by an increase in debt, with negative consequences for family farms, particularly during the Farm Crisis of the 1980s. Critiques of conventional agriculture include the problem of erosion from tilling and machinery, the pollution of groundwater from fertilizers and pesticides, the vulnerability of monocultures to pests and disease, the use of fossil fuels, and the displacement of smaller farms. The benefits of these conventional practices include increased production and efficiency and lowered costs for food.

Sustainable Agriculture

Responses to industrial agriculture include the sustainable farming and animal rights movements. Sustainable agriculture has many definitions, and the concrete practices that constitute sustainability will vary according to the specific location. In general, sustainable agriculture is described as a system that promotes ecological and human health, equity, and stable farming communities. The work of Wes Jackson, plant geneticist and cofounder of the Land Institute, exemplifies the attempt to create a sustainable agriculture based in the cultural and ecological specificities of a particular place. The centerpiece of the Land Institute's research is perennial polyculture: the development of an agricultural system that mimics Kansas prairie ecology but at the same time produces adequate yields. The benefits of this system, as Marty Bender and Wes Jackson describe, include decreased erosion, decreased use of fossil fuels and pesticides, decreased expenses for inputs and machinery, and water conservation. The work of the Land Institute is directed not only at creating sustainable agriculture based on natural systems for the bioregion but also at sustaining farming communities. The organic farming movement, which critiques the use of inorganic fertilizers, pesticides, herbicides, and, more recently, genetic engineering of crops, is frequently associated with but not identical to the concept of sustainable farming. The industrial production of meat in factory farms or "concentrated animal feeding operations" (CAFOs) is itself a source of debate and activism. Philosophers as diverse as Peter Singer, Carol Adams, and Jacques Derrida have critiqued the practice of CAFOs, sometimes advocating vegetarianism or veganism as a response. Industrial animal farming also intersects with environmental justice concerns, both for workers on the farms and because of the impacts on water, air, and aesthetic quality for surrounding neighborhoods.

Equity and Justice

The issues of migrant workers, tenant farming, and sharecropping form another crucial issue for agricultural ethics—that of farm labor and environmental justice. The displacement of "Okies" in the Dust Bowl, poignantly portrayed in John Steinbeck's novel *The Grapes of Wrath,* is a dramatic example both of the socioeconomic vulnerability of tenant farmers and of the inhumane treatment of migrant workers. Perhaps the best known of the struggles for justice for agricultural laborers is the work of the United Farm Workers (UFW), cofounded by César Chávez. Using nonviolent tactics, such as the grape boycott, the UFW negotiated for the passage of the California Agricultural Relations Act in 1975, which was intended to guarantee justice for farmworkers by ensuring the right to unionize. Workers in all areas of the food system—from agricultural labor to slaughterhouses—continue to struggle for safe working conditions, decent housing, fair wages, and benefits such as health insurance and workers compensation. The vulnerability of undocumented workers, as well of those enrolled in temporary worker programs, is of particular ethical concern.

Vandana Shiva, a leading voice for environmental justice and the rights of farmers, has argued that agricultural imperialism has extended internationally, first through the spread of green revolution technologies and more recently through genetic modification and patenting of seeds—what some have called the "Gene Revolution." In the mid-20th century, research and development organizations worked with developing nations to increase food production. The development of new varieties of wheat and rice, capable of producing higher yields from the same amount of land, helped to create what is called the green revolution. A key figure in this period is Norman Borlaug, a plant pathologist who won the Nobel Peace Prize in 1970 for his work in developing high-yield varieties of wheat. These new crops often required irrigation and fertilizer to produce their high yields, thus necessitating infrastructure and imports. Shiva's work *The Violence of the Green Revolution* describes, in contrast, the social and ecological costs of the green revolution in Punjab, India. Rather than producing abundance and peace, Shiva argues, the centralized development of the green revolution created insecurity and violence.

More recently, some advocates of farmers' rights have argued that corporations concentrate their control of agricultural production and the food supply through the use of biotechnologies that prevent seed saving, either biologically, through the production of sterile seeds, or legally, by refusing farmers the right to save seeds for replanting from year to year. The question of seeds highlights a series of ethical issues, from intellectual property rights (IPRs) to sustainable development to the precautionary principle. The patenting of seeds raises the question of IPRs when seeds that have been developed by indigenous and local farmers, and held as a common good, are transferred to private ownership. These questions overlap in ways that further complicate the ethical issues at hand. For some, genetic modification (GM) of seeds is acceptable if the seeds are placed in the public domain rather than owned by private corporations. This connects to the question of the purpose of the genetic modification. For example, creating crops that contain higher levels of nutrients might prevent disease and birth defects in regions experiencing food insecurity. Some advocates of GM technology argue that breeding insect resistance into the plant itself can decrease the use of pesticides. Genetically modified organisms (GMOs) have provoked a passionate series of critiques on ecological, ethical, and religious grounds. For some, the argument is that no one has the right to "own life." Others favor the precautionary principle, arguing that the effects of these new organisms on the environment and on human health are unknown and carry too high a risk. The argument in favor of GMOs—as with the green revolution—is based in humanitarian claims that new seeds are necessary to feed a growing human population, especially in the context of climate change. The hope is that crops can be designed to adapt to changing conditions such as drought as well as to bring marginal soils into production.

Agrarian philosophy converges with environmental justice concerns including anti-imperialism and the advocacy of farmers' rights. Wendell Berry's classic work *The Unsettling of America* argues against corporate dominance of agriculture, seeking to preserve democracy through the cultivation not only of the land but of a citizenry that is capable of living in community. Agrarian philosophers in the United States and scholars such as Vandana Shiva, Miguel Altieri, and Gary Paul Nabhan have championed the value of local and indigenous knowledges: the understanding of agriculture learned through experience and passed through culture and tradition. These philosophers of agriculture oppose the imperial science described by James Scott in *Seeing Like a State,* a view that rejects practical knowledge as unscientific and irrelevant. The revaluation of local and indigenous knowledges is a philosophical issue that requires a new understanding of epistemology and of how science should be defined.

Evolving Social Movements

In the early 21st century there are already two developments in smaller-scale agriculture: local foods and urban farming. The local foods movement is focused on the renewal of food systems. The goals of this movement include food security, economic support of farming families and cultures, environmental sustainability, food safety, and human health. Farmers markets and community-supported agriculture (CSAs) are vital elements of this movement. However, the "locavore" movement has been critiqued as simply another form of middle-class consumption, a lifestyle choice that is not accessible for many communities. The work of Gary Paul Nabhan with the organization Renewing America's Food Traditions (RAFT) is an important dimension of the local foods movement and offers a bridge to questions of food justice. RAFT seeks to support and celebrate North American food traditions and, in doing so, to promote sustainable agricultural practices. RAFT's particular focus on diversity of food traditions and heirloom species in conjunction with its attention to environmental and human health brings together questions of culture and justice with the local foods movement.

The urban farming movement includes food justice as an essential aspect of its work to construct food security. "Food justice" refers to the right of all people to safe, nutritious, and culturally appropriate food. Advocates of food justice describe many urban landscapes as "food deserts"—areas where grocery stores are sparse. Food justice is thus related to environmental justice, the movement that critiques the disproportionate burdens placed on communities of color. In this case, the issue is the denial of a benefit rather than the imposition of a burden. Residents of urban neighborhoods, who may not have cars or access to functional public transportation, are forced to buy food at gas stations and fast-food restaurants. The food at these locations is highly processed and calorie dense, contributing to obesity, heart disease, and other health problems in urban communities. Urban farming seeks to improve access to fresh foods by producing and selling garden produce within the city. In addition, urban farming can contribute to social, economic, and aesthetic revitalization of neighborhoods. These recent movements offer the hope for a safe, just, and diverse system of sustainable agricultures in urban, suburban, and rural communities.

See Also: Animal Ethics; Bailey, Liberty Hyde; Berry, Wendell; Borlaug, Norman; Ethical Vegetarianism; Genetic Engineering; Local Food Movement.

Further Readings

Berry, Wendell. *The Unsettling of America: Culture and Agriculture.* New York: Avon, 1977.

Danbom, David B. "Past Visions of American Agriculture." In *Visions of American Agriculture,* William Lockeretz, ed. Ames: Iowa State University Press, 1997.

Evans, L. T. *Feeding the Ten Billion: Plants and Population Growth.* Cambridge, UK: Cambridge University Press, 1998.
Ferriss, Susan and Ricardo Sandoval. *The Fight in the Fields: Cesar Chavez and the Farmworkers Movement,* Diana Hembree, ed. New York: Harcourt Brace, 1997.
Jackson, Wes and Marty Bender. "Investigations Into Perennial Polyculture." In *Meeting the Expectations of the Land: Essays in Sustainable Agriculture and Stewardship,* Wes Jackson, Wendell Berry, and Bruce Colman, eds. San Francisco, CA: North Point Press, 1984.
Jefferson, Thomas. *Notes of the State of Virginia.* New York: Penguin, 1999.
"Renewing America's Food Traditions." http://www.slowfoodusa.org/index.php/programs/details/raft (Accessed January 2010).
Scott, James. *Seeing Like a State: How Certain Schemes to Improve the Human Condition Have Failed.* New Haven, CT: Yale University Press, 1998.
Serageldin, Ismail. "Biotechnology and Food Security in the 21st Century." *Science,* 285/16 (1999).
Shiva, Vandana. *Biopiracy: The Plunder of Nature and Knowledge.* Cambridge, MA: South End Press, 1997.
Shiva, Vandana. *The Violence of the Green Revolution: Third World Agriculture, Ecology, and Politics.* Atlantic Highlands, NJ: Zed Books, 1991.
U.S. Environmental Protection Agency. "Ag 101." http://www.epa.gov/oecaagct/ag101/demographics.html (Accessed January 2010).

Janet Fiskio
Oberlin College

Animal Ethics

Animal ethics refers to values that should govern human relationships with domestic and wild animals. In addition to acts of deliberate gratuitous cruelty, where a person tortures an animal in order to enjoy its suffering, areas of concern include everyday activities such as the raising of animals for food (especially those raised in concentrated animal feeding operations, CAFOs), research, product testing, hunting and fishing, and sport and entertainment. Objections to some forms of these activities often focus on effects on human health or on biodiversity and environmental quality. This article is concerned solely with the interests of the animals themselves, as discussed by thinkers in the Western tradition, including modern-day advocates of an enhanced moral status for animals.

The Status of Animals in Western Tradition

Western culture is anthropocentric: Humans are regarded as superior to other life forms (and to inanimate objects) and therefore entitled to treat nature as a storehouse of resources available for our benefit. The main philosophical roots of this worldview are ancient Greek thought and the Judeo-Christian tradition. With a few exceptions, the ancient Greek philosophers believed that this superiority is natural, based on capacities

A lamb and ewe on a farm in New Zealand, where nearly all beef, lamb, and dairy products are derived from free-range animals. Most such products in the United States are produced in profitable but ethically questionable concentrated animal feeding operations.

Source: iStockphoto.com

that are unique to humans such as language and the ability to reason. Aristotle's (384–322 B.C.E.) worldview is based on the idea of unique function: the function of a knife is to cut, because that is what it does best. Artifacts such as knives are valuable only as a means and have no good of their own. Although living things were not literally designed for a function, we can understand nature as purposive, based on what each does best; thus, all living things have a good of their own. The good for plants is growth, but plants are also valuable as a means to providing food for animals and humans. Animals are sentient beings that are aware of the world through sense perception; their good is therefore sensation. They are also valuable as a means for humans—as food, in agriculture, and for transportation. The highest good for humans is happiness, both because it is the ultimate goal of all human actions and because nobody pursues happiness for the sake of anything else. Humans are endowed with reason and the capacity for virtue, and the happy life is one that is lived according to reason and virtue.

Such hierarchical thinking is often seen as inherent to the Western tradition and as manifested in exploitation, such as slavery, sexism, racism, and colonialism, as well as the treatment of animals and nature. Ecofeminists argue that inequality and injustice will never be eradicated unless people cease to think in a hierarchical way. This way of thinking is also found in Western religions, which teach that God gave humans "dominion" over the rest of creation. A widely used version of the Bible translates Genesis 1:28 as the following:

> And God blessed them: and God said unto them, Be fruitful, and multiply, and replenish the earth, and subdue it; and have dominion over the fish of the sea, and over the birds of the heavens, and over every living thing that moveth upon the earth.

Leading figures in the tradition have usually interpreted this to mean that humans are entitled to make use of animals as they see fit, and that humans have no direct duties to animals. Humans may nonetheless have duties not to mistreat animals. For instance, we have a duty to respect property rights, and therefore not to harm animals that are someone's property.

However, this duty is owed to the owner, not to the animal itself. It is also argued that cruelty to animals should be discouraged because such behavior may make one more likely to want to torture humans. This argument was developed by the Catholic theologian St. Thomas Aquinas (c. 1225–74) and has been repeated by secular thinkers such as John Locke (1632–1704) and Immanuel Kant (1724–1804).

Challenges From Within the Tradition

The first prominent Western critic of animal exploitation, well before Aristotle, was the Greek philosopher Pythagoras (c. 570–c. 490 B.C.E.). None of his works survive, so we are reliant on writers such as Aristotle and Plutarch (46–119 C.E.), himself an advocate of vegetarianism, for our knowledge of his thought. However, many authorities state that he advocated vegetarianism, perhaps because of his belief that the soul undergoes successive reincarnation (metempsychosis), which can include reincarnation as an animal.

The first advocates of enhanced moral status for animals in the modern period were Jean-Jacques Rousseau (1712–78) and Jeremy Bentham (1748–1832). Both accepted that animals are incapable of reason, but argued that the basis for moral consideration should be sentience—the ability to suffer. While Rousseau argued that eating meat is not natural for humans, it is unclear whether he was personally vegetarian. Bentham did not oppose killing animals as such, only the infliction of suffering, and there is no evidence that he was a vegetarian. Charles Darwin, who was a vegetarian, also rejected the sharp distinction between humans and other animals, arguing that they are capable of human emotions and feelings, and that the differences in mental capacity between humans and many animals is a matter of degree, not of kind.

While Bentham greatly influenced the modern animal liberation movement, his and Rousseau's views were widely regarded as preposterous, in particular, the idea of animal rights. Mary Wollstonecraft (1757–95) published what would today be called a feminist work, *A Vindication of the Rights of Women* in 1792. In reply, Thomas Taylor (1758–1835) published, anonymously, *A Vindication of the Rights of Brutes*. Wollstonecraft's arguments, he claimed, would also show that animals have rights; but this is plainly absurd and, therefore, so are her arguments for women's rights.

Nonetheless, beginning in 1822 a series of laws banning cruelty to animals were passed in England. These were not the first such laws in the Western world. In 1635 in Ireland a law was passed forbidding practices such as pulling wool off a live sheep and attaching a plow to a horse's tail, though this legislation was later repealed. The 1641 state of Massachusetts *Body of Liberties* (s. 92) forbade "any Tirranny or Crueltie" against domestic animals. However, no further legislation was passed until 1828, in New York State. The first federal law was the 1873 Twenty-Eight Hour Law, requiring that livestock transported across country be provided with water and rest at least once in every 28 hours.

The Modern Animal Ethics Movement

The earliest published book advocating an end to animal exploitation is *Animals' Rights: Considered in Relation to Social Progress* (1892) by social reformer Henry Salt (1851–1939). Salt advocated vegetarianism and an end to vivisection (dissection of living animals), blood sports, and killing animals for their skins, fur, and feathers. Other writers on the subject include George Bernard Shaw (1856–1950) and Albert Schweitzer (1875–1965). The most influential writers of the 20th century were Peter Singer (1946–) and Tom Regan (1938–). Singer's and Regan's views on the treatment of animals are almost identical—they both advocate vegetarianism. However, they come from quite different theoretical perspectives, Singer being a utilitarian and Regan an advocate of rights.

Singer, professor of bioethcs at Princeton University, is one of the founders of the Great Apes Project. He has published over 40 books, of which the most influential is *Animal Liberation: A New Ethics for Our Treatment of Animals* (1976).

Singer is a preference utilitarian. All forms of utilitarianism define the right action as whatever will produce the best outcome for all affected but they differ about the nature of "the good." Preference utilitarians identify it as whatever maximizes the satisfaction of the preferences of all involved, which may not necessarily be their own happiness. Indeed, everyone's preferences are unique, so nothing can count as "the good" except the satisfaction of the preferences of the maximum number of those affected. Everyone has an equal interest in having their preferences satisfied, and thus is entitled to equal moral consideration.

Most utilitarians have believed that only human interests are morally relevant, but like Bentham, Singer stresses that all sentient beings have an equal interest in avoiding suffering, so we are obliged to consider their interests equally with those of humans. He notes that we have already accepted that the interests of women and ethnic minorities are equal to those of white males: racism and sexism are no longer acceptable. Failure to give equal consideration to the interests of animals which, following Richard Ryder, he dubs "speciesism," should be equally unacceptable.

According to Singer, therefore, we should cease to treat animals in ways that cause them to suffer. The most obvious way in which we contribute to animal suffering is in consuming animal products, including eggs and dairy products as well as meat and fish. Since we do not need to eat meat in order to maintain good health, our only justification for doing so is that meat is tasty and convenient; but these are trivial interests and cannot outweigh the suffering inflicted on animals. In *Animal Liberation* he provided a wealth of detail about the ways in which animals suffer to provide us with gratification; over 30 years later, the picture is largely unchanged. It is true that in many countries there is now more protection for research animals, that few cosmetic companies now test on animals, and that practices such as dogfighting and cockfighting are now illegal in all U.S. states, bringing the United States into line with countries such as the United Kingdom, where dogfighting has been illegal since 1835. However, the conditions under which animals are raised for food have become worse, with most meat, dairy, and eggs in the United States and most developed countries increasingly produced in CAFOs. Moreover, with rising affluence in rapidly developing economies such as China and India, there is an increasing demand for animal products, which are most profitably produced in CAFOs.

Much research on animals, Singer argues, is of no possible benefit to humans, except perhaps to the careers of the researchers. While medical research using animals offers potential benefits to humans (though the payoff rate is low), much research is into conditions that sufferers bring upon themselves by choosing unhealthy or high-risk lifestyles. Other conditions such as diseases caused by environmental pollution and unhealthy workplaces could be avoided by changing industrial practices, waste generation and management, and transportation policies. It is also unnecessary to test new products such as cosmetics and cleaning products on animals, if only because the marginal utility value of, say, a new brand of oven cleaner is close to zero.

Animals are widely used in sport and entertainment, such as hunting, fishing, racing, and in circuses and zoos. Again, these practices cause much suffering to the animals involved, while providing only trivial benefits for humans.

Singer has been subject to criticism on several grounds. First, some have argued that we cannot know that animals suffer pain because they are unable to inform us that they are suffering. This is a weak argument, since the same may be said for infants, persons suffering from nerve damage that prevents them from talking, or indeed people who do not have a common language. But we can all recognize behavior that we take to be indicative of pain such as writhing and screaming. Moreover, most of the animals that we eat have nervous systems similar to those of humans.

Another criticism is that, in all but the smallest society, it makes no difference whether an individual becomes a vegetarian because the market is not sensitive to individual purchasing decisions. Thus, abstention from meat does not prevent any animals from suffering. Singer's answer is that if we are serious about animal liberation we should support larger-scale activities by educating friends, joining animal welfare organizations, and lobbying political representatives to support animal protection legislation. Moreover, the support of consumers for organic producers and restaurants is crucial to their profitability and even survival, so purchasing their products does make a difference.

Perhaps the most serious criticism is that if suffering is all that matters, there is nothing wrong with eating animals that have lived pleasant lives and have been killed painlessly. This is an option in countries such as New Zealand, where almost all beef, lamb, and dairy products are derived from free-range animals, free-range eggs are available in all but the most remote areas, and free-range pork and poultry products are available in many supermarkets and via the Internet. Washington University in St. Louis philosopher (and vegetarian) Adam Shriver has even argued that we should use genetic engineering to develop animals that can be raised intensively without suffering. However, for many supporters of animal liberation, the fact that these animals do not suffer seems to miss the point, which is that the rights of the animals are violated.

Tom Regan, a former philosophy professor at North Carolina State University, advocates animal rights. Regan believes that utilitarianism is seriously deficient because it fails to recognize the value of individual lives except as "vessels" for utility. The value of a life, he argues, is more than the satisfaction of people's preferences. Moreover, utilitarianism may require killing or otherwise mistreating others in order to maximize the overall good. For instance, we might be justified in forcing a person, such as a brain-damaged baby, to be the subject of a medical experiment that seems likely to yield results that will save many lives in the future. Singer himself agrees that this is indeed a consequence of his position, though it is admittedly an extremely implausible example.

Life, at least at a certain level of complexity, deserves respect, which, according to Regan, is best expressed by recognizing rights. This respect is due regardless of how many people recognize it and regardless of the contribution of the bearer of this life to the well-being of others. For Kant, respect is owed only to rational beings, but Regan argues that this standard is too restrictive. Many humans such as newborns, persons with serious brain damage, and sufferers of severe dementia are not rational beings, but we still believe that they are entitled to respect and would not subject them to medical experiments (let alone eat them). The reason why we attribute rights to humans is that our lives matter to us: we are not merely sentient but we are also conscious beings and "experiencing subjects of a life," with desires, preferences, memories, expectations, hopes, dreams, and disappointments.

Thus, our lives have inherent value and this should be recognized by acknowledging our rights. But this is true of many animals, and they are also entitled to respect and to be treated as ends, not merely as means. Even painless exploitation of animals is wrong, therefore.

Writers such as Singer and Regan draw attention to the widening of what is often called "the moral community," the class of beings that are entitled to moral consideration and respect. Over the past 200 years, at least in Western societies, slavery has been abolished, women have acquired equal status with men, and many societies provide assistance to the poor from tax revenues and accept refugees. Freedom of religion is taken for granted. For the supporters of animal liberation, their cause is just the next step in the moral progress of humanity.

See Also: Anthropocentrism; Ecofeminism/Ecological Feminism; Ethical Vegetarianism; Intrinsic Value; Singer, Peter; Utilitarianism.

Further Readings

Gruen, Lori. "The Moral Status of Animals," 2003. *Stanford Encyclopedia of Philosophy.* http://plato.stanford.edu/entries/moral-animal (Accessed January 2010).

Kraut, Richard. "Aristotle's Ethics," 2007. *Stanford Encyclopedia of Philosophy.* http://plato.stanford.edu/entries/aristotle-ethics (Accessed January 2010).

Passmore, John. "The Treatment of Animals." *Journal of the History of Ideas,* 26 (1975).

Regan, Tom. *The Case for Animal Rights.* Berkeley: University of California Press, 2004 [1983].

Singer, Peter. *Animal Liberation: A New Ethics for Our Treatment of Animals.* New York: New York Review of Books, 1976.

Alastair S. Gunn
University of Waikato

Anthropocentrism

Anthropocentrism is a philosophy that argues that human beings are the central or most significant entities in the world. This is a basic belief embedded in many Western religions and philosophies that regard man as separate from and superior to nature, and believe that human life has intrinsic value, while other entities (including animals, plants, mineral resources, and so on) are resources that may justifiably be exploited for the benefit of mankind, rather than being regarded as entities with their own intrinsic value. Many ethicists find the roots of the instrumental view of nature common in Western societies (that the natural world has value only as it benefits mankind) and of man's superiority to nature in the Creation story told in the Book of Genesis in the Judeo-Christian Bible, in which man is created in the image of God and is instructed to "subdue" the Earth and to "have dominion" over all other living creatures. However, this line of thought is not limited to Jewish and Christian theology and can be found in other sources, such as Aristotle's *Politics* and in Immanuel Kant's *Lectures on Ethics.*

Some anthropocentric philosophers support a cornucopian point of view that rejects claims that the Earth's resources are limited or that unchecked human population growth will exceed the carrying capacity of the Earth or will inevitably result in wars and famines as resources become scarce (a point of view expressed by Thomas Malthus in his 1798 *Essay on the Principle of Population*). Cornucopian philosophers (the name refers to the "horn of plenty" of Greek mythology, which could magically produce unlimited amounts of food) argue that either the projections of resource limitations and population growth are exaggerated, or that technology will be developed as necessary to solve future problems of scarcity. In either case, they see neither a moral nor a practical need for legal controls to protect the natural environment or limit its exploitation.

Challenges to Anthropocentrism

Prior to the emergence of environmental ethics as an academic field, conservationists such as John Muir and Aldo Leopold were arguing that the natural world had an intrinsic value, an approach informed by aesthetic appreciation of nature's beauty, as well as an ethical rejection of a purely exploitative valuation of the natural world. In the 1970s, scholars working in the emerging academic field of environmental ethics issued two fundamental challenges to anthropocentrism: they questioned whether humans should be considered superior to other living creatures, and also suggested that the natural environment might possess intrinsic value independent of its usefulness to mankind. The philosophy of biocentrism incorporates both beliefs: biocentrists regard humans as one species among many in a given ecosystem and believe that the natural environment is intrinsically valuable independent of its ability to be exploited by humans.

Others environmental ethicists suggested that it was possible to value the environment without discarding anthropocentrism and offered a modified approach (sometimes called prudential or enlightened anthropocentrism), which argued that humans do have ethical obligations toward the environment, but that those obligations can be justified in terms of our obligations toward other humans. For instance, one could argue that polluting the environment was immoral because it negatively affects the lives of other people now living (for instance, those who might be sickened by the air pollution from a factory) or that wasteful use of natural resources was immoral because it would deprive future generations of those resources. In the 1970s, the theologian and philosopher Holmes Rolston III argued that humans had a moral duty to protect biodiversity (i.e., to refrain from eliminating any species of animal or insect, a contested issue at the time due to debate over the legal protection of endangered species, such as that provided by the 1973 Endangered Species Act in the United States). Rolston believed that to do so would show disrespect to God's creations, and eliminate genetic diversity that could be crucial in the continued health of the environment.

More radically, some ethicists have argued that nonhuman entities such as animals or trees should have rights similar to those enjoyed by human beings. For instance, in the early 1970s, Walt Disney enterprises wanted to develop a resort complex in a wilderness area (Mineral King Valley in California), which was challenged by the Sierra Club who argued that wilderness areas were valuable for their own sake (in their undeveloped, wild condition). In conjunction with this case, Christopher Stone argued that trees and mountains should have the legal standing of a person or corporation and could therefore be represented in court and could receive compensation if they were injured. The animal rights (or animal liberation) movement that emerged in the 1970s argued that sentient beings (which would include animals as well as humans) have inalienable rights and should be granted similar protections as we grant to humans: for instance, it should be considered as immoral to kill animals for food or to breed them for use in medical experiments as it would be to use humans for those purposes.

Although the "anthro" in anthropocentrism refers to all humans rather than exclusively to men, some feminist philosophers argue that the anthropocentric worldview is, in fact, a male or patriarchal point of view, and one in which the view of nature as inferior to man is analogous to viewing other peoples (women, colonial subjects, nonwhite populations) as inferior to white Western men, and as with nature, provides moral justification for their exploitation. The term *ecofeminism* (coined in 1974 by the French feminist Françoise d'Eaubonne) refers to a philosophy that looks not only at the relationship between environmental degradation and human oppression, but may also posit that women have a particularly close relationship with the natural world, due to their history of oppression.

See Also: d'Eaubonne, Françoise; Deep Ecology; Ecofeminism/Ecological Feminism; Instrumental Value; Intrinsic Value; Leopold, Aldo; Muir, John; Rolston, III, Holmes; Singer, Peter.

Further Readings

Fleming, Michael. "Anthropocentric vs. Non-Anthropocentric Environmental Ethics." http://ocw.capcollege.bc.ca/philosophy/phil-208-environmental-ethics/non-anthropocentric.htm (Accessed August 2010).

Foltz, Bruce V. and Robert Frodeman. *Rethinking Nature: Essays in Environmental Philosophy.* Bloomington: Indiana University Press, 2004.

Keller, David R., ed. *Environmental Ethics: The Big Questions.* Hoboken, NJ: Wiley-Blackwell, 2010.

Krebs, Angelika. *Ethics of Nature: A Map.* New York: W. de Gruyter, 1999.

Steiner, Gary. *Anthropocentrism and Its Discontents: The Moral Status of Animals in the History of Western Philosophy.* Pittsburgh, PA: University of Pittsburgh Press, 2005.

Sarah Boslaugh
Washington University in St. Louis

Attfield, Robin

Robin Attfield (date of birth unknown) was a professor of philosophy at Cardiff University, Wales, from 1968 until 2010; he has also taught at universities in Nigeria and Kenya. He has published extensively in environmental philosophy as well as in the philosophy of religion. His book *The Ethics of Environmental Concern,* published in 1983, was one of the first to apply a theory of environmental ethics in a systematic way to a range of environmental issues, and he is regarded as a leading theoretician in environmental philosophy and ethics. His basic position is that all living things are entitled to respect because they have intrinsic value, and he finds support for this view in the Judeo-Christian tradition of ecological stewardship. Attfield is generally opposed to deep ecology because his belief is that it ignores the interests of individuals in favor of ecological wholes.

In *The Ethics of Environmental Concern,* Attfield examines the roots of anthropocentric and exploitative attitudes and behavior toward nature. He acknowledges the role of the Judeo-Christian belief in "Man's dominion," including Lynn White, Jr.'s account of the biblical mandate, which, he accepts, has encouraged irresponsible exploitation of the environment. However, he firmly rejects the idea of what he calls human despotism. Rather, he emphasizes the biblical tradition of stewardship, which he sees as a basis for our obligations to avoid pollution, misuse of natural resources, overpopulation, and to preserve natural areas. As stewards, we are not entitled to exploit nature as if it were merely a storehouse of resources created for our personal benefit; on the contrary, we have an obligation to consider the needs of future generations of humans, since they are equally part of our moral community. We must also recognize that humans are not the only beings with moral standing, and acknowledge that nonhumans are also entitled to respect. This is because, like humans, they have intrinsic value, though, for reasons discussed below, they are not our equals.

A second cause of our environmentally irresponsible behavior, he argues, is the secular belief in progress, which is based on the assumption that the development of new technologies will enable us to deal with environmental problems as they arise. Clearly, we cannot naively accept a myth of perpetual progress. Nonetheless, we should recognize that we can and will influence the future, and that we have obligations to provide for posterity. He recognizes that we cannot know for certain what future generations will want, especially those in the distant future, but argues that we can be sure that they will have the same biological needs that we do for food, water, shelter, and so on, and that they will not want their coastal cities to be flooded, their air to be toxic, or their children to be born with chemically induced defects. For those of us in developed countries, these obligations are not limited to our own descendants but are also owed to people in developing countries. Minimally, our obligations to the future include the avoidance of disaster scenarios such as nuclear winter, large-scale global famine leading to food wars, massive depletion of biodiversity, and uncontrolled deposits of hazardous wastes; and we should also avoid depleting resources such as fish stocks, clean water, and energy supply. He argues that we also owe it to them to reduce population growth because the worst effects of unchecked growth will be felt in the future.

The values that Attfied asserts are, he believes, rooted in Western culture and religion, but that is not why they should be accepted. Rather, they should be accepted because they can be shown to be true. Attfield is a moral objectivist and cognitivist: that is, he believes that there are moral truths (for instance, that individual lives have intrinsic value) and that humans can know these moral truths. Such approaches to ethics are vulnerable to the objection that the knowledge claims are no more than the claimed intuitions of their author. However, Attfield has defended his position by appealing to what he refers to as "biocentric consequentialism," which he explains in his book *Environmental Ethics: An Overview for the Twenty-First Century*. This position has three main features. First, the interests of all sentient beings should always be considered, though they do not necessarily all count equally: specifically, the interests of "creatures with complex and sophisticated capacities such as autonomy and self-consciousness" outweigh those of creatures that lack these capacities. Thus, the interests of humans are typically more important morally than those of other animals. Second, a basic or survival need, such as the need for food, outweighs lesser needs. Third, we should adopt practices that, overall, on balance tend to promote the basic interests of creatures with complex capacities.

Attfield's position as an environmental ethicist is complex and perhaps unique. He rejects anthropocentrism and insists that the interests of nonhuman animals must be taken into account but denies that humans and other animals are equal. He rejects holistic ethics such as deep ecology, but asserts that humans are to act as stewards of the whole of creation. His position is in some ways firmly grounded in the Judeo-Christian tradition, but he does not appeal to the authority of that tradition: on the contrary, he maintains that the fundamentals of his position are independently true, including the principle of "biocentric consequentialism."

In recent years, Attfield has published many articles on ethical theory in academic journals and several books and has also maintained his applied focus. As of 2010, he was a member of a United Nations Educational, Scientific and Cultural Organization (UNESCO) working party on environmental ethics that aims to identify possibilities for international action in this area.

See Also: Antropocentricism; Deep Ecology; United Nations Conference on the Human Environment.

Further Readings

Attfield, Robin. *Environmental Ethics: An Overview for the Twenty-First Century.* Cambridge, MA: Polity Press, 2003.

Attfield, Robin. *The Ethics of Environmental Concern.* New York: Columbia University Press, 1983.

Ten Have, Henk A. M. J. *Environmental Ethics and International Policy.* Paris: UNESCO Publishing, 2006.

Alastair S. Gunn
University of Waikato

B

BAILEY, LIBERTY HYDE

Liberty Hyde Bailey (1858–1954) was an American horticulturist, botanist, and cofounder of the American Society for Horticultural Science. He was a man of many interests and talents, a man of perception and insight. Professor Jules Janick described him as an extraordinarily successful scientist, teacher, administrator, poet, and philosopher who profoundly influenced the direction of teaching, research, and extension in horticulture in the United States. He was educated at and taught at the Michigan Agricultural College (now Michigan State University) before moving to Cornell in Ithaca, New York, where he became dean of the Cornell University College of Agriculture and Life Sciences. Most notably, he is known as the Dean of American Horticulture. He wrote scores of books, including scientific works, efforts to explain botany to laypeople, a collection of poetry, and coined the word *cultivar.* Cornell University memorialized Bailey in 1912, when Bailey Hall, the largest building on campus, was dedicated in his honor.

Bailey came from modest beginnings. Born in 1858 in South Haven, Michigan, he was the youngest child in a hard-working family that owned a fruit farm. As a child, he spent a great deal of time in the forests around his father's farm. He became interested at an early age in collecting plants and animals and in their interrelationships. His father, Liberty Hyde Bailey, Sr., a conservative puritan, was an open-minded individual and provided the family with books as well as New York newspapers. Two books—Charles Darwin's *On the Origin of Species by Natural Selection* and Asa Gray's *Field, Forest, and Garden Botany*—were to create a lasting impression on the young Bailey. At the age of 15, Bailey presented a paper titled "Birds," his first public speech, to the South Haven Pomological Society. He subsequently was elected lead ornithologist of this society. The senior Bailey and his sons were skilled and innovative farmers and their farm was known for its prize-winning apples. One of their orchards won a first premium as a model orchard for its perfection in culture, pruning, and grafting. Even as a youth, Bailey became an expert on grafting and his skills were in great demand among his neighbors.

In 1877, at the age of 19, Bailey left South Haven and began his secondary education at the Michigan Agricultural College. Here he worked with William Beal, a botanist with whom he became acquainted through the Michigan State Pomological Society. It was during this time that he became interested in plant breeding. During these student years, he organized and edited a student publication, the *College Speculum.* He became involved in the Natural History Society, the student government, and published his first articles on the identification of local florae in the *Botanical Gazette.* He began his long-term involvement with the classification of the genus *Rubus.* Academically, he was at the top of his class and the great breadth of his activities characterized Bailey's entire life.

Bailey earned his B.S. degree in 1882. After graduation, he tried working as a newspaper reporter, and then spent two years from 1883 until 1884 assisting Asa Gray at Harvard University. At Harvard, he was responsible for sorting and classifying plant specimens received from Kew Gardens in England. Gray felt that research should be conducted in a laboratory or herbarium. Horticulture was considered an ornamental art, while botany was considered the science. In the 19th century, these two disciplines were generally not considered compatible. Consequently, Gray was disappointed when Bailey returned to Michigan in 1885 to assume a newly established chair of Horticulture and Landscape Gardening. So strong was the dichotomy between botany and horticulture that Gray predicted that if Bailey returned to horticulture, he would sacrifice his growing standing as a scientist and fade away into professional obscurity.

Fortunately for the world of horticulture, Gray's prediction could not have been more erroneous. Bailey returned to Michigan State College as a professor of horticulture and landscape gardening in 1884 and stayed until 1888. In 1885, he was granted a master's degree by Michigan. Cornell University offered him the post of professor of horticulture in 1888. He accepted the position with the stipulation that Cornell sponsor a tour to Europe. This enabled Bailey to visit every significant herbarium west of Russia, including Prague, Vienna, and Uppsala. Bailey served as horticulture professor at Cornell until 1903, becoming director of the College of Agriculture at Cornell. In this same year he founded, along with S. A. Beach, the American Society for Horticulture Science. He served as president of that organization for its first four years of existence.

In 1904, the State College of Agriculture at Cornell was founded, largely through Bailey's hard work. He was appointed dean of the faculty, director of the college and its experiment station, and professor of Rural Economy. Bailey established the Department of Experimental Plant Biology at Cornell in 1907; this would later become the Department of Plant Breeding. He retired as dean and director in 1913 but continued to lead a very active life after his official retirement. When he died on December 25, 1954, in Ithaca, New York, he was described as the last living link to the Asa Gray era of American botany.

From the time Liberty Hyde Bailey retired in 1913 until his death in 1954, he remained active in horticulture, publishing prolifically and collecting and classifying plants from all over the world, including the first extensive classification of palms. The establishment of the Bailey Hortorium at Cornell University also resulted from specimens collected by Bailey during his retirement years. Bailey was a prolific writer and editor. He is credited with being instrumental in starting agricultural extension services, the 4-H movement, the nature study movement, parcel post, and rural electrification. His greatest contribution may be in the *Cyclopedia of Horticulture,* a work in five volumes that first appeared in 1914 and that is still enormously useful. Other works include *Hortus* and *Hortus Second* written with his daughter Ethel Zoe Bailey. An update, *Hortus Third,* published by the staff of the Liberty Hyde Bailey Hortorium in 1975, is a bible of horticultural taxonomy.

See Also: Agriculture; Anthropocentrism; Conservation; Land Ethic.

Further Readings

Bailey, Liberty Hyde and Zachary Michael Jack. *Liberty Hyde Bailey: Essential Agrarian and Environmental Writings*. Ithaca, NY: Cornell University Press, 2008.

Cornell University. "Liberty Hyde Bailey: A Man for All Seasons." http://rmc.library.cornell.edu/bailey/cornellu/index.html (Accessed August 2010).

Janick, Jules. *Classic Papers in Horticultural Science*. Englewood Cliffs, NJ: Prentice Hall, 1989.

Carl A. Salsedo
University of Connecticut

Berry, Reverend Fr. Thomas

Reverend Fr. Thomas Berry has helped to forge a change in the perception of the nonhuman presence in our experience. As a member of the Catholic Passionist religious order and a priest, he has recognized the influence of his learning from sciences, thinkers from China and India as much as from traditional Catholic thought. By framing all of reality from the subatomic through the human to the astronomical as subjects rather than objects, he presents a conception of an all-inclusive environmental community. In this view, all environmental subjects have moral, ethical, and spiritual significance. Concern, then, for an integrated interdependent universe and the well-being of each member of that community are inseparable. Environmental degradation and adverse impacts—whether to other human beings, oneself, or to the trees and rocks—is necessarily an ethical and moral issue. The interconnectedness of all of reality in his view, as with Matthew Fox and others, facilitates our ability to acknowledge that any violence to one has consequences for oneself and for the universe. Membership in a universal community leads to an ethical obligation to the right treatment of all members of the all-inclusive community that applies, regardless of one's approach to regulating society. For Fr. Berry, this obligation includes the imperative for a life of nonviolence and a development of a spiritual life.

Combining ecological concerns with religion and theology, Fr. Berry preferred to be known as a "geologian" or an "ecologian." In this endeavor, he sought to build or reconstruct a tradition that transcends previous thinking to include a profound concern for the relationship between humans and the Earth. Transcending the self enables an inquiry into ethics and spirituality and what Berry referred to as the "mystique of nature" and engenders a reverential attitude. Based on a tradition grounded in revelation from a personal God, this reverential attitude is appropriate to his understanding of the universe itself as the primary revelation. Further, since cosmic and socioeconomic concerns cannot be separated, the treatment and actions of humans always have an ethical issue. In his cosmic interest, Fr. Berry's thinking is related to but different from that of Teilhard de Chardin.

In Berry's reviews of several historical and cultural themes, he finds they have not only operated from a perception of nature as corrupt, in need of redemption and susceptible to exploitation for progress, but have exaggerated it. To remedy this view, he adopted a creation-centered approach. He starts from a view of the universe as mother that gives birth to humans. In this sense, the universe as a mother provides nourishment and an environment in which human life is nurtured and flourishes. Divine love and energies are expressed through a reality in which the physical, psychic, and spiritual are inseparable dimensions. The universe is material and physical as well as psychic and spiritual, where humanity is an integral part. For its part, it is in humanity that the universe is reflected and through whom the universe is consciously self-aware in a special way. Humanity is a celebration of the universe.

Fr. Berry, born November 9, 1914, in Greensboro, North Carolina, was the third of 13 children. He attended Mt. St. Mary's preparatory school in Emmitsburg, Maryland, entered the monastery of the Passionist Order in 1933, took his first vows in 1935, final vows in 1938, and was ordained a priest in 1942. During those 10 years at Passionist monasteries in New York, New Jersey, and Massachusetts, he spent his time in study and meditation. In addition to the works of early Church writers, theologians, and Western historians, he also read classical Greek and Latin thinkers and Indian and Chinese writings in translation. Between 1943 and 1947, he did graduate studies at Catholic University in Washington, D.C., reading in history and anthropology. Graduate studies made him sensitive to cultural diversity and worldviews as well as to Giambattista Vico's evolutionary approach to history, which led him to view human development in four stages. In his view, the four stages of human evolution are tribal, traditional-civilizational, scientific-technological, and now ecological, with a different consciousness in each.

He then taught, first in the Passionist seminary at Dunkirk, New York, then from 1948 to 1950 in China, before returning to teach at Dunkirk. He then served as army chaplain in Europe, when he traveled through Europe, North Africa, and the Middle East before returning to the monastery in Jamaica, New York, in 1954. He then taught at Seton Hall University, New Jersey, from 1956 and at St. John's University, New York, from 1960, and at Fordham University from 1966 to 1979. From 1976 to 1987, he was president of the American Teilhard Association and founding director of the Riverdale Center for Religious Research in New York since 1970. He died on June 1, 2009.

See Also: Animal Ethics; *Environmental Ethics* Journal; Intrinsic Value.

Further Readings

Berry, Thomas. "The New Story: Comments on the Origin, Identification and Transmission of Values." *Cross Currents,* 37 (1988).

Chapple, Christopher. "The Living Earth of Jainism and the New Story: Rediscovering and Reclaiming a Functional Cosmology." In *Jainism and Ecology,* Christopher Key Chapple, ed. Cambridge, MA: Harvard University Press, 2002.

Hope, Marjorie and James Young. "Thomas Berry and a New Creation Story." *Christian Century,* 106/24 (1989).

Lonergan, Anne and Caroline Richards, eds. *Thomas Berry and the New Cosmology.* Mystic, CT: Twenty-Third Publications, 1987.

McDougall, Dorothy C. "Toward a Sacramental Theology for an Ecological Age." *Toronto Journal of Theology,* 19/1 (2003).

Tucker, Mary Evelyn. "Thomas Berry: A Brief Biography." *Religion and Intellectual Life, 5/4* (1988).

Lester de Souza
Independent Scholar

Berry, Wendell

All movements have founders, and the greening of America is no exception. Alongside figures such as John Muir, Aldo Leopold, and Rachel Carson, Wendell Berry deserves recognition as one of the founding members, although he is sometimes overlooked in this regard. His *The Unsettling of America* (1977) is arguably as important as *A Sand County Almanac* and *Silent Spring* for promoting a green ethic in America. Published during the time of Earth Day, the Clean Air Act, the Environmental Protection Agency (EPA), and Greenpeace, Berry's message came when American agriculture was moving away from its agrarian roots toward an industrial mode of farming. Berry sounded the alarm for the mechanization and rationalization of agriculture, with its emphases on specialization, efficiency, and intensification. Berry noted that this unsettled not only rural environments and culture, but also had profoundly negative impacts more generally in the United States. Worrying about both rural and urban settings, his concern about industrialization's negative impact on soil, air, and water as well as its reduction of life to economics has been realized all too accurately in subsequent decades. His work in this regard clearly establishes him as one of the founders of the current green and sustainability movement.

Berry was born in rural Kentucky during the Great Depression. After graduating from the University of Kentucky with a B.A. and an M.A., he studied creative writing with Professor of English Wallace Stegner at Stanford, who, although known as a writer of fiction and nonfiction, was himself a conservationist (e.g., *The Quiet Crisis*). After teaching English at New York University, Berry returned to the University of Kentucky where he taught on and off over the next 30 years. Upon moving back to Kentucky, he also deliberately returned to his agrarian roots, settling on a 125-acre farm in the same county in which he was born, to reclaim land that had been neglected in previous decades. Berry, who still lives on and works that same farm, uses farming methods that he believes best exemplify what he calls "the way of agrarianism," a deliberate contrast with the industrialization that defines much of U.S. agriculture and culture today. For example, he uses horses rather than tractors for the heavy work. His way of life is meant to exemplify the green ideals embodied in his alternative vision for sustainable rural and American living.

Although Berry might be best known within environmental circles for his *Unsettling of America,* as a writer he is equally adept in poetry, novels, and essays. Spanning all three genres over the past 50 years, his oeuvre includes more than 13 novels and short-story collections, 20 volumes of poetry, and 20 collections of essays. With such range—in time and genre—one can imagine a diversity of topics. And this is true. His poetry varies from topics such as dying ("Three Elegiac Poems"), love ("The Country of Marriage"), and suffering ("The Way of Pain") to farming ("Manifesto: The Mad Farmer Liberation Front"), restoration ("The Clearing"), and dance ("The Wheel"). His essays range from religious commentary ("The Burden of the Gospels") and technology ("Why I Am Not Going to Buy a Computer") to war ("The Failure of War") and national security ("A Citizen's Response to the National Security Strategy"). His novels typically deal with the tensions between agrarian community life and unsustainable but tempting alternatives in the short term. Throughout his writings he shows uncommon insight and profound sensitivity.

However, despite the enormous variety and styles, it also appears that Berry continually circles around a single complex message, one that he approaches again and again from a myriad of angles, settings, topics, and circumstances. Berry clearly has a permanent burr under his saddle that continually draws him to write the same point in new ways and circumstances. As if responding to an ethical call that comes from elsewhere, throughout his writing and exemplified in his personal life choices, Berry has repeatedly called for an alternative vision of sustainable flourishing.

Berry's message advocates a way of ordinary living that is sustainable—something he construes extremely broadly. Although not averse to political action, he is after fundamental changes in the current direction of American culture and practice. A central concern is that our economics is upside down: we start with an economy of consumption, which drives an economy of manufacture, which drives an economy of exploitation of nature. Instead, he advocates starting with the economy of nature and nestling into it the economies of industry and consumption while guarding certain things as priceless: clean air and water, functioning ecosystems, natural beauty. More generally, Berry's arguments for sustainability concern the intertwinement of a complex of relationships—natural, social, economic, moral, religious. He has been motivated by a call to be responsibly at home in his local place and the world, which for him has meant advocating ways of living that enhance local control, personal responsibility, steadfast relationships, local knowledge, and restoration of land. In this, he connects farmland preservation with urban revitalization. He advocates a way of life that considers our boundedness and finitude as the foundation for human flourishing. These echo Jeffersonian values of small and independent landowners rather than global corporatism and economies of scale. He believes that it is only through these forms of life that human and, indeed, the flourishing of all creation will be sustainable in the long term.

See Also: Agriculture; Carson, Rachel; Conservation, Aesthetic Versus Utilitarian; Earth Day 1970; Economism; Leopold, Aldo; Muir, John; *Silent Spring*.

Further Readings

Berry, Wendell. *The Broken Ground*. New York: Harcourt, Brace & World, 1964.
Berry, Wendell. *Farming: A Hand Book*. New York: Harcourt Brace Jovanovich, 1985.
Berry, Wendell. *Life Is a Miracle: An Essay Against Modern Superstition*. Washington, DC: Counterpoint, 2000.
Berry, Wendell. *Sabbaths*. New York: North Point Press, 1987.
Berry, Wendell. *The Unsettling of America: Culture and Agriculture*. San Francisco, CA: Sierra Club 1977.
Berry, Wendell. *The Way of Ignorance: And Other Essays*. Reno, NV: Shoemaker & Hoard, 2005.
Bonzo, J. Matthew. *Wendell Berry and the Cultivation of Life: A Reader's Guide*. Grand Rapids, MI: Brazos Press, 2008.
Smith, Kimberly K. *Wendell Berry and the Agrarian Tradition: A Common Grace*. Lawrence: University of Kansas Press, 2003.

Clarence W. Joldersma
Calvin College

Biocentric Egalitarianism

While ethics' traditional concern has been with the duties among human beings, the emergence of environmental ethics has extended the scope of ethical inquiry beyond the level of human interaction, community, and nation to include nonhuman species and, in fact, the whole of nature. Fundamentally, environmental ethics has played an important part in challenging the historically dominant and deep-rooted anthropocentrism (human-centeredness) of conventional ethics. In other words, whereby the material condition for anthropocentric ethics is respect for persons, environmental ethics has begun to emphasize a respect for nature. The ethics of biocentrism, in particular, stipulates that all life forms are moral agents—entities that must be accorded moral consideration. An early proponent of this notion, Albert Schweitzer, for example, argued in 1923 that "all living beings have the will to live, and all beings with the will to live are sacred, interrelated and of equal value."

According to American philosopher Paul Taylor, anthropocentrism "gives either exclusive or primary consideration to human interests above the good of other species." Countering this notion most famously in his book *Respect for Nature: A Theory of Environmental Ethics* (1986), Taylor instead proposes an egalitarian biocentrism founded on four core principles:

- Humans are members of the Earth's Community of Life in the same way and under similar conditions as other living entities are members of that community
- The human species, along with all other species, are integral elements in a system of mutual interdependence
- All organisms are teleological centers of life in a sense that each is a unique individual attempting to pursue its own good in its own way
- Humans are not inherently superior to other living beings

Humans as Members of the Earth's Community

Taylor roots humans' membership in the Community of Life in five realities: (1) all living beings essentially face the same biological and physical requirements for survival and well-being, (2) everything that is alive can be said to have a good, (3) although only humans have free will, autonomy, and political/social freedom, more importantly, the freedom to preserve one's existence and further one's good is shared by human and nonhuman beings, (4) human beings are relative newcomers to an order of life that has otherwise been established for millions of years, and (5) we cannot do without other living beings, though they can do without us.

Interdependence

Taylor stipulates that in accepting his biocentric outlook, one understands the entire realm of life to consist of a vast complex of relationships of interdependence. As a particular change happens to one part of the system, adjustments will necessarily occur throughout the wider structure. Though interconnectedness may not always be direct, no part of the system is an isolated unit. Taylor relies here also on Aldo Leopold's work *A Sand County Almanac,* which has inspired many holistic approaches such as Taylor's.

Teleological Centers of Life

Taylor maintains that it is their *telos* (the Greek word for "end," "purpose," or "goal") that gives each individual organism inherent/intrinsic worth (things valuable for their own sake rather than things valuable only because they are valued by others, for example, by humans), and that this worth is possessed equally by all living organisms because all individual living beings have a telos and a good of their own—a good as vital to them as a human good is to a human (see also the article on Holmes Rolston III for his understanding of the notion of intrinsic value).

Denial of Human Superiority

Although we are different from other living beings in our capacities, Taylor completely rejects anthropocentric outlooks that support the idea that human beings are superior to other living things. In fact, he rejects the idea of any species being inherently superior (or inferior) to another. Just as humans place intrinsic value on the opportunity to pursue their own good in ways they deem most suitable, the realization of the good of animals and plants should similarly be valued as equally worthy. If all living beings have the same inherent worth, then the same amount of respect is due to each.

In summarizing, Taylor stipulates that if we concede the first three principles, the fourth will logically follow. He concludes: "Rejecting the notion of human superiority entails its positive counterpart: the doctrine of species impartiality. One who accepts that doctrine regards all living things as possessing inherent worth—the same inherent worth, since no one species has been shown to be either higher or lower than any other."

At a theoretical level, then, philosophers such as Taylor make bold egalitarian assumptions about the equal intrinsic worth of all entities within a chosen category. Paul Taylor does so with regard to "living things." Others, such as Tom Regan, make positive commitments to each "subject of a life" (in his case at least all mammals over one year of age). However, a sustained criticism of such egalitarian approaches is how to resolve conflict of interest cases. It has been disputed, for example, whether it is actually possible to maximize an aggregate of goods without reaching a point beyond which one entity's best interest is not pursued at the expense of another's. Taylor's (1986) response to such arguments is to maintain five "priority principles": they include giving weight to principles of (1) self-defense—permitting moral agents to protect themselves against dangerous or harmful organisms by destroying them, (2) proportionality—giving greater value to basic than to nonbasic interests, no matter the species, as competing claims arise, (3) minimum wrong—actions taken by individuals in the pursuit of ends that lie at the core of their rational conceptions of their true good must be such that no alternative ways of achieving those ends produce fewer wrongs to other living things, (4) distributive justice—providing for just distribution of interest-fulfillment when conflicting interests are all essential (basic) and hence of equal priority, and (5) restitutive justice—restoring the balance of justice after a moral subject has been wronged. These principles seemingly permit occasions that would allow a human being to get favorable treatment over some nonhuman animals or plants, though never on the grounds that the human possesses great intrinsic worth. Nevertheless, this raises concerns over the genuinely egalitarian nature of Taylor's theory. William French, for example, has pointed out a contradiction between the egalitarian principles Taylor officially endorses and the less obvious principles by which we should live.

Further criticism of biocentric egalitarianism and the intrinsic value theory it espouses has focused on a number of points (see, for example, David Schmidtz, William C. French,

James Anderson, and Tim Hayward). First, biocentrism's rejection of anthropocentrism has been accused of the practice of anthropomorphizing—simply projecting human characteristics, priorities, and needs onto other beings, whereby these might be dramatically different from what human beings can comprehend. It is therefore possible that attempts to work in another entity's best interest can do it harm, the very thing biocentric egalitarianism aims to avoid. Though Taylor points at this possible fallacy in his argument, he cannot completely dispel it.

Schmidtz has pointed out the arbitrariness of Taylor's philosophy: "if biocentrism amounts to a resolution to value only those capacities that all living things share—then biocentrism is at least as arbitrary and question-begging as anthropocentrism."

Most importantly, biocentrists such as Taylor have been accused of drawing the boundary of moral concern at living beings. Although biocentrism has shifted the moral universe away from a focus on human beings, questions remain as to why intrinsic value considerations should be limited to living beings. Deep ecologists for example, argue for intrinsic value also of rivers, ecosystems, rocks, and so on, more radically stipulating that all natural things have value in themselves, which may sometimes even exceed that of human beings.

See Also: Anthropocentrism; Biocentrism; Deep Ecology; Intrinsic Value; Taylor, Paul W.

Further Readings

Anderson, James C. "Species Equality and the Foundations of Moral Theory." *Environmental Values,* 2 (1993).

French, William C. "Against Biospherical Egalitarianism." *Environmental Ethics,* 17/1 (1995).

Hayward, Tim. *Ecological Thought.* Cambridge, UK: Polity Press, 1995.

Pyra, Leszek. "The Anthropocentric Versus Biocentric Outlook on Nature." *Analecta Husserliana,* 101/1 (2009).

Schmidtz, David. "Are All Species Equal?" *Journal of Applied Philosophy,* 15/1 (1998).

Schweitzer, Albert. *Civilization and Ethics.* London: Adam and Charles Black, 1946.

Taylor, Paul W. "The Ethics of Respect for Nature." In *Planet in Peril,* Dale Westphal and Fred Westphal, eds. New York: Harcourt Brace, 1994.

Taylor, Paul W. "In Defense of Biocentrism." *Environmental Ethics,* 5/3 (1983).

Taylor, Paul W. *Respect for Nature: A Theory of Environmental Ethics.* Princeton, NJ: Princeton University Press, 1986.

Fanny Thornton
Australian National University

Biocentrism

Biocentrism is a nonconsequentialist and nonanthropocetric ethical perspective concerned with the moral relationship between humans and nonhuman beings, animals and plants, and the Earth's natural environment, globally considered. Human beings face complex ethical dilemmas in their relationships with the natural environment and with other forms of life. These ethical problems have been addressed through different viewpoints. The article explores and discusses the place of biocentrism within the general framework of ethical approaches concerned with humankind's relationships with other living beings on Earth.

The differing ethical perspectives can be grouped into two main dimensions. The first distinguishes consequentialist from nonconsequentialist or deontological ethical theories, and the second differentiates anthropological from nonanthropological ethical theories. A consequentialist perspective sees the outcome or the consequence of an action as the criterion to assess whether such conduct or action is ethically acceptable or not, if it is good or bad; conversely, a nonconsequentalist or deontological perspective takes the view that every action or conduct has an intrinsic worth independent of its outcomes or consequences. The second dimension is concerned with the definition of the moral community and, in that context, the discussion is centered on the importance or relevance of human beings to other living beings on Earth, and the natural environment as a whole. Two contrasting viewpoints can be considered in this dimension: one is human centered, and the other is life centered or biocentered. On one side, human-centered or anthropocentric perspectives consider humans as the only species that matter for the definition of the moral community. On the opposite end, for life-centered or nonanthropocentric perspectives, nonhuman beings and the natural environment have an inherent value independent from the relevance or importance of them to humankind.

Considering both dimensions, four main groups of ethical theories can be defined: consequentialist and anthropological perspectives; consequentialist and nonanthropological perspectives; nonconsequentialist and anthropological perspectives; and nonconsequentialist and nonanthropological perspectives. In this taxonomy of ethical theories, biocentrism, being an Earth-centered or ecocentric ethic, is placed in the nonanthropological and nonconsequentialist category of ethical theories.

Biocentrism offers answers to the question of whether human beings do or do not have obligations to the Earth's natural environment, and can suggest how to balance and assess them against the interests, desires, and values of human beings, making it a nonconsequentialist and nonanthropocentric perspective. In biocentrism, humans are part of the Earth's ecosystem but are not considered above it in hierarchy. In that sense, as in other approaches within this group of ethical perspectives, a correct human ethical conduct seeks to fulfill the interests of a particular human community, without compromising moral duties in relation to nonhuman living beings and the natural environment, irrespective of the value such conduct or action might have for human development objectives.

For that reason, biocentrism is frequently contrasted with anthropocentrism, an ethical perspective that tends to see the natural environment and nonhuman living beings solely as objects for human use. While the latter bases its normative arguments on the interests, values, and desires of human beings, biocentrism attributes an intrinsic value to all plants and animals, and to ecosystems, broadly conceived.

Biocentrism Writings

Biocentrism is associated with, among others, the work of Paul W. Taylor and his book *Respect for Nature,* as well as Tom Regan and Peter Singer's writings on animal rights. It is also often associated with the biological egalitarianism of Arne Naess and the deep ecology perspective. Biocentrism is also connected to the pioneering work of Aldo Leopold, especially in his essay "Land Ethic," published in the book *A Sand County Almanac,* where he argues that a true land ethic or Earth-centered ethic is one that expands the limits of the moral community to include soil, waters, plants, and animals, or the entire ecosystem or the land. In other words, for this pioneer of the biocentric perspective, a land ethic moves humans from a dominant position to one of a simple member of the Earth community, requiring that humans respect the other members of this biotic community.

As Paul Taylor states in *Respect for Nature,* biocentrism is essentially about respecting nature and understanding that plants and animals are living beings with inherent or implicit worth in themselves, independent of any value or interest for human beings. However, the concept of biocentrism faces challenges between the respect due to nonhuman beings and the intrusive measures taken by humans within, for example, biodiversity conservation policies (e.g., the introduction of a species, or the protection of one and the potential negative impacts on other species).

Biocentrism includes different approaches or theoretical streams, some of them taking as moral references collective entities (e.g., species, ecosystems) and others taking individual living beings. For example, in the animal liberation movement, only sentient beings are considered to be worthy of ethical consideration and, for that reason, nonsentient beings (e.g., plants) are excluded from such moral consideration. For other biocentric approaches, for example, in the animal rights movement, the focus is centered on mammals. Both perspectives have been criticized for their concentration and exclusive attention on animals. Other, more moderate biocentric perspectives tend to consider that nonhuman beings have intrinsic value, but in case of conflict, the interests of humans will be considered first if these interests interfere with vital needs of human beings, although the exact definition of the term *vital needs* is a matter of never-ending discussion in different cultural and social contexts.

Biocentrism, or at least some of its approaches, have been criticized for their individualism (e.g., the consideration and respect for the life of individuals and not groups of individuals, as would be the case of an entire species or ecosystem) and for its unrealistic rationalism, in the sense that in the practical application of its ethical principles, it is the values, desires, and interests of humans that seem to prevail. In fact, the biological egalitarianism of Arne Naess, to mention just one Earth-centered ethical perspective, does not mean that human needs should never have precedence over nonhuman needs. In other words, biocentrism, while stating that human and nonhuman living beings have the same right on Earth, accepts that it is logical to affect nonhuman beings in order to meet human basic needs, as long as it remains compatible with the due respect for nature. Biocentrism has also been criticized for granting moral consideration to nonhuman beings, which for its critics can only be given on a reciprocal basis, which is not the case with plants and other animals.

See Also: Animal Ethics; Anthropocentrism; Biocentric Egalitarianism; Deep Ecology; Ecocentrism; Ecofeminism/Ecological Feminism; Intrinsic Value; Land Ethic; Leopold, Aldo; Naess, Arne; Passmore, John; Singer, Peter; Taylor, Paul W.

Further Readings

Beatley, Timothy. *Ethical Land Use: Principles of Policy and Planning.* Baltimore, MD: Johns Hopkins University Press, 1994.

Darwall, Stephen, ed. *Consequentialism.* Oxford, UK: Blackwell, 2003.

Darwall, Stephen, ed. *Deontology.* Oxford, UK: Blackwell, 2003.

Leopold, Aldo. *A Sand County Almanac.* Oxford, UK: Oxford University Press, 1987.
Naess, Arne. *Ecology, Community and Lifestyle: Outline of an Ecosophy.* Cambridge, UK: Cambridge University Press, 2003.
Regan, Tom and Peter Singer, eds. *Animals Rights and Human Obligations.* Upper Saddle River, NJ: Prentice Hall, 1989.
Singer, Peter. *Libertação animal* [*Animal Liberation,* 1st ed., 1975]. Porto, Portugal: ViaOptima, 2000.
Taylor, Paul W. *Respect for Nature. A Theory of Environmental Ethics.* Princeton, NJ: Princeton University Press, 1989.

Carlos Nunes Silva
University of Lisbon

Biodiversity

Tree ferns, banana plants, ephiphytes, bromeliads, and other plants growing in a tropical rainforest environment in the Roseau River Valley in Dominica in 2006. While 1.7 to 2 million species have been formally identified, the actual total number of species is estimated to be between 5 million and 30 million.

Source: National Biological Information Infrastructure/Randolph Femmer

Biodiversity, or biological diversity, is the variety of life found in a place on Earth or the total variety of life on Earth. Conventionally, biodiversity refers to three levels of biological organization: genes, species, and ecosystems. Genetic diversity defines biodiversity as the variation of genes within a species. Species diversity, the most common measure of biodiversity, refers to the varieties of species within a habitat or a region and also to their relative abundance. Ecosystem diversity is the result of the physical characteristics of the environment, the diversity of species present, and the interactions that the species have with each other and with the environment. The inventory of Earth's biodiversity is very incomplete. About 1.7 to 2 million species have been formally identified. Estimates of the total number of species range from 5 million to 30 million. Of the species now described, two-thirds are known from one location and many from examining only one organism or a limited number of organisms, so knowledge of the genetic variation within species is even more constrained.

Scientists around the world are warning that a substantial and largely irreversible biodiversity loss, epitomized by species extinctions, is under way. There is evidence that over the past few hundred years, humans have increased species extinction rates by 1,000 times. Biodiversity is declining due to land use change, climate change, invasive species, overexploitation, and pollution. These result from demographic, economic, sociopolitical, cultural, technological, and other indirect drivers.

The current phase of global concern over the increasing impact of human actions on biodiversity began in the mid-1980s. To be sure, the history of sporadic and often localized alarm about habitat destruction and extinction of species is much longer, with Rachel Carson's *Silent Spring* (1962) being the most prominent example. But by the 1980s, field biologists and ecologists were among the first who expressed their concern about biodiversity loss on a global scale. Their warnings were complemented by the emergence of international social movement organizations, such as World Wildlife Fund (WWF), Friends of the Earth (FOE), and Greenpeace, which continued the work of local and national conservation groups at the international level. Popularization of biodiversity threats has been closely connected with highlighting the plight of specific organisms, mainly large mammals, birds of prey, and a few groups of insects such as butterflies.

A second factor in the emergence of a global biodiversity agenda was the changing political economy of the agricultural, pharmaceutical, and chemical industries, facilitated by key innovations in biotechnology. In the so-called life sciences, the products of the new biotechniques are the prospective basis for future capital accumulation for which biodiversity represents the raw material. Bioprospecting refers to research into the "green gold" of biological resources that can be used in commercialized agricultural, pharmaceutical, or chemical processing end products. From the perspective of the developing countries where the centers of biodiversity are concentrated, this situation demands a major revision in thinking on these issues. In particular, genetic resources in these countries had previously been considered a "heritage of mankind," and Western scientists had been allowed access for research purposes on the assumption that scientific knowledge was a common property. Now, germplasm and scientific knowledge about its properties are increasingly appropriated by large transnational companies as productive resources. This applies also to traditional knowledge present in regional, local, or indigenous communities. It is increasingly recognized that cultures of these people are invaluable repositories of knowledge about useful properties of biological resources, and the companies strive to recur to such knowledge to avoid high research costs.

These developments led to the emergence of biodiversity as a high-profile topic of intergovernmental concern. An international treaty, the Convention on Biological Diversity (CBD), was adopted at the 1992 Earth Summit in Rio de Janeiro. The convention's objectives are threefold: the conservation of biological diversity, the sustainable use of its components, and the fair and equitable sharing of benefits arising from genetic resources (Article 1). The CBD process was shaped by conflicts between, on the one hand, governments of the advanced nations that were concerned with furthering the interests of their biotechnology industries by insisting on the recognition of intellectual property rights of biotechnology products and, on the other hand, the demand of the developing countries to acknowledge national sovereignty over biodiversity. The bargain in the CBD was that developing countries would conserve biodiversity and allow access to it on condition that they would share the benefits of commercial exploitation of the biological resources. Most significantly, this was the first time that "national sovereignty" was placed above natural resources, replacing the previous notion of a "common heritage of mankind" in the appropriation of biological diversity.

Why does biodiversity matter? The values of biodiversity can be divided into use values and non-use values. Use values derive from the direct role of biological resources in consumption and production (e.g., food, medicine, use in biological control, industrial raw materials). Use values of biodiversity can also be indirect, where the organisms are not used as commodities. Indirect uses include services provided by biodiversity that are crucial for human well-being, for example, ecosystem services such as the regulation of climate, floods, and water quality as well as supporting services such as soil formation and nutrient cycling. Also, non-use values are associated with biological resources if they are not directly or indirectly exploited. They include option value (for future use or non-use), educational value, and intrinsic (for example, spiritual) value of biodiversity.

In recent policy debates a new paradigm in conservation emerged that draws on arguments based on the value of ecosystem services and, hence, on the use values of biodiversity. The conservation of biodiversity is now conventionally expressed in terms of the maintenance of natural capital and the supply of ecosystem services. The economic valuation of species and ecosystems received widespread attention in the report of the Millennium Ecosystem Assessment (2005) and is now a recognized dimension of conservation science. The link between biodiversity conservation and the delivery of valuable ecosystem services adds an additional dimension to preexisting conservation concerns over habitat loss and species extinction.

See Also: Animal Ethics; Climate Ethics; Conservation; Preservation.

Further Readings

Daily, Gretchen C. and Katherine Ellison. *The New Economy of Nature: The Quest to Make Conservation Profitable.* Washington, DC: Island Press, 2002.

Gaston, Kevin J. and John I. Spicer. *Biodiversity: An Introduction,* 2nd ed. Malden, MA: Blackwell, 2004.

Hill, David, Matthew Fasham, Graham Tucker, Michael Shewry, and Philip Shaw, eds. *Handbook of Biodiversity Methods: Survey, Evaluation and Monitoring.* Cambridge, UK: Cambridge University Press, 2005.

Levin, Simon A., ed. *Encyclopedia of Biodiversity.* 5 vols., 2nd ed. Oxford, UK: Elsevier, 2007.

McManis, Charles, ed. *Biodiversity and the Law: Intellectual Property, Biotechnology, and Traditional Knowledge.* London: Earthscan, 2007.

Millennium Ecosystem Assessment. *Ecosystems and Human Well-Being: Biodiversity Synthesis.* Washington, DC: World Resources Institute, 2005.

Roe, Dilys and Joanna Elliott, eds. *The Earthscan Reader in Poverty and Biodiversity Conservation.* London: Earthscan, 2010.

Wilson, Edward O., ed. *Biodiversity.* Washington, DC: National Academy Press, 1988.

Sabine Weiland
Catholic University of Louvain

Bookchin, Murray

Murray Bookchin (1921–2006) was the first of the "New Left" writers to make the link between radical politics and environmental consciousness. His works explored his concept of "social ecology," providing a theoretical link between human and ecological frameworks. He was the author of over 20 books, and his collected works include books on ecology, history, politics, philosophy, and urban living. He was born in New York to Russian-Jewish immigrants in 1921. As a child, he participated in the Communist youth movement. Never having the opportunity to attend college, he worked in foundries and as an auto worker. While so doing, Bookchin also becoming a trade union organizer and took part in the great General Motors strike of 1946. He also served in the U.S. Army. His work in factories allowed him to become an organizer for the Congress of Industrial Organizations (CIO). As far back as the 1930s, he had broken with Stalinism in favor of Trotskyism, working for the periodical *Contemporary Issues*. Subsequently, he shifted from the Marxism of his youth and became an anarchist. As a newly found anarchist, he participated in the establishment of the Libertarian League in New York during the 1950s.

Bookchin's Works

Bookchin's first book, *Our Synthetic Environment,* was published in 1962 under the pseudonym Lewis Herber. The book addressed a broad range of ecological issues and was the first such text of the 1960s to do so (Rachel Carson's *Silent Spring* was published later that year). In his early works, Bookchin argued for a decentralized society using alternative energy sources. Throughout his writings, he developed his concept of "social ecology," stating that the solutions to environmental issues have their basis in human politics. Throughout the counterculture movements of the 1960s, Bookchin became a pivotal figure in introducing ecological politics to radical groups agitating for change. His works were influential in the emerging politics of the "New Left." Many of Bookchin's works from this time were included in his book *Post-Scarcity Anarchism* published in 1971. Throughout the 1970s, Bookchin's writings and lectures influenced the environmental movement throughout the United States and internationally. In 1974, he cofounded the Institute for Social Ecology (ISE) in Vermont, later becoming its director and lecturing on social theory, ecological philosophy, and alternative planning. Bookchin also taught at Ramapo College, New Jersey, until 1981, retiring as professor emeritus.

In 1982, Bookchin published *The Ecology of Freedom,* one of his most influential works in social thought. This book had a profound impact on the emerging ecology movement, both in the United States and abroad. In 1986 he published two books, *The Rise of Urbanization* and *The Decline of Citizenship*. These works outlined his vision for what has since been described as "town-hall" politics; local democracy with real decision-making powers.

He was active in the antinuclear movement in New England, and his lectures in Germany had an influence on the German Greens. In his book *From Urbanization to Cities,* Bookchin traced the democratic traditions that influenced his political ideology. In 1999, Bookchin moved away from the rugged individualism of libertarian-anarchist thought, moving to develop his ideas into a wider understanding of locally oriented communalism. However, Bookchin incorporated his existing concepts that focused on the need for decentralization and localism as the ideal social, economic, and industrial plan for society. Bookchin was also active in local politics. While living in Vermont, Bookchin worked with the North Vermont Green Party and served on the Vermont Council for Democracy. His later works included *Remaking Society* (1989) and

The Murray Bookchin Reader (1997). He continued to teach at the ISE until 2004. He died in 2006 at the age of 85.

Bookchin's Legacy

Bookchin's legacy to radical politics is significant, extending across the spectrum of radical politics. He is a primary figure in the fields of green politics, the antiglobalization movement, and the grassroots libertarian movement. His works on libertarian municipalism and grassroots democracy influenced the direct action movement of the 1990s, which emerged in campaigns such as Reclaim the Streets and the WTO protests from Seattle onward. While a key thinker in green politics, he remained skeptical of "new age" mysticism and was critical of biocentric philosophies such as deep ecology and the biologically deterministic beliefs of socio-biology, contributing to the divisions in the green political movement during the 1990s. Bookchin's theoretical legacy is considerable, and he developed the key concepts of social ecology, social anarchism, and dialectical naturalism that have influenced understandings of contemporary environmental and green political theory.

Bookchin's Key Works

- *Our Synthetic Environment* (1962)
- *Post-Scarcity Anarchism* (1971)
- *The Limits of the City* (1973)
- *The Spanish Anarchists: The Heroic Years* (1977, republished 1998)
- *Toward an Ecological Society* (1980)
- *The Ecology of Freedom: The Emergence and Dissolution of Hierarchy* (1982)
- *The Modern Crisis* (1986)
- *The Rise of Urbanization and the Decline of Citizenship* (1987, republished 1992)
- *The Philosophy of Social Ecology: Essays on Dialectical Naturalism* (1990, republished 1996)
- *To Remember Spain* (1994)
- *Re-Enchanting Humanity* (1995)
- *The Third Revolution: Popular Movements in the Revolutionary Era* (1996–2003)
- *Social Anarchism or Lifestyle Anarchism: An Unbridgeable Chasm* (1997)
- *The Politics of Social Ecology: Libertarian Municipalism* (1997)
- *Anarchism, Marxism and the Future of the Left: Interviews and Essays, 1993–1998* (1999)

See Also: Deep Ecology; Social Ecology; Urbanization.

Further Readings

Bookchin, Murray, Dave Foreman, Steve Chase, and David Levine. *Defending the Earth: A Dialogue Between Murray Bookchin and Dave Foreman.* Cambridge, MA: South End Press, 1999.

Callicott, J. Baird and Michael P. Nelson, eds. *The Great New Wilderness Debate.* Athens: University of Georgia Press, 1998.

Institute for Social Ecology. http://www.social-ecology.org (Accessed March 2010).

Liam Leonard
Institute of Technology, Sligo

Borlaug, Norman

Norman Borlaug studying resistance to the fungal disease wheat rust in northern Mexico in 1964. The new varieties of wheat he introduced enabled Mexico to increase its wheat production sixfold in comparison to the early 1940s, and his work eventually led to the green revolution in agriculture.

Source: Flickr/CIMMYT

Norman E. Borlaug (1914–2009) was an American agriculturalist, plant scientist, and humanitarian whose advancements in plant breeding led to significant increases in food production in Latin America and Asia during the 20th century. He was widely regarded as the father of the agricultural movement known as the green revolution, although he was reluctant to accept the title. In 1970, he was awarded the Nobel Peace Prize for his contributions to world food production and their implications for world peace.

Norman Ernest Borlaug was born on March 25, 1914, near the settlement of Saude, a community of Norwegian immigrants in northeastern Iowa. He was raised on a mixed crop and livestock family farm in nearby Protivin, outside the town of Cresco. In 1933, he enrolled at the University of Minnesota, but worked to fund his education, holding positions with the U.S. Forest Service both before and after completing his undergraduate studies. He graduated with a bachelor of science degree in forestry from the University of Minnesota in 1937. Soon thereafter, he fell under the influence of an expert in plant diseases, Elvin C. Stakman, who convinced him to pursue graduate education. Borlaug returned to the University of Minnesota and received a master of science degree in 1940 and a Ph.D. in plant pathology in 1942, conducting research with Professor Stakman. Borlaug worked briefly as a microbiologist for DuPont during 1942–44, but was persuaded by Professor Stakman to join the Rockefeller Foundation's pioneering agricultural assistance program in Mexico, a precursor to the International Maize and Wheat Improvement Center (CIMMYT). In 1944, Borlaug became the research scientist in charge of wheat improvement. He worked to solve a series of wheat production problems that were limiting wheat cultivation in Mexico. With the establishment of CIMMYT in Mexico in 1965, Borlaug assumed leadership of the wheat program. Borlaug's initial goal was to create varieties of wheat that were adaptable to Mexico's climate and could resist the fungal disease wheat rust. He achieved success within several years while also developing varieties of wheat insensitive to day length, enabling them to grow in many locations, a trait of vital significance. He developed successive generations of wheat varieties with broad and stable disease resistance and broad adaptation to growing conditions.

In 1953, Borlaug began working with a gene that created a short, compact variety of wheat. The seed heads of this variety were large, however, meaning a smaller plant could still produce a large amount of wheat. Borlaug and his team soon transferred the gene into tropical wheat varieties. The new "semidwarf" varieties produced large heads of grain without falling over from excess weight. This alteration allowed wheat output to be tripled or quadrupled on the same amount of land. By the early 1960s, many farmers in Mexico had adopted the varieties produced by Borlaug's breeding program, and wheat output in the country had soared sixfold from the levels of the early 1940s.

Following World War II, concerns about population growth and strain on food supplies spread throughout the developing world, particularly on the Indian subcontinent. Indian and Pakistani farmers adopted Borlaug's new varieties of wheat and, just as in Mexico, yields increased significantly. Soon thereafter, the Rockefeller Foundation and other donors established programs in southeast Asia to develop similar characteristics in rice.

These programs led to the creation of "semidwarf" rice varieties with significantly larger yields. The agricultural advancements developed during this time became known as the green revolution, and they are credited with saving hundreds of millions of people from hunger and starvation.

More recently, debates have arisen over the social and environmental consequences of the green revolution. Many critics argue that it displaced smaller farmers, encouraged overreliance on chemicals, and ultimately allowed greater corporate control of agriculture. Borlaug largely disagreed with these arguments, although he did acknowledge the validity of some environmental concerns. Borlaug also believed that genetic modifications and advancements in biotechnology would provide similar innovations in agricultural development in the future.

From 1984 until his death, Borlaug was the Distinguished Professor of International Agriculture at Texas A&M University, the president of the Sasakawa Africa Association, and leader of the Sasakawa-Global 2000 Agricultural Program in sub-Saharan Africa. Numerous governments, universities, scientific associations, and civic associations have honored Borlaug. He was awarded 54 honorary doctorate degrees and belonged to the Academies of Science in 12 countries. In 2006, Borlaug received the Congressional Gold Medal, America's highest civilian honor. In doing so, he became one of only five people in history to be awarded the Nobel Peace Prize, the Presidential Medal of Freedom, and the Congressional Gold Medal. He was also the driving force behind the establishment of the World Food Prize in 1985, which is awarded annually in recognition of outstanding human achievements in the fields of food production and nutrition.

See Also: Agriculture; Genetic Engineering; Technology.

Further Readings

Borlaug, Norman E. "Contributions of Plant Breeding to Food Production." *Science*, 219/4585 (1983).

Borlaug, Norman E. "The Impact of Agricultural Research on Mexican Wheat Production." *Transactions of the New York Academy of Science*, 20 (1958).

Borlaug, Norman E. "Mexican Wheat Production and Its Role in the Epidemiology of Stem Rust in North America." *Phytopathology*, 44 (1954).

William M. Collier
Independent Scholar

Bright Green Environmentalism

Bright green environmentalism is characterized by an emphasis on design, technological innovation, entrepreneurialism, and consumption practices. Bright green ideology stands in contrast to the long-held antimodernist and anti-industrial underpinnings of mainstream environmentalism. Bright green environmentalism advances the idea that economic prosperity and growth are not antithetical to environmental sustainability; the proper consumption practices and well-designed products and systems are compatible with and can generate sustainable practices.

The term *bright green* is attributed to journalist Alex Steffen, a cofounder and editor of the media organization Worldchanging. Worldchanging is an online forum, idea clearinghouse, and think tank for "bright green" ideas, ventures, and discussion. Since its inception in 2003, Worldchanging has gathered a sizable following, garnered critical praise and achieved a fair amount of public notoriety. According to Steffen, the moniker *bright green* has been appropriated by "thousands of organizations—businesses, nongovernmental organizations (NGOs), blogs, student groups, even churches." Worldchanging and Steffen have been featured in a number of mainstream media outlets like the *New York Times*.

The 2006 print publication *Worldchanging: A User's Guide to the 21st Century,* clearly advances the fundamental bright green proposition of "sustainable prosperity." Steffen tells his readers, "We don't have to destroy the planet or impoverish other people to live well. We're often told that there are trade-offs between doing well and doing right, but when we pull back and look at the big picture, those trade-offs usually prove to be illusions. In fact, we're learning more and more that doing the right thing in an intelligent way often pays off handsomely." The book then goes on to list and briefly summarize hundreds of contemporary technologies, initiatives, products, and practices as exemplars of and inspiration toward designing a bright green, prosperous future.

Worldchanging and bright green environmentalism have antecedents in the Viridian Design Movement, a project articulated by science fiction author Bruce Sterling. Through a number of public talks and papers in the late 1990s and early 2000s, Sterling proposed that solutions toward climate change and environmental degradation could be found in the production and propagation of a green aesthetic. Technology and art could be fused through design to create green products. Green social engineering, directed toward the "wealthy and the bourgeoisie," would create desires for green technology, green products, and green practices. Bright greens today similarly advocate the smart design of products, architecture, and social and economic planning as a way toward systemic change. And like the Viridian movement, bright green environmentalism embraces consumerism and prosperity as part of a sustainable future.

Bright green environmentalism also recapitulates many of the same foundational suppositions and ideas carried in the emergent field of industrial ecology. Industrial ecology values design, technological innovation, and a systems approach toward addressing ecological degradation. As explained by Robert and Leslie Ayres and elaborated in *A Handbook of Industrial Ecology,* industrial ecology draws from the cyclical and sustainable dimensions of natural processes and ecosystems in order to help design industrial systems. And industrial ecology posits that industrial activity should be developed, assessed, and managed in relation to and as interlinked with the natural environment.

Bright green environmentalism is one strand in an increasingly intricate web of green movements, ideas, and projects. Green politics has roots in a number of interrelated and overlapping social movements including environmentalism, environmental justice, feminism, and the peace movement. And green politics today are joined with and advanced by a wide range of interests including organized labor, antiracist groups, community organizers of variant political stripes, business leaders, local, state, and national governments, nonprofits, and NGOs. In sum, greens inhabit and practice a wide range of politics. In an oversimplification, some observers read environmental politics along a spectrum of green shades from light green to dark green. This spectrum tends to reflect and elicit sets of imagined, political binaries. Light green environmentalism is variously associated with reformism, individual consumer choices, anthropocentrism, and "shallow ecology." At the other end of the spectrum, dark green environmentalism can be associated with radicalism, efforts for structural and/or systemic change, revolution, ecocentrism, and "deep ecology." Bright greens appear to incorporate elements associated with both shades of green. On the one hand, bright greens treat consumerism as a salient green practice and in placing desires for consumption as integral toward environmental sustainability, bright greens clearly adopt an anthropocentric stance. On the other hand, bright greens advocate systemic change through social networking, political action, and a pragmatic openness to a full range of green practices.

Bright green ideology makes interesting if not unproblematic interventions in green political thought and practice. In asserting that green practices are compatible with economic prosperity, bright greens presaged mainstream green public discourse that now centers around green growth and the expansion of green markets as a solution to economic, social, and environmental problems, as much of the discourse in the Barack Obama administration's economic recovery plans demonstrates. And even left-of-center political writings like Van Jones's 2008 *Green Collar Economy,* as well as social movements such as the Apollo Alliance, treat green innovations, entrepreneurialism, and green jobs as fundamental to green transformation.

See Also: Ecopolitics; Steffen, Alex; Sustainability, Consumer Ethics and.

Further Readings

Apollo Alliance. http://apolloalliance.org (Accessed March 2010).

Ayres, Robert U. and Leslie W. Ayres. *A Handbook of Industrial Ecology.* Cheltenham, UK: Edward Elgar, 2002.

Cote, Ray, James Tansey, and Ann Dale. *Linking Industry & Ecology: A Question of Design.* Vancouver, Canada: UBC Press, 2006.

Jones, Van. *The Green Collar Economy: How One Solution Can Fix Our Two Biggest Problems.* New York: HarperCollins, 2008.

Robertson, Ross. "A Brighter Shade of Green: Rebooting Environmentalism for the 21st Century." *EnlightenNext Magazine,* 38 (2007).

Steffen, Alex. *WorldChanging: A User's Guide for the 21st Century.* New York: Abrams, 2006.

Sterling, Bruce. "The Manifesto of January 3, 2000." http://www.viridiandesign.org/manifesto.html (Accessed March 2010).

Sterling, Bruce. "Viridian Design." Talk given at the Yerba Buena Center for the Arts. San Francisco, California, October 14, 1998. http://www.viridiMndesign.org/notes/1–25/ Note%2000001.txt (Accessed March 2010).
Worldchanging. http://www.worldchanging.com (Accessed March 2010).

Boone Shear
University of Massachusetts, Amherst

Brundtland Report

The United Nations General Assembly (UNGA) addresses issues of global concern, which include the global environmental problems caused by developed and developing nations. In the early 1980s, concern was mounting about the increasing environmental impacts such as global warming and ozone depletion that were occurring as the world's nations were raising their standards of living.

As a response, in 1983, the UN General Assembly convened the World Commission on Environment and Development (WCED) with the intent that it would address the concern for the mounting global environmental impacts of development. The chair of this commission was the Norwegian prime minister, Gro Harlem Brundtland. The World Commission on Environment and Development is often referred to simply as the Brundtland Commission. Its report, *Our Common Future,* published by Oxford University Press in 1987, is commonly referred to as the Brundtland Report.

When the United Nations (UN) formed the World Commission on Environment and Development, or Brundtland Commission, it was charged with focusing on the following:

- To propose long-term environmental strategies for achieving sustainable development for the year 2000 and beyond
- To recommend ways concern for the environment may be translated into greater cooperation among developing countries and between countries at different stages of economic and social development and lead to the achievement of common and mutually supportive objectives that take account of the interrelationships between people, resources, environment, and development
- To consider ways and means by which the international community can deal more effectively with environmental concerns
- To help define shared perceptions of long-term environmental issues and the appropriate efforts needed to deal successfully with the problems of protecting and enhancing the environment, a long-term agenda for action during the coming decades, and aspirational goals for the world community

In response to this charge, the Brundtland Report included chapters covering, among other topics within sustainable development, the role of the international economy, population and human resources, food security, species and ecosystems, energy, industry, and proposed legal principles for environmental protection.

Of all the topics covered, the Brundtland Report is most often cited for its definition of sustainable development. The report defined sustainable development as "development that meets the needs of the present without compromising the ability of future generations to meet their own needs. It contains within it two key concepts:

- The concept of "needs," in particular the essential needs of the world's poor, to which overriding priority should be given; and
- The idea of limitations imposed by the state of technology and social organization on the environment's ability to meet present and future needs.

This definition is often credited with being the original definition of sustainable development.

The report also specifically highlighted trends within global population growth, which cannot continue indefinitely. It predicts the world population stabilizing somewhere between approximately 8 billion and 16 billion people within the 21st century. The increasing trend of urbanization was also noted, with predictions that within the 21st century, more people would live in cities than in rural areas for the first time in history. Although some of the highest growth rates are among developing countries, the report pointed out that the environmental impact of an additional individual in an industrialized nation is much greater than an individual within a developing nation. Additionally, declining birth rates of the industrialized world will mean a greater tax on the younger generations to support an aging population. For the developing world, improved health and education, especially among women, were presented as the solutions to high birth rates.

The final chapter of the report, "Chapter 12: Towards Common Action: Proposals for Institutional and Legal Change," provides a call to action to address the issue of sustainable development both within and among the countries of the Earth. It focuses on the need for agencies that are not compartmentalized in their mission, but that can address the overlapping social, economic, and environmental problems issues involved in sustainable development. It also points out the need to work cooperatively with other nations, especially when dealing with resources and ecosystems that straddle national boundaries. Six priority areas are presented in the global transition to sustainable development, including:

1. "Getting at the sources," which refers to the need for national policies and institutions to recognize and address sustainable development, the importance of regional and interregional action regarding sustainable development, and the need for global institutions and programs to address sustainability concerns.
2. "Dealing with the effects," which calls for governments to strengthen "the role and capacity of existing environmental protection and resource management agencies," as well as a strengthening of the UN Environment Programme, which should prioritize global environmental assessment and reporting.
3. "Assessing global risks," which suggests that the UN Environment Programme's Earthwatch can make sure all nations have access to global environmental information and further calls for the establishment of a Global Risks Assessment Programme.
4. "Making informed choices," which calls for an increase in the role of the scientific community and nongovernmental organizations within the global sustainability discussion as well as an increase in cooperation with industry.
5. "Providing the legal means," which refers to governments recognizing that "their responsibility to ensure an adequate environment for present as well as future generations is an important step towards sustainable development." And that states have a responsibility to both their own citizens and to other states. Finally, a recommendation is

made for the General Assembly to "commit itself to preparing a universal Declaration and later a Convention on environmental protection and sustainable development."

6. "Investing in our future," which argues for an increase in the financial investments countries make in environmental protection programs, and a reorienting of financial institutions and bilateral aid agencies to consider the environmental impacts of development projects.

The final call to action of the report includes a request that the UN transform the report into a UN Programme of Action on Sustainable Development. The Brundtland Report was the first report that investigated the causes of global environmental degradation and called attention to the links between social, economic, and environmental issues, which had been previously treated as separate issues. The report defined the term *sustainable development* and led to numerous efforts by the UN to address the interconnected causes of global environmental degradation.

See Also: Conservation; Consumption, Business Ethics and; Consumption, Consumer Ethics and; Development, Ethical Sustainability and.

Further Readings

Bugge, Hans Christian and Christina Voigt. *Sustainable Development in International and National Law: What Did the Brundtland Report Do to Legal Thinking and Legal Development, and Where Can We Go From Here?* Groningen, the Netherlands: Europa Law Publishing, 2008.

Dresner, Simon. *The Principles of Sustainability.* London: Earthscan, 2009.

World Commission on Environment and Development. *Our Common Future.* Oxford, UK: Oxford University Press, 1987.

Michelle E. Jarvie
Independent Scholar

Business Ethics, Shades of Green

Shades of green is a term coined in the 1990s to refer to various levels of commitment to environmental causes, typically as applied by businesses. Edward Freeman, Jessica Pierce, and Richard Dodd produced the first book on the subject, *Shades of Green: Business Ethics and the Environment,* in 1995. They followed this effort in 2000 with *Environmentalism and the New Logic of Business: How Firms Can Be Profitable and Leave Our Children a Living Planet.* The term remains an effective way to describe some of the nuances in how businesses extend ethical consideration to the environment. The shades, listed in increasing concern for the environment, include the legal approach, also called "light" green, whereby the company does only what is necessary to comply with the law, sometimes through force. An example is Willamette Industries of Oregon, which installed $7.4 million worth of pollution equipment in 13 factories only after an $11.2 million fine imposed by the U.S. Environmental Protection Agency. Another shade is the market approach, also called "market" green, whereby the company mostly does only what is

A farmer learning to grow Fair Trade Certified coffee in Brazil in 2008. Though Starbucks is the largest buyer of Fair Trade coffee, activists have pressed the company to increase its level of commitment, or "shade of green," with even greater focus on Fair Trade sales.

Source: U.S. Agency for International Development

necessary to satisfy customers. Freeman, Pierce, and Dodd note that "market green can apply to the industrial sector as well as to the consumer sector, and to services as well as to products."

Another shade is the stakeholder approach, also called "stakeholder" green, whereby the company meets the environmental concerns of various groups such as customers, business partners, the local community, and special interest groups. Some firms, typically after pressure from various entities, partner with groups such as the Environmental Defense Fund to protect endangered species' habitats or to reduce greenhouse gases. The highest level of green is the activist approach, also called "dark" green, whereby companies actively search for ways to protect the Earth's resources. New Leaf Paper Company, for example, is dedicated to using 100 percent consumer recycled paper in its products.

Freeman, Pierce, and Dodd use the ethical argument that conducting business in various shades of green shows correct responsibility to future generations, who, in the present tense, are our children. They refer to this issue as Our Children's Future Wager. It is built on an argument by the 16th-century philosopher Blaise Pascal, who suggested, despite some uncertainty, that it is rational to choose Christianity because afterlife consequences are severe if you don't. The authors point out that this case is similar yet different—our children do not get to choose—which should make for an even more compelling case to act with caution: "if we are to leave a *livable world* for our children and their children, we simply must pay attention to environmental matters." The authors, while acknowledging that merging business and environmental concerns is complex, turn away from traditional conflict and focus on how they can work together. Firms do not necessarily have to move from lighter to darker shades, although many have, such as DuPont, McDonald's, and United Parcel Service. Richard Daft notes that DuPont has developed a housing insulation that saves more energy than it takes to produce it, designed a new fabric made from corn, and developed biodegradable plastic silverware. McDonald's, among other practices, halted purchase of antibiotic-treated poultry, buys energy from renewable sources, and has incentives for suppliers with green practices. United Parcel Service has 2,000 vehicles that use alternative fuels.

Walmart has also recently increased sustainability practices through a stakeholder green approach that included cost savings, and a market green approach to reach new customers. In partnership with the Environmental Defense Fund (EDF), Walmart announced a goal to eliminate 20 million metric tons of greenhouse gases from its global supply chain by the end of 2015. This is equivalent to taking more than 3.8 million cars off the road, or saving 2 billion gallons of gasoline, for a year. Dominique Browning, on her EDF blog, states: "Walmart will be asking the estimated 100,000 companies that supply it to cut the amount of carbon they emit when they produce, package, and ship products. This pollution reduction goal will affect every step of the manufacturing process from raw materials to recycling. For example, suppliers could label clothes to be washed in cold water instead of hot water, or accelerate the innovation of fabrics that dry faster."

Examples of the influence of a large firm like Walmart include changing from liquid detergent in large containers to concentrates in small containers, which also reduced shipping costs; reducing wasteful packaging on DVDs; and, using displays, educating consumers about longer-lived, more energy-efficient compact fluorescent light bulbs—selling millions. Other large retailers and even manufacturing suppliers followed suit.

Compliance is seen as a major issue, as the international commodity chain often contains issues such as health and safety, illegal logging, and dishonest labeling. A clear chain of custody is needed from source to market. Walmart is working on guidelines for accountability. Bradford Plumer suggests much of Walmart's recent greening effort is to capture the higher-income market that makes up 20 to 40 percent of U.S. consumer spending and typically does not shop there.

In addition to external pressures, cost cutting, and marketing, corporate leadership can be a big factor in changing practices. Freeman, Pierce, and Dodd relate in their 2000 update that, earlier in DuPont's encounter with sustainability concerns, an executive asked plant engineers to comply with new environmental standards. They initially responded that it simply could not be done. The executive indicated the alternative was to close the plant. The engineers then came back with a proposal that, after further investigation, not only complied with standards, but saved the company money.

Jeffrey Marshall found that, of "198 medium-sized to large multinationals, most said they lacked an active approach to developing new business opportunities arising from meeting citizenship and sustainability needs." Corporate responsibility or corporate sustainability officers are being hired to address these issues.

Freeman, Pierce, and Dodd outline still more philosophical aspects of the shades of green concept: core values that link with Our Children's Future Wager. Businesses have found that articulating some bedrock, some foundation, and some basic values has enormous benefits. People, from executives to mail clerks, begin to believe in them or are attracted to the firm because of these values. In short, business strategy just makes more sense in the context of values: from huge DuPont to Ben and Jerry's, from oil and chemical companies to retail boutiques, articulating what you stand for on the environment is step one to a greener world, one that we can pass on to our children.

Freeman, Pierce, and Dodd focus on what they see as three main philosophical challenges to traditional business thinking: conservation, social justice, and ecology. While conservation of resources is the easiest to incorporate into traditional business thinking, the others are more challenging: expanding our concern to women, minorities, indigenous people is connected to expanding concern to ecology and finding a discourse outside our human-centered world.

Freeman, Pierce, and Dodd outline main barriers to such a change in worldview: the cost-benefit analysis mind-set, which focuses almost entirely on economic values; the constraint mind-set, which again focuses almost entirely on economic values and sees other values simply as constraints on profits; and the sustainable development mind-set, which sees government intervention and regulation as the primary solution. The authors repeatedly focus on—rather than cost-cutting technology—a reassessment of values as a fundamental need throughout the book.

A 2003 book titled *Shades of Green: Business, Regulation, and the Environment* took a look at how 14 pulp and paper mills in the United States, Canada, Australia, and New Zealand complied or went beyond environmental laws. Variability in environmental regulation was not as important as varying pressures from community and environmental activists, economic constraints, and differences in corporate environmental management style.

Many Shades of Green

Kathleen Morson and Bart Mongoven wrote an essay in 2007 titled "Corporate Environmental Initiatives: The Many Shades of 'Green.'" They aggregate current business environmental approaches into three (not four) categories: public relations; reacting to market transformation; and actively transforming the market. They categorize BP's recent "Beyond Petroleum" effort as primarily public relations, which is relatively easy to see through if wider program change does not occur. Other examples are programs centered only on temporary events such as Earth Day (e.g., Home Depot), without transforming wider environmental processes throughout the company.

The second, greener category is reacting to market transformation. An example is the partnership between jewelers and nongovernmental organizations (NGOs) to market conflict-free diamonds, a genuine program that avoids significant social ills in the production and marketing process. Another example is the move by several businesses to reduce or eliminate PVC packaging.

The third, greenest category is "actively transforming the market." A relatively high-profile example was DuPont's championing of the Montreal Protocol to reduce or eliminate chlorofluorocarbons (CFCs). The authors do not see a problem with the fact that DuPont also benefited financially by being first in developing alternatives to CFCs. General Electric has a similar case, its "Ecoimagination" program and strategy begun in 2004 that fosters new, less-polluting products. The authors state that "three years later, annual sales of those products topped more than $11 billion—and GE is banking on reaching $20 billion in sales by 2010. GE is now lobbying for energy-efficiency legislation and a carbon cap with other corporate and NGO members of the U.S. Climate Action Partnership. GE has a financial stake in pushing for this legislation, since it has been preparing since 2004 to make the type of machinery and products needed in a carbon-constrained world."

Morson and Mongoven mention that innovative, third-category companies will always be there, causing second-category firms to spend research and development (R&D) funds to react. The reason why this is important is this: "Although studies repeatedly show that consumers care very little about the social values of the companies from which they buy, studies also show this attitude is changing . . . the companies that find themselves in the second category—those forced to react to market transformation—are yearning for some definitions on what it means to be socially responsible."

The International Organization for Standardization hosts long-standing voluntary guidelines for environmental, health and safety, and other standards, including new standards for corporate social responsibility (ISO 26000). Standards for environmental management systems that comprehensively examine risks and energy costs throughout a firm's processes also fall under the International Organization for Standardization (ISO 14001), although they merely indicate a system is in place, not its effectiveness. The term *industrial ecology* has emerged to describe "cradle to grave" product design and whole system management of energy and material resources as inspired by the ecology of natural systems.

The larger realm of formal business ethics and environmental ethics is a relatively recent pairing. According to Richard DeGeorge, the term *business ethics* did not arise as a formal subdiscipline of academic philosophy until the 1970s. Business ethics emerged at this time in academic departments of philosophy and business, with a corresponding skepticism pitting theoretical versus applied academics.

The first anthologies on business ethics occurred in the late 1970s, including Thomas Donaldson and Patricia Werhane's *Ethical Issues in Business: A Philosophical Approach*. The first single-authored books occurred in the early 1980s and included *Business Ethics: Concepts and Cases* by Manuel Velasquez. The next decade saw academic studies combining the emerging fields of business ethics and environmental ethics. Prominent anthologies included Laura Westra and Patricia Werhane's 1998 *The Business of Consumption: Environmental Ethics and the Global Economy,* which included a chapter on "Shades of Green" by Freeman, Pierce, and Dodd.

From the same late 1990s time frame as these anthologies, Paul Hawken, Amory Lovins, and L. H. Lovins's popular book *Natural Capitalism* suggests a merging of business profitability and environmental concern that continues today with the green jobs movement. Green investing is another growing movement, especially in the arena of alternative energy. Rajat Panwar suggests this increase in ethical investing has encouraged corporate responsibility. Michael Porter supports this corporate responsibility movement in the *Harvard Business Review* in "The Link Between Competitive Advantage and Corporate Social Responsibility."

One of the more prominent summaries of this common ground between business values and environmental values is Joel Makower's 2009 book *Strategies for the Green Economy.* He highlights early successful "win–win" corporate programs. 3M was one of the antipollution pioneers who took a prevention approach in the 1970s that has saved the firm billions. Currently, large, diverse corporations such as Anheuser-Busch, General Motors, McDonald's, Nokia, and Procter & Gamble have all cut costs by reducing packaging. Others have become the largest purchasers of environmentally friendly products, including organic cotton (Walmart and Nike), renewable energy (Intel and Pepsico), landfill gas that generates electricity (General Motors), Fair Trade coffee (Starbucks), FSC-certified wood (Home Depot), and recycled products (McDonald's).

One of the key issues is standards. There are recent, reasonably clear standards for topics such as green building (LEED certification) and organic produce but not for a generally "green" business. Another issue is the gap between concern shown for the environment in consumer surveys and that shown by actual purchasing. Makower suggests that surveys need to ask, not only would the interviewee purchase green products, but under what conditions would they purchase green products? Conditions might include sold at the place where they normally shop; increase in quality or longevity; equal or lower price; or no change in habits required. This would illustrate some of the more refined decision making and trade-off values.

Makower cites numerous public surveys that attempt to organize consumers into "shades of green," typically in three to five segments regarding environmental values. One of the more prominent global surveys is Lifestyles of Health and Sustainability (LOHAS), which provides market data for both small and large companies around the world. The most concerned segment, 16 percent of the population, is very progressive, not necessarily affluent, can be influential in accepting new trends, and expects companies to act responsibly. Another recent Values and Lifestyles Survey found the following trends: (1) there is no common agreement on what environmental concern means and what actions to take, (2) libertarian views tend to trump communal ones, simplified as the environmental We're In This Together (WITT) versus the Republican Party You're On Your Own (YOYO), (3) environmental complexity can be paralyzing, (4) pocketbook environmentalism is powerful. Other value issues are ecoliteracy, nature-deficit disorder, and the tendency for "Security Moms and NASCAR Dads" to make lifestyle decisions on "what's in it for me, today."

Makower compares major-company green efforts, such as those of Levi Strauss and Coca-Cola. Levi Strauss started using more organic cotton but, so as not to build expectations, did not advertise it. Coca-Cola did advertise more recycled content in its containers but, when it couldn't meet its commitment, it was castigated by activists. Interestingly, resulting boycotts did not significantly change sales but, when the boycott affected Coca-Cola's ability to attract new recruits out of college, the company reacted. Makower notes that many activists can accept imperfection if the company understands the issues and is sufficiently concerned, but this can be difficult to get across. Even the most progressive firms, such as Patagonia, Starbucks, and Stonyfield Farms, tend to be held to higher standards. A recent campaign tried to influence Starbucks to sell more Fair Trade coffee, even though it is that product's largest purchaser.

Makower notes that many new green entrepreneurs are former dot-com "refugees." Innovators include Solar City (run by former software company owners), which encourages Silicon Valley neighbors to bond for group pricing incentives on residential solar; Sungevity, which will send a solar potential assessment for your home within 24 hours of entering your address on its Website; mkDesigns, founded by a female architect, which builds custom homes at a factory to reduce costs (the homes are further customized on-site, including energy efficiency measures); and Nine Dragons Paper, run by China's richest woman, who recycles U.S. scrap paper. Two of the author's favorite examples are (1) Nau, Inc., an outdoor clothing designer in Portland, Oregon, that offered the option to try on clothing in person, then get a 10 percent discount if you bought it online; and customer-directed giving, whereby customers could choose a nonprofit to which Nau would donate 5 percent of a sale; and (2) Greystone Bakery, founded by an astrophysicist turned Zen Buddhist priest, whose mantra was "We don't hire people to bake brownies. We bake brownies to hire people"—they would hire anyone, including former drug addicts and prisoners, and provide them with training, child care, counseling, and meaningful work.

Collaboration is also a key value. New Belgium Brewery was so successful at green innovations in the brewing industry that it provided advice for fellow brewers such as Anheuser-Busch and Coors. Materials pooling among firms using a product, often with different needs, could help offset costs to process certain products such as leather in an environmentally friendly manner. A group of at least nine major corporations linked with four major shipping operators to form a Clean Cargo Group to reduce transport of invasive species in ballast water and products.

Just as there are shades of green, there are shades of *greenwashing*, a term coined in the 1990s by Greenpeace to refer to the selling of a company's products based on image rather than actual environmental performance. Typical errors in greenwashing include "hidden tradeoffs, no proof, vagueness, irrelevance, lesser of two evils, and fibbing." Although it is getting safer for firms to advertise their green practices, Bissell took the wise approach of stating their company was becoming a "little greener," moderating expectations. A 2007 study found that 70 percent of "Americans strongly or somewhat agree that when companies call a product green, it is usually just a marketing tactic." Makower suggests there is hypocritical "green consumer washing" as well, where consumers do not act on their values when purchasing or gauging other lifestyle decisions. He challenges the consumers to take the same responsibility in their actions that they want to see in corporations.

Thus, "shades of green" fits within a large framework merging business values and environmental values. It is seen as a viable forum to define diverse, corporate, and consumer approaches toward the environment. It contrasts with the more traditional business versus environmental regulation conflict approach and continues as a summarizing theme for current corporate and environmental policy.

See Also: Consumption, Consumer Ethics and; Greenwashing; Sustainability, Business Ethics and.

Further Readings

Browning, Dominique. "Walmart Amps Up the Green Light." Environmental Defense Fund. http://blogs.edf.org/personalnature/2010/03/08/walmart-amps-up-the-green-light (Accessed February 2010).

Daft, Richard L. *Management*, 8th ed. Mason, OH: Thomson/Southwestern, 2008.

DeGeorge, Richard T. "A History of Business Ethics." Markkula Center for Applied Ethics, Santa Clara University, Santa Clara, CA (2005). http://www.scu.edu/ethics/practicing/focusareas/business/conference/presentations (accessed February 2010).

Freeman, R. Edward, Jessica Pierce, and Richard H. Dodd. *Environmentalism and the New Logic of Business: How Firms Can Be Profitable and Leave Our Children a Living Planet.* New York: Oxford University Press, 2000.

Freeman, R. Edward, Jessica Pierce, and Richard H. Dodd. *Shades of Green: Business Ethics and the Environment.* New York: Oxford University Press, 1995.

Freeman, R. Edward, Jessica Pierce, and Richard H. Dodd. "Shades of Green: Business Ethics and the Environment." In *The Business of Consumption: Environmental Ethics and the Global Economy,* Laura Westra and Patricia H. Werhane, eds. Lanham, MD: Rowman & Littlefield, 1998.

Gunningham, Neil A., Robert Allen Kagan, and Dorothy Thornton. *Shades of Green: Business, Regulation, and Environment.* Palo Alto, CA: Stanford University Press, 2003.

Hawken, Paul, Amory Lovins, and L. H. Lovins. *Natural Capitalism: Creating the Next Industrial Revolution.* Boston: Little Brown, 1999.

Makower, Joel. *Strategies for the Green Economy: Opportunities and Challenges in the New World of Business.* New York: McGraw-Hill, 2009.

Marshall, Jeffrey. "New Group Pushes 'Responsibility Officer.'" *Financial Executive,* 23/1 (2007).

Morson, Kathleen and Bart Mongoven. "Corporate Environmental Initiatives: The Many Shades of 'Green.'" Public Policy Intelligence Report. Business Ethics Network, Corporate Ethics International. April 19, 2007. http://corpethics.org/article.php?id=1003 (Accessed February 2010).

Panwar, Rajat. "Corporate Responsibility: Balancing Economic, Environmental, and Social Issues in the Forest Products Industry." *Forest Products Journal,* 56/4 (2006).

Plumer, Bradford. "How Good Is Walmart's Green?" *The New Republic* and National Public Radio. February 28, 2010.

Porter, Michael E. "Strategy & Society: The Link Between Competitive Advantage and Corporate Social Responsibility." *Harvard Business Review,* 84/12 (2006).

William Forbes
Stephen F. Austin State University

Callicott, J. Baird

J. Baird Callicott is a preeminent author in environmental ethics and philosophy. He is currently a member of the Philosophy and Religion Studies Department at the University of North Texas. Callicott served as the president of the International Society for Environmental Ethics during 1997–2000.

Callicott's works probe philosophical theory underlying environmental ethics. Much of his work builds on Aldo Leopold's land ethic, which states "A thing is right when it tends to preserve the integrity, stability, and beauty of the biotic community." The land ethic requires individuals to directly consider this land community as an independent consideration in their ethical dilemmas and debates. This approach focuses on the ecosystem. One of Callicott's texts titled *In Defense of the Land Ethic* investigates the intellectual foundations of Leopold's work and defends an ecocentric approach. In this text he examines how Western culture has led to the environmental crisis of today. Another Callicott text titled *Beyond the Land Ethic* is an extension of Leopold's position. He also edited a book titled *Companion to* A Sand County Almanac, which is a collection of works that build on Leopold's *A Sand County Almanac.*

In addition to the books listed above, Callicott is the author of many other texts including *Earth's Insights: A Multicultural Survey of Ecological Ethics From the Mediterranean Basin to the Australian Outback* and *Essays in Environmental Philosophy* and *American Indian Environmental Ethics: An Ojibwa Case Study.* He has also authored dozens of other articles and book chapters.

See Also: Biocentrism; Instrumental Value; Intrinsic Value; Land Ethic; Leopold, Aldo.

Further Readings

Callicott, J. B. *Beyond the Land Ethic: More Essays in Environmental Philosophy.* Albany: State University of New York Press, 1999.

Callicott, J. B. *Companion to* A Sand County Almanac: *Interpretive and Critical Essays.* Madison: University of Wisconsin Press, 1987.

Callicott, J. B. *Earth's Insights: A Survey of Ecological Ethics From the Mediterranean Basin to the Australian Outback*. Berkeley: University of California Press, 1994.
Callicott, J. B. *In Defense of the Land Ethic: Essays in Environmental Philosophy*. Albany: State University of New York Press, 1989.
Callicott, J. B. and M. P. Nelson. *American Indian Environmental Ethics: An Ojibwa Case Study*. Upper Saddle River, NJ: Prentice Hall, 2004.
Callicott, J. B. and M. P. Nelson. *The Great New Wilderness Debate*. Athens: University of Georgia Press, 1998.
Callicott, J. B. and F. J. R. da Rocha. *Earth Summit Ethics: Toward a Reconstructive Postmodern Philosophy of Environmental Education*. Albany: State University of New York Press, 1996.
Chappell, T. D. J. *The Philosophy of the Environment*. Edinburgh, UK: Edinburgh University Press, 1997.
Leopold, A., J. B. Callicott, and E. T. Freyfogle. *For the Health of the Land: Previously Unpublished Essays and Other Writings*. Washington, DC: Island Press, 1999.
Meine, C. and R. L. Knight. *The Essential Aldo Leopold: Quotations and Commentaries*. Madison: University of Wisconsin Press, 1999.
Nelson, M. P. and J. B. Callicott. *The Wilderness Debate Rages On: Continuing the Great New Wilderness Debate*. Athens: University of Georgia Press, 2008.
Ouderkirk, W. and J. Hill. *Land, Value, Community: Callicott and Environmental Philosophy*. Albany: State University of New York Press, 2002.

Jo Arney
University of Wisconsin–La Crosse

CARBON OFFSETS

Ethical interests exist in the operation of carbon-offset markets as well as the selection and processes involved in ethically appropriate projects and the manner in which the offsets are derived and applied.

Carbon as a principal component of greenhouse gases contributes to atmospheric loading. Emissions from human activities can be quantified and regulated either by national authorities or voluntarily by the emission sources. Voluntary regulation may be either self-regulated or by agreement in compliance with an industry standard. In its essential form, offsetting links the generation of qualifying emissions at one set of coordinates with a corresponding reduction at another. Where the offset occurs as intended, the resultant effect should be a zero increase in the loading of emissions. An overall reduction in the total environmental emissions loading could be achieved by periodically reviewing and adjusting the overall emissions limits and thresholds downward.

Emission quotas are allocated from the global and international level through the regional and domestic jurisdictions. Where the end recipient is the individual operator, the process tends to address fixed point sources of emissions more easily than nonpoint and mobile sources. As a result, point sources of emissions may bear a greater share or burden of the costs. For individual enterprises, there is a similar issue as to the assignment of emissions to various segments of a facility.

To allocate emissions within the scope a facility or even an enterprise requires management design and execution as a course of doing business. Management can achieve intended emission targets by configuring the various sources of emissions to offset reductions of emissions in some spaces in order to balance increased emissions in others. Alternatively, management can determine that it is not commercially reasonable to reduce emissions and can elect to purchase credits instead that may also qualify as offsetting in some trading markets. Thus, offsetting options include local, regional, and international arrangements, as well as offsets by carbon recapture and sequestration. Brokerages can be used to facilitate offset transactions rates at various levels of complexity from individual retail transactions involving personal activities to large, industrial-scale offsets.

Information systems can be used to map and manage carbon footprints of facilities that can be linked through software to carbon trading markets. Technological solutions can be used to flatten the supply/demand curve. Some jurisdictions have deployed "smart meters"—these are electricity supply meters that can record consumption by variables including time of day and season of the year. The information from such meters permits energy suppliers to vary rates and supplies to retail and commercial customers for supply reasons. It is even possible to assign to suppliers the remote control of noncritical individual appliances at user facilities. Such supply-based remote controls can be used to redistribute supply when demand from specified critical applications requires.

Offsets can be effectively used to limit total emissions within regions where sources of emissions are being added. At a global scale, reductions in emissions in some areas can be used to introduce other sources without exceeding the applicable limit to emissions and balancing the overall global anthropogenic atmospheric carbon burden.

New technologies can achieve reduced carbon emissions and alternative sources of energy can be valuable in maintaining energy supplies while advancing socially valuable products and services.

Kyoto and Carbon

The Kyoto Protocol at Article 17 provides a carbon trading and offsetting regime for countries to operate regulated markets. Under these provisions, regulated carbon trading occurs between signatories to the agreement with offsets between signatories and nonsignatory countries through the Clean Development Mechanism (CDM) and Joint Implementation (JI) mechanisms. Offsets can also be negotiated privately through nongovernmental markets within domestic or regional jurisdictions and globally. Nongovernmental organizations and independent self-regulated persons may also design and trade units at their own risk and discretion.

Offsetting was intended to support the introduction of new clean energy projects and technologies in industrializing economies. Under the CDM mechanism, it is possible for a party to Kyoto to obtain certified emission reductions when a qualified emission-reduction project is implemented in a jurisdiction that is not subject to Kyoto. Certified emission-reduction units are available to be applied against the Kyoto targets. Under the JI mechanism, the parties must be identified in Annex II of the Kyoto Protocol and the project must provide reductions or sinks that are additional to what would have occurred without the initiative.

For a purchaser, there are two steps to obtaining an offset. The purchaser first determines the required amount of offset for the purchaser's emissions by determining the intended carbon footprint and computing the carbon equivalent. Next, the purchaser needs to seek out and locate a vendor who will agree to implement measures to achieve the same required amount of reduction in carbon emissions to the environment. The purchaser pays the vendor to implement the reduction.

If the transaction proceeds and the required measures are implemented, the participants jointly generate no more than zero net emissions into the atmosphere. Conceptually, at the simplest level, this approach may be argued as not making the situation worse than before. The difficulty is whether existing levels should be permitted to continue due to the damage being caused. The ongoing damage from existing emission loading may already be unacceptable to the victims. Implicitly, the acceptability of existing emissions is the acceptance of damage as the cost of doing business for which insurance may be available. This results in actuarial valuations for what may be extreme health issues for humans, ecological systems, and regions. Even allowing that zero net emissions are acceptable, reality is more complex, as there are qualitatively different circumstances at the selected emission source and offset coordinates. Determining quantitative loading at the emission point is technologically easier than what may be estimated or possible offsetting reductions. The ecosystems at the respective offset points would typically be qualitatively different and even more so in an international or intercontinental offset. Factoring the dynamism of the ecosystems, the actual benefit of the practice may be difficult to isolate.

In an offset transaction, funds from purchasers of carbon equivalents serve to finance operations that minimize environmental degradation or effect atmospheric carbon neutrality. For the vendor, the same transaction can be used for capital improvements that otherwise may not have been implemented. The market and economic factors in an offset transaction may or may not include ethical considerations. For example, by securing an offsetting hydrochlorofluorocarbon (HCFC) destruction facility in China, it permitted continued production of chlorofluorocarbons (CFCs) internationally, which translated into continuing damage to the ozone as well as placing the burden of disposal of waste from the destruction facility on the local Chinese ecosystem. Similarly, nuclear energy may be presented as an offsetting facility that could benefit from additional funds from a low-grade-coal-fired facility. The result is a continuation of carbon loading at the emission location and increase in nuclear waste disposal issues at the offset point. Additionally, there would be an increase in cost of energy from the additional capital costs that could adversely impact the vulnerable in society the most.

Some Emergent Ethical Issues

With energy supply being a primary source of carbon emissions and the candidate for offsetting linked to security, a lifeboat type of ethics could be invoked. The uneven distribution of energy could be associated with a corresponding security of the parties. An ethical question arises as to the relative values to be accorded to those whose security is at risk. How the limited resources are to be distributed and the allocation of security risks to the parties and ecosystem becomes pertinent. Where a party to the supply chain is not a good actor, the ethical issues are complicated. Since energy is a necessity for human life, a right to energy as an essential human right could be postulated. The ethical issue here could be either associated with or in addition to a legal right.

As some have observed, offsetting, even where it operates as designed, legitimizes behavior that could be harmful to ecosystems. In doing so, it could provide sufficient incentive for a bad actor to operate contrary to ethical standards. Voluntary regional and industry codes of conduct could mitigate by advancing ethical considerations. Where the actors are operating in a utilitarian philosophy, offsets could be an ethically advantageous means to seek the greatest good for the greatest number.

The usual operations of business interests in not disclosing proprietary information could impede appropriate disclosure necessary to achieve actual offsets. In a competitive market, transparency between parties could suffer with ethical considerations in negotiating offsets, implementing them, and enforcing compliance where required. In practice, it has not always been possible to certify either qualification for appropriate offsets or compliance with the agreement. Failure to achieve compliance can be an ethical issue as well as an economic one. Where the recipient of offset funds diverts their use to purposes outside the agreement for offsets, there may be little the payer or the victims can do.

Continuing Issues

Carbon occurs in all organic materials including living organisms, is ingested as food, and is widely used in human products and production processes. The distinction between the use of carbon in products and in production processes is not always a simple matter. For instance, carbon-based products such as plastics derived from fossil fuels may be used directly or indirectly to generate energy. Carbon for human use (other than for direct ingestion as food) is obtained from fossil or nonfossil fuels including oil, coal, and gas, and more recently, from methanol, ethanol, propane, biomass, and oil sands or bitumen. All these sources, when appropriately fractionated and processed, may be used directly as fuels or indirectly as in the generation of electricity. Since carbon is involved in all of human life, carbon offsetting could potentially be similarly integrated into human activities. In addition to controlling its distribution in the environment, carbon offsetting may have complex, unintended, and pervasive impacts on human life.

Introducing technological solutions in an offset arrangement can be counterproductive to larger human and ecosystem interests. The valuation of a solution considered from the context of the globalized economy appears to favor capital-intensive systems over labor-intensive solutions. Valuation of a piece of equipment may be more easily visible than the work of poor, unemployed persons who may be providing the required service at lower cost. The proprietor of a new facility may have to choose between producing a favorable balance of carbon credits that represent financial gain and providing employment to a large number of subsistence-level workers. An example of this occurred in New Delhi, India, where poor people picking through waste for recyclable items could be replaced by a higher-cost, capital-intensive facility financed through offsets. Unauthenticated claims, applying past achievements for future credit, unverified implementation, enforcement of agreements, limited audit capabilities, and redistributing funding and equipment can result in the destruction of existing efficiencies and exacerbate atmospheric and social problems rather than alleviate them. Even where an offset operates as intended, there are no provisions by which the gains introduced in a project will be retained. For example, if a forest is planted as a sequestration project in an offset, there are no assurances either that the forest will not be subsequently clear-cut without further compensatory actions or enforceable consequences.

Offsets could be ethically justified where the net result is, in fact, to minimally cause no additional harm. Unfortunately, in the Kyoto context at least, the interest is independent of harm and rather in financing additional costs to perpetuate existing carbon loading.

See Also: Ecological Footprint; Social Ecology; Technology.

Further Readings

Holtcamp, Wendee E. "An Off-Setting Adventure." *The Environmental Magazine,* 18/6 (2007).

Hopkin, Michael. "Emissions Trading: The Carbon Game." *Nature,* 432/7015 (2004).

International Carbon Action Partnership. http://www.icapcarbonaction.com (Accessed April 2010).

Lashof, Daniel A. and Dilip R. Ahuja. "Relative Contributions of Greenhouse Gas Emissions to Global Warming." *Nature,* 344 (1990).

UN Intergovernmental Panel on Climate Change. "IPCC Special Report on Carbon Dioxide Capture and Storage." http://www.ipcc.ch/publications_and_data/publications_and_data_reports.htm#2 (Accessed April 2010).

Willson, Richard W. and Kyle D. Brown. "Carbon Neutrality at the Local Level: Achievable Goal or Fantasy?" *Journal of the American Planning Association,* 74/4 (2008).

Wysham, Daphne. "Carbon Market Fundamentalism." *Multinational Monitor,* 29/3 (2008).

Lester de Souza
Independent Scholar

Carson, Rachel

A nature writer and a marine biologist, Rachel Louise Carson (1907–64) is best known for her book *Silent Spring* (1962), which spearheaded the modern environmental movement in the United States by sparking public awareness about the impacts of pesticides on the environment. Carson was likely imbued with a devotion to the Earth's beauty by her mother, Maria, who had taught her as a tiny child the joy of the outdoors and the lore of birds, insects, and the creatures of streams and ponds. Her sense of wonder at the brilliance of the natural world never left her and that was a trait she guarded most dearly, as we continually see how in her writing and speeches she implores us, especially the young generations, to retain that wonderment and thrill at the Earth, our home.

Researchers Rachel Carson and Bob Hines at work on the coast of Florida in 1952, the year after the publication of her best-selling second book, *The Sea Around Us.*

Source: U.S. Fish & Wildlife Service

Rachel Carson was born on May 27, 1907, in Springdale, Pennsylvania, on a small farm. At a young age, she loved to read and enjoyed exploring with her family. She enjoyed writing stories that often involved animals, and her first story was published at the age of 11. A common thread in her reading material was the natural world. Carson graduated at the top of her class in high school and then attended Pennsylvania College for Women. She studied English, later switching her major to biology. She was admitted to graduate school at Johns Hopkins University in 1928. Carson finally earned a master's degree in zoology in 1932. In 1934, she was forced to leave Johns Hopkins to support her family. She also had to care for her aging mother when her father died suddenly a year later. At the U.S Bureau of Fisheries, Carson gave a weekly educational broadcast to generate public interest in fish biology. She was outstanding in her work, and became the second woman to be hired by the Bureau of Fisheries in a professional capacity.

After the successful completion of the third volume of her sea trilogy, *The Edge of the Sea,* her interest turned toward conservation, and she became involved in many conservation groups. The rest of her life focused primarily on the dangers of pesticide overuse. This was because she was concerned about the use of synthetic pesticides. She wanted to make public the government's spraying practices and thus devoted the next four years to the project that produced *Silent Spring* by gathering examples of the environmental damage attributed to the pesticide dichlorodiphenyltrichloroethane (DDT). By 1960, she had completed her research, but the completion of *Silent Spring* was delayed when a malignant tumor was discovered in her breast. The title *Silent Spring* was metaphorical: it suggests a bleak future for the entire natural world. Carson's main argument was that pesticides have detrimental effects on the environment, and that the effects were not limited to the pests they were created to destroy. She also accused the chemical industry of intentionally spreading disinformation, and criticized public officials for accepting the industry's claims. The book closes with a call for a biotic approach to pest control as an alternative method.

Carson's work became a powerful force for the grassroots environmental movement in the United States. She inspired some of the activism of the social movements in the 1960s, including influencing the rise of ecofeminism. The creation of the U.S. Environmental Protection Agency (EPA) was said to be directly related to Carson's work. Her shy disposition was but a polite mask for a staunch defender of the environment. She courageously and firmly opposed skeptics in the courtroom and beyond, in press releases and speeches. Carson's efforts resulted in an increased environmental awareness from laypeople to leaders, with the then Supreme Court Justice William O. Douglas, an ardent naturalist, declaring, "We need a Bill of Rights against the 20th-century poisoners of the human race."

Rachel Louise Carson died at the age of 56 in 1964, after a struggle with breast cancer. Despite a life cut short, Carson is known as one of the most influential women in the U.S. environmental movement. An editorial in the *New York Times* stated, "If her series [running in part in the *New Yorker*] helps arouse public concern to immunize government agencies against the blandishments of the hucksters and enforces adequate controls, the author will be as deserving of the Nobel Prize as was the inventor of DDT."

See Also: Environmental Policy; Globalization; *Silent Spring.*

Further Readings

Brooks, P. *The House of Life: Rachel Carson at Work.* Boston: Houghton Mifflin, 1972.

Carson, Rachel Louise. *The Sea Around Us.* New York: Oxford University Press, 2003.

Carson, Rachel Louise. *Silent Spring.* Boston: Houghton Mifflin, 1962.

Carson, Rachel Louise. *Under the Sea Wind.* New York: Simon & Schuster, 1941.

Carson, Rachel Louise, Sue Hubbell, and Bob Hines. *The Edge of the Sea.* New York: Houghton Mifflin, 1979.

Lear, L. J. *Rachel Carson: Witness for Nature.* New York: Henry Holt and Co., 1997.

"Rachel Carson Dies of Cancer; *Silent Spring* Author Was 56." *New York Times* (April 15, 1964). http://www.nytimes.com/learning/general/onthisday/bday/0527.html (Accessed February 2010).

Md Saidul Islam
Nanyang Technological University

China

Concrete housing blocks rise above a polluted river whose banks are strewn with trash in China in 1996. At least one-third of China's rivers were polluted by 2009, and pollution-related deaths in the country may be as high as 750,000 per year.

Source: World Bank

What occurs in China's environment affects all living inhabitants of the planet and its natural biophysical processes. The subject of China's environmental conditions and that nation's path toward development is of paramount importance not only to all Chinese citizens and their leaders but has profound significance for the entire world because of the many ecological, political, economic, social, and ethical consequences that will ensue from decisions made by Chinese policy makers. For example, consider the fact that China has surpassed the United States as the world's greatest emitter of the greenhouse gas and climate change catalyst carbon dioxide. This is merely one dramatic development among several that has turned the world's attention and concern toward the state of China's environment. Another is China's status as the world leader in economic growth for the 30-year period since the economic reforms of 1978, including a 10 percent annual average economic growth rate during this period—the most rapid rise ever in world history! The remarkable strength of China's economy is reflected in continuing such economic growth even during the severe global recession in 2009.

This new global significance of China's economic and environmental circumstances was indicated by its leadership role in the world's developing nations ("Group of 77 + China," which actually includes approximately 130 member nations) at the United Nations Conference on Climate Change held in December 2009 in Copenhagen. China, represented by its Premier Wen Jiabao, was the largest developing nation (in terms of its economy and population), a member of the United Nations Security Council, and, therefore, an appropriate negotiating partner with U.S. President Barack Obama at the Copenhagen Climate Change Conference.

This article aims to provide an overview of the profound developments regarding China's society and its environment; its growing environmental prominence, both globally and within its own domestic civil society; and increased priority resulting in policy shifts concerning the environment by the central government (although this may not be true on the important provincial and local levels of government, creating conflict between these governmental structures and often hindering the implementation of the central government's environmental policy directives). It intends to stimulate the reader to pursue further one of the environmental studies' most significant topics: China's environmental crisis, which is a "global crisis with Chinese characteristics" owing to the quantity, diversity, and severity of environmental problems confronting China today. Another aim of this discussion is to highlight the myriad manifestations of China's environmental crisis and to sound the alarm about the urgency of the world's environmental state of affairs. Immediate transformative action is needed in order to avert an environmental catastrophe for both China and the rest of the planet. Over its 40-year history, the global environmental movement has strived to arouse worldwide public acceptance and to mobilize governmental elites to supplant the modern industrial free market: worldview and derivative policies. These growth-maniacal values dominate virtually the entire planet, including, remarkably, China today.

China and its environmental crisis may be the proverbial "canary in the coal mine," or the demonstration project of what can happen when a society's environment deteriorates so badly as to constitute an immediate threat to its citizens' and ecology's health. A scholar of China's climate change policy shifts, Wei Liang, makes the following statement about the dire environment in contemporary China: "In recent years, severe environmental problems occur regularly. Ordinary Chinese have begun to be deprived of blue skies, clean rivers, green forests, and birds. Heart-breaking coal mine tragedies have become regular news on TV. Pollution has made cancer China's leading source of death. Nearly 500 million people lack access to safe drinking water. China is choking on its own success. The WHO [World Health Organization] found that the pollution-related death toll has now reached 750,000 a year. In comparison, 4,700 died in 2006 in China's unsafe mines."

China is the most prominent national case in the world threatened by the most menacing environmental challenges (except for Pacific Island nations possibly being eradicated by rising seas as a result of global warming). Therefore, China has the need and potential to become a world leader and role model with a green response to its environmental crisis: politically, socially, and ethically.

China's Dire Current Environmental State

Scientific reporting (both domestic and external) on the environment in China has improved greatly as a result of trepidation by Chinese citizens and leaders, as well as spurred worldwide interest because of China's status as "the world's workshop" and largest exporting economy in the world in 2009, its huge population of 1.4 billion people, and the global impact of its ecological developments. For example, sand from increasing desertification and air pollutants from industrial manufacturing sources and the burning of high-sulfur coal originating in China are propelled by wind currents across the Pacific Ocean and settle on the West Coast of the United States.

An outstanding source of the latest empirical data and scientific analysis regarding developments in China's environment is the China Environmental Forum (CEF). For the past 10 years the CEF, a subdivision of the Woodrow Wilson International Center for Scholars in Washington, D.C., established and partially funded by the U.S. Congress, has produced an annual publication, *China Environmental Series* (see Number 10, 2008–09), and a Website (www.chinaenvironmentalforum.org) and holds seminars, conferences, and symposia. It also regularly publishes papers, research notes and reports, and essays—all on the latest news and data regarding China's environment.

A thorough and comprehensive scientific description here of the environmental problems and threats besetting China is impossible given their scope, diversity, and quantity, but it is worth noting that China's environmental state includes virtually all of the ecological problems and challenges experienced by humans in Earth's biophysical environment. The components of the environmental crisis in China include (with no indication of comparative priority or severity) acid rain, air pollution, biodiversity loss, climate change, declining water resources, deforestation, desertification, dwindling food supplies, inadequate energy supplies, soil erosion, and water pollution.

The following is a brief description of China's current environmental state by newspaper reporter Michael Standaert writing in mid-2009:

> China is at a crossroads . . . Currently, one-third of China's rivers are polluted; one-fourth of its territory is desert, while another one-third suffers from severe soil erosion and drought; more than three-fourths of its forests are gone; urban residents are forced to breathe air containing lead, mercury, sulfur dioxide and other elements of coal-burning and tar exhaust. The number of cars is expected to grow from 33 million to 130 million in the next 12 years, and every 30 seconds a baby is born with pollution-related birth defects.

Underlying the bewildering scope of China's environmental problems and resulting threats, one essential point needs to be emphasized, and that is the root cause of the global environmental crisis and its particular manifestations in China: modern industrial social values and the social institutions built upon them. Every environmental threat confronting China and the planet as a whole relates to fundamental and controversial ethical and political values and policies.

The basic principle of the normative (ethical and political values) foundation of the environmental crisis in the world has been the center of the author of this article's work in the field of environmental political philosophy for the past 30 years. The environmental crisis may not be, as most scholars and the public conceive it, a challenge to technology and science, wherein breakthroughs in these fields will resolve problems. Rather, the fundamental premise of my view is that the environmental crisis is at its root a crisis in human values; a perspective that should resonate with the focus of this encyclopedia: *Green Ethics and Philosophy.* Since the environmental crisis is brought about by erroneous ethical and political philosophical values, it will take a new normative position to correct them, or at least, to ameliorate the crisis. Green ethics and political philosophy (the study of radical ethical and political environmental values) must play a primary role in responding to the modern industrial civilization that has generated so many environmental problems facing contemporary humanity, especially now in China with its severe environmental challenges.

Most scholars, and the majority of the public, conceive of the environmental crisis as a challenge to technology and science, wherein breakthroughs in these fields will resolve problems. However, missing from this view is the fundamental premise that the environmental crisis is at its root a crisis in human values. Since the environmental crisis is brought about by erroneous ethical and political philosophical values, it will take a new normative position to correct them, or at least, to ameliorate the crisis. Green ethics and political philosophy (the study of radical ethical and political environmental values) must play a primary role in responding to the modern Industrial Civilization that has generated so many environmental problems.

China's environment and its many serious issues, including the normative roots of its environmental crisis, are too globally profound to be addressed by scientific China specialists alone. Students and researchers of all geographical areas and disciplines, along with green ethical and political philosophers, should cooperate in examining the environmental crisis in China and how to respond to it. Such a combined response could be a model for the world, advancing human understanding and valuation, crisis remediation and prevention, and ultimately, social transformation to a more sustainable and just world.

Unlike the scientific authors of the founding document of the global environmental movement who overemphasized the scientific ecological limits to economic growth and omitted normative analysis of ethical and political values, contemporary students of the environment must uncover the value-based premises behind all the scientific reports and analyses of the environmental conditions in China today, as well as all over the world. For China and the world, subsequently, I shall go further and prescribe a particular set of green ethics and political philosophy "with Chinese characteristics": Confucianism.

The Normative Foundation of the Environmental Crisis in China and Elsewhere Leads to the Crucial Issue of Social Change

Looking at China's environmental crisis from the fundamental normative perspective directs attention to how social and political systems in China encompass social values. It also highlights the importance of ameliorative and preventative environmental policies that lead to two urgent questions for Chinese political leaders and citizens: (1) what social action will be required in the near- and long-term future in order to avoid environmental disaster (this is the dystopian logic of the "limits to growth" argument that has proved weakly persuasive worldwide over its nearly 40-year dominance within the environmental movement), and (2) how are we to reach the positive goals of achieving environmental sustainability and social justice through a change in social values and practices? The latter positive and normative approach is an essential one that is usually omitted from exclusively ecological analyses of China's environment (and of the global environment as well). This profound and complex point can be expressed briefly as follows: unless social justice is sought and studied along with ecological sustainability, not only will ethical and political values be ignored but ecological sustainability will also be unachievable in the practical world. This principle is vividly illustrated in China as its leaders grapple with the urgent social fact that 900 million of its poorest rural peasants seek the material betterment of their lives in the eastern coastal cities in the form of superior jobs, housing, education, medical care, and transportation. Unfortunately, these goods cannot be provided by the already overcrowded Chinese urban areas.

Once recognition of the severity of the environmental crisis in China and the world provides transformative social action, such as in China's poor rural areas, the next problem that needs to be addressed is the nature of the social changes to be made in order to respond to the crisis in order to avert catastrophe as well as to improve the ethical and political status quo. These challenges raise inescapable questions regarding social change that have been a staple of Western ethical and political philosophy of Socrates and Plato: what can the nature of these changes be in values, social practices, and institutions, and how are they to be specifically implemented given the inevitable social elements of resistance to changes of such a profound magnitude?

Regarding these vital questions, contemporary social change theory informs us that the development of social movements for change becomes critical. Therefore, in addition to considering the scientific data of China's current environmental state, the student of China's environment must consider social value and institutional changes, and how social movements and organizations might bring about such changes. The creation and maturation of the Chinese nongovernmental organizations (NGOs) sector, including ENGOs (environmental NGOs), even GONGOs (government-organized NGOs) begun in the time of the reform is important.

Although highly regulated by the central government of China since the protests of 1989, although official estimates vary regarding the numbers of civil society organizations in general, including those devoted to environmental protection in China, analysts agree that hundreds of thousands of such organizations now exist in China. The first two Chinese ENGOs, Friends of the Earth and Global Village Beijing, and thousands that have followed are devoted to environmental protection and public environmental education. Furthermore, they tend to be specific project–focused; for example, to oppose construction of an environmentally damaging industrial plant or a river dam. Such organizations could be used for broader social issues and change, as well as to strengthen the nascent but increasing environmental activist movement (the central government admits to tens of thousands of environmental protests annually) in conjunction with the recent upgrade to ministerial status of the central government's Ministry of Environmental Protection (MEP) from its previous bureaucratic incarnation as the State Environmental Protection Agency (SEPA) with no ministerial status and little political clout.

After absorbing scientific data about the various forms of China's multidimensional environmental crisis, and considering the current nature of Chinese environmental thought, the Chinese environmental movement, and the response of its political leaders to the crisis in its environment, the student of China's environment is led to the exploration of the subject of Chinese environmental politics and its many components, such as (1) the fragmented structure of the governmental structure in China between the central, provincial, and local levels with the inevitable tensions between levels, especially with regard to the controversial environmental realm (this raises serious questions about the central government's capacity to implement environmentally enhancing policies when the local governments are still stuck in the industrial growth-maniacal worldview); (2) the rise of new middle-class consumers in China seeking to live like rich Westerners with all their material excesses; and (3) the advent of new information technologies in social communications, for example, blogs, SMS (short message service), texts, Websites, Twitter, and so forth, based on the Internet and cell phone technology, affecting the dynamic relation between the Chinese people and its government (see the famous use of cell phones—China leads the world in the number in use—and texting in the successful public protest of the siting of a chemical plant in Xiamen in 2007).

Green Confucian China as a Global Catalyst for Social Change

The environmental crisis in China can provide the catalyst and platform for the renewal of modern industrial society to a just and sustainable social order by modernizing the ancient wisdom of Confucius and his followers. This result fulfills Confucius's own admonition of "both keeping past teachings alive and understanding the present" global environmental crisis, as exemplified in China. Some alternative Confucian values, as examples, that might serve as the means to take China down a different path to sustainable and just development, and be a positive role model for the world are embodied in the following Confucian ethical values:

- MIN BIN: people-oriented policy
- LI MIN: benefiting the welfare of the people
- JUN FU: equal wealth
- HE XIE: harmony between humanity and nature

The final ethical value on this list, HE XIE, or harmony between humanity and nature, may be the most important Confucian ethic from the perspective of China's and the world's environmental crisis. As Al Gore forcefully argues in his *Earth in the Balance,* it is the modern disconnection or alienation of humanity from nature that is the cause of the current environmental crisis by producing excessive and addictive consumption as a distraction from the psychic pain caused by this injurious estrangement of humanity from nature (see chapter 12 of Gore's book). In contrast, the Confucian value of HE XIE emphasizes the imperative for a harmonious relationship between nature and humanity. China's current president Hu Jintau's watchword for his administration of "harmonious society" is apt and important. It is a clear allusion to the Confucian value and emphasis upon harmony in the world, and it is an open question whether the president means harmony among human beings, or between humans and nature, or both, since the concept can be applied to ecologically crucial goals.

Understanding the ethical and political value bases of China's and the world's environmental crisis is essential not only to provide an accurate and comprehensive understanding of the nature of the current environmental state in China but to inspire potentially an effective response to the crisis that will lead China—and the world—to achieve sustainability and social justice. It is no exaggeration to conclude that the Earth's and humanity's future depends upon learning ethical and political philosophical lessons from China's environmental crisis and potentially adapting its Confucian philosophical worldview.

See Also: Carbon Offsets; Social Ecology; Technology.

Further Readings

Bell, Daniel A., ed. *Confucian Political Ethics.* Princeton, NJ: Princeton University Press, 2008.

Berthung, John. "Motifs for a New Confucian Ecological Vision." In *Confucianism and Ecology: The Interrelation of Heaven, Earth and Humans,* M. E. Tucker et al., eds. Cambridge, MA: Center for the Study of World Religions, 1998.

"Choking on Growth." *New York Times* (July–August, 2007).

Day, Kristen A., ed. *China's Environment and the Challenge of Sustainable Development.* Armonk, NY: M. E. Sharpe, 2005.

Economy, Elizabeth C. "The Great Leap Backward? The Costs of China's Environmental Crisis." *Foreign Affairs,* 86/5 (2007).

Economy, Elizabeth C. *The River Runs Black: The Environmental Challenge to China's Future.* Ithaca, NY: Cornell University Press, 2004.

Garner, Jonathan. *The Rise of the Chinese Consumer: Theory and Evidence.* West Sussex, UK: John Wiley, 2005.

Gore, Al. *Earth in the Balance: Ecology and the Human Spirit,* 2006 ed. New York: Rodale Publishers, 2006.
Kassiola, Joel Jay and Sujian Guo, eds. *China's Environmental Crisis: Domestic and Global Political Impacts and Responses.* New York: Palgrave Macmillan, 2010.
Leslie, Jacques. "The Last Empire: Can the World Survive China's Headlong Rush to Emulate the American Way of Life? *Mother Jones* (February 2008).
Myers, Norman and Jennifer Kent. *The New Consumers: The Influence of Affluence on the Environment.* Washington, DC: Island Press. 2004.
Shao, Bin. "Consumerism, Confucianism, Communism: Making Sense of China Today." *New Left Review,* I/222 (March/April, 1997).
Shapiro, Judith. *Mao's War Against Nature: Politics and the Environment in Revolutionary China.* New York: Cambridge University Press, 2001.
Song, Ligang and Wing Thye Woo, eds. *China's Dilemma: Economic Growth, the Environment and Climate Change.* Washington, DC: Brookings Institution Press, 2008.
Standaert, Michael. "China Turns to Clean Tech to Stimulate Its Economy." *San Francisco Chronicle,* May 10, 2009.
Tu, Wei-Ming. *Confucian Thought: Selfhood as Creative Transformation.* Albany: State University of New York Press, 1985.
Tucker, Mary Evelyn and John Berthrong, eds. *Confucianism and Ecology: The Inter-Relation of Heaven, Earth, and Humans.* Harvard University Center for the Study of World Religions. Cambridge, MA: Harvard University Press, 1998.
Turner, Jennifer L., ed. *China Environment Series.* China Environment Forum of the Woodrow Wilson Center for Scholars, Issue 9 (2007); Issue 10 (2008–09).
Xie, Lei. *Environmental Activism in China.* London: Routledge, 2009.

Joel Jay Kassiola
San Francisco State University

Civic Environmentalism

Civic environmentalism is an approach to environmental policy that emerged in the 1980s as a divergence from the first generation of environmental policy dominated by the top-down regulation of the central authority. This article first defines the concept and details its emergence and historical roots, then contrasts it with the traditional environmental policy and politics, explores the concepts and ideas it is related to, reviews the role of different levels of government in civic environmentalism, and closes with an examination of its criticisms and limitations and situations in which it is not appropriate for use.

Definition

Civic environmentalism is an ad hoc process of custom-designing answers to complex environmental problems in a specific location involving the stakeholders. It attempts to reconcile the economic development versus environmental protection dichotomy by replacing the traditional top-down, federal command-and-control regulation with a new decentralized system that combines the most effective elements of command-and-control and free-market approaches to environmental protection. It strikes a new balance between national standards and local solutions. This type of social action has limited government involvement, encourages bottom-up initiatives, emphasizes accountability at the local level, and claims to achieve better environmental outcomes more rapidly at a lower cost. Its cornerstones include public-private partnerships and collaboration, citizen participation, flexibility and custom-tailored solutions particular to location, and market-based incentives. At its best, it blends legal measures with fiscal policy, good science, governance mechanisms, and civic engagement. However, it is important to note that civic environmentalism is not a substitute for but a complement to federal regulation.

Historical Roots

The shift in the approach to environmental policy is rooted in two phenomena that occurred concurrently: the changing nature of the environmental problems and the state and local governments' stepping into the void in environmental policy created by the federal funding cutbacks in the 1980s. The first generation of national environmental policies were developed to deal with earlier problems caused mostly by large industrial polluters. Each addressed one problem at a time, focusing on individual species, pollutants, or pollution forms and usually independently of the circumstances of a particular place. Their top-down, prescriptive nature and fragmentation across policy areas proved ineffective against the newly emerging, more complex, and diffuse issues such as nonpoint source pollution, pollution prevention, habitat protection, and ecosystem management. These issues follow from land use decisions and bring forth the conflict of public interest with private property rights. Concurrent with the recognition of these different sets of problems, the 1980s saw the loss of momentum of the previous decade of environmental regulation and funding cutbacks, particularly in the United States. This caused a shift in focus of activity from the federal to the state level, and a new approach to environmental policy and politics was born.

While these uncoordinated bottom-up efforts were arising all over the United States, by the early 1990s, traditional environmental policy and politics was widely criticized for its inability to balance economic and environmental objectives and a consensus emerged on the need for a new approach in environmental policy. Alternatives discussed centered on more local and bottom-up initiatives, market-based approaches, and economic incentives such as fees, taxes, subsidies, and tradable permits as well as nonregulatory tools including public education, technical assistance, voluntary government programs, voluntary agreements negotiated through stakeholder participation, and new ways of investing in and operating public works, all tailored to local conditions. In his 1994 book *Civic Environmentalism: Alternatives to Regulation in States and Communities,* DeWitt John labeled the emerging alternative as "civic environmentalism." In fact, civic environmentalism is not one specific way of problem solving; rather, it can be better described as a heterogeneous collection of varied civic collaborative problem-solving approaches.

Civic environmentalism proposed using new tools and politics to address these newly recognized environmental problems with an integrative comprehensive approach. It differs from traditional command-and-control approaches in four ways: it focuses on a different set of environmental problems, uses different tools, seeks to overcome fragmentation, and searches alternatives to confrontation between the regulator and the regulated. Furthermore, it is more planning based and preventative. Many important ecosystem restoration and management initiatives of the time, such as the Everglades restoration in Florida,

CALFED Bay-Delta program in California, or the interstate Chesapeake Bay program involving the states of Maryland, Pennsylvania, Virginia, and the District of Columbia have been presented as successful examples of civic environmentalism by researchers. Many similar transboundary environmental problems cannot be solved by one-size-fits-all regulatory approaches. Collaborative watershed councils, species conservation, or estuary protection are other examples of cases deemed amenable to civic environmentalism approach. These complex ecosystems require an adaptive management strategy and civic environmentalism is the proposed approach to achieve it.

Civic environmentalism is advocated for its synthesis of the strengths of the federal government in environmental policy making to set the national standards and the ability of state and local governments to provide flexible and cost-minimizing solutions. Recognizing that geography of environmental problems hardly follows existing political boundaries, civic environmentalism proposes to achieve these solutions through collaboration of a diverse spectrum of stakeholders including citizens, community leaders, property owners, polluters, corporate executives, developers, and environmental action groups as well as local, state, and federal agencies. Solutions are not the product of a single agency or leader. These processes are marked by creativity, dynamism, and innovation to harness market forces to solve an environmental problem. This is sometimes achieved by the creation of economic value for landowners who voluntarily participate in environmental programs for habitat conservation or water quality protection. At other times, the economic incentive may be developed for a polluter to minimize their negative externalities.

Related Concepts and Ideas

While the term *civic environmentalism* is comparatively new, the idea has close ties with many familiar environmental concepts and approaches that have been around for a long time. Its focus on place-based solutions ties it to the concept of "ecological place." Solutions based on a specific place and its unique characteristics relate it also to "community-based environmental protection." In fact, civic environmentalism is proposed as a way of resolving the tension of the dual environmental system in which the centralized authority is setting standards and overseeing on one side while bottom-up community-based environmental protection programs are seeking pragmatic decisions on the other. "Community-based environmental protection" is the collective name for a variety of techniques and approaches government agencies use to encourage and support civic environmentalism. Involvement of all stakeholders is the common point between collaborative problem solving and civic environmentalism. Its recognition of interconnectedness of environmental problems and the holistic approach of taking the whole ecosystem with all of its elements rather than focusing on a single element like air or water is what it shares with "ecosystem management."

Environmental education is an important component of civic environmentalism as a prerequisite to effective citizen engagement and participation. At the root of the desire to include all stakeholders in environmental decisions is the goal of environmental justice. Civic environmentalism has also been called "third way" in environmental protection and is closely associated with the "third-way politics" that gained popularity in the 1990s and tries to synthesize right-wing free-market capitalist and left-wing democratic-socialist economic policies. With its attempt to find a balance between economic and environmental well-being while at the same time achieving environmental justice, civic environmentalism is also closely related to sustainability. Some civic environmentalists suggest industrial ecology—the emulation of the ecosystems in industrial processes of extraction, production, distribution, consumption, and waste—as the model for economic development. Civic environmentalism shares a number of elements with "ecological modernization" as well: their emphasis on innovative solutions, market-driven strategies, and changing role of government to allow for a more flexible, decentralized, cost-effective, and collaborative policy making.

Role of Federal, State, and Local Governments

Civic environmentalism is presented as the cornerstone of the second generation of environmental policy that redefines responsibilities among levels of government. Federal government's role in civic environmentalism is envisioned as providing top-down support for bottom-up environmental initiatives through provision of legal tools, technical assistance, and financial resources. Many agree that civic environmentalism without a strong federal framework establishing minimum standards is destined to fail. The federal government can also provide local collaborators with information, expertise, and small grants. Federal experts from outside the state can participate in the technical working groups convened locally to design the settlements or can work on loan to local agencies and organizations. Federal agencies may be able to help jump-start local collaborative processes by working as a catalyst. Through joint training, swapping staff, and helping to construct databases, federal managers can foster the competence of state and local agencies and build their capacity for effective involvement.

States can have a variety of roles in civic environmentalism. As repositories of federal environmental funds, they typically provide and organize the financial and technical sources that are available to local government. In extensive transboundary issues, they may get more involved in the process. Their capacity to capture the economies of scale within the university and business communities enables direction of their resources toward place-based environmental problem solving. Local initiatives rely on legal tools, technical assistance, and financial resources provided by the state and federal governments in crafting their locally driven solutions.

Challenges

Civic environmentalism is not without its challenges and critics. The criticisms can be grouped into three general areas: participation and effectiveness issues, reconciliation of top-down and bottom-up components, and limited applicability. Participation critiques argue that devolution of environmental policy from federal governments to lower levels may not necessarily increase the influence of citizens; rather, local elites and interest groups may take over domination of decision making. A related point of dispute is on whether civic environmentalism actually produces better processes and outcomes. Lack of evaluation leaves this issue unresolved.

As an emerging paradigm, civic environmentalism embodies a variety of different emphases and methods. Distinctions are made between "narrow focus" civic environmentalism that stresses the complex, interdependent, and place-based nature of contemporary environmental problems and argues communities will craft superior solutions if they have sufficient technical information and "broad focus" civic environmentalism that stresses the interdependency of contemporary environmental, social, and economic issues and calls for a civic renewal and power sharing between the stakeholders. In its narrow form, civic environmentalism is regarded as weak participation. Different forms of the model are criticized for favoring top-down or bottom-up elements or being self-contradictory about their relationship, unable to reconcile the two.

Finally, not all problems and situations are appropriate for resolution through civic environmentalism. Global problems such as acid rain or climate change may require a blend of civic, national, and international approaches. Civic environmentalism will thrive in places with stronger pro-environmental policies, whereas in historically industry-dominated areas traditional regulatory programs may achieve better outcomes. Some have argued that it only works at local levels or as long as the cause of the environmental problem is near the negative effects but cannot handle large geographic areas and temporally complex environmental problems. Three factors affect the success of civic environmentalist problem solving: the transboundary nature of environmental problems conflict with the idea of placed-based solutions; collaborative problem solving is perceived to be lengthy and time consuming, thereby discouraging participation; and voluntary programs might not work without financial incentives.

Despite its limitations and criticisms, its compatibility with many other approaches of the day make civic environmentalism a potentially valuable tool. However, as the civic environmentalists themselves agree, it needs better articulation to be put into practice more effectively.

See Also: Democracy; Ecological Restoration; Environmental Justice; Environmental Policy; Sustainability and Distributive Justice.

Further Readings

John, DeWitt. *Civic Environmentalism: Alternatives to Regulation in States and Communities*. Washington, DC: CQ Press, 1994.

Landy, Marc K., Megan M. Susman, and Debra S. Knopman. *Civic Environmentalism in Action: A Field Guide to Regional and Local Initiatives*. Washington, DC: Progressive Policy Institute, January 1999. http://www.ppionline.org/documents/Civic_Enviro_Full_Report.pdf (Accessed March 2010).

Montague, Peter. "The Environmental Movement—Part 3: Civic Environmentalism." *Rachel's Democracy & Health News,* 735 (2001). http://www.rachel.org/en/node/5418 (Accessed March 2010).

Shutkin, William A. *The Land That Could Be: Environmentalism and Democracy in the Twenty-First Century.* Cambridge, MA: MIT Press, 2000.

Sirianni, Carmen and Lewis Friedland. "Civic Environmentalism." Civic Practices Network, July 1995. http://www.cpn.org/topics/environment/civicenvironA.html (Accessed March 2010).

Aysin Dedekorkut
Griffith University

Climate Ethics

Climate change has been characterized as the "greatest moral challenge of our time." While the causes and impacts of climate change are now broadly understood, comparatively little attention has been directed toward the ethical challenges posed by climate change or to how the values, beliefs, and ethics of human societies drive the various actions that contribute to climate change. Nor has much attention been given to the goodness or badness of policies designed to remedy or adapt to climate change. Indeed, most of the ethical discussion to date has been relatively theoretical. Nonetheless, discussions about the rightness or wrongness of human (in)action around climate change and its impacts—in other words, about climate ethics and climate justice—have opened a veritable "Pandora's box" of issues. Resolving these issues also presents a series of dilemmas, contradictions, and paradoxes. Not surprisingly, more has been written about technocratic feasibility of interventions to minimize (mitigate) or adapt to climate change than about the actual moral correctness of those actions. This article considers some of the emerging ethical considerations surrounding anthropogenic climate change. After first briefly outlining the scope of expected impacts, the article then discusses the various ethical aspects of the climate change crisis and its potential remedies.

Existing and Expected Impacts of Climate Change

According to the Intergovernmental Panel on Climate Change (IPCC), climate change will bring numerous biophysical changes to the Earth, with environmental, social, political, and economic consequences. Rising temperatures, for instance, are expected to lead to more frequent and severe heat waves in many places. Associated urban "heat island" effects will likely result in the deaths of tens of thousands of people across the world's cities—especially in the megacities of the industrializing global south (developing nations concentrated predominantly in the Southern Hemisphere). Higher-intensity storm events and associated rainfall are predicted to bring widespread flooding and concomitant property and infrastructure damage. Frequent droughts will threaten food and water supplies and are expected to severely disrupt the livelihoods of peasant farmers in Africa, parts of South America, and southeast Asia. Wildfires, desertification, and concomitant soil erosion, combined with unreliable water supplies and crop failures, will lead to large-scale abandonment of nonviable agricultural land, higher food prices, and large-scale migrations to cities, in turn heightening the exposure of increasing numbers of people to impacts such as extreme temperatures and flooding. Soaring electricity costs, declining housing affordability, and escalating food costs will further harm vulnerable populations (e.g., poor, elderly, sick, indigenous people). A higher prevalence of insect-borne diseases will also likely harm both crops and people, especially impacting people living in subtropical cities (as tropical disease ranges expand). And rising sea levels will likely devastate many low-lying coastal settlements (from hamlets to megacities) by directly inundating built environments or severely disrupting the livelihoods of people who live within them (e.g., flooding coastal farmlands, severing transport linkages, destroying critical infrastructure, contaminating aquifers). Very large numbers of people (potentially millions) will likely be displaced as "climate refugees."

Biodiversity is also expected to decline significantly as the expected impacts of climate change take hold. Climate change will affect the physiology of individual plants and animals, the geographic distribution of species (as climatic zones shift), and the diversity of species (as individual species become extinct due to changes in predator-prey relations, vegetation composition, fire regimes, tolerance thresholds, and associated disruptions to ecosystem processes). Specialist species will be more susceptible to impacts as they are adapted to unique ecological niches and may be unable to adjust behavioral patterns such as mating, foraging, nesting, and so forth, and/or genetic predispositions may limit their ability to adapt to new environmental conditions (e.g., tolerance limits to temperature, pH, relative humidity, chemical exposure, pathogens) within the relatively short time frames in which climate changes are expected to occur. Changes to some species will also likely flow onto other species, resulting in cascading extinction episodes. For example, the bleaching of corals as sea temperatures rise and seawater increases in acidity (as oceans soak up carbon) will result in massive species extinctions (invertebrates and fish dependent upon coral will no longer have food and shelter). Alpine species are also especially vulnerable as they will not be able to simply migrate to other places, given they are isolated on mountaintops.

Ethical Considerations

Discussions of ethics center upon notions of what is right or wrong, good or evil, fair or unfair, and so on. Ethical precepts that have been applied to climate change include utilitarian (greatest good for the greatest number), Kantian, contractarian (surrendering individual liberties for a civilized society), doing no harm, respecting others, obligations to the vulnerable and marginalized, freedom from coercion, and taking responsibility for personal actions, among others. But there are two broad ethical dimensions to climate change problems and their potential remedies. The first—deontological ethics—pertain to the rightness of a particular action; the second—consequentialist ethics—relate to the rightness of the effects or impacts of a particular course of action. Important questions include "who is responsible for greenhouse gas emissions—individuals, communities, nation states, producers, consumers?"; "who is responsible for reducing emissions and by how much?"; "what are our obligations to future generations and nonhuman species, if any?"; "what level of sacrifice should current generations make?"; and "do developed nations have a moral responsibility to accept climate refugees and to help developing nations to mitigate and adapt to climate change?" These questions and many others address ethical concerns pertaining to distributive justice, compensatory justice, procedural justice, and equity.

Scholars generally recognize that there are several basic expressions of equity insofar as climate change is concerned, including (1) equitable distribution—all humans should have uniform exposure to benefits or harms; (2) compensatory equity—the most disadvantaged and vulnerable populations who are disproportionately exposed to climate change impacts should be compensated to offset inequalities, either in monetary, aid, technological, or developmental assistance; (3) procedural equity—all individuals (irrespective of class, race, gender, religion, etc.) should have equal capacity to participate in climate change negotiation and decision making. Many scholars also argue that actions that contribute to climate change violate the Universal Declaration on Human Rights, and much of the ethical literature on climate change considers the human rights impacts of climate change. Knowing how these different versions of equity are mobilized in climate change discussions can potentially enable us to better evaluate policies, actions, and remedies.

Scholars have also identified a number of ethical conundrums that arise from climate change. First, the people least responsible for climate change (e.g., vulnerable populations in developing countries of the industrializing South and within the developed North) will likely bear the greatest burden of anticipated impacts. Second, while the industrialized North is responsible for most greenhouse gas (GHG) pollution to date, limiting further emissions will have a disproportionately harsh impact on developing counties seeking to industrialize. Third, technological and market-based solutions for climate change can further entrench inequalities between the north and south, may disproportionately benefit the elites who are most responsible for the problem (e.g., venture capitalists and transnational corporations), and may exacerbate the problem itself (developing new technologies, e.g., coal gasification, can increase emissions). Fourth, abrupt and catastrophic climate change could result in a mini ice age in Europe, raising the specter of increased emissions to adapt to this new problem, further jeopardizing the fate of future generations. But not adapting would condemn thousands to death. Fifth, the future generations of humans and other species will disproportionately suffer from climate change impacts even though they have not caused them. Sixth, some of the potential solutions to climate change may, in fact, exacerbate existing problems (e.g., replacing fossil fuels with biofuels to mitigate carbon emissions can increase the price of grains and other foodstuffs and exacerbate biodiversity loss as fragile and/or threatened habitats are converted to crops for food and fuel). Finally, mitigating climate change to benefit future generations may place unbearable costs upon the most vulnerable people within the current generation and, conversely, taking action to adapt the current generation to climate impacts (e.g., installing air conditioners or relocating vulnerable settlements) may exacerbate climate change in the future.

Principles of Climate Ethics (Climate Justice)

Conceptions of justice typically address notions of what is right and what is fair. Climate justice has two dimensions—a social justice element and an ecological justice component. When considering the social dimension of climate ethics, commentators urge that attention be given to how multiple axes of difference (e.g., gender, class, race, ethnicity, and disability) configure vulnerability or resilience to climate change impacts, the alternative knowledge and perspectives that traditionally marginalized groups might offer, and, in turn, how strategies to combat or adapt to climate change might empower marginalized and vulnerable communities to take their own action against climate threats.

Several principles have been posed to resolve ethical tensions arising from climate change to achieve just actions and outcomes. Many stem from the so-called Bali Principles. There is insufficient room to describe them all here, but they include the following:

- *Global commons:* the atmosphere is a global commons to which all species are equally entitled
- *Do no harm:* climate change must be mitigated and harmful greenhouse gas levels reversed
- *Polluter pays:* the people most responsible for climate change should fix it
- *Beneficiaries pay:* the beneficiaries of past GHG emissions (e.g., industrialized nations and elites in developing nations) should bear the burden of responsibility for mitigation and adaptation—also termed *historical responsibility*
- *Common but differentiated responsibility:* all people bear a common responsibility to halt climate change, but the greatest burden falls to those with the ability to pay and to those who benefit most from greenhouse gas–producing activities
- *Compensatory equity:* powerless, disadvantaged, and socioeconomically vulnerable people who are worst affected by climate change should be compensated by those who have benefited
- *Precautionary principle:* uncertainty around the exact causes and impacts of climate change is no excuse for forestalling mitigation and adaptation action
- *Intergenerational equity:* current generations who have benefited from climate-harming actions have an ecological debt to future generations and must take action to ensure that future generations are not harmed by (in)actions of present generations and are able to meet their own needs (e.g., sustenance, shelter)
- *Intragenerational equity:* steps should be taken to ensure that marginalized and vulnerable people within the current generation do not disproportionately suffer from climate change impacts

- *Environmental justice:* people of color and the poor should not bear a disproportionate impact of climate change
- *Meeting basic needs:* people in the developing world must be allowed to generate a certain level of emissions to meet their basic requirements for shelter, food, transport, etc.
- *Affordable clean energy:* all people have a right to affordable and sustainable energy
- *Participatory democracy:* vulnerable and traditionally marginalized groups have a "right to represent and speak for themselves" in policy and decision making
- *Ecological justice:* nonhuman species have moral considerability and must be represented in climate change decision making as well as benefit from efforts to mitigate and adapt to climate change

It is telling that few scholars have considered the ecological justice imperative of climate change. Important questions here include "who represents nonhuman life forms at the policy table?"; "what are our obligations and responsibilities toward other species?"; and "should we compensate other species who are affected by our actions, and if so, which ones, how many, and by what means?" Concepts such as ecological citizenship and Buddhist economics are beginning to address such questions—by asserting that all species are tied together in a complex web of socio-ecological relations and that humans have both an obligation and a responsibility to ensure that other species are not imperiled by our actions.

Potential Remedies

Potential climate change remedies take two forms: mitigation strategies and adaptation strategies. Adaptation options include actions like developing drought-resistant crops, storm-resistant housing, resilient infrastructure, and food and water security (e.g., recycling water and growing food locally). Mitigation proposals include reducing emissions by altering lifestyles and behavior (e.g., driving less, reducing air travel, switching to renewable energy sources, and geo-sequestering carbon). But to be ethical and equitable, many of these potential solutions will have little effect unless they also entail transfers of wealth and knowledge from elites and developed nations to impoverished people and developing countries. Some mitigation actions can have pernicious consequences. For example, some developed nations have begun purchasing forest tracts in developing countries to function as carbon sinks. But if indigenous communities are forced out of such reserves, it will only heighten inequalities. And in some ways, developed nations are simply transferring their emissions problem elsewhere instead of adjusting excessive consumption behaviors and fossil fuel–dependent technologies and infrastructures that caused the problem in the first place. Moreover, preserving carbon sinks in the developing world could limit those countries' development options—amounting to neocolonialism. More ethical options may include developing new technologies like renewable energy and transferring them to the developing world; retraining and reskilling people currently working in carbon-intensive industries for "green jobs"; and even radically reworking the way governments, corporations, and individuals interact with each other and with the biosphere.

From an adaptation perspective, we will require better medical services, alternative sources of water (such as recycling effluent), planned retreat from flood-prone land, and distributing critical infrastructure—so we are less reliant on single sources for electricity, water and waste treatment, and the like. But such solutions are expensive and will do little to redress inequities unless they are made available to poorer communities and individuals. Developed nations must also be prepared to change their immigration policies to accept peoples displaced by climate change.

Solutions for conservation biology are also fraught with difficulties. Establishing protected corridors between isolated habitat patches may improve the flow of genetic material and enable threatened plant and animal species to move, but such corridors also facilitate the movement of weeds and pest species and/or may increase the movement of wildfire. Revegetating degraded areas with local native species may foreclose the opportunity to plant species that will be better adapted to future climatic conditions. And warehousing vast numbers of seeds in "seed banks" and establishing captive-breeding programs in zoos does little to redress the cause of climate change. Nor will it enable the restoration of ecosystems that will be decimated by climate change.

The ethical issues raised by climate change are complex. Clearly, there is no one single ethical position that helps us to address all aspects of climate change and its impacts. Moreover, competing ethical positions can create paradoxes and dilemmas. While there is a dire need for a globally agreed set of ethical principles to address climate change, arriving at such an agreement will be fraught with challenges. Few, however, would disagree that we must find a more socio-ecologically sustainable way of life.

See Also: Anthropocentricism; Biocentricism; Biodiversity; Carbon Offsets; Ecocentrism; Ecological Footprint; Ecopolitics; Energy Ethics; Environmental Justice; Environmental Values and Law; Future Generations; Human Values and Sustainability; Intergenerational Justice; Kantian Philosophy and the Environment; Land Ethic; Precautionary Principle; Preservation; Religious Ethics and the Environment; Sustainability and Distributive Justice; Tragedy of the Commons; Urbanization; Utilitarianism.

Further Readings

Adger, W. N. "Scales of Governance and Environmental Justice for Adaptation and Mitigation of Climate Change." *Journal of International Development,* 13 (2001).

Caney, S. "Justice and the Distribution of Greenhouse Gas Emissions." *Journal of Global Ethics,* 5 (2009).

Daniels, P. "Climate Change, Economics and Buddhism—Part 1: An Integrated Environmental Analysis Framework." *Ecological Economics,* 69 (2010).

Gardiner, S. M. "Saved by Disaster? Abrupt Climate Change, Political Inertia and the Possibility of an Intergenerational Arms Race." *Journal of Social Philosophy,* 40 (2009).

Goodman, J. "From Global Justice to Climate Justice? Justice Ecologism in an Era of Global Warming." *New Political Science,* 31 (2009).

Hillerbrand, R. and M. Ghil. "Anthropogenic Climate Change: Scientific Uncertainties and Moral Dilemmas." *Physica D,* 237 (2008).

McNamara, K. E. and C. Gibson. "'We Do Not Want to Leave Our Land': Pacific Ambassadors at the United Nations Resist the Category of 'Climate Refugees.'" *Geoforum,* 40 (2009).

Page, E. A. "Distributing the Burdens of Climate Change." *Environmental Politics,* 17 (2008).

Rosales, J. "Economic Growth, Climate Change, Biodiversity Loss: Distributive Justice for the Global North and South." *Conservation Biology,* 22 (2008).

Westra, L. *Environmental Justice and the Rights of Ecological Refugees.* London: Earthscan, 2009.
Wilby, R. L. and G. L. W. Perry. "Climate Change, Biodiversity and the Urban Environment: A Critical Review Based on London, UK." *Progress in Physical Geography,* 30 (2006).

Jason Byrne
Griffith University

Club of Rome (and Limits to Growth)

The idea of limits to growth has been among the most influential and controversial notions in environmental debates. For many greens, infinite growth on a finite planet is clearly impossible. Yet governments, business, most economists, and even mainstream environmental groups have tended to resist such conclusions, favoring the idea that economic growth can be decoupled from rising environmental impacts.

The possibility of boundless expansion was a widely shared, modern article of faith until the Club of Rome called it into question. Founded in 1968 by Italian industrialist Aurelio Peccei and Scottish scientist Alexander King, the club's original members consisted of a small, international group from the worlds of science, industry, diplomacy, academia, and civil society. Their concerns over the "predicament of mankind" included the dominance of short-term thinking and unlimited resource consumption in an ever-more interdependent world. The club soon commissioned systems modelers, led by Donnella and Dennis Meadows at the Massachusetts Institute of Technology, to produce *The Limits to Growth* report of 1972.

The report focused on five key elements: population, industrialization, pollution, food production, and nonrenewable resource depletion. Various runs of its World3 computer model led to the conclusion that if growth in these areas persisted, limits to growth would be reached sometime within 100 years, with the most probable result being a sharp decline in population and industrial capacity. The standard world model run, based on business-as-usual-trends, showed a collapse in the first half of the 21st century resulting from nonrenewable resource depletion—one of several scenarios examined.

The often-misunderstood consequences of exponential growth were a key theme of the report, notably the potential for sudden, catastrophic impacts. Another central message was that technology alone could not push back limits for long: models assuming technological solutions that lifted one restraint on growth simply grew until another limit was reached, merely delaying overshoot and collapse.

Despite the claims of some critics, the authors of *The Limits to Growth* did not predict impending doom; in fact, they emphasized the possibility to alter such growth trends. They argued that a stable population and constant stock of capital would allow an "equilibrium state" that could be sustained far into the future—in which improved technology would be welcome and some forms of ecologically benign growth, including increased leisure, could continue.

The report was a publishing success, selling 30 million copies in more than 30 languages, but also provoked much criticism. Economists denounced the failure to capture the price mechanism's role in stimulating greater efficiency, substitution of alternate resources, and investment in improved technology. Promethean writers such as Julian

Simon and Herman Kahn countered the club's neo-Malthusianism by celebrating the unlimited capacity of human ingenuity, liberated by free markets, to find technological solutions to any short-term resource and pollution problems. Southern countries resisted any suggestion of limits on economic development to reduce poverty, while some on the political left similarly saw talk of limits as a way to cement existing social inequalities and a threat to rising living standards for the majority. Although the oil price shocks of the 1970s vividly illustrated the threat of resource scarcity, energy shortages and commodity inflation were later followed by declining resource prices and even, as the *Economist* put it in 1999, a world "drowning in oil." Such developments appeared to support the critics who downplayed depletion concerns. Meanwhile, by the 1980s, ideas of limits to growth were being pushed aside by sustainable development, which, in most interpretations, downplayed the conflict between endless economic expansion and ecological sustainability. Similarly, the idea of ecological modernization, which became the dominant approach to environmental matters in many countries, rejected a discourse of limits and promised "win-win" opportunities for "green growth."

While the Club of Rome's analysis was clearly out of fashion for a time, some observers have seen renewed relevance in light of growing concern over climate change, peak oil, food and water scarcity, species extinctions, and decline of the world's fisheries, among other ecological challenges. Two updates of *The Limits to Growth,* in 1992 and in 2004, confirmed the original message, while adding that humanity had squandered the opportunity to change course and already overshot environmental limits. Studies using ecological footprint analysis have similarly concluded that total human demands now exceed the capacity of the biosphere. A study by Australian physicist Graham Turner found that in the 30 years following *The Limits to Growth* model, real-world data has been in line with the unsustainable trajectory of the report's standard run or business-as-usual scenario. Meanwhile, in light of the lack of evidence to date to support hope for decoupling economic growth and environmental impacts, a 2009 report by the United Kingdom Sustainable Development Commission argued that there was an urgent need to find ways to achieve "prosperity without growth."

Such conclusions remain controversial—for reasons including the challenge they represent to deeply ingrained cultural notions and an economic system that appears to require endless expansion—but proponents of the limits to growth thesis maintain that we avoid them at our peril.

See Also: Consumption, Consumer Ethics and; Ecological Footprint; Western "Way of Life."

Further Readings

Club of Rome. http://www.clubofrome.org (Accessed February 2010).

Hails, Chris, et al. *Living Planet Report 2008*. Gland, Switzerland: WWF International, 2008.

Jackson, Tim. *Prosperity Without Growth? The Transition to a Sustainable Economy.* London: Sustainable Development Commission, 2009.

Meadows, Donella H., Dennis L. Meadows, and Jørgen Randers. *Beyond the Limits: Global Collapse or a Sustainable Future.* London: Earthscan, 1992.

Meadows, Donella H., Dennis L. Meadows, Jørgen Randers, and William W. Behrens III. *Limits to Growth: A Report to the Club of Rome,* 2nd ed. New York: Universe Book, 1974.

Meadows, Donella H., Jørgen Randers, and Dennis L. Meadows. *Limits to Growth: The 30-Year Update.* White River Junction, VT: Chelsea Green, 2004.
Simon, Julian L. and Herman Kahn, eds. *The Resourceful Earth: A Response to Global 2000.* New York: Blackwell, 1984.
Turner, Graham M. "A Comparison of the Limits to Growth With 30 Years of Reality." *Global Environmental Change,* 18 (2008).
Victor, Peter. *Managing Without Growth: Slower by Design, Not Disaster.* Cheltenham, UK: Edward Elgar, 2008.

Anders Hayden
Dalhousie University

Conservation

A park ranger maintaining a sign for the Togiak National Wildlife Refuge in southwestern Alaska, where 2.3 million acres have been conserved in the Togiak Wilderness, one of the largest such areas in the country.

Source: U.S. Fish & Wildlife Service

Conservation thought has been part of the Western sociocultural landscape for more than two centuries. Its practice and ethic have acted as something of a counterbalance to capitalistic ideology and action related to industrialism, growth, and development, which have altered the face of the planet. In some ways, the evolution of conservation as an ethical system has always lagged behind the development of capitalism. Because the widespread understanding of environmental damage was slow to unfold, conservationists were often put in the position of responding to damage wrought by increasing demands on the Earth by the expanding economy.

Conservation as a philosophy, ethic, and form of action has been expressed widely and in many different ways in art, music, literature, and life sciences, as well as in individual lifestyles and the political and economic arenas. Yet the conservation ethic has not been predominant in the West. In the United States, especially, natural resources have been viewed as essential for growth of the capitalist system based on the creation of individual, corporate, and national wealth in what many have historically viewed as an unlimited abundance of free natural gifts there for the taking.

One significant barrier to conservation thought is its hazy time horizon that offers deferred rewards. While capitalism can be fragmented because of competition and social class interests, its unifying principle is profitability, measured in shorter-run periods such as quarters or years. In addition, capitalism, in its core areas, promises increased social well-being, most evident in consumer comforts.

Conservation, however, is both a short-run and long-run idea that is fragmented in its ethics and practice. First, the idea of conservation has challenged and, to some extent, mitigated the predominant ideas and practices of unlimited growth and rapacious use of natural resources. As a result, capitalist enterprises have adopted conservation practices, sometimes grudgingly. This is not because conservation is always seen as a good end in and of itself, but because it saved money for the extractors and processors of natural resources and increased bottom-line profitability. Capitalism, because of its profit orientation, is, at its core, materialist.

Second, conservation's deepest roots are idealistic, based on mutually cooperative relationships between humans and nature. The ethic translates into action through wise use of resources to leave a better world for future generations, that is, using natural resources for the benefit of everyone. While the practice of a conservation ethic can have clear material outcomes, this deferred gratification is a distant promise, not as tangible or rewarding as immediate profits.

Since the predominant capitalist system has tended to co-opt significant areas of conservation thought, conservation's ideological fault lines lie along complicated ethical terrain. They include varying ethical views of the following:

- Anthropocentrism and envirocentrism, including differing religious and philosophical views
- Economic and environmental rationality
- Pragmatic use values and aesthetics, with nature as a value in and of itself
- Masculine and feminine views of nature
- Individual and corporate ownership and commonwealth, with an emphasis on government's role in conservation
- Free markets and planned markets
- Development and so-called undeveloped places that have different cultures and views of the natural, human, and cosmological world
- Local communities and national, regional, and global economies, with an emphasis on economic systems and geography
- Locally grounded traditional knowledge and science for analysis and management
- Smaller economies and larger economies of scale
- Growth and limits to growth, including population, consumption, and development of physical infrastructure
- Efficiency and waste management
- Individual and corporate freedom and social control

Given the complicated map of conservation thought, ethics, and action, it is important to study some of the its evolution from the Enlightenment, Romanticism, during the last half of the 19th century, and its application during the Progressive period around the turn of 20th century. A brief look at these time periods helps provide a framework for understanding some of the nuances of conservation ethics that underpin current understandings of sustainability.

The Enlightenment

The great minds of the Enlightenment generally were not directly interested in conservation of nature in and of itself. In fact, the powerful Enlightenment heritage that spawned both democracy and modern capitalism ultimately wove the economy, technology, political processes, education, and sciences into a sociocultural fabric to support growth, competition, and consumerism that exploit natural and human resources while heightening personal satisfaction for many.

French philosopher Jean-Jacques Rousseau (1712–78) was something of an exception when he linked healthy human development to interactions with nature. In *Emile* (1762), he wanted to create an environment that cultivated a child's desire to learn from nature. Rousseau's individualism was based on growing awareness of sensations triggered by nature that ultimately allow us to grow outside ourselves and think for ourselves. With a basis for comparison, individuals develop a sense of morals or ethics that allow us to participate more fully in our human relationships.

Thomas Jefferson (1743–1826) envisioned a nation of individual liberty with government for the good of all—commonwealth. Jefferson's notions of commonwealth and agrarianism link individuals, communities, and government with conservation. His scientific approach to agriculture and philosophy of democracy are key elements in some threads of the conservation ethic.

Human history is, in the view of some, interpreted as the struggle between individuals and authority or community. This is certainly a crucial aspect of history. Enlightenment thinkers elevated that stature of humans. Individuals are rational beings capable of building a better world. Government by the people builds democracy for communities and nations.

But there is another facet of human history that sheds light on the emergence of conservation. Our efforts to gain and apply knowledge—spurred on by the Enlightenment—also have focused on survival and protecting ourselves from natural risks. Conservation picks up ideas from the Enlightenment that suggest human beings can make rational, scientifically based decisions about their future based on common interests. Conservationists, however, do not rely solely on the notion of rational decision making in competitive, individualistic economic markets, another product of Enlightenment thinking. Conservationists are mainly concerned about ecological rationality, which stresses more cooperation among individuals and focuses on respect for the ecology as crucial to human interactions. In this sense, conservationists place high value on the ability to act positively on common concerns about the long-run impacts of human activities on the environment. Government is a unifying force for protecting the environment. More recently, market activity has begun to consider environmental impacts. Conservation as a science measures the impacts of these activities and its practitioners seek to manage the environment for short- and long-run human benefits.

Romanticism

Conservation roots are particularly deep in the Romantic period that followed Enlightenment from around the turn of the 19th century. The expression of conservation in this period was strongly emotional and implied strong feelings toward nature and places, and, sometimes, a sense of nostalgia and loss. It was at least partly a reaction to the onset of the Industrial Revolution and the increasing pace of sociocultural change in both the United States and Europe.

As a major element of the complex Romantic reaction against the Industrial Revolution, the transcendentalism of Henry David Thoreau and Ralph Waldo Emerson idealized times past, emphasizing an ethic of simplicity to elevate individual humans to a spiritual plane within nature, beyond the dirtiness of factories. Openness to nature tempered greedy self-interest—the extreme of Adam Smith's premise of individuals' insatiable wants and needs—moving individuals to view human relationships with nature in new ways. Yet, the early New England transcendentalists either did not understand or dismissed the flagrant waste of their time. If they did understand the rapacious activity, they did not dissent. At one level, they were believers in progress. In fact, Emerson and many other writers and artists of the time seem to have embraced the machine age as compatible with nature. In addition, writers such as Walt Whitman praised American democracy, fanning community aspirations for democracy and conservation in later times.

In the midst of the Romantic period, nature-focused art and writing flourished. On one hand, wildlife painters such as John James Audubon (1785–1851) chronicled the country's birds, leaving a realistic record of species that have persisted or become extinct in their wild settings. Others, such as those of the Hudson River School, chronicled the nation's rich landscapes, capturing mysterious qualities of their subjects and inspiring viewers to see the beauty and natural wonders of the American land. Besides the journals of early explorers of the continent, writers such as Thoreau, John Burroughs (1837–1921), and John Muir (1838–1914) opened the way for a distinctly American style of nature writing that was both popular and influential in political and economic circles.

Muir's preservationist stance, which became the basis for national parks, was uncompromising. In the Romantic tradition, he wanted vast wilderness areas, scarcely touched by human activity, to be places where people could escape exploding urbanization. They would remain wilderness forever. This was not a notion of preservation rooted in utilitarian and pragmatic economics. It was rooted in cosmovision, aesthetics, and ethics. The value of wilderness lay in its ability to help humans transcend their daily existence to learn more about their oneness with nature. The practical application of his ethic became the basis for national parks.

The Last Half of the 19th Century

As the pillaging of natural resources intensified with rapid economic growth during and after the Civil War, the conservation movement gradually gained strength. The last half of the 19th century could be characterized as a slowly growing scientific and public consciousness about growing environmental problems in the United States. The emergence of research-based scientific management for forests, in particular, underpinned the rational branch of the new conservation movement and affected its ethical characteristics.

Science began to play an important role in documenting the impacts of human activities on the American landscape, especially in the developing field of forestry, but also in other areas. The 1864 publication of *Man and Nature* by George Perkins Marsh, for example, warned of the need to protect natural resources. Frederick Starr, in an 1865 report for the new U.S. Department of Agriculture, predicted a timber famine within 30 years and urged research on forest management and setting up forest plantations. In an 1867 report to the Wisconsin legislature, I. A. Lapham showed the relationship of forests and stream flow, stressing the need for tree planting to protect watersheds. In 1877, Franklin B. Hough, first chief of the Division of Forestry in the U.S. Department of Agriculture, presented a report that demonstrated the effects of forests on climate, stream flow, and soil. In 1878, John Wesley Powell authored *Report on the Lands of the Arid Region,* which dealt with water problems in the West.

Meanwhile, a group of scientists around the world, including Joseph Fourier of France (1859); John Tindall of Ireland and England (1859); Svante Arrhennius and Arvid Högbom, both of Sweden (1896); and T. C. Chamberlin of the United States (1897) were developing an understanding of the role of greenhouse gases in climate change.

As scientists began to develop new understandings of the environment, the federal government also gradually began to increase its role in natural resources management. Activities included establishment of the U.S. Department of Agriculture (USDA) and land grant college system for agricultural education (1862); Wilderness Act (1864); Mineral Land Act (1865); Mining Act and establishment of Yellowstone National Park (1872); Timber Culture Act (1873); USDA Division of Forestry (1876); Desert Land Act (1877); Timber and Stone Act and Timber Cutting Act (1878); USDA Division of Forestry (1881); Yosemite National Park (1890); and the Forest Reserve Act (1891).

Progressive Politics

Government, which had begun to take a more active role in resource management after the Civil War, ratcheted up its activities after the turn of the 20th century. The zenith of Progressive Republican politics, with President Theodore Roosevelt (1858–1919) at the lead, redefined and strengthened the federal government's role in conservation. Roosevelt, with John Muir (1838–1914) and Gifford Pinchot (1865–1946), built the processes of thought and action that formed the basis for the Conservation Movement into the 1960s and 1970s.

As Pinchot stated any number of times, the nation's natural resources were to be used wisely for the benefit of everyone. Muir, who wanted to preserve wilderness for the sake of wilderness, disagreed with Pinchot's philosophy. Roosevelt befriended both of them and pushed both of their causes in the face of opposition from business and political interests. Roosevelt's, Muir's, and Pinchot's positions all depended on public ownership of land for the enjoyment of future generations.

The adoption of conservation practices, with their core concepts of preserving resources for future use, took several generations to gain wide acceptance. As the federal and state governments became more involved in conservation, research and practical expertise moved beyond forestry into areas such as soil erosion, watershed management, and air quality, as well as healing damaged environments. While pragmatic and utilitarian in practice, the predominant side of the conservation movement showed an increasing recognition of the importance of nature in shaping human endeavors, evolving gradually into the sustainability movement.

Different Approaches to Conservation Today

J. Rodman analyzes ecological consciousness rooted in historic Western values as taking the four following forms:

- Resource Conservation, in the tradition of Gifford Pinchot and Theodore Roosevelt
- Wilderness Preservation, in the tradition of John Muir and the Sierra Club
- Moral Extensionism, a diverse tradition that claims intrinsic value and rights for nature and challenges anthropocentrism
- Ecological Sensibility, in the tradition of Aldo Leopold, which recognizes nature's intrinsic value, accounts for individuals' interactions with the land's ecosystems, and is based on an ethical code

Alternatively, P. Hay proposes the modern conservation ethic manifests itself in such areas as green sympathy, ecophilosophy, ecofeminism, spirituality, critiques of science and knowledge, sense of place, political thought, liberalism, socialist traditions, and human ecology.

Rodman's and Hay's approaches have common goals, but suggest why the conservation movement has been fragmented, subject to differing interpretations that sometimes find themselves creating friction for each other.

See Also: Conservation, Aesthetic Versus Utilitarian; Forest Preservation Laws; Green Liberalism; Muir, John; Pinchot, Gifford; Thoreau, Henry David.

Further Readings

Collins, T. "Changing Forests, Changing Forest Service: Chronology and Quotes." ASPI Critical Issues Series No. 4. Livingston, KY: Appalachia—Science in the Public Interest, 1992.

Collins, T. "Community Capitals and Land Use: Values for Assessing Sustainability." North Central Regional Center for Rural Development. Community Capitals Framework Workshop, 2007.

Hay, P. *Main Currents in Western Environmental Thought.* Bloomington: Indiana University Press, 2002.

Hays, S. P. *Conservation and the Gospel of Efficiency: The Progressive Conservation Movement, 1890–1920.* New York: Atheneum, 1980.

Marx, L. *The Machine in the Garden: Technology and the Pastoral Ideal in America.* Oxford, UK: Oxford University Press, 2000.

Miller, C. *Gifford Pinchot and the Making of Modern Environmentalism.* Washington, DC: Island Press, 2001.

Muir, J. *The Wilderness World of John Muir,* Edwin Way Teale, ed. Boston: Houghton Mifflin, 1954.

Rodman, J. "Four Forms of Ecological Consciousness Reconsidered." In *Ethics and the Environment,* D. Scherer and T. Attig, eds. Englewood Cliffs, NJ: Prentice-Hall, 1983.

Rousseau, J.-J. *Emile, or On Education.* New York: Columbia University Institute for Learning Technologies. http://www.ilt.columbia.edu/pedagogies/rousseau/contents2.html (Accessed November 2005).

Udall, S. L. *The Quiet Crisis.* New York: Holt, Rinehart and Winston, 1963.

Weart, S. "The Discovery of Global Warming. Introduction: A Hyperlinked History of Climate Change Science." http://www.aip.org/history/climate/summary.htm (Accessed November 2005).

Timothy Collins
Western Illinois University

Conservation, Aesthetic Versus Utilitarian

John Muir (1838–1914) and Gifford Pinchot (1865–1946) mark two significant points on the complex map of conservation and environmental ethics. While both men shared the common goal of natural resource conservation, their approaches represent polar opposites with different ethical foundations.

The Scottish-born Muir developed an aesthetic view of nature. After traveling across the country, he ended up on the West Coast. An engineer, naturalist, and writer, he was

Conflicts over use of conservation lands and national parks have involved development, recreation, logging, plant diversity, and wildlife habitat. The photo shows a large resort development built within sight of the Bon Secour National Wildlife Refuge in Alabama in 2009.

Source: U.S. Fish & Wildlife Service

idealistic, an advocate of preserving wilderness for the sake of wilderness to leave vast areas of land untouched by development.

Pinchot, a two-term Progressive Republican governor of Pennsylvania, was an ardent political and scientific advocate of conservation through wise use. His worldview grew partly out of his French training as America's first professional forester. He was a pragmatist and a utilitarian, interested in sustainable timber yields at a time when clear-cutting had devastated the nation's forests.

Differences between Muir and Pinchot define conflicting fields of thought, ethics, and action that fragmented the late 19th century's growing conservation movement. The historic context included rapidly increasing population and industrialized exploitation of the nation's natural bounty. The conservation movement was partly a reaction against industrialization, but it also was an effort to control industrialization. Muir and Pinchot—both considered political Progressives—offered clear alternatives for rebuilding and preserving the nation's natural heritage. They were articulate conservation advocates with important connections and sympathetic ears in the nation's political and business circles.

The sociocultural context of this time included struggles over access to and control of public lands to assure low-cost natural resources. Sales, gifts, and leases of public lands to private interests caused considerable controversy. Muir and Pinchot, with their political and writing skills, both played a role in helping to crystallize government's role in managing and conserving the nation's rich commonwealth of resources. Progressive Republicanism—expressed by Pinchot and Muir, with help from other politicians including President Theodore Roosevelt (1901–09)—ultimately strengthened federal control over natural resources planning.

Muir and Pinchot came to the fore of the conservation movement as naturalists and scientists began raising alarms about environmental destruction, especially after the Civil War. Two well-publicized disasters helped bolster the fledgling conservation movement, increasing public awareness about irresponsible forestry practices and the need for conservation:

- Several massive forest fires resulted from wasteful harvesting. In and around Peshtigo, Wisconsin, for example, at least 1,200 persons died and about 1.2 million acres burned across the upper Midwest in October 1871.
- Deforestation damaged watersheds and caused major flooding in downstream communities. Clear-cutting in central Pennsylvania's Appalachian Mountains was a contributing factor in the May 1889 Johnstown flood that killed about 2,200 persons.

Pinchot's pragmatic ethic of forest conservation was built on scientific management of the nation's forests as renewable resources for multiple use to assure a future supply of wood and wood products for construction and other uses; protect watersheds and water supplies from pollution and soil erosion; and maintain recreation and wildlife areas. His multiple use philosophy—implemented while he was head of the U.S. Forest Service—has been a guiding, but contradictory and sometimes controversial philosophical tenet for the service since its inception in1905 during Roosevelt's term. The position was essentially an attempt to build a compromise that recognized national growth, but was intended to use the forests for the widest possible national benefit.

Muir's conservation ethic was spiritual and possibly romantic. He emphasized the aesthetic beauty of forests, especially in the western United States, eventually shunning Pinchot's pragmatic stance. Muir's nature writings were inspirational; he had an ability to convince some politicians and business leaders to support his cause. His activism helped secure congressional passage of the 1899 national parks bill that added Yosemite and Sequoia to the nation's permanently protected list of lands. The National Park Service, with its mission of protecting the nation's natural heritage, was formed in 1916 shortly after Muir's death. It is clearly part of his legacy. Muir cofounded the Sierra Club in 1892. The group quickly became a political force in land preservation; he served as president for 22 years until his death. The Sierra Club's political activism and wide-ranging educational efforts continue to shape conservation thinking and politics today.

The fault line between Muir's and Pinchot's philosophical and ethical traditions remains active. The U.S. Forest Service, for example, has struggled to balance multiple use imperatives in its management of complex ecosystems built around mixed public–private land ownership. Heated scientific, political, and economic debates over forest management practices, timber sales, and conflicts involving recreation, wildlife, plant diversity, and logging have challenged the practical application of Pinchot's philosophy as diverse stakeholders have placed myriad demands on the system. The Forest Service's efforts to preserve wilderness areas have been a response to pressures from constituents whose views are more like Muir's "nature for nature's sake" position.

Meanwhile, national parks also face challenges to their mandate to protect the natural and historic heritage. Development in and near parks, coupled with issues of park access and levels of use, affects fragile ecosystems. In fact, national parks, while protected from natural resources extraction, still must deal with disparate multiple use demands. As a result, they are not usually the untouched, lightly traveled natural areas Muir envisioned.

The division between the pragmatic (utilitarian) and the aesthetic conservation ethics marks contradictory terrain in the shifting map of American conservation that Muir and Pinchot helped draw. Leadership by individuals with widely differing ethical perspectives, such as Muir and Pinchot, helps deepen our understanding of the various facets of the conservation movement more than a century after its emergence. Their efforts laid the groundwork for later environmental reforms that broadened federal powers in the 1930s and the 1960s and 1970s under both Democratic and Republican administrations.

See Also: Conservation; Muir, John; Pinchot, Gifford; Preservation; Utilitarianism.

Further Readings

Collins, T. "Between a Rock and a Mandate: The Philosophy of the U.S. Forest Service." ASPI Critical Issues Series No. 2. Livingston, KY: Appalachia—Science in the Public Interest, 1992.

Frome, M. *The Forest Service*. Boulder, CO: Westview Press, 1984.

Hay, P. *Main Currents in Western Environmental Thought.* Bloomington: Indiana University Press, 2002.
Miller, C. *Gifford Pinchot and the Making of Modern Environmentalism.* Washington, DC: Island Press, 2001.
Muir, J. *The Wilderness World of John Muir,* Edwin Way Teale, ed. Boston: Houghton Mifflin, 1954.

Timothy Collins
Western Illinois University

Consumption, Business Ethics and

The relationship of business ethics and consumption of natural resources may be thought of as a relatively recent pairing in environmental studies. This perception is understandable because, according to Richard DeGeorge, the term *business ethics* did not arise as a formal subdiscipline of academic philosophy until the 1970s. Yet "ethics in business" has been a forum of discussion outside formal academic philosophy since the Industrial Revolution. Rogene Buchholz suggests that the Protestant ethic provided moral limits on consumption by individuals during the early stages of industrialization in Western Europe and the United States. Despite moral concerns also outlined by foundational free market economist Adam Smith in the 1700s, this ethic eventually weakened during the development of a consumer society.

Aldo Leopold, perhaps best known for his seminal 1948 essay "The Land Ethic," wrote a 1928 piece called "The Homebuilder Conserves," illustrating this concern over a growing consumer society: "A public which lives in wooden houses should be careful about throwing stones at lumbermen, even wasteful ones, until it has learned how its own demand [for] lumber helps cause the waste it decries." Consumers are still a primary moral focus of many discussions about overconsumption, including Douglas MacCleery's article suggesting Leopold's land ethic is incomplete without a corresponding consumption ethic—otherwise we simply export our environmental impacts to other nations.

Yet Harvard professor John Kenneth Galbraith, in his 1958 book *The Affluent Society,* highlighted (as did earlier Marxist critiques of capitalism) the strong influence of production, marketing, and advertising in determining consumption preferences and behavior. At the other end of the 1960s environmental movement, the Club of Rome and its "Limits to Growth" effort in 1972 offered a prominent evaluation of capitalist society's thresholds regarding consumption.

Business ethics emerged at this time in academic departments of philosophy and business, with a corresponding skepticism pitting theoretical versus applied academics, not unlike that which occurred during the emergence of environmental philosophy in the same time frame (see *Environmental Ethics* Journal). The first anthologies on business ethics appeared in the late 1970s, including Thomas Donaldson and Patricia Werhane's *Ethical Issues in Business: A Philosophical Approach.* The first single-authored books emerged in the early 1980s and included *Business Ethics: Concepts and Cases* by Manuel Velasquez. Ethical codes are another typically nonacademic branch of ethics in business that originated before the 1980s and grew in focus at this time. Paul Camenisch provided a key source, *Grounding Professional Ethics in a Pluralistic Society.*

Thus, the next decade saw academic studies combining the emerging fields of business ethics and environmental ethics. Prominent anthologies were edited by Dan Miller in 1997 (*Acknowledging Consumption: A Review of New Studies*), David Crocker in 1997 (*Ethics of Consumption: The Good Life, Justice, and Global Stewardship*), and especially Laura Westra and Patricia Werhane in 1998 (*The Business of Consumption: Environmental Ethics and the Global Economy*). Miller's effort emphasized from a social scientist perspective the concept of consumption as the "vanguard of history" and the consumer as "global dictator." It noted the tendency to focus on household consumption while there were also important realms of industrial, commercial, and public (government) consumption. Crocker is a senior research scholar at the Institute for Philosophy and Public Policy and the School of Public Affairs at the University of Maryland. He is also a founder and former president of the International Development Ethics Association. His anthology drew from eminent scholars from many disciplines—philosophy, economics, sociology, political science, demography, theology, history, and social psychology.

The Business of Consumption is the anthology that most mixes business ethics and environmental ethics. Biologist John Lemons analyzed north–south inequity in per capita consumption rates and draws from the ethics of risk analysis. Policy analyst Donald Brown critiqued structural adjustment in the global south. Planning professor William Rees highlighted the concept of an "ecological footprint," comparing environmental impacts of different nations. Rees suggested gradual implementation of taxes and quotas on consumption levels. He importantly argued that a solution cannot be sought through the new information, rather than a product-based, economy, as new buying power simply maintains high consumption levels.

Former World Bank economist Herman Daly considered "natural capital," including the dollar value of ecosystem services. Economist Julian Simon argued for free-market solutions, while philosopher Ernest Partridge countered with concern for future generations. Philosopher Mark Sagoff further critiqued the narrow scope of economic valuation. Editor and philosopher Laura Westra critiques anthropocentrism. Business ethics concludes this anthology. George Brenkert addressed effects of Western business practice on diverse local cultures. Donald Mayer suggests corporate marketing, along with domestic and international law, has institutionalized overconsumption. Rogene Buchholz and Sandra Rosenthal examined roots of consumptive behavior, while Andrea Larson provides examples of innovative businesses and the "Natural Step" program.

Around the same, late 1990s time frame of these anthologies, P. Hawken, A. Lovins, and L. H. Lovins's book *Natural Capitalism* suggested a merging of business profitability and environmental concern that continues today with the green jobs movement, highlighted by more corporate-oriented books such as Joel Makower's *Strategies for the Green Economy.* The emergence of environmental management systems and corresponding energy audits provide common ground through cutting both energy consumption and costs. R. Panwar suggested this increase in ethical investing has encouraged corporate responsibility. Green investing is a growing movement, especially in the arena of alternative energy.

A prominent example of evolving common ground between business and environment is the emerging field called "industrial ecology." Originating with work at MIT in the 1970s, the field coalesced around R. Frosch and N. Gallopoulos's 1989 *Scientific American* article "Strategies for Manufacturing" and the National Academy of Science's 1991 Colloqium on Industrial Ecology. Life cycles of specific products are analyzed with respect to the materials, energy, and wastes used and produced in the process. Technological improvements offer hope through construction of products that incorporate recycled and recyclable materials. Wastes from one product are ideally used in construction of other products, even by other firms, creating more of a closed loop that reduces movement of pollution into the more open natural ecosystem.

While largely technological in focus, the field has a foundational ideology, attempting to reinvent industrial processes to align with more closed rather than open ecological systems. Frosch in his 1992 article "Industrial Ecology: A Philosophical Introduction," suggests that "ecologists talk of a food web: an interconnection of uses of both organisms and their wastes. In the industrial context we may think of this as being use of products and waste products."

Paul Hawken explains, in his well-read 1993 book *The Ecology of Commerce,* how industrial ecology involves more than simply product life cycle analysis: "Industrial ecology provides for the first time a large-scale, integrated management tool that designs industrial infrastructures as if they were a series of interlocking, artificial ecosystems interfacing with the natural global ecosystem . . . applying the concept of an ecosystem to the whole of an industrial operation, linking the 'metabolism' of one company with that of another."

The field's origins are closely tied ideologically to the concepts behind sustainable development. Thus ethical critiques of the sustainable development paradigm, such as those related to reliance on neoclassical economics or anthropocentrism, may transfer to this field. Industrial ecology became so prominent by the late 1990s that Yale University established a Center for Industrial Ecology. Yale now hosts the International Society for Industrial Ecology and its *Journal of Industrial Ecology,* which reflects an interdisciplinary nature. One of the more highly read recent articles is "The Quantitative and the Qualitative in Industrial Ecology" by Reid Lifset. A new journal in the field is *Progress in Industrial Ecology.*

Michael Porter supported the related, broad corporate responsibility movement in a recent issue of *Harvard Business Review* in "The Link Between Competitive Advantage and Corporate Social Responsibility." C. P. Egri suggests that proactive corporate leadership is an essential component of this movement. Some companies simply attempt to maintain an image of environmental responsibility. J. Marshall found that of "198 medium-sized to large multinationals, most said they lacked an active approach to developing new business opportunities arising from meeting citizenship and sustainability needs." Corporate responsibility or corporate sustainability officers are being hired to address these issues. G. Enderle found common grounding for this corporate movement within Christian environmental stewardship.

Despite these initiatives, the sheer size of the global economy, along with emerging consumer populations such as China, illustrate the seriousness of the issue. Paul Krugman won the 2008 Nobel Prize in Economics for his work in economic geography, illustrating the ability of various nations to compete in areas of specialization formerly held by specific nations through comparative advantage, due to the increasing size of the global economy, transfer of technology, and reduced transportation costs. Douglas MacCleery pointed out that, since the first Earth Day in 1970, the average family size in the United States dropped by 16 percent, while the size of the average single-family house being built increased by 48 percent. R. Goodland and H. Daly pointed out the need for restrictions on free trade. Robert Frodeman noted how environmental ethics has recently emphasized a more pragmatic, policy-oriented approach. A continued, corresponding effort in business ethics highlighted by M. E. Porter will help problem-solving efforts in this critical arena of society.

See Also: Business Ethics, Shades of Green; Consumption, Consumer Ethics and; Ecological Footprint; Greenwashing; Sustainability, Business Ethics and.

Further Readings

Buchholz, Rogene A. "The Ethics of Consumption Activities: A Future Paradigm?" *Journal of Business Ethics,* 17/8 (1998).

Camenisch, Paul. *Grounding Professional Ethics in a Pluralist Society.* New York: Haven Press, 1983.

Crocker, David A. and Toby Linden, eds. *Ethics of Consumption: The Good Life, Justice, and Global Stewardship.* Philosophy and the Global Context Series. Lanham, MD: Rowman & Littlefield, 1997.

DeGeorge, Richard T. "A History of Business Ethics." Markkula Center for Applied Ethics, Santa Clara University (2005). http://www.scu.edu/ethics/practicing/focusareas/business/conference/presentations (Accessed February 2010).

Egri, C. P. "Leadership in the North American Environmental Sector: Values, Leadership Styles, and Contexts of Environmental Leaders and Their Organizations." *Academy of Management Journal,* 43/4 (2006).

Enderle, G. "In Search of a Common Ethical Ground: Corporate Environmental Responsibility From the Perspective of Christian Environmental Stewardship." *Journal of Business Ethics,* 16/2 (2006).

Frodeman, Robert. "The Policy Turn in Environmental Ethics." *Environmental Ethics,* 28 (2006).

Galbraith, John Kenneth. *The Affluent Society.* Cambridge, MA: Riverside Press, 1958.

Goodland, R. and H. Daly. "If Tropical Log Export Bans Are So Perverse, Why Are There So Many?" *Ecological Economics,* 18 (1996).

Hawken, P., A. Lovins, and L. H. Lovins. *Natural Capitalism: Creating the Next Industrial Revolution.* Boston/London: Little Brown, 1999.

Leopold, A. "The Homebuilder Conserves." *American Forests and Forest Life,* 34/413 (1928).

MacCleery, Douglas W. "Aldo Leopold's Land Ethic: Is It Only Half a Loaf Unless a Consumption Ethic Accompanies It?" *Journal of Forestry,* 98/10 (2000).

Makower, Joel. *Strategies for the Green Economy: Opportunities and Challenges in the New World of Business.* New York: McGraw-Hill, 2009.

Marshall, J. "New Group Pushes 'Responsibility' Officer." *Financial Executive,* 23/1 (2007).

Miller, Dan, ed. *Acknowledging Consumption: A Review of New Studies.* New York: Routledge, 1995.

Panwar, R. "Corporate Responsibility: Balancing Economic, Environmental, and Social Issues in the Forest Products Industry." *Forest Products Journal,* 56/4 (2006).

Porter, M. E. "Strategy & Society: The Link Between Competitive Advantage and Corporate Social Responsibility." *Harvard Business Review,* 84/12 (2006).

Westra, Laura and Patricia H. Werhane, eds. *The Business of Consumption: Environmental Ethics and the Global Economy.* Lanham, MD: Rowman & Littlefield, 1998.

William Forbes
Stephen F. Austin State University

Consumption, Consumer Ethics and

This Indian home, where laundry dries outdoors and women carry small amounts of water in buckets, likely consumes vastly fewer resources than a typical American household. Such disparities make calculating sustainable levels of consumption for the world difficult.

Source: World Bank

In recent decades, particular consumer choices as well as larger patterns of consumption have come under ethical scrutiny. Critics argue that some acts or rates of consumption are morally wrong and should be avoided, either because the commodity being consumed should be protected or because the processes by which it is manufactured, brought to market, or used causes harm to others. Social movements for various forms of ethical consumerism—whether based in objections to specific consumption choices like ethical vegetarianism or larger consumption patterns like movements for sustainable living—depend on this claim that consumer behavior and personal consumption choices should be subject to ethical rather than merely prudential considerations.

Consumer and consumption ethics thus represent a challenge to a certain form of classic liberal constructions of privacy. Conventionally, nearly all consumer choices have been regarded as private actions that should aim only to maximize the interests or satisfy the preferences of individual consumers, rather than as public actions of ethical or political concern. Exceptions to this privatization of consumption have historically been rare, such as the prohibition against owning human slaves, with moral limits placed on how commodities are produced or used but relatively few on what or how much can be consumed. How or what we consume is each person's private business, according to this view, and is properly subject to prudential but not ethical considerations. Unless they cause harm to others through their acts of consumption, classic liberals like John Stuart Mill insist, consumer behaviors transpire within a protected private sphere, immune from the intrusive and conformist gaze of public pressure or coercive law. Consumer and consumption ethics only partly challenges this received view about privacy, maintaining that some consumer choices and consumption patterns are, in fact, public and ethical rather than private and merely prudential, with extrinsic harmful effects being the criterion for setting the boundary of the private sphere. This line of ethical critique seeks to shed light on the manifold ways in which one's consumption can harm another, raising objections when that harm is serious and avoidable, whether through the consumption of alternative commodities or reduced overall consumption.

Scholars have taken two general approaches to developing an ethics of consumer choices and consumption. First, some have subjected particular consumer choices to ethical interrogation, seeking to link certain acts of consumption to harmful consequences that result from them. For examples, some tuna consumption has been linked to fishing practices that kill dolphins and sea turtles, certain automobiles have been linked to heightened risk to other drivers or increased health consequences from pollution, and some garments have been linked to inhumane sweatshop conditions for the workers who assemble them. In each case, the harmful consequence lies at a spatial and temporal distance from the consumer purchase or act of consumption and so is invisible to the consumer, but ethicists locate that act within the causal chain that produces the bad outcome such that making alternative consumer choices becomes an ethical imperative. Consumer or consumption ethics accepts standard moral prohibitions against causing avoidable harm and seeks to draw attention to commonly neglected causal links between ostensibly innocent choices and such harm.

Second, critics also point to larger patterns of consumption rather than discrete consumer choices as culpable for various avoidable harmful consequences. Examples of such aggregate analysis include claims that excessively large carbon footprints cause the harms associated with anthropogenic climate change, that fuel-inefficient sport utility vehicles are harmful in a way that fuel-efficient hybrid cars are not, and that the moral problem with the acquisition of luxury goods lies not in their mere consumption but in their overconsumption, understood as some level beyond a threshold of justified shares or an ostentatious display of contemptible overindulgence. In each case, moral permission could be granted to persons for the consumption of any particular commodity or bundle of commodities up to some level, but beyond that level the further consumption of that particular good or bundle of goods would trigger the production of some bad outcome. Here, it is assumed that there exists a finite capacity to yield the resources necessary for commodity production and to absorb the wastes associated with their manufacture and use, such that aggregate consumption within the bounds of that capacity are benign, but that consumption levels that exceed this capacity become culpable for harm in a way that like consumption within those limits does not.

In the remainder of this article, the basis for consumer or consumption ethics is critically examined, with several challenges to the causal linkage between individual consumer choices or aggregate consumption patterns and the harm with which they are sometimes associated critically assessed. In addition, some preliminary responses to these challenges are offered on behalf of the objective of clarifying the applicability of ethical analysis to consumer choices and behavior.

Ethics of Particular Acts of Consumption

When some consumer's consumption choices cause harm to others, those choices become subjects of ethical criticism. Such criticism is often contentious and may be unjustified, however, when the causal links between the consuming act and the alleged harm is indirect or diffuse, when the moral status of the affected subject is disputed, when the harmful outcome is unknown by the consumer, and when benign alternative commodities or consumption options are unavailable. In each case, some doubt is cast on whether one person's consumption does, in fact, cause morally relevant avoidable harm to others in a way for which the former can be held responsible.

Links between an act of consumption and some alleged harm are indirect when the former sets in motion some causal chain that, together with other causes and often temporally and spatially distant from the agent, the harm eventually obtains. An example of indirect harm that invites ethical critique is the demand for commodities manufactured by child labor, which such consumption arguably supports. These links are diffuse when a small number of similar consumption acts would be insufficient to cause the harm in question, but when many similar acts take place within a bounded space and time that harm would result. Harm that results from pollution is typically diffuse in that any one person's consumption typically fails to cause discernable harm, given the small amounts of harmful pollutants that any one person can emit by themselves and the environment's capacity to assimilate those wastes, but many similar consumption acts yield levels of pollution that adversely affect many. For each individual polluter, their consumption acts are neither necessary nor sufficient to cause harm, and no pollution-related harm can be traced back to particular actions or persons. As a result, the responsibility for harm is widely diffused among a great many persons and actions such that some fallaciously conclude that no person's consumption acts or choices played any culpable role in harming others. Derek Parfit in *Reasons and Persons* (1986) terms this fallacy a "mistake in moral mathematics" in that it erroneously conflates very small contributions toward larger harmful outcomes with zero causal contribution to those outcomes. Ethical consumerism demands that consumers refrain from making consumption choices that contribute toward indirect and diffuse as well as direct harm, since it is primarily in the former manner that consumption causes harm, but as causal links become more indirect and diffuse the culpability of offending acts becomes less clear.

Where causation for harm can be established but where those suffering the harm are not recognized as moral subjects, the wrongness of that consumption is challenged. For example, many deny that nonhuman animals have morally relevant interests or can be the objects of any sort of moral obligation, even the duty to refrain from causing avoidable and serious harm. Consumption that seriously harms nonhuman animals, such as that motivating some forms of ethical vegetarianism in the interest of avoiding animal cruelty, is seen as morally wrong by those granting moral status to nonhuman animals but not by those who deny that the mistreated animals warrant ethical consideration. Indeed, the key premise for scholarly work on animal ethics has involved the moral status of nonhuman animals, as seen in Peter Singer's *Animal Liberation* (2001). Consumption choices that require or support animal cruelty are only controversially unethical in that some deny that nonhuman animals have any morally relevant interests in avoiding cruel treatment. Were it not for this denial of moral subjectivity, such consumption would be straightforwardly viewed as wrong, at least insofar as benign alternatives existed.

Consumers who cause harm through their consumption acts without anticipating that harm in advance or recognizing it after the fact are often viewed as having not acted wrongly, given the presumption that the harmful outcome in question was accidentally caused. Many controversially unethical consumption choices involve this sort of harm, where intent to injure is wholly lacking and even minimal recognition of potential harm does not accompany the offensive act. This problem is exacerbated by the difficulty in obtaining accurate information about the conditions under which many commodities are produced that often frustrates efforts to practice ethical consumerism. For example, some diamonds are mined through the use of slave labor and profits from their sale supports brutal conflict, but other gems are not implicated in this causal chain that leads to human suffering. Consumers may inadvertently purchase a "conflict diamond" while mistakenly believing it to have been mined and brought to market through ethically defensible labor and business practices. This mistaken belief may be the result of some certificate that falsely claims it to be so, or it may simply be impossible to distinguish diamonds mined under objectionable conditions from those coming from benign ones. In such cases, the wrongness of such purchases is disputable, at least insofar as consumers could not reasonably have known how their consumption was related to the bad outcome to which it contributes. However, that wrongness is less disputable where consumers could have obtained accurate information about the harmful commodity but chose not to, or failed to engage in due diligence to avoid ethically objectionable consumption.

Finally, it is sometimes thought that no consumption choice can be condemned as wrong unless benign alternatives are available. While it is ordinarily wrong to kill and eat another human when alternative sources of sustenance are available, the wrongness of cannibalism arguably becomes less clear when eating one person is the only way to save the lives of others. Lifeboat cases in ethics turn on this form of emergency, where some worse outcome would result unless someone opted to engage in what would in nonemergency settings be regarded as an unethical consumption choice. But this judgment is clouded in cases where the act of consumption is not required for bare survival or when benign alternatives are merely difficult to find. Suppose, for example, that someone dressed only in light clothing entered a department store during a snowstorm, looking for a coat to protect them against the cold, and was confronted only with rack after rack of garments produced under conditions of brutal worker exploitation. Does this count as the sort of emergency for which the above exemption might apply? Some might claim that it does, but this analysis depends on how necessary purchasing a coat from that particular store was for the cold consumer. What if there were benignly produced coats, but they cost beyond what the cold consumer could afford? The second instance describes circumstances under which many consumers opt to consume commodities that they know or suspect are produced under ethically dubious or objectionable conditions, since benign alternatives to harmful products are often more expensive options. Invoking the idea of emergency in cases where benign alternatives are merely more costly would be to excuse nearly all of what ethical consumerism urges: namely, to consider more than merely the price and intrinsic qualities of some commodity in comparing it to alternatives, but also to ascertain and consider the extrinsic effects of its consumption on others.

Ethics of Patterns of Consumption

While some discrete acts of consumption might be benign on their own, they can cause significant harm when combined with many similar acts or when consumption bundles exceed some threshold. Unlike the diffused harm that results from many individuals making similar consumption choices, however, these patterns of consumption can result from some person's aggregate consumption of qualitatively different commodities. For example, many people simultaneously driving automobiles that are powered by the combustion of fossil fuels can create dangerous airborne pollution, but their individual culpability when others suffer illness as the result of this pollution would be diffused among the many other polluters. But one person or a small group can over time and as the consequence of their various consumption acts cause harm for which responsibility is not widely diffused among a large group of culpable agents, even when none of these discrete consumption acts is sufficient to cause harm on its own. For example, a person or small group might over time and through many different consuming acts deplete the financial reserves that were earmarked for the higher education of a family member, harming them. While no particular frivolous consumer decision depletes that college fund on its own, many frivolous ones over the course of time could easily do so. In such cases, it is the larger patterns of consumption rather than any particular act that would be viewed as harmful, and which would therefore invite ethical criticism.

As with particular acts of consumption, ethical criticism of larger patterns of consumption becomes contentious when those harmed are not viewed as moral subjects, where the harm is unanticipated and unknown and where benign alternatives are not available. Larger patterns of consumption necessarily involve indirect harm, insofar as no particular consumption act or choice can be viewed as uniquely culpable for the harm in question, though as noted above they need not entail diffused responsibility. In many respects, ethical criticism of consumption patterns follows the same analysis of causal contribution to harm and thus invites similar challenges to its imperatives, but patterns of consumption differ from specific consumption acts in that ethics prescribes a relatively clear injunction against unethical consumption acts but is less clear about how to remedy unethical patterns of consumption. This conceptual difficulty can best be seen in the way that critics understand and prescribe responses to overconsumption.

The most comprehensive metric for measuring aggregate consumption rates is the ecological footprint, which calculates the physical area of biologically productive land needed to accommodate the resource consumption and waste production associated with an individual's or group's consumption. Since all forms of consumption require physical resources and yield wastes that must be assimilated by the environment, while the Earth's capacity to yield such resources and assimilate wastes is finite, the ecological footprint offers a metric for comparing consumption rates among persons and groups as well as for determining whether aggregate consumption patterns are ecologically sustainable. As Mathis Wackernagel and William Rees point out in *Our Ecological Footprint* (1996), the total human ecological footprint now well exceeds the ecological capacity of the planet. Reserving 20 percent of the planet's biological capacity to support all life forms other than humans, Wackernagel and Rees calculate, there are approximately 1.3 hectares of biologically productive land per person to support human consumption, and the average ecological footprint of humanity is 1.8 hectares per person. This leaves an "ecological deficit" of 0.5 hectares per person between supply and human demand for ecological goods and services, and such ecological deficits cannot be sustained over the long term. Overall, this analysis suggests, humanity must reduce its consumption by 40 percent or it will cause serious environmental harm as a consequence of exceeding ecological limits.

The controversy over the ecological footprint as it applies to ethical consumerism lies not so much in this analysis of unsustainable aggregate human consumption patterns, however, as in what this implies for specific individuals and groups. Wide disparities between the ecological footprints of different groups, where the average U.S. citizen requires 5.1 hectares of land to support their consumption while the average Indian requires only 0.38 hectares, complicate the ethical analysis of how much each might be allowed to consume in order to bring about a sustainable human impact on the planet. While footprint data show that humanity must reduce its aggregate consumption by 40 percent in order to avoid exceeding ecological limits, it is less clear what follows for the consumption patterns of the average American or Indian. Expecting both to reduce their consumption by that same proportion would be unreasonable, some argue, given that the average Indian consumes at a level that is well below what is sustainable on a global per capita basis while the average American footprint is nearly four times that sustainable level. In making interpersonal and intergroup comparisons of aggregate consumption patterns, ecological footprint analysis reveals the wide disparity among consumption rates and the gap between current and sustainable consumption, but must be joined with an account of justified entitlements to ecological goods and services before it can offer the ethical guidance that is needed to identify and correct indefensibly large footprints. Consumption ethics in such cases identifies morally objectionable states of affairs but requires a theory of distributive justice in order to provide a defensible remedy to them.

See Also: Carbon Offsets; Climate Ethics; Ecological Footprint; Ethical Vegetarianism; Greenwashing.

Further Readings

Crocker, David. A. and Toby Linden. *The Ethics of Consumption: The Good Life, Justice, and Global Stewardship*. Lanham, MD: Rowman & Littlefield, 1997.
Dauvergne, Peter. *The Shadows of Consumption: Consequences for the Global Environment.* Cambridge, MA: MIT Press, 2008.
Parfit, Derek. *Reasons and Persons.* New York: Oxford University Press, 1986.
Princen, Thomas, Michael Maniates, and Ken Conca, eds. *Confronting Consumption.* Cambridge, MA: MIT Press, 2002.
Schor, Juliet B. *The Overspent American: Why We Want What We Don't Need.* New York: Harper Perennial, 1999.
Singer, Peter. *Animal Liberation.* New York: Harper Perennial, 2001.
Vanderheiden, Steve. *Atmospheric Justice: A Political Theory of Climate Change.* New York: Oxford University Press, 2008.
Wackernagel, Mathis and William Rees. *Our Ecological Footprint: Reducing Human Impact on the Earth.* Gabriola Island, British Columbia, Canada: New Society Publishers, 1996.

Steve Vanderheiden
University of Colorado, Boulder

Cornucopians

Cornucopian is a label given to individuals who assert that the current environmental problems faced by society either do not exist or can be solved by technology and/or the free market. Cornucopians reject population growth projections and the notion of finite resources and carrying capacity. They also discard arguments for increased government regulation to protect natural resources and the environment. Cornucopians tout capitalism and human intervention as the essential features of human progress. Many arguments in support of the cornucopian position can be traced back to the work of Julian Simon and Herman Kahn.

The cornucopia is a mythical horn of plenty. According to Greek legend, this horn could magically provide humans with anything that they wanted or needed. The title *cornucopian* is derived from the ancient mythical cornucopia and is based on the conviction that technology and the free market can also provide humans with anything that they need or want. This combines Adam Smith's invisible hand with an unfettered belief that technology has the ability to overcome all human problems. Smith used the metaphor of the invisible hand to argue that individuals who pursue their self-interest in the market are, at the same time, contributing to societal interests by increasing revenue for the society as a whole. Cornucopians are confident that technology will rise to meet the demand of individuals and society.

Population

One of the cornerstone arguments of the cornucopian position is a denial of the Malthusian assertions that the world's population will grow rapidly and exponentially. Malthus suggested that a growing population would have negative consequences for humanity. He believed that the population would become unsustainable in the sense that humanity would run out of natural resources to meet the growing needs of the human race. The Malthusian position led Paul Ehrlich to call for population control in the 1970s. In his book *The Population Bomb,* he predicted a disaster for humanity due to overpopulation. Specifically, Ehrlich predicted that millions of people would starve to death because of what he called a "population explosion." He believed that this tragedy would take place in the late 20th century. His assertion did not come to be.

Many cornucopians challenged the notion of a pending Malthusian catastrophe. There are two basic parts to this dissent. The first centers on a denial of exponential population growth. Although the population of the planet did grow rapidly after 1800, due to medical and technological advances, cornucopians have maintained that growth would slow and even out over time. To bolster this claim they often reference statistics from international organizations such as the United Nations, which have adjusted and lowered their world population predictions since the turn of the last century.

The second part of the cornucopian dissent has to do with the effects of population growth. As noted above, individuals in the Malthusian tradition suggest that population growth left unchecked would lead to the ruin of humanity. Contrary to this position, cornucopians note that as the population of the world has grown steadily since 1800, so has the standard of living increased. The standard of living refers to the number and quality of goods and services available for humans. Taking this one step further, cornucopians often assert that population growth might actually improve the human condition because the number of goods and services have increased over time. For this reason, cornucopians reject the environmentalists' arguments about the detrimental effects of unlimited growth. Rather, from a cornucopian perspective, it is limiting economic growth that would prove detrimental for the human race. In the book *Population Matters: People, Resources, Environment, and Immigration,* Julian Simon argues that growing prosperity and technology will only increase the amount of resources available, which in turn will raise that standard of living for all.

Resources

Cornucopians also reject the notion that Earth has finite resources. This directly relates to their stance that technology can be utilized to regenerate or replace any resources under pressure. One often-used example is the rise of fiber-optic cable as a replacement for metals, especially copper, in communication lines. In their rejection of finite resources, cornucopians also challenge the concepts of carrying capacity and Garrett Hardin's tragedy of the commons. Carrying capacity refers to the number of individuals an environment can support without serious detrimental impacts to the system. Garrett Hardin related carrying capacity to environmental social goods, such as clean air and clean water, which are shared by all but owned by no one. According to Hardin, without regulation of our common resources, such as by government, individuals set to maximize their own utility would destroy the common goods we share.

Cornucopians reject many of the claims that underlie environmentalist arguments for environmental protection and government regulation. While there are many different types of environmentalism, almost all forms call for some type of government intervention to save and/or protect natural resources. Those in the anthropocentric, or human-centered, tradition only believe in regulation to protect human interests while other types of environmentalists may value and want protection of resources not directly related to human well-being. Cornucopians stand in opposition to all of these positions. In reference to ideology and government regulation, cornucopians fall nearest the libertarian tradition. They value minimal government intervention and place high importance on individual liberty. It is individual liberty and freedom that allows the market to grow and technology to be invented. The only legitimate role that the government has in reference to resources is to protect private property. Private property is needed for the market to flourish.

Criticisms

There are several criticisms that have been levied at the cornucopian worldview. The most common criticism is that cornucopians simply ignore evidence that is contradictory to their position and that they only choose methodologies that already support their perceptions. For example, one argument levied against their position charges cornucopians with ignoring the effects of population growth on ecosystems, which, environmentalists argue, are essential for the survival of humans. Moreover, an environmentalist would argue that cornucopian assertions about the standard of living are overstated and that cornucopians ignore the living conditions of the poor around the planet. In defense of their position, cornucopians maintain that their position is logical and based on sound evidence. Rather, they argue that it is the environmentalist positions that are illogical, emotive, and based on doomsday scenarios that never come to pass.

See Also: Ecocentrism; Future Generations; Tragedy of the Commons.

Further Readings

Ehrlich, Paul. *The Population Bomb.* Cutchogue, NY: Buccaneer Books, 1971.

Hardin, Garrett. "The Tragedy of the Commons." *Science,* 162/3859 (1968).

Simon, Julian. *Population Matters: People, Resources, Environment, and Immigration.* New Brunswick, NJ: Transaction Publishers, 1990.

Simon, Julian, et al., eds. *The Resourceful Earth: A Response to Global 2000.* Oxford, UK: Blackwell Press, 1984.

Smith, Adam. *Wealth of Nations,* edited by C. J. Bullock. Vol. X. The Harvard Classics. New York: P. F. Collier & Son, 1909–14.

Jo Arney
University of Wisconsin–La Crosse

Cost-Benefit Analysis

Cost-benefit analysis (CBA) is one of the most favored and used techniques to estimate the economic value of the nonhuman world in order to inform economic and environmental policy. Its origins lie in welfare economics and it is perhaps the most dominant economic valuation technique used in policy making around the world, especially in the areas of environmental policy, transport, and healthcare.

The essential theoretical foundations of CBA are very simple. Benefits are defined as increases in human well-being (utility)—usually reduced to monetary values; and costs are defined as reductions in human well-being—again usually reduced to monetary values. For a project, proposal, or policy to qualify and be supported on cost-benefit grounds, its benefits must exceed its costs. And herein lies the simplicity (and problem) with CBA: it operates on a "profit-making" logic, where once we have estimated the monetary costs and benefits of a particular use of the environment—and therefore have established a common denominator for decision making—we simply weigh the options and, on the basis of monetary calculation (does the proposal create more profit/money/wealth/jobs/economic growth than it costs?), make the decision.

Some environmental groups have been lobbied against CBA for decades. Perhaps the main argument is that the application and use of monetary measures and values as the common metric systematically misrepresents, distorts, and corrupts the decision-making process in environmental issues in "crowding out" noneconomic valuations, and worse, forcing noneconomic valuations to commit a category mistake by having to express themselves as monetary units or valuations. As F. Ackerman and L. Heinzerling note, "To say that life, health, and nature are priceless is not to say that we should spend an infinite amount of money to protect them. Rather, it is to say that translating life, health, and nature into dollars is not a fruitful way of deciding how much protection to give them. A different way of thinking and deciding about them is required." In other words, CBA represents a form of economism, or economic imperialism, whereby all noneconomic valuations and preferences are not merely "crowded out," according to J. Barry, but to add insult to injury, are forced to present and translate themselves in monetary forms. Related to this is that the trades involved in CBA are from the citizen's perspective, entirely involuntary. Just as it is assumed all values and valuations can be or ought to be reduced and reducible to monetary units, so it is simply and axiomatically assumed that citizens are willing to accept money in exchange for an added health risk or a blighted landscape, whether or not this is true. In this sense CBA can be said to "make offers you cannot refuse" in a language and idiom you cannot refuse.

There are significant and often overlooked political implications of such neoclassical economic techniques as CBA. The monetizing and quantification at the heart of neoclassical economics, as expressed in CBA, is perfect (indeed, functional) for a liberal political order based in large part on having utilitarianism as its "public philosophy." One of the advantages of reducing values and political or ethical debate to monetary units is that it enables the orderly administration of public policy to be maintained while also permitting (some degree) of "permissible" ethical and political plurality. That is, you can have any ethical or political views you like, but if you want them to "count" in public policy deliberation, they have to be translated into monetary terms or be capable of being so translated.

Critics of CBA, such as ecological economists and many greens, argue that what is needed is the following:

- Openly political, participatory, and democratic forms of decision making such as "citizens' juries," "consensus conferences," roundtables and other democratic decision-making innovations that include citizens more in the policy-making process, according to G. Smith
- Policy-making systems that recognize the full variety and range of valuations people have in relation to the environment—through such processes as multicriteria decision analysis (MCDA), according to G. Munda

CBA has an important function in environmental and economic policy and decision making, but it only usefully and properly fulfills this function if it is kept to its appropriate sphere of competency. CBA and monetary valuation techniques require continuous improvement to better assess the human values ascribed to the environment. Even with such improvements, CBA is limited by its use of money as its unit of analysis, and should be accompanied by other nonmonetary valuation approaches in environmental policy and decision making.

See Also: Democracy; Instrumental Value; Intrinsic Value; Utilitarianism.

Further Readings

Ackerman, F. and L. Heinzerling. *Priceless: On Knowing the Price of Everything and the Value of Nothing.* New York: New York Press, 2004.

Barry, J. *Rethinking Green Politics: Nature, Virtue and Progress.* London: Sage, 1999.

Munda, G. *Social Multi-criteria Evaluation for a Sustainable Economy.* New York: Springer, 2008.

Smith, G. *Beyond the Ballot: 57 Democratic Innovations From Around the World.* London: The Power Inquiry, 2005.

John Barry
Queen's University Belfast

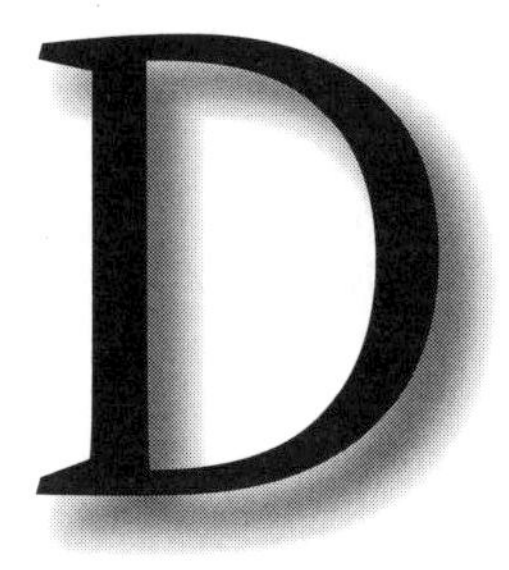

"Death of Environmentalism, The"

"The Death of Environmentalism: Global Warming Politics in a Post-Environmental World" is the title of a long article written by Michael Shellenberger and Ted Nordhaus in 2004 and published with the contributions of the Breakthrough Institute and the Evans/McDonough research firm. The article states that the environmental movement is no longer able to deal with the world's ecological crisis because of its exclusive interest in technical policies, like pollution controls or standard setting that weakens its popular inspiration without incrementing its political power.

The authors adopt a corporate marketing research approach to analyze opinions, values, and beliefs of environmental communities in the United States and Canada. They build upon a large number of interviews with several environmental thinkers, community leaders, and funders in order to investigate the reason why modern environmentalism is increasingly unable to deal with the most pressing ecological challenges, first of all global warming, notwithstanding the investment of hundreds of millions of dollars into combating them. From the collected interviews, the authors deduce that those responsible for the loss of the environmental movement's authority are the environmentalists themselves. Environmentalists affirm they are on the right track but, despite the efforts toward converging on an international "green deal," the current results do not fit the intent.

According to the authors, this happens because environmentalists provide a very narrow definition of their issues as merely pertaining to the environment, nature, and the ecosystems outside the human domain. This literal interpretation of the object, the purpose, and the identity of environmentalisms turns it into a special interest and debilitates the popular understating of its general relevance. Environmental groups consider themselves defenders of endangered nature and refuse any critical reinterpretation of their aims, strategies, or achievements: this is the reason why their campaigns are doomed to failure.

"The Death of Environmentalism" is articulated in two parts. In the first and longer part, the authors argue that, after the successful environmental campaigns that led to a number of environmental laws approved in North America during the 1960s and 1970s, modern environmentalism regards ecological issues as a sectional interest and not as the core of politics. This consideration induces a sense of shallow enthusiasm for the protection of a thing called "the environment" but no mobilization toward the real fulfillment of an alternative and articulated worldview.

The strategic framework environmentalists generally adopt requires three stages: the definition of a problem as "environmental," the provision of a technical remedy, and the selling of this remedy to legislators by using a variety of political tactics (e.g., lobbying, public relations, advertising). This strategy has been applied to global warming but it led to unsuccessful international political initiatives; their failure was a direct consequence of the environmental movement's reductive view about the deep causes of global warming and the lack of an inspiring vision for the future.

In order to give an account of the environmentalist strategy to contrast carbon emission, the authors outline the story of fuel efficiency regulation in the United States. On several occasions since the 1970s, the environmental movement missed the opportunity to form effective alliances with industry and unions by adopting a narrow view of what the environmental struggle was about. As a consequence, environmentalists failed to win legislative agreement on carbon, to determine the fuel efficiency standard, and to regulate the U.S. auto industry.

Shellenberger and Nordhaus argue that the roots of the environmental movement's failure can be found in the way in which the movement itself categorizes certain problems as environmental and others as nonenvironmental. The supposed nonenvironmental problems are disregarded and no comprehensive political strategy is crafted in order to form larger alliances with other groups, which are justified only to the extent these groups can be of some help to the environmentalists' cause.

Together with the establishment of more inclusive alliances, the authors state the environmental movement should also be concerned with the elaboration of a strategy to help developing countries in adopting environmentally friendly behaviors, ethical trade agreements, and technology transfer.

In the second and shorter part of the article, the authors affirm that environmentalists have been too timid in raising the alarm about global warming and that their focus on technical solutions leads to the adoption of very short-term policies with no remarkable consequences. The excessive attention to the policies downplays the relevance of the politics that support them; but, as long as its failures are understood as essentially tactical and the solutions essentially technical, it will not be possible to change the fate of the environmental movement.

The authors suggest that a new kind of political project (like the New Apollo Project they describe) is necessary to face the world's problems with a single comprehensive strategy. These projects can be elaborated by means of the synergic effort of several interest groups and foundations whose aim is to create new jobs, to free the United States from oil, to improve the investments in clean energy, and to offer an inspiring vision of civil and unions' rights, business, and the environment.

In the concluding section, Shellenberger and Nordhaus suggest the necessity for the environmental movement to turn its apocalyptic alarms into hopeful scenarios and transformative visions framed around the core American values. New bridge values can lead toward a more coherent and collective project able to attract consensus and greater investments, so that different groups can align their interests by contemporary advancing action on global warming.

See Also: Climate Ethics; Environmental Policy; Environmental Values and Law.

Further Readings

Latour, B. "It's Development, Stupid! or: How to Modernize Modernization." http://www.bruno-latour.fr (Accessed November 2009).

Latour, B. *Politics of Nature: How to Bring the Sciences Into Democracy.* Cambridge, MA: Harvard University Press, 2004.

Nordhaus, T. and M. Shellenberger. *Break Through: From the Death of Environmentalism to the Politics of Possibility.* New York: Houghton Mifflin, 2007.
Nordhaus, T. and M. Shellenberger. "The Death of Environmentalism: Global Warming Politics in a Post-Environmental World." http://www.thebreakthrough.org/images/Death_of_Environmentalism.pdf (Accessed November 2009).
Young, J. *Post-Environmentalism.* London: Belhaven Press, 1990.

Chiara Certomà
Sant'Anna School of Advanced Studies

d'Eaubonne, Françoise

Françoise d'Eaubonne (1920–2005) was a French radical feminist, an activist, and the author of a vast body of writings that include historical and science fiction novels, memoirs, biographies, and philosophical treatises about women and the environment. Outside France, she is best known for coining the term *ecofeminism* in her 1974 book *Le Féminisme ou la Mort (Feminism or Death).*

Born in Paris, d'Eaubonne spent her childhood in Toulouse in the company of her mother, the child of a Carlist revolutionary (a legitimist political movement in Spain that called for the placement of a different line of the Bourbon family on the Spanish throne); and her father, a member of the religious Sillon movement (a French religious and political movement to bring Catholicism closer in line with French Republican and socialist ideals) and an anarchist sympathizer. A witness to the physical decline of her father (who was exposed to poison gases in the trenches during World War I) as well as the arrival of Republican exiles from Spain and freed Jews returning from Nazi death camps, d'Eaubonne's experiences during her childhood and in her youth molded her perceptions of the world and shaped her into a militant radical and feminist.

After studies at the University of Toulouse and the School of Fine Arts of Toulouse, Françoise d'Eaubonne joined the Resistance. At the time of France's liberation, she was a member of the Communist Party. She would subsequently leave upon learning about the Moscow Trials, Joseph Stalin's series of trials of his political opponents during the Great Purge.

In 1947, she published her first novel, *Comme un Vol de Gerfauts (A Flight of Falcons).* Two years later, she read Simone de Beauvoir's *Le Deuxième Sexe (The Second Sex)* and was introduced to feminism. During the decades of the 1950s and 1960s, she worked as a reader in major publishing houses, all the while publishing her own novels, biographies, and essays. On the political front during this period, she protested against the war in Algeria and campaigned for minority rights.

Cofounder of the Mouvement de Libération des Femmes (MLF; Women's Liberation Movement), d'Eaubonne was among 343 French women who signed the famous manifesto "*Je me suis fait avorter*" ("I had an abortion"), which was published in a 1971 issue of *Le Nouvel Observateur,* a major French periodical. In that same year, she was a key participant in the founding of the Front Homosexuel d'Action Révolutionnaire (FHAR; Revolution Homosexual Front), a radical queer group that identified its enemies as *hétéroflics* (heterocops) and the entire structure of bourgeois capitalism.

In 1978, she also created the Mouvement Ecologie et Féminisme (Ecology and Feminism Movement), which would have more of an impact in Australia and the United States than in France.

Although a full translation of d'Eaubonne's work *Le Féminisme ou la Mort* does not currently exist in English (most of her works are hard to find and have not been translated into English), excerpts from this essay can be found in Elaine Marks and Isabelle de Courtivron's *New French Feminisms,* published in the early 1980s. In this work, d'Eaubonne coined the term *ecofeminism* as a way of linking the devaluation of women to the devaluation of the Earth.

In subsequent works such as *Ecologie, Féminisme: Révolution ou Mutation?* (1978) and "What Could an Ecofeminist Society Be?" d'Eaubonne locates the beginnings of the oppression and exploitation of women and nature at the end of the Neolithic period. It is in this period, she believed, that we first witness the appearance of patriarchy thanks to men's discovery of their role in procreation and a change in cultivation practices from dry agriculture to farming with the plow and irrigation. Following both of these events, men greatly increased their wealth and the human population. From these developments, she argued, originate the two ecological disasters that currently threaten our species: overcrowding and the exhaustion of resources. Thus, according to d'Eaubonne, the only hope of salvation for humanity can be found in women's liberation and the death of patriarchy.

With respect to the future, she believed that the most pressing danger facing humanity stems from what she calls the "era of data processing," where the manipulation of genes and reproduction and the use of robotics to replace human workers are seen as remedies to the exhaustion of resources. Linking women and ecology to this equation, she maintained that women are the keepers of the values of life since they are more involved in the ecological problem than men due to their role in giving life to future generations. Thus, in order to overcome this crisis, women must be allowed to regulate their own fertility and they must commit themselves to the dismantling of the current patriarchal system. Furthermore, capitalism, a by-product of patriarchy that is now in the imperialist stage, and market economies as well must disappear and be replaced by an ecological solution to the current production-consumption problem that has given rise to much of the environmental crisis.

See Also: Ecofeminism/ Ecological Feminism; Human Values and Sustainability; Warren, Karen.

Further Readings

d'Eaubonne, Françoise. *Ecologie, Féminisme: Révolution ou Mutation?* Paris: Editions A.T.P., 1978.

d'Eaubonne, Françoise. "Feminism—Ecology: Revolution or Mutation?" *Ethics and the Environment,* 4/2 (2000).

d'Eaubonne, Françoise. *Le Féminisme ou la Mort.* Paris: P. Horay, 1974.

d'Eaubonne, Françoise. "The Time for Ecofeminism," trans. Ruth Hottell. In *Ecology: Key Concepts in Critical Theory,* Carolyn Merchant and Roger S. Gottlieb, eds. Atlantic Highlands, NJ: Humanities Press International, 1994.

d'Eaubonne, Françoise. "What Could an Ecofeminist Society Be?" *Ethics and the Environment,* 4/2 (2000).

Marks, Elaine and Isabelle de Courtivron, eds. *New French Feminisms.* New York: Schocken, 1981.

Danielle Roth-Johnson
University of Nevada, Las Vegas

Deep Ecology

Norwegian philosopher Arne Naess coined the term *deep ecology* in 1973 to describe a philosophical method of inquiry that sought deeper relationships to the natural world that views humans as embedded in nature rather than separate from it. Influenced by Rachel Carson's *Silent Spring* that documented the negative human intervention in the environment through the use of pesticides, he sought to combine ecological awareness with nonviolent action based on Gandhian principles. Naess developed his Eight-Point Deep Ecology Platform as a method of considering an interconnected and reciprocal relation between humans and nature, rather than the anthropocentric or human-centered view that inserted a dualistic split between the two. Seeking the answers to how we should live in the world, often situated in a cabin high in the mountains of Norway, Naess considered deep experience, deep questioning, and deep commitment as imperative for an ethical approach to human thought and action. He developed what is termed an *ecosophy* that embodies ecological wisdom as a way of being, thinking, and acting in the world.

Naess focused on the intrinsic value of all life, including nonhuman life, that he believed could be cultivated through a process he called "self-realization." The experience of Aldo Leopold's recognizing the life force in the eyes of a dying wolf that brought him to the understanding that humans did not have the right to control other members of the biotic community is an example of this revelation. Such an experience brings about empathy and concern that assists humans in understanding that they are participants in the web of life, but not masters of it. This realization leads to an understanding of the interconnection of all life, not dissimilar to systems theory that indicates that an action or event in one part of the web affects all the others. Recognizing human dependence on other life forms for our physical and psychological well-being, the framework of relationships then extends to ecosystems and bioregions and leads to a desire to protect nonhuman life.

Instead of cultivating self-serving, egotistical interests, Naess believed that human identity should be nested within a larger framework he called "the ecological self" that encompassed all the diversity of life forms that inhabit the planet. Naess recommended experiential activities that would assist humans in aligning with their nonhuman counterparts and contributed to a classic book, *Thinking Like a Mountain,* which described a frequently used exercise called "The Council of All Beings." In this activity, humans take on the representation of a nonhuman life form such as a dolphin or a butterfly and speak for those without language capacities in a ritual format.

The Eight Points

Along with George Sessions, Naess developed the Deep Ecology Platform as a device to demonstrate how people of different spiritual backgrounds or cultural belief systems might agree on the principles of protecting life. The eight points include (1) all life has value in itself, independent of its usefulness to humans, (2) richness and diversity contribute to life's well-being and have value in themselves, (3) humans have no right to reduce this richness and diversity except to satisfy vital needs in a responsible way, (4) the impact of humans in the world is excessive and rapidly getting worse, (5) human lifestyles and population are key elements of this impact, (6) the diversity of life, including cultures, can flourish only with reduced human impact, (7) basic ideological, political, economic, and technological structures must therefore change, and (8) those who accept the foregoing points have an obligation to participate in implementing the necessary changes and to do so peacefully and democratically.

Although Arne Naess died in 2009, his legacy lives on in a social movement that bears the moniker *deep ecology.* His ideas on ecology and ecosophy were developed in numerous books and articles, notably *Freedom, Emotion and Self-Subsistence* (1975); *Ecology, Community and Lifestyle* (1989); and *Life's Philosophy: Reason and Feeling in a Deeper World* (2002). Among those who have carried on his mission are Thomas Berry (*Dream of the Earth*), David Abram (*Spell of the Sensuous*), Bill Plotkin (*Soulcraft*), Bill Devall and George Sessions (*Deep Ecology: Living as If Nature Mattered*); Warwick Fox (*Toward a Transpersonal Ecology*), Paul Shepard (*Nature and Madness*), Gary Snyder (*Practice of the Wild*), Joanna Macy (*World as Lover, World as Self*), Theodore Roszak (*Voice of the Earth*), David Orr (*Earth in Mind*), and many others too numerous to mention.

Offshoots of Deep Ecology

Deep ecology has spawned numerous offshoots that include ecopsychology (Chellis Glendinning, Andy Fisher), social ecology (Murray Bookchin), and ecofeminism (Carolyn Merchant, Susan Griffin). It also has critics including Michael Zimmerman, who has expressed concerns about ecofascism, comparing deep ecology to the Nazi movement's ideas of "blood and soil," and Bron Taylor, who finds that the movement has strayed too far from its philosophical roots and romanticizes hunter-gatherer ideology while shunning agriculture; distances itself from Abrahamic religions while embracing indigenous, pagan, or Eastern religious worldviews; and believes in a decentralized democracy. Problematic aspects of Christian approaches that view humans as stewards or caretakers of the Earth with an attitude of dominance are presented in a classic article by Lynn White titled "The Historical Roots of Our Environmental Crisis."

Environmental ethics has used deep ecology principles as an alternative to utilitarian models. Rather than relying on reason and individual responsibility, a deep ecological view aspires to lift up virtues such as compassion and forgiveness and to allow for emotional responses instead of only abstract thinking. Deep ecologists are concerned with the disconnection between head and heart and seek to develop a holistic approach to problem solving.

Deep ecology is often linked back to the transcendentalists Ralph Waldo Emerson and Henry David Thoreau and considers some of the early environmentalists like John Muir among its founding fathers. Over the past several decades, it has gained prominence with followers who admire Native American principles and those who follow the tenets of Buddhism's "do no harm" and interrelationship principles. Joanna Macy has used Buddhist teachings, systems theory, and grief and despair work to create workshops around the "Work That Reconnects" and "The Great Turning." Thai monk and social activist Sulak Sivaraksa embodies many deep ecological principles in his efforts to reduce materialistic tendencies and Western models of consumption as part of his Buddhist approach to protecting the natural world. Vandana Shiva and the Chipko Movement to stop the senseless cutting of trees in India are other voices from the Asian subcontinent that are often referenced by deep ecologists.

It has also contributed to the Green Party ideas and been advocated by those who seek to reform economic systems, such as David Korten, Bill McKibben, and Paul Hawken. Some of the more radical political movements that are often associated with it include Dave Foreman's Earth First, the Animal Liberation movement, and activist Derrick Jensen. The movement also finds its way into new spiritual ideas like the creation spirituality advocated by Matthew Fox and is linked with new science luminaries David Bohm and

Fritz Capra. Proponents often resonate with Brian Swimme's Universe Story as a way of reframing the narrative by which humans relate to the planet.

Environmentalists who support green energy including solar and wind as a way to reduce dependency on fossil fuels find meaning in deep ecology principles. Sustainability practices such as organic farming methods, recycling and zero waste, voluntary simplicity, and back to the land enthusiasts find agreement with many deep ecology proposals.

See Also: Anthropocentrism; Carson, Rachel; Intrinsic Value; Naess, Arne; Social Ecology.

Further Readings

Drengson, Alan. *The Deep Ecology Movement.* Berkeley, CA: North Atlantic Books, 1995.

Naess, Arne. *Ecology, Community and Lifestyle.* Cambridge, UK: Cambridge University Press, 1993.

Seed, John, et al. *Thinking Like a Mountain.* Gabriola Island, British Columbia, Canada: New Society Publishers, 1988.

Sessions, George, ed. *Deep Ecology for the 21st Century.* Boston: Shambhala Publications, 1995.

Stephanie Yuhas
University of Denver

Deep Green Theory

In contrast to the mainstream environmental movement's emphasis on lifestyle changes and technology, deep green theory focuses on entire systems of power, including industries and institutions. Copsa Mica, Romania, shown here, became one of Europe's most polluted towns mainly because of factory emissions from government-owned heavy industry.

Source: iStockphoto.com

Deep green theory (DGT) is a critique of the ethics of "deep ecology" put forward by writers such as Richard Sylvan and David Bennett. Distinctions are made between "nonethical living," which includes anthropocentric exploitation of the environment, and three ethical positions: "shallow," "intermediate," and "deep" green ethics. Deep green theory develops Arne Naess' concept of deep ecology or "ecosophy," which rejects the anthropocentric for ecocentric models of living, with human life and ecology coexisting equally. However, deep green theory develops the concepts of deep ecology, incorporating alternative ideological understandings of the possibilities of deep green politics. Such changes emerge from the individual conscience changing of "self-realization" put forward by Bill Devalls and George Sessions, which then can be developed into personal and ultimately communal politics.

The central critique of deep green theory to deep ecology centers on the absence of a challenge to existing capitalist structures by the latter. In addition, deep green theorists argue that the absence of a framework to put deep ecological concepts into action lacks a framework. Deep green theory incorporates a political praxis that includes animal rights, antiglobalization, feminism, and anticapitalism within its activist structures, thereby creating a theoretical ideology for adherents of deep green politics that is absent in the philosophy of deep ecology. Deep green theory allows for a consideration of socialism, which is rejected as anthropocentric in deep ecology.

This framework has been influenced by Murray Bookchin's concept of "eco-socialism," creating a political basis for an ideological critique that is congruent with the tenets of deep green theory, and that provides for a more extensive critique of the destructive and exploitative nature of capitalism in conjunction with other radical perspectives such as ecofeminism. Sylvan called his own theoretical perspective "deep green theory" (DGT), making a distinction with the religious and philosophical basis of Naess's deep ecology and deep green theory's pragmatic political framework for dealing with existing environmental issues. Distinctions are also made between the personalized basis of deep ecology (based on holism, biospherical egalitarianism, and self-realization) and bringing these concepts into the wider political realm through the deep green theory activism. This pragmatic perspective allows for the creation of a broad green coalition with pacifists, feminists, and anarchists, but one that holds to deep green thinking as its core ideological concept.

This primal response is the basis for understandings of "rural sentiment," which can be seen as part of what Arne Naess (1972) originally called "eco-centricism," the valuing of the hinterland over the self. The dichotomy between deep green and ecomodernist paradigms has its basis in R. Eckersley's definition of an "anthropocentric/ecocentric cleavage." The distinction is made clear from the following quote by Eckersley:

> The first approach is characterized by its concern to articulate an eco-political theory that offers new opportunities for human emancipation and fulfillment in an ecologically sustainable society. The second approach pursues the same goals in the context of a broader notion of emancipation that also recognizes that moral standing of the non-human world.

While both approaches are concerned with the environment, it is the emphasis placed on "human emancipation" over "the non-human world" that demarcates the anthropocentricism of the sustainable development culture from an ecocentric perspective. Eckersley also cites the broadly similar distinctions found in the ecological theories of Naess ("shallow and deep ecology"), O'Riordan ("technocentricism and ecocentrism"), Bookchin ("environmentalism and social ecology"), and so on. The positioning of humankind in relation to other species and ecosystems is pivotal in regard to this theoretical contextualization of two main distinct features of current environmental thought. While not aligned with a traditional understanding of the left/right divide within political ideology, the distinction between anthropocentric and ecocentric does have its basis in humankind's technical and industrial capabilities, which have become the basis for the type of environmental destruction evident in contemporary society. The "eco-socialism" of

Rudolf Bahro and the "left-biocentricism" of David Orton have also been cited as contributors to the development of DGT.

While traditionally the Left pinpointed control of the means of production as the crucial issue of political contestation, deep green theory is more concerned with how the means of production impact upon the environment and to what extent this is acceptable in society. Deep green theorists have responded to this by addressing the technical nature of industrial development, and the need to critique that development through deep green politics, or alternatively, to regulate industry.

In other words, the root of technocentrism lies in social and political compromise between the Earth's resources and human development with technology as the cutting edge of this manipulation of the Earth's resources. Technocentric approaches are determined with no overhaul of human social systems envisaged and despite recognition of the inherent ecological problems of this analysis. Ecocentrism conversely places humankind not to the fore of the global ecosystem, but rather sees humanity as part of an organic whole, with a moral imperative to restrain activity and growth and to interact and cooperate with the greater ecosystems that populate the Earth. This view holds a respect for a pristine, natural world in its own right before any aspect of human economy and development is considered with human beings living in a spirit of cooperation and ecumenism with the environment.

The deep green view of environmentalism had its roots in the ecological, feminist, and other new social movements of the 1960s and 1970s and has challenged the hierarchical hegemony of political dominance and technological development over social and ecological systems around the globe. Deep green ideology goes beyond old left-wing attempts at "controlling the means of production" or of deconstructing class systems and sets its point of origin before the era of revolution to the beginning of modernity and the age of Enlightenment. By questioning the concept of social order based on expansive development that had its roots in the Enlightenment project, present-day environmental protests have rejected the concept of a technologically driven modernity in itself, radically moving beyond the position of "sustainable development" by questioning the validity of development from an ecocentric perspective.

Deep green theory views mainstream environmental activism as having various levels of impact in the struggle for sustainability. The deep green theory holds that industrial civilization is now unsustainable and must be actively challenged at all levels in order to secure a sustainable future for all species on the planet. The DGT perspective argues that the dominant culture will not undergo a voluntary transformation to a sustainable way of living. Deep green theory purports that industrial civilization must be extensively challenged in order to save as many species and bio-organisms as possible, noting that the Earth's sustainable capacity is further diminished as the spread of industrial society continues exponentially.

Within deep green theory, lifestyle or personal changes are not considered effective methods of creating meaningful change. The mainstream environmental movement is seen as being distracted by its emphasis on lifestyle changes and technological solutions instead of confronting systems of power and holding individuals, industries, and institutions accountable.

Deep green views technological solutions, no matter how well intentioned, as inadequate, and possibly leading to accelerated ecological destruction and pollution. The deep green movement looks to preindustrial and precivilization, land-based cultures as models for sustainable ways of living.

The main principles of deep green theory as put forward by Sylvan and Bennett (1994) include the following points:

- Traditional ethics are inadequate for dealing with the ecological crisis.
- Deep green theory takes an ecocentric perspective on all issues.
- The intrinsic value of all natural items overrides human interests.
- No single species deserves special ethical treatment.
- Human use of the environment is tolerated, with only "use of too much and too much use" forbidden.
- Indigenous hunting for sustenance is tolerated; hunting for trade is not.
- The burden of proof lies with those who advocate interference with the environment. This interference now needs to be justified in advance, rather than waiting for reasons not to interfere with the environment to become commonplace.
- The economics of sufficiency should replace the economics of profit.

In the main, deep green theory sets out a manifesto for green citizenship. At the heart of the DGT approach is a desire to make deep ecological ethics less virtuous and more commonplace in order to facilitate the adoption of ecological practices by the greater majority in society. For DGT adherents, this approach is necessary in order to meet the challenges of environmental degradation that threatens all species and the planet. In other words, the metaphysical and enlightened pathway set out through deep ecology needs to be applied to living in a practical manner, without diluting the significant aspects of ecological ethics. Nonetheless, deep green theory allows for an adaptation of shallow and intermediate green practices in the ultimate pursuit of deep green ethics and provides a broad framework for this to be achieved throughout the endeavors of humankind to make the transition to deep green lifestyles.

See Also: Deep Ecology; Ecocentrism; Ecofeminism/Ecological Feminism; Ecology; Future Generations; Naess, Arne.

Further Readings

Devall, W. and G. Sessions. *Deep Ecology: Living as If Nature Mattered.* Salt Lake City, UT: Gibbs M. Smith, 1985.
Leonard, L. *The Environmental Movement in Ireland.* Dordrecht, Netherlands: Springer, 2008.
Leonard, L. and J. Barry, eds. *The Transition to Sustainable Living and Practice.* Advances in Ecopolitics, vol. 4. Bingley, UK: Emerald, 2009.
Naess, A. *Ecology, Community and Lifestyle: Outline of an Ecosophy.* D. Rothenberg, trans. Cambridge, UK: Cambridge University Press, 1989.

Liam Leonard
Institute of Technology, Sligo

Democracy

"Democracy" literally means "rule by the people" and history suggests that it should be thought of as a work in progress. Although its origins have been attributed to ancient Athens, the philosophy, institutions, and political culture that underpin modern democracies were substantially transformed by the Enlightenment and the American Revolution. Philosophically, democracy argues that people have both the capacity and the right to rule themselves, so governments must gain their citizens' consent to exercise power legitimately. Democracy is institutionalized in any system of government that either enables citizens to participate directly in the policy-making process or allows them to elect representatives to make policies on their behalf. Democratic political cultures encourage citizens to participate in the formal policy-making process, persuade them to accept the outcomes of this process, and discourage them from threatening democratic institutions. Democracy is currently being further transformed by historic changes in the late 20th century such as globalization, the rise of the Internet, and the emergence of environmentalism.

One of the first instances of democratic government was that of Athens 2,500 years ago where citizens met regularly to vote on decisions affecting the running of their city. They also elected a council to represent the various regions of Athens and set the agenda for the assembly's deliberations. The right to participate was restricted to adult male citizens, so women, slaves, and foreigners were excluded. A few centuries later, Rome (which went through phases of republic and empire) experimented with systems involving an appointed senate, assemblies of elected representatives, and a forum open to all adult male citizens.

By the end of the first millennium, various European city-states had adopted systems where the nobility periodically elected leaders from among themselves. As these city-states merged to form larger countries (often through wars of conquest), parliaments were created as a way for the nobility to be consulted by their monarchs about major policy decisions. In 1215, for example, the English barons forced King John to sign the Magna Carta to formalize some limited civil rights such as the protection of private property and the right to a fair trial. The English Civil War of the 17th century saw an expansion of the role of the parliamentary system and the temporary abolishment of the monarchy. In 1679, the parliament passed the Habeas Corpus Act that required people to be brought before the courts rather than be imprisoned without trial. Even after the restoration of the king, power continued to shift from the monarch to the parliament, and the number of people allowed to vote progressively expanded.

The first modern democracy emerged after the American Revolution. In 1789, the Constitution of the United States of America created a federal republic where adult male citizens periodically elected people to represent them in Congress. A president was elected every four years via an electoral college to lead the government. Women and slaves were initially not allowed to vote, but slavery was abolished in the 19th century and the right to vote or run for office was eventually expanded to include all adult citizens in the 20th century (achieving what is known as universal suffrage). The first 10 amendments to the Constitution formed the Bill of Rights, which included protection of freedom of speech, freedom of association, freedom of the press, and the right to a fair trial, among other things. The Constitution and Bill of Rights were heavily influenced by the Magna Carta and Habeas Corpus.

During the 19th and 20th centuries modern democracies multiplied and spread around the world, but they faced several challenges. Some monarchs were reluctant to relinquish powers and some empires were disinclined to grant their colonies independence. Fascism confronted democracy in World War II, socialism/communism competed with it during the cold war, and military dictatorships posed serious challenges in some regions. The long-term historical trend, however, appears to be toward democracy and approximately two-thirds of the world's countries are now governed by democratic political systems of some kind.

The Role of a Constitution

The dynamic history of democratic government has been tracked by an equally innovative development in philosophy. In ancient Athens, Aristotle introduced the idea of a constitution to set out the various roles of the different governing bodies, but he saw democracy as simply one of three kinds of possible government (the other two being monarchy and aristocracy). In 1690, John Locke discussed the idea of individuals having rights to support liberty, equality, and majority rule. He introduced the notion of a social contract between citizens and their government, where people consent to give up some of their naturally endowed liberties and abide by majority decisions in return for the benefits of living in a democratic society. These ideas were picked up by later philosophers such as Jean-Jacques Rousseau and were eventually enshrined in the U.S. Constitution and Bill of Rights. The idea of men having rights was championed by Thomas Paine during the American Revolution, and Mary Wollstonecraft argued that these rights should be extended to women. In the mid-19th century, John Stuart Mill stressed the importance of preserving individual liberty for both men and women and developed the idea of representative government as a way to overcome the difficulties associated with organizing large numbers of people.

The philosophy behind modern democratic government continued to develop in the 20th century. During the 1950s, the school of pluralism emerged in the United States as a reaction to Marxist claims that government was being dominated by the interests of big business. Pluralists such as Robert Dahl conducted a series of studies that indicated business did not always get what it demanded from democratic government. In the 1970s, scholars like Stephen Lukes challenged this conclusion by demonstrating that corporate power had hidden dimensions (such as the ability to set the agenda and to engineer a favorable ideology) that can constrain democratic decision making. Jürgen Habermas argued that if representatives from all positions on a contested issue negotiated until they reached a true consensus, the narrow individual interests would be stripped away, leaving only the common interests of all in the final agreement. John Rawls revived the idea of designing a social contract that would avoid extremes of injustice and poverty.

Just as the philosophy behind democracy has developed and diversified over time, so have the institutions that are designed to support democratic government. Whereas the citizens of Athens were a small-enough group to practice direct democracy, it is not practical for larger societies. This has led to the predominance of representative democracy. These systems limit popular voting to the periodic election of governments and the occasional referendum on major policies. Representative democratic institutions include the periodic election of an executive and a legislature; a constitution that separates powers between the government, the judiciary, and enforcement agencies; the rule of law; individual rights; and an independent media. Modern democracies are also usually accompanied by mixed economies that blend private, public, and community sectors in varying proportions.

There are still significant variations in modern representative democracies. The British system has been labeled "parliamentary" because the executives (prime minister and cabinet) have to be members of the legislature. The U.S. system, on the other hand, is known as "presidential" because a president is elected separately from the legislature and appoints the cabinet. The U.S. system is also known as a liberal democracy because its institutions rely very much on the ideas of 17th- and 18th-century liberalism (i.e., its philosophy and political culture is focused very much on the freedoms of individual citizens, and governments strive to limit the size of the public sector in comparison to the combined private and community sectors). Social democracy, on the other hand, focuses more on social goods and collective rights, as opposed to liberalism's focus on individuals. It involves a large public sector to provide social welfare and services. Such systems prevail in many northern European countries.

Requirements for Success

To be successful, democratic institutions need supportive political cultures (i.e., a constructive way for citizens to view and interact with the political system). First, citizens need to see the whole system as legitimate (i.e., they have to believe that elections are free and fair, that they are being treated equally under the law, and that no one is being unfairly favored). Second, they have to accept the outcome of elections or major policy decisions, even if they don't like the result. Third, while they may campaign for the reform of the system, they refrain from advocating its overthrow. Finally, some citizens need to be willing to get involved in government by forming pressure groups and political parties. Pressure groups consolidate the concerns of the citizens, aggregate their demands into a coherent set, and provide leaders that can negotiate with the government. Political parties link citizens to the executive by recruiting, training, and supporting candidates for election. Without supportive political cultures, democratic systems may collapse (as happened to the Weimar Republic in Germany between the two world wars).

Democracy has transformed itself over 2,500 years, and in this time, the school of philosophy on which it rests has grown, democratic governing institutions have diversified, and new supportive political cultures have developed. A number of dramatic historical changes in the late 20th century, however, have generated some further challenges and transformations. These changes include rapid globalization, the rise of the Internet, and the emergence of environmentalism.

Globalization is the increased technical, economic, social, political, and environmental interlinking of societies around the world. While it can be traced back to empires, colonization, and trade, globalization has been rapidly accelerating in the last few decades. The process has expanded democratic institution building into the international arena. At the regional level, the European Union has established a parliament that covers 27 member countries with political parties and pressure groups operating across national borders. At the global level, the United Nations General Assembly gives the government of each member country a vote on major resolutions and has brokered several international agreements supporting individual rights. These include the Universal Declaration of Human Rights and the International Covenant on Civil and Political Rights. Such agreements embrace the kinds of rights and freedoms that were elucidated by thinkers such as Locke, Paine, and Wollstonecraft.

The rise of the Internet is part of the globalizing process and has also offered new ways of extending democracy between and within countries. Governments now routinely release draft policies, plans, legislation, or reports and call for public comment/submissions online before a final decision is made. Pressure groups use Websites to tell their story directly to the public and to contact politicians. They also use e-mail to keep their members informed, organize protests, and mobilize volunteers. Political parties use Websites to raise money and to release their policies, particularly in the lead-up to elections. For those citizens who have access to the Internet and are politically active, this has greatly improved their ability to interact with government, which has led some scholars to suggest that we are moving toward a new age of "digital democracy."

The third transformation of democracy has been triggered by the emergence of global environmental issues and the rise of the green movement. Concerns about environmental damage led to the creation of local pressure groups in the 1960s that grew into a network of international organizations in the 1980s. The green movement used all the institutions of democracy, particularly their rights to freedom of speech, freedom of assembly, and periodic elections. When they were dissatisfied with the response of the established political parties, they ran candidates for election themselves and formed their own green political parties. These green parties have gained some measure of success around the world, getting elected to parliaments from Australia to Europe. John Dryzek suggests that since society relies on the environment for resources, services, and waste disposal, we should redesign our institutions to mimic natural ecosystems. Robyn Eckersley has pointed out that if the decisions we make collectively today concern not just the current population but future generations and other species, how is legitimate consent obtained? These ideas have led to the concept of ecological democracy that requires a restructuring of democratic institutions, a change to conventional political cultures, and a philosophical shift to broaden the scope of democratic governance to encompass the environment.

History suggests that democracy is not a single fixed destination—it is a journey that started with the direct democracy of ancient Athens, overtook the emergence of European parliaments, was spurred on by the American Revolution, and is currently visiting various modern liberal and social democratic systems. The school of philosophy, set of institutions, and range of political cultures are still evolving and recent developments suggest that the journey is far from over.

See Also: Civic Environmentalism; Ecopolitics; Environmental Pluralism; Green Party, German; United Tasmania Group.

Further Readings

Aristotle. *The Politics*. London: Penguin, 1981 [c. 335–323 B.C.E].

Dahl, Robert. *Who Governs? Democracy and Power in an American City*. New Haven, CT: Yale University Press, 1961.

Dryzek, John. *Rational Ecology: Environment and Political Economy*. Oxford, UK: Blackwell, 1987.

Eckersley, Robyn. *The Green State: Rethinking Democracy and Sovereignty*. Cambridge, MA: MIT Press, 2004.

Habermas, Jürgen. *Theory of Communicative Action*. Trans. Thomas McCarthy. Boston: Beacon Press, 1981.

Hague, Barry and Brian Loader, eds. *Digital Democracy: Discourse and Decision Making in the Information Age*. London: Routledge, 1999.

Locke, John. *Two Treatises of Government*. Cambridge, UK: Cambridge University Press, 1990 [1690].

Lukes, Stephen. *Power: A Radical View*. New York: Macmillan, 1974.

Mill, John Stuart. *Utilitarianism: On Liberty and Considerations on Representative Government*. London: Everyman, 1987 [1859–61].

Paine, Thomas. *The Rights of Man*. London: Penguin, 1987 [1791].

Rawls, John. *A Theory of Justice*. Cambridge, MA: Harvard University Press, 1971.

Rousseau, Jean-Jacques. *The Social Contract.* London: Penguin, 1986 [1762].
Wollstonecraft, Mary. *Vindication of the Rights of Woman.* London: Penguin, 1985 [1792].

Michael Howes
Griffith University

Development, Ethical Sustainability and

The activities of a community are sustainable if it converts resources into waste no faster than its ecological support systems can convert waste into resources. While this ecological definition seems straightforward, sustainability also has a complex ethical dimension because it involves moral choices. Ethical sustainability, in the sense of morally justifiable sustainable living, is represented by the sum of our choices and policies designed to render our living sustainable that are most defensible under generally accepted theories of justice and ethics. Currently the activities of humanity as a whole and those of most human societies are far from sustainable. Ethical development describes how we can achieve the transition to sustainable living, as opposed to other interpretations of development that are often informed by quite contrary aims.

The most widely known definition of *sustainability* comes from the 1987 Brundtland Report, which framed the concept in terms of meeting the basic needs of presently existing people without compromising the ability of future generations to meet their own needs. While this definition acknowledges intergenerational obligations, it leaves open how many people can make such demands before the total exceeds the capacity of the biosphere to meet them and how far into the future this can be calculated. Subsequent definitions, such as the one in the Hannover Declaration, referred to the continued satisfaction of basic human physical needs and of higher-level social and cultural needs, such as security, freedom, education, employment, and recreation, and combined those with the continued productivity and functioning of ecosystems. Although those definitions describe more accurately the necessary condition of how human needs can be met sustainably, they still do not resolve the question of limits.

The meaning of sustainability has become even fuzzier with the efforts of sociologists, economists, and anthropologists to introduce diverse definitions within their respective disciplines. However, because all human activities take place within ecosystems, and are supported by them, the ecological definition given in the first sentence circumscribes all others. Ecosystems consist of communities of species and their nonliving physical environment; they serve as environmental support structures for human enterprises that deliver raw materials and energy and recycle wastes. By linking sustainability to ecological cycles and their capacities, the ecological definition provides the sufficient condition for sustainability, not just the necessary one of human needs being met. This focus on what is scientifically sufficient seems rather important if we are to have any chance at attaining sustainability.

Chances of arriving at a sustainable modus vivendi for humanity within the next decade or two are not exactly encouraging. Mounting evidence has now convinced most people of the reality of a human-caused global environmental crisis. The rates of resource depletion are increasing as the global human population and its consumption patterns continue to grow out of control; pollution continues, with its effects on climate, habitat quality, and public health; species continue to disappear at an alarming rate as humans introduce competitors, modify ecosystems, deplete habitats, and modify landscapes and climates. As for causes, five self-reinforcing processes have been identified: economic growth, population growth, technological expansion, arms races, and growing income inequality.

Moral choices about sustainable living have become more important as the environmental impact of human populations has approached the sustainable maximum that is dictated by the capacities of the supporting ecosystems. This environmental impact (I) represents the total demand on ecosystem services (raw materials, energy, and waste processing). It is defined as $I = PAT$, where P is the population size, A is the affluence per person, and T is the environmental impact created per currency unit spent (which depends mainly on what kinds of technology are involved). The environmental impact of a population is often expressed as its ecological footprint. The $I=PAT$ formula applies unconditionally to the global human population, meaning there are no other variables that influence the environmental impact. Since humanity's impact began to exceed the sustainable maximum of the biosphere in the mid-1980s, we have remained in the condition of global ecological overshoot.

A population can only sustain itself for as long as it remains below the overshoot threshold, that is, its total environmental impact stays below the sustainable maximum. Findings from animal populations suggest that the consequences of prolonged overshoot for the population and the entire ecosystem tend to be severe: the population is likely to crash or fluctuate wildly, or the entire ecosystem can collapse into a less-complex state. The longer humanity stays in overshoot, the more severe the consequences for the entire planet will be. In this context, working toward ethical sustainability means making every morally justifiable effort to decrease our collective impact or footprint to a level below the sustainable maximum.

Another aspect of ethical sustainability that gives rise to moral choices emerges directly from the $I=PAT$ equation. The maximum sustainable impact can be reached by a large population of poor people with each making small demands, or by smaller populations of more affluent people demanding more environmental services per individual. As attractive as any form of sustainability appears in comparison to its alternative, not all sustainable solutions will be regarded as equally "ethical" (i.e., morally justifiable), nor will all people agree on their order of preference. Historically, most societies have developed ways to regulate the growth and consumption of their populations. The ethical challenge lies in the means and ends of collective decision making.

At the regional or local level it is usually more difficult to determine what constitutes the maximum sustainable impact because resources flow in from other regions and wastes are transferred out. In the age of global trade there is hardly a region or locality left on Earth whose human inhabitants rely entirely and exclusively on the services of the local ecosystems. Nevertheless, those inflows and outflows of resources and wastes can be measured and the net impact of the population determined. In that instance, it is particularly useful to express the environmental impact as the population's ecological footprint because the latter is measured in hectares of productive land area. A region's footprint can exceed its available area of productive land for one of two reasons, and each is accompanied by important ethical challenges. Either the region imports resources and exports wastes to an extent that is probably unjust toward people elsewhere; or the region's population is in overshoot. In each case, the dictates of ethical sustainability present the population with some difficult choices.

The range of moral choices is further diversified by the fact that human survival comes in various modes. A human population can survive merely, miserably, idealistically, unjustly, or acceptably. Mere survival means that its members are barely alive, whereas miserable survival allows them a small amount of autonomy; idealistic survival allows for considerable luxury and is only possible if the population rigorously controls its own numbers and activities; unjust survival involves an inequitable situation where a minority survives in luxury while a majority suffers; most other modes of survival are considered acceptable. As the individual citizens make choices about their reproduction, spending habits, and lifestyle, they contribute to the collective choice of survival mode of subsequent generations, or they may be opting against their survival entirely. By choosing to what extent we drift into overshoot mode, and for how long we are to remain in that mode, present-day humanity decides the fate and well-being of future generations.

An entirely different kind of moral choice involves how people decide to behave toward nonhuman nature, which includes all individual life forms and the species, communities, and ecosystems to which they belong. Under an anthropocentric ethic, it is legitimate to utilize all of nonhuman nature as means to our ends, that is, for maximizing the well-being of humans. Such limitless exploitation is not justifiable in the views of people who subscribe to biocentric or ecocentric ethics, because to them, nonhuman life forms possess intrinsic value, interests, and rights. Gaian ethics, a specific form of ecocentrism, focuses particularly on the responsible and respectful treatment of species, ecosystems, and the planet as a whole. Under those environmental ethics, every life form, human and nonhuman, deserves our respect and must not be used solely as means to our ends. Thus, a certain mode of sustainable survival may or may not be acceptable to a person, depending on what environmental ethic he or she holds. However, the current mass extinction of species, a direct result of human overshoot, is of concern to people of all ethical convictions, albeit for different reasons.

As outlined so far, ethical sustainability involves acting on justifiable moral choices and making decisions on the basis of acceptable standards of behavior. This is only possible if the individual actually has the freedom and power to make those decisions. Some argue that in order for any sustainable form of living to be "ethical," it must have been achieved in a democratic way. Ethical sustainability in that view includes the requirement for universal democratic rule. Opponents to this argument claim that after two decades of overshoot, waiting until a sufficient number of people make the right decisions on their own accord is no longer an option. Some go so far as to regard democracy as unsustainable unless it is mitigated in an appropriate cultural context that influences people's values, attitudes, and beliefs toward an ethic of sustainable living.

Ethics

In the foregoing discussion, sustainability was referred to as ethical if the means by which one tries to achieve it are morally justifiable, meaning that we can show that such means or strategies do not contravene any standards of moral behavior to any undue extent. This begs the question of what kind of ethics are to be applied for such a justification. Moral philosophers distinguish between three broad kinds of ethics: deontology, utilitarianism, and virtue ethics. Each kind of ethic can contribute its own justification, and most provide objections to some of the possible strategies that could be employed in our quest for sustainability. In addition, each of the three ethics comes in an anthropocentric and ecocentric/biocentric variety. Their respective contributions to sustainability are as follows.

Deontological ethics is all about rules. Those rules are to be applied rigorously to guide decisions and actions, regardless of circumstances, conditions, or other mitigating considerations. The most important one, often referred to as the Golden Rule, asks to always treat others in ways that one wishes to be treated. Under this prime directive, any measures toward sustainability must be applied equitably to all people. One might think that this requirement would attract few overt objections; however, the political realities around the world suggest that much fewer people are willing to act according to principles of equity and reciprocity than are espousing those ideals in words.

As far as other rules are concerned, the field divides into anthropocentric deontologists and the ecocentric/biocentric kind. The former include the devout followers of most major religions. Most world religions offer surprisingly little guidance when it comes to sustainable living, perhaps because they evolved under conditions where sustainability was taken for granted. Secular anthropocentric deontologists tend to regard human rights and dignity as their prime source of rules. This, too, offers little help for sustainability because those rights are usually interpreted as only applying to presently living people. Some rights appear difficult to grant while others, such as the one that calls for clean drinking water for every person, actually turn from ungrantable to immaterial under conditions of overshoot. In contrast, ecocentric deontologists have quite a lot to say about sustainability, as exemplified by Aldo Leopold's dictum "a thing is right when it tends to preserve the integrity, stability, and beauty of the biotic community. It is wrong otherwise."

Utilitarians would approach the question of whether or not an action is justifiable by weighing its potential benefits against its potential harms. It is, in fact, impossible to estimate the benefits of perfect sustainability in terms of the sum total of human well-being and flourishing, because it is not clear how many generations we can look forward to under those conditions. At best, our sun is expected to provide the Earth with an adequate supply of energy for another 6 billion years; on the other hand, paleontologists estimate the average lifetime of a species as about 7 million years. Compared to such massive benefits, almost any seemingly immoral act appears justifiable in the eyes of the utilitarian, including the totalitarian enforcement of draconian measures for population reduction, radical redistribution of resources, public repossession of the means of production, and the like. For the utilitarian, sustainability could be achieved in two ways. One involves maximizing the total well-being of humanity, which would require that the population becomes as large as could be sustainable under any conditions; this would result in mere to miserable survival for billions in an equitable way. The other way is to maximize the average well-being of humanity, which would mean that the population size be fixed by whatever means necessary to a suitable level at which all individuals can live comfortably.

While some may regard such options with less than absolute dismay, the mere fact that most utilitarians would be likely to indiscriminately approve just about all strategies makes them poor judges. Furthermore, ecocentric utilitarians would find themselves resolutely opposing anthropocentric utilitarians because the latter would approve strategies for their benefit to humans while the former would condemn them for their harm to other life forms. Some ecocentric utilitarians have gone so far as to deny any moral justification to human civilizations beyond the gatherer-hunter stage.

Many advocates of virtue ethics base their approach on the teachings of Aristotle, who postulated four classical virtues as standards of behavior: courage, temperance, prudence, and justice. Numerous cultures and religions have advocated their own lists of virtues and vices. The principle of virtue ethics is the idea of achieving a balance between excess and deficiency of any virtue on one's list. That balance is regarded as the standard for leading a virtuous life. The standard does not prescribe any specific decisions or actions; it merely requires that they be virtuous under the given circumstances. From the perspective of virtue ethics, strategies toward sustainability are evaluated by the kinds of actions that they demand under the circumstances. Anthropocentrists would ask that human life and well-being be respected, while biocentrists and ecocentrists would demand that the same or similar considerations be extended to other life forms and ecosystems. The anthropocentrist's primary focus on human flourishing puts her at a disadvantage because this moral "tunnel vision" invariably leads to mistakes where species and ecosystems are lost because they appeared unimportant at the time. In contrast, ecocentric virtues appear to go a long way toward defining an effective ethic for sustainability. Together with ecocentric deontology, ecocentric virtue ethics has the potential to define the moral standards that inform sustainable living.

Development

The term *development* has been used in diverse meanings that have given rise to heated controversy. At its center lies the idea of improving something in a planned approach along a predetermined timeline. In a socioeconomic context, many have come to regard development as simply quantitative growth of anything that is considered good. More specifically, it often describes the deliberate modification of a local piece of natural environment to facilitate its use as a means of production, for example, as a parking lot. At the international level, development often refers to the deliberate modification of existing social, cultural, and economic structures to render a country more economically productive and accessible to the global market. These notions formed the basis of traditional concepts of international development, of economic growth, and of progress in general under the modern paradigm. In its most dogmatic form, this notion of progress relies on the assumption, referred to as cornucopianism, that there are no real limits to the growth of economies, populations, industries, and capital. Of course, this assumption clashes head-on with the concept of sustainability, not to mention the Laws of Thermodynamics and all our scientific understanding of nature. Hence, the term *sustainable growth* represents an oxymoron, at least in the sense in which it tends to be used. Needless to say, notions of sustainability are not part of the ideology of growth, which is also referred to as the Dominant Social Paradigm.

In response to increasing criticism of the growth ideology and its pervasive influence, alternative economic models were proposed during the past three decades, along with the umbrella concept of sustainable development. To what extent development can be sustainable depends, of course, on one's definition of development. Evidently, it cannot mean quantitative growth, but qualitative improvement seems perfectly sustainable if one puts up with the effect of diminishing returns. In fact, no obvious limits exist to personal development in the sense of increasing one's maturity, knowledge, skills, moral responsibility, and judgment, or any kind of professional competence. Although it is now widely used, the term *sustainable development* is of little conceptual use unless it is accompanied by clear definitions of development and sustainability, which is rarely the case in the popular literature.

The idea of sustainable development has had a particularly strong influence on the policies of the United Nations and United Nations Educational, Scientific and Cultural Organization (UNESCO), which gave rise to the vision of the Millennium Development Goals (MDGs). Unfortunately, they, too, do not take into account limits of growth, nor do they even acknowledge our overshoot. As noted above, without such precise specifications, any proposition or undertaking that is advocated under the banner of sustainable development remains less than ethical, lacking moral justification.

The concept of ethical development seems to be less encumbered by such difficulties. Because it includes by definition a moral justification, it can neither deviate from the goal of sustainability, nor can it rely on counterproductive notions of growth and technological potential. As argued above, the "ethical" in this term refers primarily to ecocentric virtue ethics and deontology. Development is understood in the sense of transition measures, paradigm changes, reexamination of values, structural reform, measures toward adaptation, and restraint, all aimed at the goal of sustainable living for humanity.

See Also: Anthropocentrism; Club of Rome (and Limits to Growth); Ecocentrism; Ecological Footprint; Environmental Justice; Future Generations; Intergenerational Justice; United Nations Millennium Development Goals.

Further Readings

Diamond, Jared. *Collapse: How Societies Choose to Fail or Succeed.* London: Viking Penguin, 2005.

Ehrlich, Paul R. and Anne H. Ehrlich. *One With Nineveh: Politics, Consumption and the Human Future.* Washington, DC: Island Press, 2004.

Lautensach, A. "The Ethical Basis for Sustainable Human Security: A Place for Anthropocentrism?" *Journal of Bioethical Inquiry,* 6/4 (2009).

Leopold, Aldo. *A Sand County Almanac.* New York: Ballantine, 1966 [1949].

Meadows, D., J. Randers, and D. Meadows. *Limits to Growth: The 30-Year Update.* White River Junction, VT: Chelsea Green Publishing, 2004.

Potter, Van Rensselaer. "A Response to Clements' Environmental Bioethics: A Call for Controlled Human Fertility in a Healthy Ecosystem." *Perspectives in Biology and Medicine,* 28/3 (1985).

United Nations Environment Programme (UNEP) and Millennium Assessment Board. *Living Beyond Our Means: Natural Assets and Human Well-Being.* London: UNEP-WCMC, 2005. http://www.millenniumassessment.org/documents/document.429.aspx.pdf (Accessed February 2009).

United Nations Millennium Project. *Investing in Development: A Practical Plan to Achieve the Millennium Development Goals.* New York: United Nations, 2005.

Wackernagel, Mathis and William Rees. *Our Ecological Footprint: Reducing Human Impact on the Earth.* Oxford, UK: John Carpenter, 1996.

Wenz, Peter. *Environmental Ethics Today.* Oxford, UK: Oxford University Press, 2008.

World Commission on Environment and Development. *Our Common Future: The Brundtland Report.* Oxford, UK: Oxford University Press, 1987.

Alexander K. Lautensach
University of Northern British Columbia

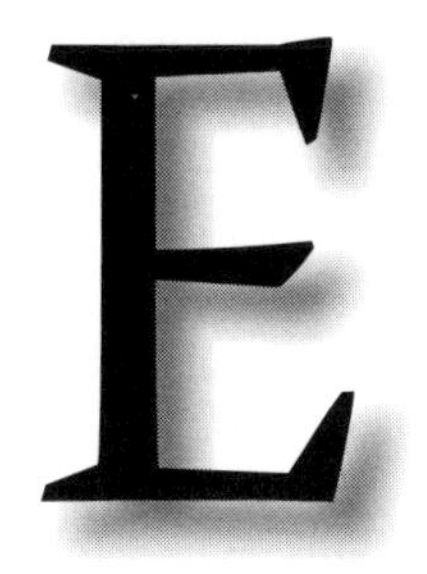

Earth Charter

The Earth Charter initiative is a particularly notable juncture both in environmental ethics and in the attempts to formulate a global ethic. It represents a significant step not only in the content of the final document, but also in its drafting process and the ongoing activity it has sparked. The development of the charter entailed nearly a decade and involved international consultation with tens of thousands of participants. The intent was the creation of a "people's treaty" that expressed a consensus of values and principles among diverse cultural, economic, religious, and scientific interests. Its content is comprehensive; it connects ecological integrity and sustainability to social and economic justice and the establishment of enduring peaceful relations. The charter and its supporters are not without their critics, however. And despite a decade of efforts since the launch of the Earth Charter, the initial aspiration of formal adoption by the United Nations General Assembly, similar to the Universal Declaration of Human Rights or the Declaration on the Rights of Indigenous Peoples, has not been fulfilled, although it has been endorsed by United Nations Educational, Scientific and Cultural Organization (UNESCO) and many other bodies. Nevertheless, not only does the charter provide a framework for ongoing dialogical efforts in crosscultural ecological ethics, the efforts for full adoption continue, and the charter has become the focus of numerous educational, political, and social projects at all levels, from grassroots and municipal bodies to international initiatives. More than a mere document, the Earth Charter is perhaps best understood as a movement.

Development of a "People's Charter"

The proposal for a new charter or universal declaration that would deal with ecological protection and sustainable development came from the work of the United Nations World Commission on Environment and Development (WCED), *Our Common Future,* also known as the Brundtland Commission report, in 1987. Although creation of such a charter was attempted at the 1992 Rio Earth Summit, an intergovernmental agreement could not be reached. The Rio Declaration, while containing important principles, did not represent the cohesive vision many, particularly those representing nongovernmental organizations (NGOs), had sought.

In 1994, Jim McNeill, secretary general of the WCED, along with Queen Beatrix and Prime Minister Ruud Lubbers of the Netherlands, brought together the secretary general of the Rio Earth Summit, Canadian Maurice Strong, with former Soviet Union president Mikhail Gorbachev to lead a new civil society initiative. These two leaders worked at first through organizations they each had founded (Earth Council and Green Cross International, respectively) with initial funding from the Netherlands. This strategy led to the initiative of creating a charter to a more varied set of participants, taking it out of the domain of official representatives of nations. It would be in the hands of interested organizations and individuals, and consultations were sought at all levels of society. The intent was to develop a "soft law," or set of guidelines that could be used at many levels of community and international deliberations, rather than an official, legally binding treaty. There was still an objective of international adoption, as the document would theoretically represent a broadly based consensus of the world's peoples.

In 1995, Ambassador Mohamed Sahnoun of Algeria became the first executive director of the project, which then embarked on several years of consultation that drew on hundreds of international documents and representatives from all parts of the world, including scientific and religious leaders, professionals, schoolchildren, and interested citizens, to determine what should be included in such a charter. An Earth Charter secretariat under the leadership of Maximo Kalaw of the Philippines was created at the Earth Council in Costa Rica. Mirian Vilela of Brazil took on the coordination of activities relating to the charter at the council in 1996.

Toward the end of that year the Earth Charter Commission was formed to oversee the drafting process. It consisted of 23 members from all over the world. Gorbachev and Strong served as cochairs of the commission, and Stephen C. Rockefeller, professor of religion and dean of Middlebury College in Vermont, was invited to chair the drafting committee. This committee first examined hundreds of intergovernmental documents and civil society declarations and statements of global ethics, including the Universal Declaration of Human Rights of 1948, the Stockholm Declaration of the Human Environment in 1972, the World Charter for Nature from 1982, *Caring for the Earth* from 1991, the declaration of the Parliament of World Religions of 1992, and the Draft International Covenant on Environment and Development. In March 1997, this committee produced the first benchmark draft, released at the Rio+5 Summit. It was sent all over the world for feedback and suggestions for revision.

Over the three-year time frame of the drafting process, 45 national committees were formed, and dialogues and regional conferences took place in all major regions of the world as well as online. The draft was revised and a second draft issued in 1999, after which further consultations took place. The final text was approved at a meeting of the Earth Charter Commission in Paris in March 2000, at UNESCO headquarters. It was formally launched at the Peace Palace in The Hague in June 2000.

The Earth Charter Document

The Earth Charter consists of a preamble followed by a statement of principles and subprinciples grouped into four areas, and a forward-looking conclusion. The preamble is intentionally evocative. Responses to the first draft indicated that the charter should be more than a statement of principles; it should be a visionary proclamation that spoke to the predicament of the present time and stirred the spirit of human aspiration with the power of poetic and even spiritual expression. The preamble became the primary vehicle for this, setting the principles in a context evocative of home, family, community, diversity, solidarity, and common destiny, emphasizing the interconnectedness of life. It sounds a note not only of alarm but of hope. The current predicament of this global community with its trends of depletion and devastation, of poverty, violence, and consequent suffering are identified as "perilous—but not inevitable." The evidence of the inclusion of sensibilities of religious groups is evident but not identifiable with any particular traditions. The ethical concept of responsibility, for instance, is intensified by the naming of a "reverence for the mystery of being" and the "gift of life."

The statement of principles that follows is divided into four sections as follows:

I. Respect and Care for the Community of Life

1. Respect Earth and life in all its diversity.
2. Care for the community of life with understanding, compassion, and love.
3. Build democratic societies that are just, participatory, sustainable, and peaceful.
4. Secure Earth's bounty and beauty for present and future generations.

II. Ecological Integrity

5. Protect and restore the integrity of Earth's ecological systems, with special concern for biological diversity and the natural processes that sustain life.
6. Prevent harm as the best method of environmental protection and, when knowledge is limited, apply a precautionary approach.
7. Adopt patterns of production, consumption, and reproduction that safeguard Earth's regenerative capacities, human rights, and community well-being.
8. Advance the study of ecological sustainability and promote the open exchange and wide application of the knowledge acquired.

III. Social and Economic Justice

9. Eradicate poverty as an ethical, social, and environmental imperative.
10. Ensure that economic activities and institutions at all levels promote human development in an equitable and sustainable manner.
11. Affirm gender equality and equity as prerequisites to sustainable development and ensure universal access to education, health care, and economic opportunity.
12. Uphold the right of all, without discrimination, to a natural and social environment supportive of human dignity, bodily health, and spiritual well-being, with special attention to the rights of indigenous peoples and minorities.

IV. Democracy, Nonviolence, and Peace

13. Strengthen democratic institutions at all levels, and provide transparency and accountability in governance, inclusive participation in decision making, and access to justice.

14. Integrate into formal education and life-long learning the knowledge, values, and skills needed for a sustainable way of life.

15. Treat all living beings with respect and consideration.

16. Promote a culture of tolerance, nonviolence, and peace.

What is immediate evident from this list of the main principles is that these four categories are not distinct and separate from one another. Under the first category, "Respect and Care for the Community of Life," for instance, principle I.3 is "Build democratic societies that are just, participatory, and peaceful," a statement fleshed out as a category itself (IV) further in the document. Similarly, subprinciple I.3.b is "Promote social and economic justice, enabling all to achieve a secure and meaningful livelihood that is ecologically responsible," expanded later as well under category III. This explicit recognition that the ecological sustainability and integrity of the rest of life on the planet are deeply interconnected with the realities of human well-being in terms of social justice, economic equity and opportunity, political voice and peace (and vice versa) is one of the great achievements of the document as a work of ethics.

It was the frustration of many, particularly voiced by nongovernmental organizations (NGOs) at the Rio Summit, that the existing discourse of "sustainable development" was too narrowly focused on adjustment of an existing economic system to limits of resources and waste, rather than comprehensively understanding this deep interconnectedness of justice and the flourishing of life, that prompted the movement to a civil initiative in the first place. So in that respect, the Earth Charter has been a remarkable success.

This is not to say that the value of other-than-human life is understood only in these terms. Up front, in subprinciple I.1.a., there is an explicit affirmation of the value of every form of life independent of its worth to humankind. Yet evidence of a bias toward instrumentality (where the biosphere is valuable only insofar as it is useful for humankind) in the document has been identified by critics. It is a debatable question whether the initial statement is sufficient and whether it indeed informs the vision as a whole as expressed in the rest of the principles and subprinciples.

Its recognition of interconnectedness of ecological, social, economic, and political well-being is not the charter's only strength as a work of ethics. It also expands international discourse about rights. It incorporates rights both implicit and inconceivable in the 1948 Universal Declaration and reinforces those in the Declaration on the Rights of Indigenous Peoples. Subprinciple III.9.a. (under Social and Economic Justice, Eradicate poverty) not only names as objects of specific universal rights potable water, clean air, food security, uncontaminated soil, shelter, and sanitation, but calls for a guarantee of equitable distribution of such necessities of life among as well as within nations. More striking, perhaps, is the inclusion of a right (IV.13.a.) of people to clear and timely information on environmental and development matters that would affect them. And though rights to own, use, and manage natural resources are affirmed (I.2.a.), these rights and the possession of freedom, knowledge, and power (I.2.b.) are tempered with responsibilities to promote the public good and prevent environmental harm.

There is no mention, however, of rights with regard to other-than-human beings, particularly animals. Given that the goal of the drafting process was to create a document with values as widely shared as possible, this is perfectly understandable. The concept of animal rights is not widely shared, even within the cultures that gave rise to it, much less around the globe. Indeed, how to phrase regard and care for animals became one of the issues of contention during the consultation process.

Finn Lange, representing the Greenland government and Arctic hunting cultures at a consultation in Assisi, Italy, objected to draft language that called for "respect and compassion" in the treatment of animals. Respect was not a problem, he said; their hunters respected their prey, but compassion seemed to imply feeling sorry for animals that were killed for food. That did not respect their culture. This led to an hour-long discussion, introduced by Bawa Jain of India, of the ideas of "compassion" in the Jain and Buddhist religions. While productive for understanding among the participants, this discussion did not persuade Lange that such wording would be acceptable to Arctic peoples. An alternative word, *reverence,* was proposed by a Christian, (Franciscan) Charlie Spencer. It was openly and respectfully discussed, but ultimately rejected in favor of "consideration" (IV.15).

This example is one illustration of the difficulty of creating a single, comprehensive global ethic. Values and principles generated under some conditions—culture being intricately connected with the physical realities of climate, topography, and biotic populations—may not be appropriate in all the diverse biozones of Earth. This example is particularly stark in that regard, as the Asian concept of compassion in relation to animals comes out of a lush landscape with many alternatives for human sustenance, whereas local resources for human survival in the Arctic region are limited.

The cultures of this region are varied, but no doubt each has an understanding of proper relations between humans and other animals that fostered sustainable communities until the social and environmental impacts of industrial societies intervened. The concern for the welfare of animals in a culture involved in industrial factory farming is not necessarily transferable to a hunting culture. Predation, after all, is an essential natural process. Keeping in mind the overall objective of sustainable communities, the participants were able to reach an agreement satisfactory to the indigenous peoples of the Arctic, who endorsed it at an Inuit Circumpolar Conference.

This example, however, raises the question put to all attempts at a global ethic: is it possible to find agreement comprehensive enough to be truly representative, and at the same time potent enough to be meaningful? This example by itself might indicate that the process of reaching agreement always entails a compromise that softens the intent or narrows the scope, but there is at least one example where the opposite occurred. In the first draft's discussion of "promoting a culture of peace" the emphasis was on elimination of weapons and conversion of military resources to peaceful purposes. Hindu representatives found this too limited. They suggested that there needed to be more emphasis on conflict prevention, and so language about comprehensive strategies to prevent violent conflict and the use of collaborative problem solving to manage environmental and other conflicts was included.

A different reference to the military points up yet another issue: whether the aspirations embodied in such a document are achievable in reality. Subprinciple II.6.c states, "Avoid military activities damaging to the environment." This begs the question of whether it is possible to maintain any military activity at all on that basis. No matter how desirable, military activities are unlikely to be forsaken or transformed to such an extent that they are not "damaging to the environment."

One further issue is how representative the document actually is. There is no doubt that the intent was to be as inclusive as possible, and that the consultation process was unprecedented in that regard. Still, it had a beginning and it is possible that much of the terms of the later consultations were effectively, if not intentionally, determined to some degree by the worldviews of those who were responsible for the first draft and organization of the project. Questions about who was invited and who actually came to the table could be raised. Those who responded to such an invitation may have been those who were already amenable to the kind of cosmopolitan approach to ethics that the Earth Charter (and its predecessors) represents. This may be true, but it is also the case that the consultations provided an opportunity for participants to see a larger vision than they may have had before. Those motivated to discuss an "earth charter" out of concern for environmental degradation could learn about the links to issues of poverty. Some already motivated by those two issues but not as initially receptive to statements about gender equality or democracy could be persuaded at least to accept them as part of an overall comprehensive vision. The insistence within the charter on democratic and participatory processes in governance and decision making should make evident, however, how misdirected are any allegations by fringe elements that the charter is part of a conspiracy to enforce a totalitarian or radical agenda, or a new religion, although such ridiculous claims abound on the Internet.

As a work of global ethics, the Earth Charter is global in a sense that it deals with concerns of global impact and unites issues in a global perspective, and it is a singular achievement on that account. No ethic is likely to be global in the sense of representing universal consensus, no matter how extensive the consultations in its formation. What is not yet clear is whether the Earth Charter is or with effort and time will become a "global ethic" at least in the terms to which its supporters have aspired, that of a widely shared one. The measurement of the worth or overall success of this enterprise may not be in the end an all-or-nothing matter. Incremental success can be significant in various ways. Its supporters generally describe the charter as a "living document" rather than one chiseled in stone, one whose principles not only continue to fund further conversation but already inspire and guide actions in a wide range of endeavors.

The Continuing Earth Charter Initiative

It is far beyond the scope of this article to document in any complete and comprehensive way the organizations that have adopted or otherwise indicated their support of the charter since its official launch or the extensive activities connected to the Earth Charter. The following is meant to be suggestive only, by showing examples of the geographic spread and broad range of fields of endeavor that have found the charter meaningful and helpful, and the creativity that it has stimulated.

Areas of activity include formal endorsements by governments at many levels, NGOs, religious groups, educational institutions, and other institutions and associations, as well as by individuals. There are myriad examples of educational activities, activities undertaken by religious groups, and productions by artists inspired by the charter. Business and professional groups have been incorporating principles from the charter into guidelines and standards. Youth groups are particularly active.

Over 5,000 organizations are listed on the Earth Charter Initiative Website as having endorsed the charter. These include over 450 businesses and close to 800 schools, universities, and other educational organizations. The faith communities that have publicly endorsed the charter range all over the globe and religious traditions. The hundreds listed include the national bodies of the Anglican Church in Southern Africa and New Zealand;

pages of congregations of Roman Catholic Sisters, Friends' meetings around the world, Baha'i groups and Soka Gakkai communities; a Hindu temple in Spain, a Christian Institute in Taiwan, the Muslim Community of the Quad Cities in the United States, the United Sikhs of India, and many others. The diversity indicates a large degree of success in the efforts during the framing process of listening to the concerns of religious communities, of incorporating spiritual insights and values, without privileging one group over others. Religious communities all over the world evidently have been able see their values expressed in the Earth Charter. Government endorsements have come from central governments and their ministries in Bolivia, Brazil, Costa Rica, Egypt, Ecuador, Honduras, Mexico, and the Republic of Niger. Numerous municipal governments—Jordanian cities and Spanish cities, and towns in Vermont—have endorsed it as well, as have over 2,000 NGOs.

The Earth Charter Initiative and the Dutch National Committee for International Cooperation and Sustainable Development (NCDO) have cowritten a report to guide businesses that want to become more environmentally and socially responsible, relating the charter to other leading global initiatives such as the UN Global Compact and the Global Reporting Initiative (GRI). The staff of the Arlington Hilton in Washington, D.C., are implementing several initiatives to put the Earth Charter into action in their place of work. Japan Classic Live for the United Nations group performed a charter-inspired musical, *Our Blue Planet,* in New York and Washington. Little Animation, Inc., of Montreal, Canada, has created a children's version of the Earth Charter, condensing it to eight principles: Life, Interconnected, Family, Past, Earth, Peace, Love, and Future. Oscar-nominated director Alexei Kharitidi, the Little Animation team, and coproducers Rosie Emery and Studio Kinetika have created a musical video for children in several languages. Accompanying classroom activities are on their Website. One of the eight episodes, "One is Life," was posted on YouTube in September 2009. These are only glimpses into business initiatives, artists' activities, and educational resources being developed around the Earth Charter.

Youth has been one of the primary foci of the Earth Charter Initiative. From tiny Sierra Leone, where there are 10 Earth Charter Youth Groups using the Earth Charter as a guide to ways in which they can create a culture of peace in their troubled country, to the very active network of youth in Germany, young people all over the world are learning about the charter and considering their own responsibility in creating a sustainable, flourishing world.

The United Nations General Assembly has not adopted the Earth Charter as of this writing. It may one day happen. Perhaps by then it will not be the spur to change envisioned by its initiators, but rather a recognition of the change brought about through the efforts of a persistent movement. For there is a vibrant community within small villages and across oceans who have taken the Earth Charter and turned it into song, poetry, imagery, policy, into hope.

See Also: Brundtland Report; Precautionary Principle; Religious Ethics and the Environment; Sustainability and Distributive Justice; United Nations Environment Programme.

Further Readings

Corcoran, Peter Blaze, Mirian Vilela, and Alide Roerink, eds. *The Earth Charter in Action: Toward a Sustainable World.* Amsterdam: KIT Publishers, 2005.

Earth Charter. http://www.earthcharterinaction.org (Accessed January 2010).

Habel, Norman. "The Inaugural Earth Charter Forum." *Ecotheology: Journal of Religion, Nature & the Environment,* 5/7 (1999).
Little Animation, Inc. "The Little Earth Charter." http://www.littleanimation4kids.com/LEC_home.html (Accessed January 2010).
Lynn, William S. and Ronald Engel, eds. Special edition. *Worldviews: Environment Culture Religion,* 8/1 (2004).
MacGregor, Sherilyn. "Reading the Earth Charter: Cosmopolitan Environmental Citizenship or Light Green Politics as Usual?" *Ethics, Place & Environment,* 7/1–2 (2004).
Roberts, Jan. "Earth Charter." *Yes!* Magazine (Winter 2000–01). http://www.yesmagazine.org/issues/a-new-culture-emerges/401 (Accessed January 2010).
Shiva, Vandana. *Earth Democracy: Justice, Sustainability, and Peace,* Cambridge, MA: South End Press, 2005.
Tucker, Mary Evelyn. "World Religions, the Earth Charter, and Sustainability." *Worldviews: Environment Culture Religion,* 12/2–3 (2008).
Vilela, Mirian. "The Earth Charter Endeavour: Building More Just and Sustainable Societies Through a New Level of Consciousness." *Social Alternatives,* 26/3 (2007).

Nancie Erhard
Saint Mary's University, Halifax

Earth Day 1970

This crowd of about 20,000 gathered near Independence Hall in Philadelphia on April 21, 1970, the day before the first Earth Day, to listen to a performance by the cast of the Broadway musical *Hair* and hear speeches by Ralph Nader and U.S. Senate Minority Leader Hugh Scott.

Source: Earth Week Committee of Philadelphia

Senator Gaylord Nelson (1916–2005), Democrat from Wisconsin, announced a day of environmental protest and awareness raising to be held on April 22, 1970. Earth Day is often referred to as the start of the modern environmental movement, though initially it was a purely American event. It is widely stated that 20 million people (a figure for which there is no reliable evidence) participated in a range of activities across the nation, and that 2,000 universities and colleges and 10,000 schools were involved. Celebrities included folk singer and activist Pete Seeger and actors Paul Newman and Ali McGraw. Earth Day became international in 1990, with an estimated 200 million participants, and today involves up to a billion people, according to some sources, with almost all countries represented. Earth Day may thus be the largest secular gathering in the world. The United Nations (UN) celebrates Earth Day on a date between March 20 and 22, the spring equinox in the Northern Hemisphere, and many countries also participate at this time. The UN observance includes a symbolic ringing of the UN Peace Bell (a gift to the United Nations from Japan in 1954) at the exact time of the equinox. The focus of Earth Day has shifted from protest to year-round programs including the promotion of environmentally responsible behavior such as recycling, as well as environmental education in schools and in the wider community.

Earth Day was originally called the National Environment Teach-In, this being a 1960s term for a mass gathering designed to raise awareness of politically controversial issues, with an emphasis on grassroots participation leading to antiestablishment action. The first teach-ins, beginning in 1965, were organized by radical groups—notably, the University of Wisconsin–based Students for a Democratic Society (SDS), and the University of California–based Free Speech Movement and Vietnam Day Committee—as a means of galvanizing mass opposition to the Vietnam War. However, organizations such as SDS had political and social concerns beyond ending the Vietnam War and shared a strongly left-wing agenda, though they were prone to fragment over ideological issues; their teach-ins were sometimes a prelude to student occupation of university administrative buildings. Earth Day was also intended as a focus for protest, but it was not designed to be divisive or confrontational. Rather, it was intended as an opportunity for individuals and groups that had many different environmental agendas such as air and water pollution, pesticides, endangered species, and the loss of open space and wilderness to recognize that they shared a common concern for the planet and therefore had a common agenda in working for ecological protection—*ecology* being, at the time, an unfamiliar term. Mainstream politicians supported Earth Day: Congress adjourned in order to allow members to address the nation, while New York Mayor John Lindsay closed much of Fifth Avenue for several hours for Earth Day activities, including lectures, concerts, and street theater, in which tens of thousands of people participated.

It is easy to exaggerate the significance of the first Earth Day. For example, the Earth Day Network Website states that it led to the creation of the U.S. Environmental Protection Agency and the passage of the Clean Air, Clean Water, and Endangered Species Acts. This is overly simplistic: the environmental awareness that made Earth Day such a spectacular success can be traced to earlier seminal works such as Rachel Carson's *Silent Spring*. Moreover, the passage of environmentalist legislation was not built merely on the enthusiasm of millions of people but on the regular political activities that determine the outcome of any legislative proposals. Environmental organizations such as the Audubon Society and the Sierra Club lobbied for the passage of conservation and environmental cleanup legislation. Business, especially the automobile industry, lobbied intensively against the Clean Air Act and opponents of the Endangered Species Act such as utility companies and rancher associations argued that it would infringe on private property rights and unduly restrict economic growth.

Many of the claims made by contributors to Earth Day seem extravagantly apocalyptic from the perspective of 2010. *The Environmental Handbook,* edited by Garrett De Bell, was a collection of articles compiled as an educational resource for Earth Day. Authors included biologists Barry Commoner and Paul Ehrlich. The main concern expressed was global famine due to overpopulation, but there were also warnings of toxic air pollution in U.S. cities, the extinction of 75 percent or more species globally, pervasive water pollution leading to the total collapse of all fisheries globally, the exhaustion of all nonrenewable resources, and even a new ice age. Some of these disasters were predicted to occur as soon as 1975, and certainly by 2000; at the time of writing, none of them has come to pass. Climate change skeptics use these mistaken predictions as ammunition against claims of anthropogenic climate change, arguing that the fact that so many scientists were wrong in 1970 is a reason for distrusting the equally calamitous predictions of today. However, it can be argued that environmental science, which was in its infancy in 1970, is now much more advanced.

Nevertheless, Earth Day 1970 is widely regarded as the most significant single event in raising environmental awareness in the United States and as bringing together a wide variety of people who were concerned about specific issues to work together in a common cause. It forced politicians to recognize that millions of Americans were deeply concerned about environmental health and quality. Perhaps the most extraordinary feature of Earth Day is that it was the largest mass gathering in U.S. history—perhaps 10 percent of the U.S. population was involved—and yet, other than a small number of activities, it was not in any way centrally organized. Nelson and his associates had a very small budget and almost all of the activities were organized by volunteers. Moreover, unlike much of the mass protest of the Vietnam era (and some recent antiglobalization activities), Earth Day was entirely peaceful, despite the huge numbers involved. These were certainly factors in influencing the federal government to pass the legislation named above, along with the Resource Recovery Act, the Resource Conservation and Recovery Act, the Toxic Substances Control Act, the Occupational Safety and Health Act, the Federal Environmental Pesticide Control Act, the Federal Land Policy and Management Act, and the Surface Mining Control and Reclamation Act—all passed during the 1970s, within a decade of the first Earth Day.

See Also: Carson, Rachel; Civic Environmentalism; Endangered Species Act.

Further Readings

Bailey, Ronald. "Earth Day, Then and Now." *Reason* (May 2000). http://reason.com/archives/2000/05/01/earth-day-then-and-now (Accessed February 2010).
Carson, Rachel. *Silent Spring*. Boston: Houghton Mifflin, 1962.
De Bell, Garrett. *The Environmental Handbook*. New York: Ballantine, 1970.
EarthDay Network. http://www.earthday.net (Accessed February 2010).
Lewis, Jack. "The Spirit of the First Earth Day." *EPA Journal* (January–February 1990). http://www.epa.gov/history/topics/earthday/01.htm (Accessed February 2010).
Nelson, Gaylord, with Susan Campbell and Paul Wozniak. *Beyond Earth Day: Fulfilling the Promise*. Madison: University of Wisconsin Press, 2002.

Alastair S. Gunn
University of Waikato

Earthlife Africa

Earthlife Africa (ELA) is an environmental organization founded in 1988 in Johannesburg, South Africa. It was an important actor in the development of the South African environmental movement, notable especially for its anti-apartheid position that helped to define the concept of environmental justice. Today, the organization continues its work on fair access to electricity; the dangers of nuclear energy, toxics, and acid mine drainage; and issues surrounding climate change. It has expanded to include branches in Cape Town, eThekwini (formerly Durban), Tshwane (formerly Pretoria), and Windhoek, Namibia. Each of its branches is an autonomous member-driven volunteer organization that conducts research and publicity campaigns.

Many of the environmental organizations that existed before Earthlife Africa were focused on wildlife protection and worked within the framework of apartheid. Modeled after the activist U.S. organization Greenpeace, Earthlife instead rejected apartheid and sought to address environmental issues in relation to people. Along with the Koeberg Alert (a peace-promoting, antinuclear group) and the Cape Town Ecology Group (which reappropriated the Dolphin Action and Protection Group motto "Dolphins should be free" to fashion its own motto, "People should be free"), Earthlife pursued an agenda that framed ecology in terms of social justice. Its stated goal is to achieve "a better life for all people without exploiting other people or degrading the environment." The prominent South African Environmental Justice Networking Forum (EJNF), an alliance of community-based nonprofit organizations that promotes values of ecological sustainability, nonracism, nonsexism, and democratic governance, was incubated at Earthlife Africa's 1992 conference titled What Does It Mean to Be Green in South Africa? ELA and EJNF have worked together since then to stand up for marginalized groups facing environmental and ecological stresses.

One of the first campaigns on which the two organizations collaborated was the 1992 exposure of pollution by Thor Chemicals, which had been shipping hazardous waste from the United Kingdom, the United States, Brazil, and Italy to its reclamation plant in Cato Ridge, South Africa. Though it was supposed to treat the waste there, it stored mercury and other heavy metals in warehouses instead, where they leaked into the soil and groundwater and sickened several employees. ELA and EJNF worked with the Chemical Workers Industrial Union and the Legal Resources Center as they put pressure on the South African government to take action. As a result of their initiatives, Thor Chemicals later compensated the affected workers. Additional Earthlife Africa toxics campaigns have focused on preventing the siting of new hazardous waste plants, in which the organization has shown the ability to engage a number of stakeholders. With its coalition of unions, political parties, nongovernmental organizations (NGOs), and government supporters, ELA successfully halted plans for a toxic waste dump in Chloorkop in 1996. It has also stopped the construction of numerous incinerators with the help of its nongovernmental partners.

Earthlife Africa's other activities centered on pollution have raised awareness about the environmental effects of mining and other heavy industries. The mining of coal, gold, diamonds, platinum and palladium, and chromium have made South Africa the wealthiest nation on the continent, but it has had serious ecological consequences as uncontained acid seepage from mining sites has contaminated river and lake systems. ELA has pointed to these problems as well as to the problem of air pollution generated by the country's other industries in order to support its advocacy for a more diverse South African economy. ELA's 1998 protest against air pollution earned it a reputation for creative media tactics when it placed gas masks on three prominent Johannesburg statues.

True to its philosophy of peace and nonviolence, ELA remains firmly opposed to the use of nuclear power for South Africa's electricity needs. Experience with poor pollution control at waste treatment plants, incinerators, and mines has led the organization to be skeptical about the probability of safe storage for nuclear waste in South Africa. Concerns over high levels of radiation emitted by nuclear plants also inform ELA's position, as does the memory of the 1986 disaster at the Chernobyl nuclear power plant in Ukraine, which occurred just two years before the organization was founded. For these reasons, Earthlife Africa has fought against the South African utility Eskom's proposed plans for a Pebble Bed Modular Reactor (PBMR) nuclear plant.

The organization has been critical of plans for the development of the country's energy sector in general, citing heavy reliance on nuclear and coal power, lack of investment in alternative energy projects, significant increases in electricity tariffs, and uneven distribution of energy to the poor as major failings of the utility, government, and international funders. Coal, South Africa's most abundant fossil fuel resource, supplies 95 percent of the country's energy and makes it the largest carbon dioxide (CO_2) emitter on the African continent. Because 30 percent of the population lacks access to reliable basic electricity, meeting these energy needs is a high priority for development. ELA has contested the World Bank's proposed $3.75 million loan to Eskom for an additional coal-fired power plant, advocating instead a shift to renewable energy.

With its research and media campaigns and its connections to several other nonprofit groups in South Africa, Earthlife Africa and its several branches will continue to be a force in the country's growing environmental movement and an important part of the movement's history.

See Also: Climate Ethics; Development, Ethical Sustainability and; Ecological Footprint; Environmental Justice.

Further Readings

Cock, Jacklyn. *Connecting the Red, Brown, and Green: The Environmental Justice Movement in South Africa.* Durban, South Africa: University of KwaZulu-Natal, 2004.

Earthlife Africa. http://www.earthlife.org.za/index.php (Accessed August 2010).

McDonald, David A., ed. *Environmental Justice in South Africa.* Athens: Ohio University Press, 2002.

Sophie Turrell
Independent Scholar

Earth Summit

The term *Earth Summit* refers to the 1992 Rio de Janeiro Summit; however, when used in plural, it describes the series of summits: Stockholm 1972, Rio de Janeiro 1992, and Johannesburg 2002, which focused on human environment and sustainability.

Rio de Janeiro, 1992: United Nations Conference on Environment and Development (UNCED)

This two-week event is also known as the Rio Summit, Rio Conference, Earth Summit, and, in Portuguese, Eco '92. The summit was convened by the United Nations and was attended by representatives of 172 countries, including 108 heads of state. Over 2,400 nongovernmental organization (NGO) representatives and 17,000 individuals took part in the parallel Global NGO Forum, which had consultative status. The aim of the conference was to reconcile the mutual impact of human socioeconomic activities and the environment. A large number of sustainability-related topics were discussed including production patterns and residual waste, alternative energy sources, carbon dioxide (CO_2) emissions reduction, and the growing scarcity of water.

The conference produced the Framework Convention on Climate Change, which was the basis of the 1997 Kyoto Protocol on carbon emissions. The Convention on Biological Diversity, a legally binding document that promotes the conservation of biological diversity and the sustainable use of its components, was adopted and opened for signature at the Earth Summit. As of this writing, 168 countries have signed and ratified this convention. The summit also adopted 27 principles to help guide environmental and economic responsibility under the name of the Rio Declaration for Environment and Development.

The blueprint of the Agenda 21 program was presented at the summit at the end of four years of political and scientific consultations. Agenda 21 is a comprehensive action framework to be implemented at international, national, and local levels and addresses the interaction between humans and the environment. It is divided into four main sections dealing with social and economic dimensions, conservation and management of resources for development, strengthening the role of major groups, and mean of implementation. Agenda 21 is implemented by the countries on a voluntary basis; however, the United Nations (UN) created the Commission on Sustainable Development within the UN Economic and Social Council, which monitors and supports its implementation. In 1997, the UN General Assembly held a special session celebrating five years of Agenda 21 and evaluated its impact. The conclusion was that implementation remains "uneven."

At the Rio Summit, 12 cities received the Local Government Honors Award for innovative local environmental programs. Among them were Sadbury, Canada, for implementing a program to rehabilitate the environment affected by local mining; Austin, Texas, for green building; and Kitakyushu, Japan, for education and training in pollution control. The Rio Earth Summit was preceded by the Stockholm 1972 Summit and followed by the Johannesburg 2002 summit.

Stockholm 1972: United Nations Conference on Human Environment

This first major conference on environmental issues marked a radical change in international environmental politics. It is credited with the development of the basic principles of international environmental law. The conference was convened by the UN General Assembly and was attended by representatives of 113 countries, 19 intergovernmental agencies, and over 400 NGOs. A variety of environmental issues were discussed, ranging from chlorofluorocarbons (CFCs) and their impact on the ozone layer to global warming.

After three days of debates, the conference issued the Declaration on the Human Environment, known as the Stockholm Declaration, containing 26 principles and 109 recommendations. It attempted to emphasize the importance of protecting both species and habitats. This declaration was the first international document to officially recognize the right to live in a healthy environment. The conference also approved the creation of the United Nations Environment Programme (UNEP), tasked with assisting developing countries in implementing sound environmental policies and sustainable development. Its mandate has since been expanded to coordinate the development of environmental policy consensus. The conference inspired several later documents such as the Brundtland Report, published in 1987, which provided the first definition of sustainable development.

Johannesburg 2002: World Summit on Sustainable Development

The Johannesburg Summit was convened by the United Nations in order to address sustainable development and to evaluate progress on Agenda 21. Official participation was more restricted than at previous summits; however, the event enjoyed wide popular participation from nongovernmental groups and industry. The United States did not send an official delegation to the meeting; Secretary of State Colin Powell attended briefly and read a short statement to the delegates.

The summit concluded with the Johannesburg Declaration that, although more general than the Rio Declaration, commits the nations to sustainable development. The declaration contains a comprehensive list of barriers to sustainable development that need to be addressed. During the summit, over 300 partnership initiatives were established and centralized by the UN Department for Economic and Social Affairs. The initiatives database can be accessed on the Internet, facilitating ongoing cooperation and action.

Civil society groups are requesting an Earth Summit to be held in 2012.

See Also: Biodiversity; Brundtland Report; Development, Ethical Sustainability and; Environmental Law; United Nations Environment Programme; United Nations Millennium Development Goals.

Further Readings

Agenda 21. http://www.un.org/esa/dsd/agenda21 (Accessed February 2010).

Baylis, John and Steve Smith. *The Globalization of World Politics*, 3rd ed. Oxford, UK: Oxford University Press, 2005.

Commission on Sustainable Development. http://www.un.org/esa/dsd/csd/csd_csd16.shtml (Accessed February 2010).

Earth Summit. http://www.un.org/geninfo/bp/enviro.html (Accessed February 2010).

Earth Summit 2012. http://www.earthsummit2012.org/index.php?id=708 (Accessed February 2010).

Stakeholders Forum. http://www.stakeholderforum.org (Accessed February 2010).

United Nations. "Earth Summit." http://www.un.org/esa/earthsummit (Accessed January 2010).

United Nations. "Report of the United Nations Conference on the Human Environment." http://www.unep.org/Documents.Multilingual/Default.asp?documentID=97 (Accessed January 2010).

United Nations Department for Economic and Social Affairs. http://www.un.org/esa/sustdev/documents/WSSD_POI_PD/English/POI_PD.htm (Accessed February 2010).

United Nations Environment Programme (UNEP). http://www.unep.org/Documents.Multilingual/Default.asp?DocumentID=43 (Accessed February 2010).

United Nations Framework Convention on Climate Change (UNFCCC). http://unfccc.int/kyoto_protocol/items/2830.php (Accessed January 2010).

United Nations General Assembly Resolution S-19/2.

Daniel Tomozeiu
University of Westminste

Ecocentrism

Within environmental philosophy, ethics, politics, and activism, a general perception pervades that human beings are the direct cause of the ecological crisis that our planet faces. Ecocentrism underscores the notion that the worldviews that humans hold profoundly contribute to their misuse and abuse of the natural world. One of the most important belief systems or worldviews supporting forms of environmentalism that are dedicated to taking full account of human contributions to ecological degradation, ecocentrism reconceives nature and the planet Earth as primary and humankind secondary. Ecocentrism offers a strong critique of any worldview modeled upon a view of the nonhuman natural world as having less intrinsic value because it is an object to be controlled and whose purpose resides outside itself. Ecocentrism is thus one current of environmentalism whose justifications are both ontological, denying the existential divisions upon which human beings' primacy is often asserted, and ethical, refusing to assign greater intrinsic value to human beings, and instead promoting egalitarianism.

Ecocentrism has been described by Stan Rowe as a fundamental shift in values from human beings to planet Earth. With this shift, the preeminent metaphor of life, the organism, is replaced with the Earth. The ecosphere, understood holistically, is seen as a Being whose importance transcends that of any one species, including humans. Environmentalists who subscribe to ecocentrism maintain that, until the ecosphere itself is recognized as the common ground of all human activities, human beings will continue to put their interests first, thereby undermining more healthy and sustainable ways of living. As an environmental ethos, ecocentrism holds that human civilization and urbanization have obscured our intimate connection with the natural world and its ecological purposes.

Despite a number of hotly contested views concerning the origins of the ecocentric perspective, it is accurate to state that the majority of early hunter-gatherer cultures were rooted in nature-centered belief systems that involved a sacred sense of the Earth and its inhabitants. The traditional Native American philosophy of the sacred "Circle of Life" is one example of an ecocentric perspective depended upon to order human activity and determine values. Ecocentrism was imperiled by the rise of agricultural social orders whose aims included humanizing and domesticating the wildness of nature. Some writers have also argued that the rise of Western religions helped to obscure interdependence with nature, promoting forms of thought that see nature as raw material to be controlled and shaped by human will. This instrumental view of nature was challenged by a number of thinkers who became disillusioned with the prevailing economic and technological domination of nature. Nineteenth-century philosophers, artists, and naturalists such as Henry David Thoreau, George Santayana, Walt Whitman, and John Muir exemplified an ecocentric view. Ecocentrism in the 20th century was supported by writers and ecologists such as D. H. Lawrence, Robinson Jeffers, and Aldo Leopold. Leopold influenced those who contributed to the transition from conservation to ecology in the 1960s and 1970s—ecological writers and thinkers such as Aldous Huxley, Loren Eiseley, Rachel Carson, Lynn White, Jr., Paul Shepard, Edward Abbey, and Gary Snyder. Also tracing their philosophical origins to Leopold, the deep ecology movement, as advanced by Arne Naess and George Sessions in the early 1980s, developed a set of principles advocating ecocentrism.

In terms of its ethical positions, ecocentrism is often compared and contrasted with other "isms," the most famous of which is anthropocentrism, or a human-centered view. The chief problem with anthropocentrism from an ecocentric point of view is that the protection and enhancement of nonhuman nature is fully contingent upon the demands of utility and human values. Ecocentrism, by contrast, develops a noncontingent basis for the protection of nonhuman nature since humans are held to hold intrinsic value just as animals, rivers, lakes, and mountains do. Ecocentrism thus possesses a radically egalitarian ethics. Technocentrism is also contrasted with ecocentrism because the values of human mastery of the nonhuman world are inconsistent with the respect and humility toward nature that are reflected in an ecocentric view. Biocentrism, a view that takes the Earth's organisms to be of central importance, is often compared to ecocentrism, which is the more inclusive approach because it recognizes the deep interdependence and interactivity of living and nonliving entities.

Because ecocentrism involves the emancipation of people and nature, the two litmus issues for ecocentric activism and policy are human population reduction and wilderness preservation for its own sake rather than for human interests. Other action principles of ecocentrism include the reduction of human consumption of the ecosphere and the promotion of ecocentric governance. The legal implications of an ecocentric paradigm are potentially profound since current environmental protection law is largely implemented to protect humans, not the environment. An example of the anthropocentric orientation of environmental law is the restriction of pollution, which is a law of protection from pollution designed for humans only. Law derived from the values of ecocentrism instantiates natural equilibrium as a fundamental and common legal value and its maintenance as a human right.

Certain forms of ecological thinking have been critical of ecocentrism, especially as practiced within the scope of its strong anti-anthropocentrism. Ecofeminists, for example, have argued that it is not necessary to eliminate anthropocentrism but instead to strip it of its gender-biased articulation. Other criticisms have warned that the radical egalitarianism of ecocentrism is perilously close to nihilism and can possibly be used as the platform for oppressive ideologies, such as sanctioning measures aimed at quickly reducing the human population to ecologically sound levels.

See Also: Anthropocentrism, Biocentrism; Bright Green Environmentalism; Carson, Rachel; Deep Ecology; Leopold, Aldo; Naess, Arne.

Further Readings

Brown, Charles. "Anthropocentrism and Ecocentrism: The Quest for a New Worldview." *The Midwest Quarterly*, 36/2 (1995).

Eckersley, Robyn. *Environmentalism and Political Theory: Toward an Ecocentric Approach.* Albany: State University of New York Press, 1992.

Mosquin, Ted and J. Stan Rowe. "A Manifesto for Earth." *Biodiversity*, 5/1 (2004). http://www.ecospherics.net/pages/EarthManifesto.pdf (Accessed April 2010).

Rowe, J. Stan. "Ecocentrism: The Chord That Harmonizes Humans and Earth." *The Trumpeter*, 11/2 (1994).

Rowe, J. Stan. *Home Place: Essays on Ecology*, 2nd ed. Edmonton, Canada: NeWest Press, 2002.

Sessions, George. "Ecocentrism and the Anthropocentric Detour." *ReVision*, 13/3 (1991).

Michael Uebel
University of Texas at Austin

Ecofascism

To understand ecofascism, it is important to know what is meant by "fascism" and "eco-," and to gain a cursory understanding of ecophilosophical debates that lead to the development and continued use of the term.

Fascism is a political philosophy/political studies term that defines an ideological form of governance. It was first used to describe Mussolini's Italian government in the 1930s and 1940s; Hitler's government in Germany and Franco's in Spain are also early exemplars of fascist forms of governance. In general, fascism is associated with extreme right-wing single-party rule based on a violent and totalitarian police state, where all government policies and actions are geared toward upholding hegemonic party power. Fascism is intolerant of competing points of view, whether political, social, or economic, and tends to subjugate and destroy (using extreme violence and propaganda when necessary) any competing views or minority opinions (including supporters of such minority views). Last, fascism tends to be associated with a vision of an ideal, pure national identity and a romanticized, apotheosized era and/or race; for example, Hitler's romantic idealization of the Aryan race and German motherland. Fascist governments tend to be willing to commit any action, no matter how harmful or atrocious, to help realize and bring about this return to a glorified past and to promote the well-being of the assumed glorified race/people/motherland.

The prefix *eco-* is added to fascism when dealing with environmental debates that have developed over the last decades. Those levying the charge of ecofascism attempt to highlight that philosophies—and, by extension, policies—advocated by ecofascists are intolerant of other views and have the potential to lead to violent and totalitarian actions taken on behalf of an idealized nature and/or its inhabitants. These debates may include charges of fascism in general, and ecofascism in particular, surrounding the following issues.

Human Population Numbers

Garrett Hardin and Paul Ehrlich have both been charged with fascism regarding their work dealing with the biosphere's carrying-capacity potential. Both scientists argue that human population numbers must be reduced in order to avert natural and human-induced tragedies that they claim exponential population growth will, in all likelihood, precipitate. The premises of their arguments and their proffered solutions tend to be misrepresented and they historically have been charged with being misanthropic and with advocating fascist, totalitarian national and global forms of governance that will be in charge of how many children families can have.

Charges Against Deep Ecology

Some ecofeminists argue that the deep ecology ideal of expanding individual self-awareness to include identifying one's self with the biosphere is patriarchal and can degenerate into fascist-like policies and lifestyle changes advocated by these deep ecologists. These ecofeminists claim that such an expansion of the self-centered ego subsumes the interests and identity of other species and individuals in the biosphere to the needs and urges of the self-thinking (male) ego of the deep ecologist, thus generating a form of solipsistic fascism regarding environmental issues.

Ramachandra Guha offers another charge of ecofascism against those deep ecologists who advocate wilderness protection. Guha argues that wilderness is a global north, and especially North American, concept and when exported to the global south results in fascist actions taken in the defense of nature. For example, he points out that wilderness reserves in India and Africa are created at the expense of indigenous peoples who have lived there for decades or centuries. Such indigenous peoples have historically been forcibly removed so that a global north, deep ecological image of "pure" wilderness free of humans can be created and maintained.

A corollary of this charge from the global south is the emerging debate about climate destabilization and various proposed international treaties that will regulate greenhouse gas (GHG) emissions. Global south countries are charging global north countries with a form of climate-fascism, as global north countries are making demands on global south countries to severely restrict their GHG emissions for the benefit of the whole planet. Global south countries respond by pointing out that global north countries consume and pollute in disproportionate numbers and are able to do so because of a history of colonialism. In effect, as with wilderness preserves, the global north is using environmental science to regulate and govern global south politics and lifestyles based on global north conceptions of nature, resulting in charges of ecofascism.

Radical Environmentalism

Many governmental officials, spokespeople for those on both the conservative right and in wise-use movements, and some ecophilosophers claim that the tactics and goals of radical environmentalists are a form of ecofascism. These critics claim that the actions and ideals of those in Earth First!, the Earth Liberation Front, and the Animal Liberation Front are fascist and commit acts of ecoterrorism because members of these radical environmentalist groups utilize violence and sabotage to illegally protect flora and fauna that radical environmentalists claim are intrinsically valuable and sacred. Some also charge that these radical environmentalists, as well as some deep ecologists, glorify a primal, pure state of the planet and are willing to commit any action in order to bring this original, pure state back into being.

The Land Ethic

Aldo Leopold's famous dictum that a thing is right when it preserves the integrity and functioning of a biotic community and is wrong otherwise has gained followers within ecophilosophy. However, those in the animal rights movement, as well as other ecophilosophers critical of the land ethic, point out that if the goal is to maintain harmony (i.e., health) of the larger biotic whole, then the rights and interests of individuals who may threaten this presumed health will be sacrificed. In short, a "fascism of the whole" will trump the interests of the individuals within a biotic community and will also lead to a slippery slope of potentially fascist governmental policies designed to protect the larger biotic whole. Land ethicists have responded by claiming that animal rights philosophers are actually ecofascists if their arguments are taken to their logical extreme, in that animal rights philosophers are willing to uphold the rights of individual animals even at the expense of an endangered species or larger ecosystem that may be threatened by the presence of an overabundance of a single species.

See Also: Animal Ethics; Anthropocentrism; Biocentric Egalitarianism; Callicott, J. Baird; Climate Ethics; Deep Ecology; Land Ethic.

Further Readings

Callicott, J. Baird. *In Defense of the Land Ethic: Essays in Environmental Philosophy.* Albany: State University of New York Press, 1989.

Ehrlich, Paul. *The Population Bomb.* Minneapolis, MN: Rivercity Press, 1975.

Guha, Ramachandra. "Radical American Environmentalism and Wilderness Preservation: A Third World Critique." In *Varieties of Environmentalism: Essays North and South,* Ramachandra Guha and Juan Martinez-Alier, eds. London: Earthscan, 1997.

Hardin, Garrett. "Lifeboat Ethics." *Bioscience,* 24 (1974).

Ivakhiv, Adrian. "In Search of Deeper Identities: Neopaganism and 'Native Faith' in Contemporary Ukraine." *Nova Religio: The Journal of Alternative and Emergent Religions,* 8/3 (2005).

Katz, Eric, Andrew Light, and David Rothenberg. *Beneath the Surface: Critical Essays in the Philosophy of Deep Ecology.* Cambridge, MA: MIT Press, 2000.

Plumwood, Val. "Nature, Self and Gender: Feminism, Environmental Philosophy, and the Critique of Rationalism." *Hypatia,* 6/1 (1991).

Sagoff, Mark. "Animal Liberation, Environmental Ethics: Bad Marriage, Quick Divorce." In *Environmental Philosophy: From Animal Rights to Radical Ecology,* Michael Zimmerman, ed. Upper Saddle River, NJ: Prentice Hall, 1993.

Taylor, Bron, and Todd LeVasseur. "Ecotage and Ecoterrorism." In *Encyclopedia of Environmental Ethics and Philosophy,* J. Baird Callicott and Robert Frodeman, eds. Detroit, MI: Macmillan Reference, 2008.

Zimmerman, Michael. "Ecofascism: A Threat to American Environmentalism?" In *The Ecological Community: Environmental Challenges for Philosophy, Politics, and Morality,* Roger Gottlieb, ed. New York: Routledge, 1997.

Todd LeVasseur
University of Florida

Ecofeminism/Ecological Feminism

Ecofeminism is unique among ecological theories in its placement of patriarchy at the center of analysis of social and ecological oppression. Ecofeminism derives an epistemology, an ontology, and sociopolitical project out of the everyday lives of women that challenges the phallocentric framework that shapes Western thought (philosophically, economically, and politically). Beginning with a perspective "from below," the point of view of the oppressed, ecofeminists contest the dominant logic of capitalist patriarchy as independence *from,* and domination *over,* nature via reason, science, and technology. Furthermore, ecofeminists argue that global inequality and ecological destruction are not the by-products of individual actions, but are structural outcomes of three unequal sets of power relations: north over south, men over women, and humans over nature.

First coined by Françoise d'Eaubonne, the term *ecofeminism* contends that the domination of women and the domination of nature are interconnected processes. Consequently, ecofeminism articulates a politics that embraces the coliberation of women and nature from "modernization" and "development." Karen Warren underscores these connections, claiming that ecology must have a feminist theory/practice and feminism must have an ecological theory/practice.

Women carrying bundles in Malawi, where 53 percent of the population lived in poverty in 2004. Ecofeminism finds a link between the domination of women and the domination of nature and emphasizes the direct impact of environmental pollution on female bodies and reproductive processes.

Source: World Bank

Ecofeminists promote social relations based on reciprocity, cooperation, mutual aid, respect, love, and care as the foundation of ecologically just and sustainable alternatives. They therefore reject capitalist-patriarchy values of *homo economicus,* which favors economic self-interest, competition, short-term profit-oriented contractual obligations, and abstract equality through the state. Rather than seeking the liberal goal of equality of opportunity through the "public" realm of politics and economics on capitalist and patriarchal terms, ecofeminists put forth a gynocentric framework to implode the hierarchal dichotomies of public–private, male–female, culture–nature, civilization–wilderness, mind–body, and reason–emotion.

As a social movement, the historical roots of ecofeminism derive from the 1960s and 1970s feminist, peace, and ecology struggles against megadam projects, nuclear power plants, agribusiness, deforestation, and toxic waste dumps. Ecological issues became feminist issues, and manifested in activism based on women's positionality in the social division of labor and their relation to environmental degradation. Women are overwhelmingly responsible for the production and reproduction of life, and therefore positioned to be directly concerned with preserving environmental integrity to ensure food, water safety, and security, along with environmentally healthy spaces for their children and family. Additionally, it is through women that environmental health and human health become linked, as environmental conditions directly affect female bodies and their capacity for reproduction. Increased rates of breast cancer, birth defects, and miscarriages, as well as toxification of breast milk through polychlorinated biphenyls (PCBs) and other industrial contaminants are directly associated with environmental pollution. For these reasons, ecofeminists critique industrialization for its toxification of land and body, which falls predominately upon women.

Contending that science and technology are gendered rather than neutral processes, ecofeminism critiques the historical roots of patriarchy in specific object relations *with,* and enlightenment productions *of,* nature. Patriarchy emerges from men's object relations to nature, organized around tools specialized for lethality and plunder. Patriarchy's control over technological development manifests itself through government and corporate funding of the military-industrial complex. Furthermore, technological development is shaped by the elusive quest to conquer and "develop" nature more efficiently through extractive industries, dam projects, genetic engineering, and the mass production of toxic chemicals. In fact, these practices effectively destroy the capacity of ecological systems to support human and nonhuman life.

Patriarchal science is implicated in the production of the ecosocial crisis as well. Francis Bacon, Isaac Newton, and other founding fathers of modern science articulated a mechanistic and reductionist conception of nature where humans were no longer seen as part of nature, nor was nature seen as sacred and a subject with agency. Enlightenment thought plays a central part in the codomination of women and nature. Newtonian science was premised on a linear conception of space–time, which ruptured and rejected any understanding of the cyclical flows and rhythms of nature. The enlightenment produced a nature devoid of organic relations of interdependence and holism. Nature was dead matter that could be disaggregated and reconnected at will through basic mathematical formulas and abstract and universal laws.

Moreover, rather than nature being conceived as a provider of plenty and therefore a rich surplus, enlightenment thought depicted "mother nature" as stingy. Nature was "feminized" and seen as nonproductive; its only guarantee was the provision of scarcity. Rational science could force nature to yield plenty by constructing it like a "fruitful womb for the use and benefits of a man's life." These linguistic conflations between nature and woman emphasize patriarchy's reduction of both to an object for male gain.

Ecofeminists contend that capitalist activities can only be construed as "productive" through an extremely narrow and illusory conception of productivity: the reduction of nature into a quantity of money. In actuality, such activities destroy and degrade the productivity of nature, whose processes of cleaning the air and water, forming the soil, cycling and fixing nutrients in the soil, sequestering carbon, regulating climates, and pollinating plants are increasingly imperiled by industrialism's planetary destruction. Capitalism's logic of growth obscures the true costs of a reductionist conception of productivity: declining agricultural yields due to soil erosion, salinization, waterlogging, and toxification; declining rainfall and biodiversity due to deforestation; and declining carbon absorption by forests and oceans.

Furthermore, capitalist-patriarchy depends upon the colonization of the most productive members of society: women. Women's role in social reproduction (most fundamentally, giving birth) is the basis for all further production. Under patriarchy, women's social reproduction practices are naturalized and turned into a "free gift" through its reduction to biological fact. The same process occurs for the *labor* and *land* of "foreign peoples." Patriarchal economic and political systems are therefore conceived as nonproductive predatory modes of production where the naturalization of women, foreign people, and land turns them into "raw" materials, and colonies to be freely appropriated and restructured at will.

The coliberation of women and nature from capitalist-patriarchy becomes vital to rethinking alternative economies premised on the production of life rather than death. To legitimate such a political project, ecofeminists reorient the language and focus of economic production to account for the ways in which the productivity of women and nature are primary for life and the foundation for all further "development." This shift is dependent on the concepts of ecological reproduction and social reproduction (the care work of child bearing/rearing, the provision of food, clothing, shelter, emotional/sexual labor, and knowledge/cultural transmission).

In opposition to enlightenment logic, ecofeminism insists that social liberation must be contextualized within ecological flows that form the preconditions for life. For ecofeminists, the path to freedom does not lie in catch-up development and universalizing the privileges of white males—the path of liberal feminism—as this is ecologically unsustainable and requires neocolonialism. Subsequently, ecofeminist movements attempt to restore integrity to the cycles of life that are increasingly disconnected, fragmented, and broken by "modernization" and "development" through the reempowerment of women's knowledge and economic production systems.

An alternative path to embracing capitalist-patriarchy and the universalization of the Western "monoculture of the mind" can be found in the turn to "wild politics." To rebuild the planet's biodiversity and ecological integrity, ecofeminists embrace lifeways that are premised on strengthening indigenous and non-Western cultures, not assimilating them into "modernity." The ecofeminist turn is toward rebuilding nature's economy and the sustenance economy rather than the market economy. These empowerment strategies are interconnected with building direct democratic practices and organizations to localize economic and political power in order to build healthy and vibrant community economies. Collectively, these moves prioritize women's experience and knowledge in the construction of ecologically sustainable and socially just worlds.

In the quest to move beyond patriarchy, ecofeminists often reinforce the essentialism and dualism of the former by reinscribing the nature–woman connection. Claiming that women's reproductive capabilities make them inherently closer to nature than men may give rise to biological reductionism. As a result, women alone are viewed as the historical subjects that will create an ecologically sustainable and nonoppressive society. This essentialism not only produces an overly simplistic binary of women are good and men are bad, but also infuses a politics of female separatism. In doing so, ecofeminists seek to replace patriarchal science with goddess worship and reject rationality outright, rather than particular historical formations of rationalization. Additionally, by gendering nature as a nurturing mother, ecofeminism merely turns the worldview from male-centric to woman-centric, reinforcing the dualistic polarity of male–female. Critics contend that ecofeminism needs to move beyond essentialism and dualism in order to create a liberatory politics that is truly postmodern.

See Also: d'Eaubonne, Françoise; Ecopolitics; Warren, Karen.

Further Readings

Biehl, Janet. *Rethinking Feminist Politics.* Boston: South End Press, 1991.

Hawthorne, Susan. *Wild Politics: Feminism, Globalization, Bio/Diversity.* North Melborne, Australia: Spinifex Press, 2002.

Merchant, Carolyn. *The Death of Nature: Women, Ecology, and the Scientific Revolution.* San Francisco, CA: Harper & Row, 1980.

Mies, Maria and Vandana Shiva. *Ecofeminism.* London: Zed Books, 1993.

Salleh, Ariel. *Ecofeminism as Politics: Nature, Marx, and the Postmodern.* London: Zed Books, 1997.

Warren, Karen. *Ecofeminist Philosophy: A Western Perspective on What It Is and Why It Matters.* Denver, CO: Sagebrush Press, 2000.

Justin Myers
The Graduate Center, City University of New York

"Ecological Crisis, The Historical Roots of Our"

Lynn White, Jr.'s (1907–87) "The Historical Roots of Our Ecological Crisis," published in 1967, triggered controversy by connecting rampant environmental destruction globally with the linking of Western science and technology and the Latin branch of the Judeo-Christian heritage that grants humans dominion over nature. The piece, coming as it did during the rise of the environmental movement in the late 1960s, set off a debate that still simmers today.

Tracing the history of the world's ecological crisis requires connecting dots that relate human activities and values that comprise the sociocultural system's patterned activities. White focused on the relationship of religion, science, and technology as factors in environmental degradation. His contribution to the growing environmental literature added a cultural dimension to our understanding of ecological history. Consciously or not, he tapped into an intellectual vein opened by German scholar Max Weber, whose 1905 book *The Protestant Ethic and the Spirit of Capitalism* documented relationships between religion and the evolution of capitalism.

Critics of White's work usually accuse him of misinterpreting the Bible. His work stimulated the development of a Christian ecological movement to counter his thinking, which was, in fact, partly misinterpreted by his critics. For example, White, himself a Christian, cautioned readers about Christianity's complexities and how sociocultural context shapes its consequences.

As White noted, ecological crises resulting from human activities are part of our history, predating impacts of the continuing Industrial Revolution. Changes in agricultural practices during the Middle Ages, as well as harnessing wind and waterpower, increased human capacity to manipulate the environment. This laid the foundation for the later Industrial Revolution, whose Western scientific and technological roots corresponded with the Latin thread of Christianity. White provided supporting evidence to show that development in the East, with its Greek roots, followed a different course, with less environmental impact.

From White's idealistic perspective, Eastern and Western views of sin shaped sociocultural evolution in different historical contexts. For Greeks, sin was an intellectual blindness, where orthodoxy (clear thinking) could illuminate the soul of the sinner, leading to salvation. For the Latin branch of Christianity, sin was a moral evil. Salvation was found through right conduct. White compared the intellectualism and contemplation of Easterners to voluntarism and action orientation of Westerners, concluding that the conquest of nature was a logical possibility, given the Western context.

In essence, White captured the contradictory nature of both the Bible and Christianity as part of the seemingly infinite complexities of the human sociocultural milieu. His point that the Bible gives humans dominion over the Earth sparked discussions of the different meanings of dominion in relationships of humans with each other and with their Creator. This raises the philosophical problem of text reading and interpretation, along with the nature of belief and levels and types of doubt and unbelief. This philosophical/metaphysical/theological discussion raises a question of whether the quality of the text reading and faith are more important than historic facts.

White's historic approach to the practical application of Christianity in different sociocultural settings shed new light on faith and religion, making them more relativistic than his critics would like them to be. This is the essence of abstract ideals held by the faithful meeting concrete everyday activities by believers and nonbelievers alike. White presented an evidence-based challenge to narrow Christian orthodoxy or dogmatism.

Proponents of eco-Christianity can find ample examples of stewardship, ecofeminism, and creation spirituality in Scripture, which emerged in a pastoral and agrarian context; they often find their viewpoints challenged by Christianity's hierarchy and fundamentalists. Unfortunately, they have a more difficult time finding historical applications of Christian environmentalism. In fact, the Christian contributions to environmentalism and sustainability today tend to flow more from asceticism and social justice in both the Bible and other traditions that have a historic basis.

With his arguments from historic records that show, for example, how missionaries tried to stamp out paganism by chopping down sacred groves, White provided ample evidence of environmental destruction wrought in the name of Christ. Likewise, he offered the radical humility of St. Francis of Assisi as an alternative form of Christianity, with the seemingly wry note that Francis, unlike many of his left-wing followers, became a saint and was not burned at the stake. Francis advocated humility of the human species, placing us on an equal footing with all of God's creatures. His thinking, once thought banished, is one wellspring for creation spirituality today.

As an early foray into environment and religion, White's article is a benchmark. It was problematic, however, that he did not take Weber's cue to link the sociocultural changes Christianity underwent in the face of global markets' evolution from mercantile to industrial capitalism and their relationship to environmental problems, all of which have roots in the Middle Ages.

The emergence of mercantile capitalism in the 16th and 17th centuries opened the way for widespread domination of the global ecology and subjugation of some populations. Western nations established colonies to provide them with low-cost natural resources and labor (often slave) to create vast amounts of wealth for an evolving middle class. This process occurred with little protest from institutionalized Christian denominations that were part of the governing structure.

The evolution of capitalism and the Enlightenment broadened human capacity to influence the Earth's environment, and religious institutions adapted along the way, even though they viewed many Enlightenment thinkers as a faithless lot. Enlightenment thinkers separated church and state. They saw the older moral philosophies as bankrupt and confining. Instead, they sought to empower individual human agents in the economy, politics, and other aspects of the sociocultural system. New philosophical premises emerged to support the development of individual freedom, laissez-faire democracy, and free markets that ultimately emerged as a highly complex, industrialized, wage-based, consumer economy that girdled the globe figuratively and literally. Christian denominations again adapted to fit the changing world.

White's contention that what we do about ecology depends on our ideas about the man–nature relationship is crucial to his argument about the role of religion in development and environmental degradation. His idealist approach to the problem taps him into a main branch of sociological theory that human thinking creates perceptions about the world that shape our actions. Sociologists, of course, also recognize the materialist bases for human activities, and White could have taken advantage of this conceptualization by pointing out more clearly how science and technology became the tools not only of religion, but also of capitalist entrepreneurs and their drive to seek wealth in the name of God.

See Also: Deep Ecology; Religious Ethics and the Environment; Sustainability and Spiritual Values; Western "Way of Life."

Further Readings

Carnoy, M. *The State and Political Theory.* Princeton, NJ: Princeton University Press, 1984.

Christian Ecology Foundation. "Christian Ecology." http://www.christianecology.org (Accessed November 2009).

Domhoff, G. W. *Who Rules America Now? A View for the '80s.* New York: Simon & Schuster, 1983.

Grey, Mary. "Ecofeminism and Christian Theology." *The Furrow*, 51/9 (2000).
O'Connor, James. *Accumulation Crisis*. New York: Basil Blackwell, 1984.
Sears, R. T. and A. J. Fritsch. *Earth Healing: A Resurrection-Centered Approach*. Livingston, KY: ASPI Publications, 1994.
Weber, Max. *The Protestant Ethic and the Spirit of Capitalism* (1905). http://xroads.virginia.edu/~HYPER/WEBER/cover.html (Accessed November 2009).
White, Lynn. "The Historical Roots of Our Ecological Crisis." *Science*, 155/3767 (1967). http://aeoe.org/resources/spiritual/rootsofcrisis.pdf (Accessed November 2009).

Timothy Collins
Western Illinois University

Ecological Footprint

The ecological footprint (EF) is a measure of a population's wide-ranging demands on global natural resources. It has become one of the most widely used measures of humanity's environmental impacts, one that has been used to highlight the apparent unsustainability of current practices and the wide inequalities in resource consumption between and within nations.

The EF estimates the biologically productive land and sea area needed to provide the renewable resources that a population consumes and to absorb the wastes it generates—using prevailing technology and resource management practices of world average productivity. It measures the requirements for cropland, grazing land for animal products, forested areas to produce wood products, marine areas for fisheries, built-up land for housing and infrastructure, and forested land needed to absorb carbon dioxide emissions from energy consumption. One can estimate the EF, measured in "global hectares" (gha), at various scales—for individuals, regions, nations, and humanity as a whole. The resulting figures can also be compared to how much productive area—or biocapacity—is available.

Bill Rees of the University of British Columbia's School of Community and Regional Planning was the originator of the EF, which Mathis Wackernagel further developed in his dissertation under Rees's supervision. Together, Wackernagel and Rees published the book *Our Ecological Footprint* in 1996. Rees's motivation in developing the EF was to reopen questions of limits to growth and human carrying capacity. However, rather than trying to determine how many people a given land area or the entire planet can support, the EF asks the question in reverse order, calculating how much area is necessary to support a certain population, given current consumption levels and technology.

EF calculations have called into question the sustainability and equity of current consumption and production practices. Recent figures from the Global Footprint Network (GFN) show that the per capita global footprint of 2.6 gha exceeds available biocapacity of 1.8 gha. In other words, 1.4 planets would be needed to sustain current resource demands or, alternatively, it takes the Earth a year and five months to regenerate what is used in one year. The implication of such ecological overshoot—which began in the mid-1980s, according to the GFN—is that life-supporting biological resources are being depleted. Results include collapsing fisheries, declining forest cover, and the accumulation of pollution and waste—including the greenhouse gases that drive climate change.

Using the Ecological Footprint

EF analysis can also show whether a nation is living within the biocapacity of its own territory or if it is an "ecological debtor" drawing on the ecological capital of other parts of the world. Meanwhile, per capita EFs show a wide divergence in the demands on nature from people in different societies, ranging from the United Arab Emirates at the high end (10.3 gha/person) to Haiti at the low end (0.5), with the United States (9.0), Germany (4.0), China (1.8), and others in between (2006 data). Such figures are the basis of claims that if all of humanity consumed like the average American, five planets would be needed. Ecological footprints also vary greatly within countries according to level of affluence. One Canadian study found, for example, that the average EF of the richest 10 percent of the population was nearly 2.5 times greater than that of the poorest 10 percent.

Researchers have combined footprint analysis with measures of human development to assess whether nations are on track toward sustainable development—defined as a per capita EF lower than the available per capita biocapacity and a high rating on the UN Human Development Index (above 0.8). Such analysis has produced the finding—an inconvenient one for defenders of a capitalist market economy—that Cuba is the only nation yet to meet these minimal conditions for sustainable development.

Environmental educators and activists have used the EF to raise awareness of unsustainable consumption patterns, often with the goal of encouraging lifestyle change and, less frequently, to promote awareness of wider structural forces driving such patterns. Many online footprint calculators have appeared on nongovernmental organization (NGO) Websites with such goals in mind. These allow people to calculate their personal EF and to make sometimes eye-opening comparisons to estimates of available biocapacity or to average footprints of other people locally and globally. Meanwhile, social scientists have used the EF as a comprehensive indicator of nations' ecological impacts in order to test empirically different social theories of the forces driving those impacts.

Some organizations have used the EF to question the degree to which high rates of natural resource consumption contribute to improved well-being. The United Kingdom–based New Economics Foundation includes the EF in a measure of the efficiency with which resource consumption is converted into long and happy lives. Its Happy Planet Index (HPI) consists of the product of the mean national average of life satisfaction (as recorded on international surveys) multiplied by average life expectancy, divided by per capita EF. Countries ranking highest on the HPI in 2009 were Costa Rica, Dominican Republic, and Jamaica, while richer developed nations fell well back in the rankings. According to this index, most affluent nations have become much less ecologically efficient since the 1960s as subjective well-being has increased little, while per capita footprints have risen significantly.

Although EF analysis can lead to a radical critique of the current economic and social order, it has found increasing mainstream acceptance among business and governments. Multinational corporations such as Walmart and BP have introduced programs to shrink their footprints, while the World Business Council on Sustainable Development has begun exploring pathways toward achieving a "one-planet economy" by 2050.

A number of governments have also embraced the concept. Wales has adopted the EF as a sustainability indicator, while the European Union has begun to examine doing the same. Switzerland is incorporating EF data into its Sustainable Development Plan and Japan has adopted the EF as one way to measure success of its Basic Environment Plan. In 2006, Britain's Environment Secretary David Miliband proclaimed that enabling a move to "one-planet living" was his department's mission, while he acknowledged being struck by calculations showing that Britons were living as if there were three planets to support them. Meanwhile, the United Arab Emirates launched an Ecological Footprint Initiative in 2007 in response to gaining the dubious distinction of having the world's largest per capita EF. In addition, more than 100 cities around the world use the EF measurement.

The growing use of the EF illustrates the power of a compelling metaphor. Wackernagel and Rees originally used the term *appropriated carrying capacity;* however, the concept did not take off until Rees coined the term *ecological footprint,* reportedly inspired by his new computer's smaller footprint on his desk. Since then, the idea of the ecological footprint—and offshoots such as the carbon footprint and water footprint—has become ubiquitous in environmental debates.

Critiques of the Ecological Footprint

Despite its rapid ascent and widespread use, the EF has faced a wide range of criticism. One of the EF's attractions is that it provides a single aggregate indicator of ecological impacts; however, such aggregation requires simplification of a complex reality. One of the most controversial simplifications is the way the EF deals with energy consumption by estimating the land area needed to grow trees capable of absorbing the carbon dioxide released from fossil-fuel combustion. Meanwhile, EF calculations prior to 2008 assumed nuclear power to have the same impact as electricity produced with a world average mix of fossil fuels, which critics argued did not reflect nuclear power's much lower carbon emissions. As of 2008, the GFN no longer included nuclear power in national footprint calculations. Given that energy consumption accounts for nearly one-half of the global footprint, the methodology chosen to account for this type of resource use has a large impact on the overall measure and conclusions drawn from it.

Critics also argue that the EF methodology rewards more intensive production methods that increase yields per unit of land in the short term, but might actually be less sustainable in the long run by, for example, accelerating land degradation. Similarly, organic farming methods with lower yields than conventional agriculture could appear to have a bigger footprint, despite other ecological benefits. Others argue that EF analysis is overly anthropocentric, focusing only on land and sea area that is useful to the human economy and failing to allocate space for the needs of other species. Indeed, the EF does not measure changes in biodiversity, for which other indicators are needed.

EF advocates acknowledge that it cannot include all significant environmental impacts given the lack of data for some issues and the difficulty of converting some types of ecological demands, for which no regenerative capacity exists, into a measure of land area. Among the key impacts not reflected in the EF are those related to toxic substances, greenhouse gases other than carbon dioxide, and water consumption. While not all ecologically significant impacts are included, supporters of the EF maintain that the measure describes a minimum condition for sustainability, namely, that footprints must be smaller than available biocapacity.

Some critics also object to what they perceive to be the implication that a population ought to live within the biocapacity of its own territory and not draw on outside resources through trade. The idea that cities or countries with an EF that exceeds their territory's biocapacity are running an "ecological deficit" can leave the impression that some form of self-sufficiency or autarchy should be the goal, although the GFN maintains that such ideas are not inherently antitrade. Indeed, in a world in which populations and scarce biocapacity are not uniformly distributed, being an ecological debtor or creditor is not necessarily a useful indicator of sustainable practices. Small, densely populated areas may be very ecologically efficient, but will almost inevitably be ecological debtors drawing on outside resources, while large, sparsely populated countries such as Canada and Australia appear to be living well within their territories' ecological means, even though their per capita footprints are among the highest in the world.

Proponents of the EF have acknowledged some of the concept's limitations and have worked to refine it. The first set of standards for proper calculation and communication of the EF was produced in 2006, and continual revisions are occurring. Meanwhile, a review in 2008 produced for the European Commission concluded that, although complementary sustainability indicators and further improvements in data quality and methodology are needed, the EF is a useful indicator of sustainable natural resource use that is easy to communicate and understand.

Despite the inherent difficulties in establishing an aggregate environmental indicator, the EF has quickly emerged as a widely used measure of humanity's demands on the Earth—and a concept capable of communicating to a wide public the idea that significant change is needed to reduce those demands.

See Also: Club of Rome (and Limits to Growth); Consumption, Consumer Ethics and; Western "Way of Life."

Further Readings

Global Footprint Network: http://www.footprintnetwork.org (Accessed February 2010).

Moran, Daniel D., et al. "Measuring Sustainable Development—Nation by Nation." *Ecological Economics,* 64/3 (2008).

Van den Bergh, Jeroen C. J. M., and Harmen Verbruggen. "Spatial Sustainability, Trade and Indicators: An Evaluation of the Ecological Footprint." *Ecological Economics,* 29/1 (1999).

Venetoulis, Jason, and John Talberth. "Refining the Ecological Footprint." *Environment, Development, Sustainability,* 10/4 (2008).

Wackernagel, Mathis, and William E. Rees. *Our Ecological Footprint: Reducing Human Impact on the Earth.* Gabriola Island, British Columbia, Canada: New Society Publishers, 1996.

World Wildlife Fund, Global Footprint Network, and Zoological Society of London. *Living Planet Report 2008.* Gland, Switzerland: World Wildlife Fund, 2008.

York, Richard, Eugene A. Rosa, and Thomas Dietz. "Footprints on the Earth: The Environmental Consequences of Modernity." *American Sociological Review,* 68/2 (2003).

Anders Hayden
Dalhousie University

Ecological Restoration

Ecological restoration is a generic term for projects to remediate environments that are perceived as having been damaged, typically by human activity. Such projects fall on a spectrum ranging from attempts to re-create, as closely as possible, the original environment, to "clean up" projects where, for instance, an opencast mine that was formerly a forested area is

The 136,000-acre Pilanesberg Game Reserve in South Africa, shown here in 2006, was used for farming and mining before undergoing the largest restoration project in the world, including the reintroduction of thousands of game animals.

Source: Wikipedia/Joonas Lyytinen

replaced by a lake surrounded by a park. This article explains different concepts of restoration and discusses the controversy over whether natural environments, once destroyed, can ever be fully restored.

To refer to environmental changes as "damage" is to make a value judgment. Many people see the felling of forests and the draining of wetlands for agriculture—whether to grow tobacco in North Carolina or to plant oil palms in Indonesia—as "improvement," whereas others see them as destruction of natural systems. Perhaps the more neutral term *transformed* is preferable. However, almost everyone agrees that if a natural environment has been affected to the point where it is unproductive, polluted, and ugly, then it would be good to remediate it to a productive, unpolluted, and attractive state. This could include a productive industrial enterprise, a housing area, a recreational complex, or a more or less natural area. This article discusses only the last named.

The term *restoration* is most commonly used for projects that seek to return an area to a state that resembles the original, such as cleaning up a polluted river or lake or replanting a forest. One of the earliest restoration projects in the world (beginning in 1941) is the University of Wisconsin–Madison Arboretum program. From small beginnings, the prairie restoration movement now encompasses a total of 5.7 million hectares (14 million acres) of degraded agricultural land that has been taken out of production and replanted with native vegetation.

One of the largest single restoration projects in the world is Pilanesberg Game Reserve near Sun City, South Africa. This park of over 55,000 hectares (136,000 acres) was formerly farmed and mined; in the 1970s, the government of what was then Bophuthatswana (a black "homeland" created in apartheid-era South Africa) decided to relocate the farmers and remove almost all traces of the town and mine workings; thousands of native animals were reintroduced in what is still the largest-ever translocation of game animals.

In New Zealand, island restoration ecology—including mainland "island" restored areas—has been very successful. Because of the number and variety of introduced pest fauna and flora, most of which in some way compete with or prey on native species, areas must first be cleared of such pests (and, on the mainland, surrounded by pest-proof fencing) before native species can be reintroduced. Because of the need to kill thousands of pests, restoration projects are often criticized by animal liberationists.

Ongoing Management of Restoration

In all these cases, it is recognized that natural systems cannot simply be restored and then left to function by themselves: some degree of management will always be required.

Restored prairies need controlled burning and control of exotic plants such as goldenrod; a sustainable balance between predators, prey, and vegetation in Pilanesberg requires culling of overpopulated species; in New Zealand, there is constant vigilance to ensure that pests do not reintroduce themselves, as well as to protect the islands against new pests such as the Argentine ant.

Alternatively, where a natural environment has been irreversibly changed, the goal could be to create a substitute environment that would also have a high degree of natural value, for instance, to transform an opencast mine into a lake. In the United Kingdom, for instance, companies are legally required to restore old quarries in this way and the British Aggregates Association operates a Restoration Guarantee Fund. As a result, the area of wetlands—most of which have been drained over the years—is now increasing, providing both wildlife habitat and recreational opportunities.

Not all damaged environments can be restored to their original state. Often this is for geological reasons, for instance, mountaintop removal in West Virginia and Kentucky. Moreover, in all the cases cited above, some of the original species cannot be introduced because they have become extinct. This does not, of course, mean that since restoration will never be "perfect" it should not even be attempted. But what if it were technically possible to restore an environment to the point where a visitor, or even an expert, could not distinguish it from the original?

The classic example in the environmental ethics literature is dune mining for rutile (titanium oxide) on Fraser Island, Queensland, Australia, in the late 1970s. In reply to objections from local people, the developer (and the state government) admitted that parts of the dune system would be seriously damaged, but claimed that when the mining ended the system could be rebuilt to be identical to the original. Many coastlines in the world, especially dune systems, are dynamic and subject to massive changes as a result of extreme weather events, but dunes are typically restored by the same processes that created them in the first place. Thus the value of the system would be reduced only temporarily, and this loss would be more than compensated for by the economic benefits of the project. Robert Elliot, at the time a lecturer at the University of Queensland, dubbed this argument "The Restoration Thesis."

How "The Restoration Thesis" Works

Elliot, who was one of the main opponents of the mining, argued that even if the restoration of the dunes were technically possible (which he disputed), the intrinsic value of the dunes would be lost forever. Intrinsic value is the value that something has as an end, for its own sake, not merely as a means to some other end. It is quite a different kind of value from instrumental values such as economic or recreational value. He argued that the value of natural areas derives from the fact that they are products of nature rather than of humans, and that they maintain continuity with the naturally originating environment.

Elliot is careful to point out that he is not arguing that everything that is natural is superior to everything that is made or modified by humans. Smallpox is natural but it is better to interfere with nature to prevent people suffering from smallpox. However, he and others who share his view argue that areas such as coastlines and forests are valuable precisely because they are products of nature—we stand in awe of nature's creative ability, diversity, and beauty, beside which human artifacts seem puny and contrived.

Elliot draws an analogy with the value of original works of art versus the value of fakes and forgeries. Consider, for example, the Dutch painter Johannes Vermeer (1632–75), who is widely accepted as one of the greatest Dutch painters of the period. Only around 30 of his paintings exist. The forger Han van Meegeren (1889–1947) was a painter whose own work was rejected by art critics, and, in revenge, he produced paintings that he successfully passed off as recently rediscovered works of Vermeer and other famous Dutch artists: his most famous Vermeer forgery, *Supper at Emmaus,* was regarded by some critics as "Vermeer's" finest painting. The point of the analogy is that just as we value works of art partly because of their origins (it was painted by Vermeer himself, not a forger), so we value the products of nature itself, not a restoration.

It is true that original works of art are valued more than fakes (or copies), and while this is partly because of the market value of an original, it is also the case, at least in Western culture, that we attach particular value to originality and creativity. Vermeer developed his own style; van Meegeren merely painted in the style of Vermeer. It is also true that an object's value may be largely a function of its origin and associations, for instance, an inexpensive wedding ring or the first painting produced by one's child. Nonetheless, the appropriateness of the analogy with works of art has been questioned. We value creativity and originality in art because works of art are intentional artifacts, but nature does not have intentions, let alone intentions to produce originals or fakes.

Perhaps more importantly, to claim that restorations are unnatural because they were created by humans is to locate humans outside nature. Anthropocentric ethics draws a sharp distinction between humans and nature, which is seen as no more than a storehouse of resources that humans are entitled to exploit for their own good. However, Linda Graber has noted that the opposite end of the environmental spectrum is populated by what she calls "wilderness purists," who also believe that humans are in no way part of nature. Neither of these views is consistent with the worldview that sees humans as part of nature, exemplified in Aldo Leopold's land ethic, where we are required to see ourselves as "plain members of the biotic community." In Leopold's view, environmental restoration would be as much a duty as cleaning up hazardous waste sites or removing graffiti from public buildings.

See Also: Animal Ethics; Intrinsic Value; Leopold, Aldo.

Further Readings

Atkinson, I. A. E. "Opportunities for Ecological Restoration." *New Zealand Journal of Ecology,* 11 (1988). http://www.nzes.org.nz/nzje/free_issues/NZJEcol11_1.pdf (Accessed January 2010).

Elliot, Robert. *Faking Nature: The Ethics of Environmental Restoration.* London and New York: Routledge, 1997.

Graber, Linda. *Wilderness as Sacred Space.* Washington, DC: American Society of Geographers, 1976.

Kucharik, Chris. "Assessing the Progress of Prairie Restorations in Southern Wisconsin" (2009). http://www.sage.wisc.edu/in_depth/kucharik/prairie/prairie.html (Accessed January 2010).

Pilanesberg Game Reserve. http://www.pilanesberg-game-reserve.co.za (Accessed January 2010).

Alastair S. Gunn
University of Waikato

Ecology

When he coined the term in 1866, Ernst Haeckel defined *ecology* as "the total science of the connections of the organism to the surrounding external world." Today, the word *ecology* is commonly used within green philosophy in two different senses. The first sense defines ecology as the study of relationships among biological and nonbiological processes making up ecosystems and the biosphere (as well as referring to the systems themselves being studied). The second sense defines ecology as a philosophical position emphasizing the interconnectedness of human societies with the natural world, as well as the social movement dedicated to putting that philosophy into practice.

Development of Scientific Ecology

The idea of the world as an interdependent system exists in varying forms in a variety of cultures, but the primary influence on the modern science of ecology has come from Western intellectual tradition. Ecological science grew out of classical natural history. Ancient Greek and Roman thinkers, notably Aristotle and Pliny the Elder, proposed classification systems for the organisms and physical regions of the world that integrated them into a conceptually coherent scheme. During the European Enlightenment, a priority was placed on empirical testing of claims about the workings of the natural world. This led to discarding major portions of classical natural history and formalizing scientific societies and research institutions such as botanical gardens. During this period, several thinkers promoted the idea of the natural world as an integrated and interdependent whole. Notable among them was Carl Linnaeus. Linnaeus is most famous for inventing the modern classification system for organisms, into which he inserted humans as just one more species. The necessity of each to the harmony of the whole indicated that the entire system was created as a unit, and that new species could not appear, nor could existing ones go extinct.

The European colonial experience made a significant mark on the development of ecology. Colonialism brought Europeans into contact with a wider variety of ecosystems than they had previously been familiar with. Intensive exploitation of natural resources, particularly in circumscribed areas such as the island colonies of the Caribbean, forced colonial rulers to develop better understandings of the relationships among such factors as forest cover, climate, and erosion. Naturalists working in the colonies often drew on ecological ideas from colonized peoples, such as traditional Hindu and Muslim botany in India. Among the notable authors of this period was Alexander von Humboldt, who documented relationships between abiotic and biotic factors in his travels through South America. One of Humboldt's most famous findings was that the plant communities of the Andes formed a regular series arranged from the lowlands up to the peaks, governed by temperatures at different altitudes. He emphasized both the meticulous collection of empirical data as well as faith in the underlying unity of natural processes.

Western ideas about the natural world were revolutionized by Charles Darwin's theory of evolution. Darwin built on the findings of naturalists who went before him, but his insistence that species appear, change, and go extinct over time directly challenged ideas of divinely ordained permanent harmony. Instead, evolution through natural selection posited a different mechanism for establishing the interrelations among all species as well as the abiotic factors in their environments. Ecosystemic relationships could now be seen as the unplanned product of mutual adaptation of various species to each other's presence in the environment. In the early portion of the 20th century, ecological science emphasized a contrast between natural harmony and exogenous disturbance. According to thinkers such as Frederic Clements, each region of the Earth has a natural climax ecosystem. If left alone long enough, plant and animal communities will progress through a series of stages to reach the climax—a process called "succession." Once reached, the climax would be stable and maximize biodiversity. However, exogenous disturbances such as human interference are capable of pushing conditions back toward earlier successional stages.

In the middle portion of the 20th century, ideas from general systems theory—originally promoted in order to manage the massive military efforts of World War II—were applied to ecology. Ecologists focused on closed model ecosystems, such as islands, and investigated how pathways of nutrient and energy flow bound the various ecosystem components together. During this period, ecologists had high hopes of deriving a set of universal laws of ecology that could be validly applied to any ecosystem. Proposed laws included the proportionality of an ecosystem's size and isolation to its biodiversity, the correlation between biodiversity and ecosystem stability, and mathematical models of equilibrium between predator and prey populations (the so-called Lotka-Volterra equations).

The most ambitious hopes of systems ecologists hit a bump in the late 20th century with increasing recognition of the inherent disequilibrium in ecological systems. Ecologists are divided over whether these findings constitute a revolution in the discipline (moving from an equilibrium-based "old ecology" to a disequilibrium-embracing "new ecology") or whether they simply extend older work. In either case, the "new ecology" emphasizes that disturbances are an inevitable and necessary part of ecosystem dynamics, and that plant and animal communities are often temporary and ad hoc combinations rather than perfectly balanced wholes. For example, it was once believed that as the ice sheets retreated at the end of the last ice age, ecological communities moved northward as packaged wholes, because each species was dependent on the others in its ecosystem to survive. But it is now believed that species moved northward at differing paces, creating new combinations that continued to rearrange themselves under changing climatic conditions.

Basic Ecological Principles

Several basic principles unite work on scientific ecology. The most central concept is the ecosystem. An ecosystem is an assemblage of populations of organisms, as well as abiotic factors (e.g., geology, climate) sharing a geographical space and influencing each other. Ecosystems can be studied as a series of nested scales, from extremely local to continent-spanning. These broadest scale ecosystems are often referred to as biomes. All of the Earth's biomes then make up the biosphere, a single overarching system encompassing all life on Earth as well as its interactions with the lower atmosphere and upper crust (lithosphere). The boundaries of ecosystems and biomes are defined by (1) the falloff in the intensity of interaction among factors and (2) changes in the species composition and/or physical structure (e.g., forest versus grassland). Ecosystems rarely have neat boundaries. Transition zones between ecosystems are referred to as "ecotones." Within an ecosystem, each species is said to occupy a niche. The same niche can be occupied by different species in different ecosystems, as with groundhogs in North America and wombats in Australia, which both fill the "mid-sized burrowing herbivore" niche.

One key characteristic of any ecosystem is its biodiversity. Biodiversity refers to the amount of variability within and between organisms in the ecosystem. Higher biodiversity is generally seen as desirable; for some philosophical positions it is an intrinsic good, while others simply point to the greater resilience of high-biodiversity ecosystems. The organisms and abiotic components of an ecosystem are bound together by a series of biogeochemical cycles. These cycles move energy and nutrients through the system. For example, the carbon cycle can be traced beginning with the stock of carbon in the atmosphere (in the form of carbon dioxide). Carbon dioxide is taken in by plants and transformed into various carbohydrates through photosynthesis. The carbon is passed to herbivorous animals when they eat the plants. These animals' respiration returns some of the carbon to the atmosphere, while the remainder is absorbed by carnivorous animals and/or decomposing microorganisms. Oxygen, nitrogen, phosphorus, and sulfur also travel through critical biogeochemical cycles. Most of these cycles pass through the food web, defined as the network of the "who-eats-who" relationships in an ecosystem. At each step of the food web, the ecosystem can support fewer organisms. Thus, it is typical for an ecosystem to have a large number of plants, fewer herbivores, and just a few carnivores.

The overall composition of an ecosystem tends to change over time. Ecologists once held that for each place on Earth, there was a stable natural climax condition that ecosystems would reach if given enough time. Later findings suggested that ecosystems move through an "adaptive cycle." The cycle has four phases: exploitation, conservation, release, and reorganization. During the exploitation phase, organisms move quickly to take advantage of opportunities in their environment. As all of the niches begin to get staked out, the ecosystem moves into a conservation phase centered on the maintenance of stable relationships among species. But the conservation phase cannot last forever; the structure of the ecosystem becomes increasingly rigid, and conditions begin to invite disturbances, for example, by stockpiling fuel for a wildfire or food for a beetle outbreak. When the disturbance hits, the ecosystem enters the release phase. The complex structure that had built up over the previous two phases is torn down, and nutrients and opportunities are thrown open. This release makes reorganization possible during the fourth phase, and a new ecosystem can potentially be built up that is quite different from the old one. Eventually the uncertainty of the reorganization phase is resolved in favor of some course of development, setting in motion a new exploitation phase.

Attention to scale—both in time and in space—is critical to understanding ecosystem processes. What looks like a major disturbance at a local, short-term scale may be a necessary part of a stable cycle when examined at a regional, long-term scale. Moreover, processes occurring at one scale can influence those at other scales. Broader-scale processes often act as a check or flywheel stabilizing narrower-scale processes (e.g., by influencing the direction the system takes during a reorganization). At the same time, narrower-scale processes can cascade up to revolutionize broader-scale dynamics, as when a local event triggers a wider disturbance and systemic release phase.

Human Ecology

The majority of ecological research has focused on natural systems in isolation from human activity, or has framed human activity as an exogenous disturbance. However, the name *human ecology* has been adopted by a number of loosely related attempts to integrate human societies into the ecological perspective. One variant of human ecology has sought to simply apply models and metaphors from nonhuman ecology to social dynamics. One such example is the human ecology of Robert Park's "Chicago school" of urban sociology. The Chicago school had little to say about the natural environment as such, but it treated human communities as populations undergoing competition and succession much like populations of other species would.

Other variants of human ecology have aimed to treat humans as just one more organism in the ecosystem, subject to the same laws and processes. Proponents argue that separation of humans and nature is based on an unjustified idea of human specialness or a Cartesian mind–body dualism. Opponents of integration point out features unique to human social systems that are not found in other ecological systems, such as language, foresight, and reflexivity (theories about human systems are part of those systems, whereas nonhuman nature contains no theories about itself). A leading resolution of the tensions within human ecology is the idea of a "coupled human–environment system." The coupled system view holds that human and nonhuman systems each have their own unique dynamics, but neither can be understood apart from its connections to the other. Coupled systems research has aimed to understand how human activities are driving changes in the biosphere, such as climate change and tropical deforestation.

The Ecology Movement

Ecological science was based on a view of natural systems as inherently interconnected wholes. This view was quickly translated (by ecologists and nonecologists) into a philosophical standpoint that went beyond analyzing the workings of ecological systems. It is common to divide the environmental movement into two general tendencies—a moderate, reformist ("light green") side and a radical ("deep green") side. When used to refer to a philosophy and a movement, "ecology" typically denotes the radical branch. Reformist environmentalism may make extensive use of ecological science in order to guide resource use and conservation strategies. But radicals charge that the goals being pursued by the reformists—such as improvement of human welfare and manipulation of nature—are based on a worldview in which humans are separate from, and superior to, the rest of nature.

Ecology as a philosophy has roots as deep as those of scientific ecology. Its modern form is usually traced most directly to the late-19th- and early-20th-century wilderness movement (e.g., John Muir) and urban hygiene movement (e.g., Jane Addams). Aldo Leopold laid out a "land ethic" capturing many of the main concerns of philosophical ecology after becoming disillusioned with the prevailing reformist approach to environmental conservation. The origin of the modern manifestation of the ecology movement is typically dated to the 1962 publication of Rachel Carson's *Silent Spring*. In her book, Carson linked scientific ecological findings about the effects of toxins on ecosystems to a larger concern about human society's hubris in delinking itself from natural systems.

Philosophical ecology is comprised of a large number of specific subphilosophies, but they are conventionally grouped into an opposition between "deep ecology" and "social ecology." The deep ecology branch, which includes deep ecology proper as well as bioregionalism and related viewpoints, emphasizes the moral equality of all species and ecosystems with humans. Deep ecologists typically place the blame for human destruction of nature on anthropocentric (i.e., human-favoring) cultural or philosophical outlooks that have become dominant especially in the West.

Social ecology encompasses green socialist, anarchist, ecofeminist, and environmental justice perspectives. While agreeing that destruction of nature and a loss of harmonious human–nature relationships are serious problems, social ecologists point to structural injustices within human society as the main culprit. Capitalism, patriarchy, and other oppressive systems are said to produce highly unequal impacts on people from environmental problems. Varieties of social ecology that are heavily influenced by postmodernism have been particularly interested in the "new" scientific ecology, seeing in its embrace of disorder and contingency parallels with postmodernism's rejection of social coherence and order. Nevertheless, most practicing scientific ecologists deny that their findings support postmodernism, as the "new ecology" is still rooted in a scientific search for reliable, systematic truths about how the world works. Critics of philosophical ecology cite promethean and critical theories. Promethean critics assert the moral primacy of humans over the rest of the biosphere, as well as the feasibility of dominating and manipulating nature for human ends. Critical theory questions the idea that human societies could or should be modeled on an ideal of an objectively knowable, harmonious ecosystem. In many cases, the line between social ecology and critical theory is blurred.

See Also: Biodiversity; Ecocentrism; Ecological Restoration; Haeckel, Ernst.

Further Readings

Begon, M., C. R. Townsend, and J. L. Harper. *Ecology.* Oxford, UK: Blackwell, 2006.

Dobson, A. P. *Green Political Thought.* London: Routledge, 1995.

Gunderson, L. and C. S. Holling. *Panarchy: Understanding Transformations in Human and Natural Systems.* Washington, DC: Island Press, 2001.

Odum, E. P., et al. *Fundamentals of Ecology.* Florence, KY: Brooks Cole, 2004.

Stentor Danielson
Slippery Rock University

Philippe Boudes
Paris 7 University, LADYSS-CNRS

Economism

"Economism" means the dominance of "economic" ways of thinking and acting and denotes economics, specifically neoclassical economics—the dominant and most powerful and widespread form of economic thinking in public policy and political debate—as a form of ideology.

The imputed value-free (and therefore nonethical and nonpolitical) character of modern neoclassical economics has perhaps done most to establish the predilection for economism within much contemporary political and environmental political debate. Simply put, economics is not and simply cannot be an "ethics free zone." It is not value-free, objective, and scientific. In fact, much of the criticism of economics is motivated by a desire for supporters of neoclassical economics to "come clean" about the value judgments and ethical positions that underpin it. Modern economics is as ideological and value based as feminism, conservatism, nationalism, socialism, or green politics. This is why *political economy* is a more accurate term to describe any approach to the economy, and this explicit political and ethical aspect is something that was understood in the history of economics by great economic thinkers such as Adam Smith, Thomas Malthus, John Stuart Mill, and

John Maynard Keynes. However, unlike these political positions, modern orthodox economics has evolved to a "mythic" status in its almost "full spectrum" domination, widespread acceptance, and infiltration of many aspects of modern life and ways of thinking, within and far beyond the academy. The power of economism, therefore, lies in its everyday, commonsensical quality in that neoclassical economic thinking is viewed as "the" way to think about the economy and the "economic"—that is, other ways of viewing and thinking about the economy are often not ever considered or if they are, they are considered as novel, strange, dangerous, or merely of historical interest. Economism therefore is an ideological reading of how we should think about economic issues.

This ideological reading of modern economic thinking shows it to construct a fantasy world of monetary valuation and human productive activity that has only tenuous links to the real source of all economic activity—namely, the ecological dynamics of nature and the biological and social needs of human beings. The global and globalizing economy operates in a fantasy world of "24-hours-a-day/7-days-a-week" work and production patterns, and quarterly returns that fly in the face of the ecological time of the seasons and regenerative capacities of natural resources as well as the biological and psychological time human beings need not only to function efficiently but to flourish. Canadian political theorist John McMurtry would go further and claim that the orthodox economics taught in universities and used to structure and inform state and business decisions in investment, production, employment, and so on, is an economics in the service of, in the title of one of his books, the *Cancer Stage of Capitalism*. The argument here, and one that echoes elements of the *Limits to Growth* analysis, is that growth for growth's sake is the logic of the cancer cell, and that is precisely what we have today in orthodox economics promoting growth that does not add to quality of life or long-term sustainable development.

In addressing the status of economics as a science, critics argue that we have to say loudly and often, "the emperor has no clothes." That is, the dominant economic paradigm is not a science but has been falsely trading on its reputation as a "hard social science," in contrast to the "soft" social sciences of sociology, cultural studies, and politics and the non–social scientific disciplines of the humanities such as philosophy, for example. This lack of predictive capacity (that we are taking as constitutive for a body of knowledge to count as a "science" in the manner of the natural sciences) does not mean that we view the knowledge produced by conventional economics as useless. Conventional economics has produced knowledge and findings that have been useful. The point here is that economics cannot persist in the pretense that it is a value-free and predictive social science and that, consequently, its usefulness and rationale needs to be grounded upon other principles.

That economics is not value free, objective, or neutral is a standard argument leveled against orthodox or neoclassical economics. The normative character of conventional neoclassical economics can be gleaned from any undergraduate textbook on the subject. These textbooks speak of a world of "rational consumers" and "utility maximizing individuals" with determinate "consumer preferences" in a world of "perfect competition" and so on. The *homo economicus* that forms the bedrock of much neoclassical economic thinking (particularly microeconomics) is not only a fiction (a fantasy to return to one of the features of autism discussed above) but a deeply normative and not to say ideological fiction. The human subject of economics that is, is not described or simply reflected in economics, but conventional economics actively creates and prescribes this human subject as an ideal to be attained. In other words, what economics is doing is not describing how the world and human beings are (facts) but mandating or prescibing how the world and human beings should be (values or ideology).

Now, while all forms of human knowledge from physics to philosophy make generalizing and simplifying assumptions, within neoclassical economics there is a clear sense that the *homo economicus* and the associated human economic experience discussed are simply how human beings are and how the economic world is. That is, these models are not simplifying assumptions so much as capturing the essential character, motivation, and modus operandi of human beings when they enter economic relationships. Human beings as revealed by neoclassical economics qua *homo economicus* or "economic man" (the gendered connotations of this are noteworthy and, of course, criticized by ecofeminists) are essentially selfish, individualistic, hedonistic, and possessed of desires that both explain their behavior and these desires can never be satisfied. And herein lies the essence of economism—it does not objectively describe the world as it is, but rather, is actively engaged in a project to prescribe how the world ought to be and work to bring that world into being.

See Also: Ecofeminism/Ecological Feminism; Ethics and Science; Green Altruism; Human Values and Sustainability.

Further Readings

Barry, J. *Environment and Social Theory*, 2nd ed. London: Routledge, 2007.
McMurtry, J. *The Cancer Stage of Capitalism*. London: Pluto Press, 1999.
Nutter, G. W. "On Economism." *Journal of Law and Economics*, 22/2 (1979).

John Barry
Queen's University Belfast

Ecopedagogy and Ecodidactics

Ecopedagogy describes education that is designed to provide learners with the skills, knowledge, values, and attitudes required to make the transition toward sustainable living. Even though this particular kind of education is sorely needed in the light of the global environmental crisis, it has only recently gained acceptance into mainstream curricula. Education systems at all levels have so far largely failed to prepare learners for the challenges arising from the crisis. Ecopedagogy has three overarching aims: to compensate for the transmission of harmful and counterproductive learning outcomes that commonly occur through conventional education; to elicit more productive learning outcomes, including ecological concepts and the values, beliefs, and attitudes that provide the basis for sustainable living; and to liberate the learner from exploitative dependencies and systemic constraints. The strategies employed for teaching and learning toward those aims is referred to as ecodidactics.

In spite of intense public discussion and political initiative at all levels from global to local, humanity has not made much progress toward making its tenure on Earth sustainable. Resources are depleted ever faster as the global human population and its consumption patterns continue to grow out of control; pollution continues with its effects on climate, habitat quality, and public health; species continue to disappear at an alarming rate as humans introduce competitors, modify ecosystems, deplete habitats, and modify landscapes and climates. Five self-reinforcing causes have been identified: economic growth, population growth, technological expansion, arms races, and growing income inequality. Recognizing that education can make a decisive difference in preparing the next generation of decision makers to perform better, the United Nations declared 2005–2014 the Decade of Education for Sustainable Development. Numerous regional programs are beginning to make a difference, albeit at a distressingly slow pace.

Part of the explanation for this slow progress lies in the failure of mainstream education to elicit the requisite understanding of what ecological sustainability means and to instill the appropriate values and attitudes. In fact, the curricula (in their explicit as well as their hidden forms) of leading educational institutions, as well as the decisions and actions of the majority of their graduates, suggest that formal education has contributed more to the problems than to their solutions. Detrimental learning outcomes include such beliefs as economic growth as a good in itself, denial of limits to growth, freedom from "nature" and dominion over it, omnipotence of science and technology, and such attitudes as complacent optimism, moral nihilism and materialism, consumerism, and neoliberal individualism.

The first overarching aim of ecopedagogy, therefore, is to compensate for the harm already done by formal education, by the global media and entertainment industry, and by general cultural influences. This may include the deliberate unlearning of factually wrong information, and it invariably necessitates a change in personal values and the acceptance of moral responsibility. The second aim includes the promotion of neglected learning outcomes that bring significant benefits toward sustainability. Besides a foundation in human ecology and basic sustainable life skills, they include an awareness of our dependence on "nature," of our integration within the natural environment, and of the limits to consumption and to technological solutions. It is important that the learner recognize the transdisciplinary significance of those concepts instead of following the tradition of relegating them to the field of environmental science (and leaving them there unexamined). Some cognitive skills are to help the learner conceptualize ecological overshoot, such as the ability to perceive one's environment in a holistic way, to extrapolate to global dimensions, and to the long term. This group of outcomes has been referred to as "ecoliteracy." Requisite values and attitudes include a concern for future generations, respect for nonhuman nature, and an ecocentric environmental ethic.

Ecopedagogy pays particular attention to concepts of progress. The learner is encouraged to critically analyze dominant concepts of progress for contradictions arising from their ideological content of consumerism, laissez-faire, cornucopianism, technologism, and scientism. It is expected that the learner adopt a personal ethic of sustainability by extending existing values of justice toward future generations and ecosystem restoration.

Efficiency, restraint, adaptation, and structural reform to facilitate sustainable practices are to become moral norms. The goal is that the learner adopt a concept of progress that is informed by sustainability, and that he or she acquires a vision for and awareness of the future that includes change and sustainable solutions.

Learners who achieve the outcomes described above are equipped for the transition to sustainable living. Yet one significant group of obstacles remains that tends to prevent the learner from turning thought and sentiment into action. Often these obstacles manifest themselves as feelings of powerlessness, of disenfranchisement, and the absence of choices. Ecopedagogy therefore includes as its third overarching aim the empowerment of the learner to take action and to make a difference. Strategies from other liberation pedagogies are adapted and implemented in three stages. First, "conscientization" of the learner allows him or her to recognize ideological hegemonies (e.g., corncucopianism) and unjust dependencies (e.g., consumerist habits and market pressures), to develop an ecological conscience and a concept of self-worth that defines itself less on peer approval and social status and more on explicit and internal moral criteria. Conscientization then leads to the empowerment of the learners toward critique, action, and accepting responsibility. Empowerment, in turn, allows the learners to liberate themselves from exploitative dependencies and the influences of counterproductive ideologies.

From this brief summary of ecopedagogy, four recurrent themes emerge. It takes a holistic view of the individual and humanity; it emphasizes the decisive role of values in education and in life decisions; it encourages the open critique of ideologies that usually exert their influence implicitly; and it seeks to empower the learner toward action.

See Also: Consumption, Consumer Ethics and; Human Values and Sustainability; Intergenerational Justice.

Further Readings

Fien, Jon. *Education for the Environment: Curriculum Theorising and Environmental Education.* Geelong, Australia: Deakin University Press, 1993.

Freire, Paolo. *Pedagogy of the Oppressed.* New York: Continuum, 1986.

Lautensach, Alexander and Sabina W. Lautensach. "A Curriculum for a Secure Future: Agenda for Reform." In *The Havoc of Capitalism: Educating for Environmental and Social Justice,* Juha Suoranta, Donna Houston, Gregory Martin, and Peter McLaren, eds. Rotterdam, Netherlands: Sense Publishers, 2009.

Orr, David R. *Ecological Literacy: Education and the Transition to a Postmodern World.* Albany: State University of New York Press, 1992.

Orr, David R. "Rethinking Education." *The Ecologist,* 29/3 (1999).

United Nations. *Learning Our Way to Sustainability* (2009). http://www.unesco.org/en/esd/ (Accessed November 2009).

Wackernagel, Mathis and William Rees. *Our Ecological Footprint: Reducing Human Impact on the Earth.* Oxford, UK: John Carpenter, 1996.

Alexander K. Lautensach
University of Northern British Columbia

Ecophenomenology

Ecophenomenology is a less-known branch of environmental philosophy compared to land ethics, deep ecology, and ecofeminism. Despite this status, it is an environmental philosophy that is gaining in popularity with theorists and is quickly maturing in its sophistication and argumentation. This article briefly looks at how ecophenomenology has developed, what it is, a few of its theorists and practitioners, and gives a quick summary of various ecophenomenological standpoints.

Ecophenomenology builds upon and adds to the phenomenological method found within the humanities, and especially within the analytical method of philosophy. Edmund Husserl (1859–1938) developed and advocated modern phenomenology as a

"Philosophy of Rigorous Science" in response to the empirical naturalism and logical positivism of Europe that was gaining ascendancy in the early 1900s. *Phenomenology* as a term is derived from the Greek *phainomenon,* meaning appearance, thus phenomenology is the study of appearances. Husserl argued that humans are able to understand the physical, material world around them (the phenomenal world) through intuition and sense experience. This view was subsequently challenged by Martin Heidegger (1889–1976) who added the concept of *dasein,* or "being-in-the-world," to the phenomenological tradition. Heidegger nonetheless agreed with Husserl that phenomenology is both hermeneutic and interpretive. Furthermore, both maintained phenomenology must be qualitative and descriptive, meaning the natural, sense-and-corporeally experienced phenomenal world is accounted for with descriptive statements. For Husserl, this is the "lived world," knowable and experienced in a present "now" of "lived time." Maurice Merleau-Ponty (1908–61), one of the foundational figures in ecophenomenology, also came out of this phenomenological tradition. Other early philosophers and theologians who have helped create the analytical ground from which ecophenomenology has grown include the Jewish monist philosopher Baruch Spinoza (1634–77) and the process theologian Alfred North Whitehead (1861–1947).

One of the goals of ecophenomenology is to do away with the inherited philosophical tradition (dating back to Socrates via Plato) of positing the mind (and especially the capacity to reason) as a separate, even transcendent, phenomenon from the natural world and upon which human exceptionalism is based. Rather, ecophenomenologists argue that our bodies and minds are actually part of the natural world within which they are embedded. Ecophenomenology argues that as we are evolved mammals—with senses, bodies, and minds that both shape and are shaped by and which are part of our surrounding environments and ecosystems—any epistemological and ontological understanding of nature, and thus understanding of our relationship to nature, must be built upon this insight. The ecophenomenological method therefore includes describing our embodied, reciprocal relationships with nature, including what this relationship entails for our own self-understanding and idea of what counts as meaning in an embodied, phenomenal world. From this position, our reciprocal, embedded, and embodied experiences with and within nature are seen to be loci of value, a priori of theoretical and philosophical abstraction and examination. Ecophenomenology argues that this value-laden, a priori world of sense experience assumes ontological and epistemological priority in the philosophical tradition.

Early hints of subsequent ecophenomenological standpoints can be seen in Aldo Leopold's land ethic, when he writes that humans can only value things that they first feel and love. These insights were also shared by John Muir, who developed a pantheist love of nature and wilderness based in large part on embedded sense experience with flora and fauna in their various ecosystems. Both Leopold and Muir challenged humans of their eras to learn to value nature from a place of relationship and phenomenological embeddedness.

Another early example of ecophenomenological work, with attendant repercussions on how humans could potentially perceive and thus change their habitual relationships with the environment, is seen in anthropologist Keith Basso's well-known article "Stalking With Stories." Based on ethnographic fieldwork with Apache Native Americans, this article describes how, in the Apache worldview, places actively interact with and thus shape human self-understanding and values, including those values and duties that humans have toward the natural world.

Such an animistic perception of the world and the repercussions of this worldview were taken up in philosopher David Abram's epochal book *The Spell of the Sensuous.* In this neoanimist/neoshamanic book, Abram argued that the "more-than-human" world that humans are in constant relationship and dialogue with is made in large part from a collection of intelligent subjects worthy of moral reasoning and care. He further argued that the natural world is a phenomenal world of mystery, so that approaching this world through compartmentalized, analytic logic based on written, phonetic language does not open humans to the fullness of relationship that is possible with the natural world and thus within our embodied selves. Parts of this relationship with the wider natural world will always remain beyond human knowledge and understanding, yet this does not detract from the value that the more-than-human world entails. The publication of this book (which went on to become a best seller) by Abram has since given the bourgeoning field of ecophenomenology an established classic.

Other environmental philosophers have emphasized embeddedness and direct sense experience of nature, including Edward Abbey, Henry Thoreau, Rachel Carson, Joseph Wood Krutch, and Gary Snyder. These authors and philosophers maintain that our sense of self must include our physical, sense-based phenomenal experience of nature and they all can be considered to have helped lay the foundation for today's ecophenomenology. Equally, the ecofeminist move toward the "erotic" and recognition of intersubjectivity and interexperientiality can also be said to have affinities with the ecophenomenological standpoint. Lastly, an ecophenomenological approach has developed in the field of architecture, looking at how constructed and lived spaces interact with and shape human sense perceptions and environmental values.

It should be noted that for many, ecophenomenology carries an underlying normative claim. From this perspective, one of the goals of ecophenomenology is to correct an inherited dualist, hierarchical value of the human capacity for reason over the nonhuman world at large, where such a privileging of human rationality has helped lead to the current ecocrisis. Rather, ecophenomenologists advocate an environmental philosophy based on nondualist, iterative, sense-based perception that will lead to valuing nature more fully as a cocreator of our embedded epistemologies.

See Also: Carson, Rachel; Conservation, Aesthetic Versus Utilitarian; Deep Ecology; Ecofeminism/Ecological Feminism; Land Ethic.

Further Readings

Abram, David. *The Spell of the Sensuous: Perception and Language in a More-Than-Human World.* New York: Pantheon Books, 1996.

Basso, Keith. "'Stalking With Stories': Names, Places, and Moral Narratives Among the Western Apache." In *The Nature Reader,* Daniel Halpern and Dan Frank, eds. Hopewell, NJ: Ecco Press, 1996.

Brown, Charles and Ted Toadvine, eds. *Eco-Phenomenology: Back to the Earth Itself.* Albany: State University of New York Press, 2003.

Casey, Edward. *The Fate of Place: A Philosophical History.* Berkeley: University of California Press, 1997.

Howes, David. *Sensual Relations: Engaging the Sense in Culture and Social Theory.* Ann Arbor: University of Michigan Press, 2003.

Husserl, Edmund. *General Introduction to Pure Phenomenology.* Boston: M. Nijhoff, 1982.

Ingold, Tim. *The Perception of the Environment: Essays on Livelihood, Dwelling and Skill.* London: Routledge, 2000.
Leiter, Brian and Michael Rosen. *The Oxford Handbook of Continental Philosophy.* New York: Oxford University Press, 2007.
Seamon, David. "A Way of Seeing People and Place: Phenomenology in Environment-Behavior Research." In *Theoretical Perspectives in Environment-Behavior Research.* Seymour Wapner et al., eds. London: Kluwer Academic/Plenum Publishers, 2000.
Toadvine, Ted. *Merleau-Ponty's Philosophy of Nature.* Evanston, IL: Northwestern University Press, 2009.

Todd LeVasseur
University of Florida

Ecopolitics

Ecopolitics, more commonly known as "environmental" or "green politics," is, along with feminism, the newest political ideology and associated social and political movements of the postwar era. One finds in green politics an extremely broad understanding of the scope of "politics" and the "political" that encompasses almost everything one does, including one's choices about consumption, transport, waste, fertility, food, job, and so on. Often these are presented in such broad terms that they can be viewed as "pre-political" or as simply too big, urgent, and important for normal "politics." This is presented in either the sense that green politics is about issues of ecological survival that make "politics" irrelevant or that the real "cause" of the ecological crisis is "beyond politics" and has to do with "deeper" dynamics of the ecologically destructive consciousness and ignorance of humanity and/or the spiritual malaise of modernity. Greens thus vacillate between a "common sense" account of politics as the dirty machinations for state power and thus to be avoided and one in which "politics" is an all-encompassing facet of life.

In part, this neglect is also because of the "neither left nor right" element in green movements, which is "anti-political" and eschews serious discussion of central elements of politics in favor of ideological exhortations concerning the "common interests of humanity"—humans should be seen not so much as citizens, workers, and so on, but as "plain members" of the "land community," according to A. Leopold. Greens are thus, for the most part, instinctive global thinkers, and "think globally, act locally" has become one of their best-known catchphrases. But this naïve exhortation concerning common global interests has been progressively harder to sustain. In green theory, there has thus been a loss of innocence marked by a stepping back from an anarchist rejection of the state. As R. Grove-White argues, modern environmentalism needs to be seen as having evolved not only as a response to the damaging impacts of specific industrial and social practices but also, more fundamentally, as a social expression of cultural tensions surrounding the underlying ontologies and epistemologies that have led to such trajectories in modern societies.

Green politics has some unique features compared with nongreen conceptions of politics, some of which are outlined below. The first issue concerns the temporal frame of green politics: as expressed in its central concern with ecological sustainability and sustainable development, it suggests the integration of a concern for the future and for future generations. From a green political point of view, the future needs to be included as an explicit, rather than implicit, dimension of contemporary politics to ensure decisions today do not detrimentally affect those yet to be born. The second issue has to do with the fact that ecological problems do not respect national or cultural boundaries. This is an issue of scale. Pollution problems—such as, most dramatically, climate change—are transnational and global in scope. Thus politics and thinking and acting politically must also be transnational and global in scope and approach. It can no longer (as if it ever could) be solely concerned with a particular society independent of other countries and international processes. This facet of green politics is captured in the famous green movement slogan of "acting locally, thinking globally." The third issue, perhaps most unique to green politics (and contentious), is that political thinking and considerations of what is "politics" can no longer remain within the species boundary, that is, being solely or primarily concerned with human social relations and phenomena. Thus, green politics is unique among all other approaches to understanding and practicing politics in its challenge to the "human-centeredness" or anthropocentrism of dominant conceptions and practices of politics (and ethics). For example, from a green political point of view, "harm" is not something that is limited either in reality or its normative scope to only humans. If we care about reducing or minimizing harm (and suffering), then we are compelled to extend our political and moral thinking beyond humanity to include the nonhuman world.

Green politics involves the necessary and desirable bridging of the gap between society and nature. T. Hayward stated the following:

> The most distinctive green idea is that of natural relations. These are of numerous kinds: there are natural relations of biological kinship between humans, on which familial and social relations are supervenient; between humans who are not kin, too, relations are naturally mediated, for instance in the sense that reproductive and productive activities occur in a natural medium; such activities normally involve modifying the natural environment in some way, and all humans, individually and collectively, have relations to their environment.

This suggests that green politics has a naturalistic perspective that has two main components: first, a sensitivity to the significance of the natural environmental contexts, preconditions, opportunities, and constraints on human activity; second, a recognition of the centrality of internal human nature, of seeing humans as natural beings with particular modes of flourishing, like other natural beings. The integration of biological and ecological insights into political theory would produce a form of politics that began from acknowledging human biological embodiedness and ecological embeddedness. The most advanced thinking in this aspect of green politics is within certain strands of ecofeminism. In terms of ecological embeddedness, the greening of social theory involves the acceptance of ecological limits and parameters to collective human activity. As K. Lee puts it, given the ecological facts of the world, and our dependent relationship with it, "any adequate social/moral theory must therefore address itself to these characteristics [of the world] and the character of the exchange [between humans and nature]. If it does not, whatever solution it has to offer is of no relevance or significance to our preoccupations and problems."

Ecology: Connecting the Natural and Social Sciences?

Ecology, or the "ecological paradigm," has, for some, since its emergence in the last century, promised a "unified science of nature and society." Ecology and its aim of a "naturalistic" account of the human condition, while originating as an empirical natural science dealing with the relationship between species and their environments, has also become a form of social and moral theory.

However, the science of ecology (dealing with facts and the way the natural world is) has tended to go hand in hand with normative claims (dealing with values and how the world should be and how we ought to treat and use nature) and has found it difficult to maintain a strict and lasting separation between facts and values. Eroding this strict distinction has placed ecology in a unique position as a "science," as a form of knowledge that seems to bridge the natural and social sciences. This can be seen in how environmental studies or environmental management programs in universities necessarily have to straddle the social and natural sciences, reflecting the interdisciplinary character of the forms of knowledge appropriate to articulating social–environmental relations. This scientific basis is yet another unique feature of green politics, meaning it is the only political ideology with a firm basis in modern science.

Time and Future Generations

One of the central claims of green politics is its concern to extend the temporal dimension of our thinking and action to include a concern and recognition of the needs, interests, and rights of those yet born. This reorientation of politics toward the future is at the heart of the idea of sustainable development, with its explicit recognition that present decisions about how we treat and use the environment cannot be made without considering the likely impact of these decisions on the type of environment that we leave to future generations, and thus the effect the latter may have on the welfare of the future or on the ability of the future to meet its needs. Thus the greening of politics requires lengthening its temporal frame, as B. Adam suggests in her 1998 work on the relationship between environment and social theory. A sense of the future dimension involved is the Native American saying that "We don't inherit the Earth from our parents, but borrow it from our children." How far into the future is, of course, an open and important question, the answer to which will have important implications for society. However, the key issue is that greening social theory implies extending its temporal range into the future, in particular, making the likely ecological impacts of present courses of action an explicit object of analysis. As Adam puts it:

> We learn about and relate knowledgeably to a multidimensional space, but our understanding of the temporal dimension of socioenvironmental life is pretty much exhausted with knowledge about the time of calendars and clocks. Nature, the environment and sustainability, however, are not merely matters of space but fundamentally temporal realms, processes and concepts . . . without a knowledge of this temporal complextity . . . environmental action and policy is bound to run aground, unable to lift itself from the spatial dead-end of its own making.

She goes on to develop the centrality of adopting new modes of conceptualizing time and their associated modes and patterns of living and interacting with the environment.

She suggests "industrial time" based on a simplistic linear perspective and Newtonian physics is at the heart of environmental problems and our lack of concern for future generations. For her, the following is true:

> Industrial time is centrally implicated in the construction of environmental degradation and hazards; second, as a panacea it worsens the damage. Industrial time, in other words, is both part of the problem and applied as the solution. As long as time is taken for granted as the mere framework within which action takes place and is used in a pre-conscious, pre-theoretical way . . . it will continue to form a central part of the deep structure of environmental damage wrought by the industrial way of life.

This suggests not only the need for social theoretical research into the industrial conception of time and how it is connected to environmental destruction, but also nonindustrial and non-Western conceptions of time and futurity.

Beyond the Nation-State and Globalization

Green politics requires the adoption of an international and global perspective, since social–environmental relations are not always contained within a specific geographical area and the effects of social intervention in the environment do not always respect national boundaries. In this way, the greening of social theory requires that social theory be self-consciously international and global in its outlook and analysis. As S. Yearley points out, "environmental dangers pose supranational problems; these need solutions to which national governments are not well suited."

While all conceptions of politics and political thinking have always had an international dimension (though often the various links between societies were assumed, rather than made an explicit object of study), the main focus of politics was on the internal dynamics of society. With the advent of global and transnational environmental problems, highlighting the ecological interconnections between geographically separated parts of the Earth, the study of the origins and effects of and alternatives to these various transnational environmental problems necessarily requires that one must look beyond (and sometimes below) the nation-state. The interconnectedness of the world's economic and ecological systems means that what happens within a society increasingly has its origins in processes outside that society.

This means that one must look not just beyond the boundaries of the nation-state but also at international actors other than the nation-state. These include powerful global nongovernment actors such as transnational corporations (TNCs); weaker but still significant environmental nongovernment organizations (NGOs); and transnational political and economic institutions such as the United Nations and the World Bank. Put another way, theorizing about the environment is a particular aspect of theorizing about globalization. This link between globalization and the environment can be seen in M. Paterson's (1996) view that "if globalization is environmentally problematic . . . the politics of those concerned with environmental problems lies in resisting globalizing forces, such as multinationals, banks and governments, in their attempts to negotiate new international regimes on the environment."

Green politics here often incorporates the ideas of green thinkers and activists who argue that the global political and economic system maintains an exploitative relationship between the affluent northern countries (the so-called developed world) and the poor southern countries (the so-called developing world). Here the greening of social theory suggests analyzing the cultural, political, economic, and ecological effects of globalization within the context of north–south relations. For example, while global environmental threats such as climate change affect everyone on the planet, this does not mean that humanity as a whole is in the same boat. First, it is not "humanity" as a whole that is to blame for global environmental problems; those with economic and political power, mainly developed, industrialized nations, are mostly to blame for causing environmental problems by, for example, being responsible for the vast bulk of global carbon pollution from the use of fossil fuels. Second, environmental problems do not have the same effects on everyone. Thus, for example, affluent countries and groups are better able to protect themselves from environmental problems than poor countries and groups. As J. Seabrook (1998) notes:

> Globalization is not an organic growth, but a carefully wrought ideological project. . . . If the perpetuation of privilege is to be the guiding force of the world, let it be identified as such. . . . If the abuse of the resource-base of the earth and the intensifying exploitation of its people is the supreme civilisational pursuit of the culture of globalisation, no matter what the consequences, why shrink from saying so?

Rejecting the claim that globalization is not something natural or organic—that is, something that is both "given," "inevitable," and beyond human control and also somehow "good"—has recently played a major part in critiques of parties and governments under neoliberalism. In this way, viewing or describing something as a force of nature, a "fact of life," or "natural" has tremendous ideological power in prescribing particular forms of action or inaction. Since environmental degradation is caused by poverty and socioeconomic inequality, stopping environmental destruction requires alleviating poverty, which in turn requires the creation of a fairer global economic system.

Beyond the Species Barrier

According to scholar Brian Baxter, "it is now intellectually unacceptable to develop political theories in which the sole focus of concern is human well-being and values, ignoring the issues which greens have pushed to the fore concerning the well-being of other species, and the biosphere in general." Baxter makes this claim on the basis that there are compelling moral arguments (from within green theory) that mean that it is illegitimate not to extend moral considerability and concern beyond humans to include (at least parts of) the nonhuman environment (animals, plants, ecosystems). This normative approach to including the natural environment and its interests in human social theory has much to commend it, even though it goes against many settled assumptions about our attitudes toward and treatment of the nonhuman world. However, one can also advance reasons why the nonhuman world should be part of the agenda of social theory on the more factual grounds that social–environmental relations are constitutive of human societies. That is, we cannot fully grasp or understand any human society without also understanding the types and meanings attached to the ways in which society views, values, treats, and uses its natural environment.

At the same time, there are those like K. Milton, whose ideas suggest that social theory, at least that part of social theory that is based on reflecting on culture, cannot remain concerned with human culture alone. As she puts it, "As we learn more about both human and nonhuman animals, *it becomes increasingly difficult to sustain the view that culture is uniquely human*" (1996, emphasis added). In this effort she is joined by some recent moral philosophers who hold that morality, in the sense of acting or behaving in accordance with moral principles, like "culture," is not something that is unique to humans. Politics and theorizing about politics can be (indeed, it must be) human based, but does not have to be human centered. That is, while social theory must begin from an analysis of human society, it does not necessarily have to be exclusively concerned with human affairs, interests, and events. For example, given both the coevolutionary history of humans and animals, as well as the various material ways in which human societies use and have relationships with animals, social theory can reorient its aims and objects of study to include these nonhumans as fellow members of "society."

Conclusion

It is rather paradoxical that while green politics seems to be from an examination of the external environment, that it should lead us to look inward toward an examination of "human nature." In this way, one could say that by its very nature (excuse the pun), when the word *environment* is used, one almost automatically moves in the direction of speaking of "external nature," and from there it is difficult to prevent this discussion from spilling over into debates about "internal nature." We begin by analyzing the natural world around us (that which environs us) and our relation to and with it, and find that this cannot be done without reflecting on "us" as embodied beings, a particular species of animal evolved from social primates, and our own "nature." And green politics places before us the ultimate paradox of humanity: that the "human condition" (which at the same time is our "natural condition") is one in which humanity is both a part of and apart from the environment.

See Also: Animal Ethics; Ecofeminism/Ecological Feminism; Leopold, Aldo; "Should Trees Have Standing?"; Sustainability, Seventh Generation.

Further Readings

Adam, B. *Timescapes of Modernity. The Environment and Invisible Hazards*. London: Routledge 1998.

Barry, J. *Environment and Social Theory*, 2nd ed. London: Routledge, 2007.

Benton, T. *Natural Relations: Ecology, Animal Rights and Social Justice*. London: Verso, 1993.

Hayward, T. *Ecological Thought: An Introduction*. Cambridge, MA: Polity Press, 1995.

Lee, K. *Social Philosophy and Ecological Scarcity*. London: Routledge, 1989.

Milton, K. *Environmentalism and Cultural Theory: Exploring the Role of Anthropology in Environmental Discourse*. London: Routledge, 1996.

Paterson, M. "UNCED in the Context of Globalisation." *New Political Economy,* 1 (1996).
Seabrook, J. "A Global Market for All." *New Statesman* (June 26, 1998).
Singer, P., ed. *Ethics.* Oxford, UK: Oxford University Press, 1994.
Yearley, S. *The Green Case: A Sociology of Environmental Issues, Arguments and Politics.* London: Routledge, 1991.

John Barry
Queen's University Belfast

Ecotourism, Responsible

These tourists staying at an ecotourism ranch near the Chaco-Pantanal Reserve in Paraguay in 2004 were transported by truck to a birdwatching site. Issues like this suggest the difficulties in determining whether many ecotourism operations are truly ethical or sustainable.

Source: National Biological Information Infrastructure/ Andrea Grosse

Responsible ecotourism is ecotourism that is ethically justified, based on the assumption that in return for the benefits that tourists receive from travel, they ought to contribute to environmental conservation and to improvement in the economic situation of poor people, and to show respect for culture. Ecotourism is commonly presented as providing substantial economic social and environmental benefits to host countries, especially poor countries, as a sustainable alternative to extractive industries such as the logging of native forests. However, there is widespread concern that unscrupulous operators engage in "greenwashing," using false or misleading claims about the impacts of their operations.

Conventional Tourism

Tourism may be defined as leisure travel for recreational purposes. Tourism, domestic and foreign, is one of the world's largest industries: in 2008, the 922 million international tourists generated revenue of $944 million. Mass tourism is a 20th-century phenomenon, but the earliest evidence of tourism is from ancient Babylon and Egypt. People traveled to visit antiquities, famous buildings, and religious festivals as well as for educational and commercial reasons. By Roman imperial times, travel had become straightforward with no borders from Britain to the Middle East, a common coinage and language, safe travel by sea, and well-maintained roads in many areas. Early modern tourism was also largely cultural and urban, notably the "Grand Tour" of European antiquities, art galleries, museums, cathedrals, and other elements of the Classical and

Renaissance periods. The Grand Tour was regarded as the culmination of the education of well-to-do young English gentlemen.

With industrialization, rising incomes, and improved working conditions, more people were able to travel, especially with the advent of the railway. Cheap air travel in the 1960s made all-inclusive holidays affordable, and hordes of people from northern Europe began to descend on the Mediterranean coast. The vast majority of these tourists were interested only in sun, sea, romance, and cheap wine; ancient villages rapidly became mass tourism destinations, with seriously adverse effects on the environment and local cultures. In the 1970s, resorts sprang up throughout the tropics, notably in the Caribbean, southeast Asia, and the Pacific as well as in parts of Africa, South America, and India. These developments have caused environmental, social, cultural, and economic disruption, especially in small island nations. Fortress-like resorts displace local people and traditional economic activities, destroy forests and coral reefs, and pollute lagoons. Resorts are usually owned by international consortia so the profits go offshore, while rich Western (and, increasingly, Asian) tourists demand air-conditioning, limousines, and the like, which have to be imported along with luxury foods and liquor. As a result, international tourism may actually have a negative impact on the economies of some nations. Resorts do create jobs for local people but most of these are low skilled—gardener, driver, maid, cleaner, porter, sales assistant—poorly paid, and seasonal.

Elements of Ecotourism

The idea of ecotourism can probably be traced back to the environmental movement of the early 1970s, and its growing importance was recognized by the United Nations designation of 2002 as International Year of Ecotourism and the 2007 Oslo Global Ecotourism Summit.

According to the International Ecotourism Society, ecotourism is "responsible travel to natural areas that conserves the environment and improves the well-being of local people." Martha Honey, cofounder of the Center on Ecotourism and Sustainable Development, states that ecotourism involves the following:

- Involves travel to natural destinations
- Minimizes impact
- Builds environmental awareness
- Provides direct financial benefits for conservation
- Provides financial benefits and empowerment for local people
- Respects local culture
- Supports human rights and demographic movements

Honey's definition is widely quoted and is used in this article.

The first criterion, travel to natural destinations, should be interpreted broadly so as to include more than just wilderness considered as an area that has not been modified in any significant way by humans. Indeed, leisure travel to such areas as Antarctica or remote valleys of Papua New Guinea is probably incompatible with the goals of ecotourism. Moreover, the U.S. model of national parks as essentially uninhabited areas, which has been largely followed in countries such as Canada, Australia, New Zealand, and much of Africa and Asia, implies a sharp divide between humans and nature. By contrast, the British model, widely followed in much of Europe, recognizes that humans and natural systems can coexist in a sustainable way and is in line with Aldo Leopold's land ethic, based on the notion of humans as members of the land community.

Minimizing impact includes social and cultural as well as environmental impacts. The ecotourist's environmental impact begins with travel: destinations such as Costa Rica, Nepal, New Zealand, the Galapagos, and Botswana can be reached only by air from the developed countries where the overwhelming majority of ecotourists live. For instance, a passenger on an 80-percent-occupied 747 flight from Los Angeles to Kathmandu will be responsible for the emission of nearly three tons of carbon dioxide (CO_2), plus other harmful emissions. Many ecotourists purchase carbon offsets to neutralize this, but the effectiveness of offsets (especially tree planting) is disputed, and, of course, one could choose not to take the trip at all but to instead donate money to an environmental organization. Local travel and accommodation, including energy use and waste management, also need to be considered.

It may be argued that minimizing impacts excludes all forms of environmental modification except for those required for any form of tourism other than primitive camping. Minimally, this includes construction of accommodation and related facilities including water supply and waste management. But there is disagreement about how much further modification is permissible. Accessibility to areas beyond the immediate vicinity may require the construction of tracks and bridges; the alternative is to limit visitors to the superfit, or to run the risk that visitors making their own way may end up damaging fragile environments. Impacts on wildlife also raise issues. Even passive activities such as bird watching and marine-mammal watching may alter behavior, especially if tourists are allowed to get close to sensitive species.

The goal of building environmental awareness applies to the tourists themselves but also to local people. If ecotourism is successful in generating revenue and creating jobs, it will be publicized by local media and may encourage local entrepreneurs to move into this area. Tourists such as journalists and teachers will also be able to build awareness in their home countries by reporting on their experiences.

Ecotourism provides direct financial benefits for conservation via entrance fees to protected areas and donations. In some cases, foreigners pay substantially higher fees than locals: for instance, in Tanzania, foreigners pay $25 one way for the ferry from Dar-es-Salaam to Zanzibar, whereas locals pay less than $5.

Ecotourism can provide financial benefits and empowerment for local people if tourists patronize locally owned businesses providing accommodation, food and drink, local transport, and other goods and services.

Respecting local culture includes dressing appropriately and avoiding offense, for instance, not using the left hand to offer food or payment in Muslim societies, and not pointing the feet, when sitting, toward a person or a Buddhist image in Thailand.

Supporting human rights is somewhat problematic. Foreign tourists are not in a position to influence government policy and in former European colonies, Western human rights advocacy is often seen as neocolonialist rhetoric. Human rights organizations are divided over whether people should visit countries such as Myanmar, a repressive military dictatorship whose main opposition leader Aung San Suu Kyi opposes tourism, which she sees as support for the regime. Others argue that tourism enables cultural exchange and that tourists can target their spending to locally owned (as opposed to government-owned) businesses; that local people have told travel guide authors, such as writers for the *Lonely Planet Guide,* that they want foreigners to visit; and that human rights abuses are less likely to happen when foreign tourists are around.

Examples

Bicycle touring, which became increasingly popular from the mid-19th century, is a paradigm of ecotourism. The first national tourism organization in the world, the Bicycle Touring Organization, was founded in Britain in 1878. Writings of early bicycle tourists are full of descriptions of the delights of birdsong, wildflowers, and colorful sunsets. The cyclist consumes no fossil fuels (except in traveling to the destination area), meets local people, eats at local restaurants or markets, and can easily avoid dealing with multinational companies.

Island nations of the South Pacific are increasingly promoting ecotourism, though standards vary. The industry is most developed in Western Samoa, where the emphasis is on small-scale, locally owned businesses. Accommodation is typically in simple beach cabins or in family homes, and visitors are welcome to participate in village life.

Some operators allow sustainable consumption of wildlife. In New Zealand, for example, where there are no native terrestrial mammals, tourists include trophy and recreational hunters who wish to hunt the several introduced species of deer, as well as goats and wild pigs. These animals are environmentally destructive and it is often argued that these hunters are paying to do what conservation agencies would have to do anyway. Introduced game birds such as pheasants and two introduced species of trout may also be hunted, and although these species are not pests, "harvesting" them is not environmentally destructive.

Similar arguments are presented for trophy hunting in some southern African parks where populations of species such as lions, elephants, buffalo, and wildebeests have overextended the carrying capacity of the park. Thus, a trophy hunter who spends $50,000 or more on a two-week hunt may fulfill at least some of the criteria of ecotourism, especially if the operator trains local people to carry out more than menial roles.

Problems

A major problem for would-be ecotourists is determining whether a provider offers genuinely ethical services. This is compounded by the fact that there is no internationally accepted and trusted accreditation body as there is, for example, for Fair Trade coffee and other products. "Greenwashing" is thus widespread. Service providers often make vague claims about sustainability, recycling and reuse, purchasing locally, providing fair wages and working conditions, and contributing to the revival of local culture.

Tourists can check some of these claims in advance by consulting local nongovernmental organizations (NGOs), reliable guidebooks, personal contacts, and operators' Websites. Dishonest marketing can provide no more than temporary, short-term profit, and tourism operations need to endure long-term profitability in order to recoup investments so, it may be argued, there is no point in making false claims on Websites. For example, visitors to an operation that promotes itself as providing accommodation constructed by local people from traditional, renewable resources, locally sourced food and drinks, sustainably generated electricity to power in-room appliances, electric vehicles and boats, composting toilets, spring water, and the like will be quickly disillusioned if they find imported meat and liquor in the restaurants and bars, bottled water in the in-room refrigerator, and a Jeep Grand Cherokee to meet them at the airport.

However, understanding infrastructure requires research, and most people would have no idea how to evaluate water and waste systems or how to identify the timbers used in constructing the structures, or whether they are derived from unsustainable forestry practices. It is also effectively impossible for the visitor to know the wages and conditions of employment for workers, let alone to evaluate them. For instance, Malaysia is a middle-income country that has a successful conventional tourist industry and is currently promoting itself as an ecotourism destination. At the time of writing, Malaysia had full employment and indeed relied on immigrants to fill jobs that locals do not wish to do such as unskilled manual worker in the construction industry and domestic worker. The potential tourist is not in a position to evaluate what is a fair wage for a worker such as a maid or a gardener and has no idea of the conditions in which such workers and their families live.

A further problem is the impossibility of determining the ownership of businesses. Nobody expects a Hilton or a Sheraton to be owned by an average local family but, other than operations run by local or state governments, the tourist has no idea who might own what appears to be a local business.

Despite these difficulties, responsible tourism is possible, provided one is prepared to spend some time on research.

See Also: Conservation; Green Altruism; Greenwashing; Leopld, Aldo; Preservation.

Further Readings

Honey, Martha. *Ecotourism and Sustainable Development: Who Owns Paradise?* Washington, DC: Island Press, 1999.

International Ecotourism Society. http://www.ectourism.org (Accessed January 2010).

Mozer, David. "Less-Developed Countries and Bicycle Tourism" (2009). International Bicycle Fund. http://www.ibike.org/encouragement/travel/eco-tourism.htm (Accessed January 2010).

South Pacific Ecotourism Guide. http://www.pacific-travel-guides.com south-pacific/ eco-tourism/index.html (Accessed January 2010).

Alastair S. Gunn
University of Waikato

Electronic Manufacturers and Recycling Laws

Electronics play an essential role in today's rapidly moving society. Utilization of electronics can be found in products such as information and communication technologies, household entertainment systems, refrigerators, dishwashers, washing machines, and clothes dryers. Knowledge improvements in advanced manufacturing technologies over the years have resulted in the miniaturization of some electronic appliances and increased production. Compounding the increased proliferation of electronic goods and a shorter life span is the significant cost of recycling them at the end of their lives. These challenges typically result in the dumping of obsolete electronics without recycling. Mandatory recycling laws help avert the dumping of old electronics and ensure that significant amounts of obsolete electronics are utilized as secondary raw materials in the manufacture of new electronic goods. Additional benefits of recycling laws are the ability to help stimulate innovation in the design of electronics and in the manufacturing process.

A combination of knowledge improvements in technological capabilities in manufacturing on the one hand, and the optimization of market opportunities by the growing consumption by humans of more electronics on the other hand poses significant concerns about the quantities of electronics being produced globally. The primary reasons for these

U.S. consumers discarded about 3 million tons of obsolete consumer electronic products in 2010, making the country the world leader in annual electronic waste production.

Source: iStockphoto.com

concerns include the consequential effects of such mass production of electronics on the Earth's limited natural resources and the fate of these electronics at their end of life. This phenomenon whereby beneficial gains achieved through technological improvements are offset by increased human demands for more electronics per capita is a classical example of the "rebound effect," sometimes referred to as the "Jevons Paradox."

Recycling-Related Laws for Electronics

Recycling laws have been mandated in several jurisdictions (predominantly in developed countries) to curb the dependence on natural resources by electronics manufacturers, and to minimize any potential hazardous impact in the life cycle of electronic products.

Recycling laws provide minimal requirements for safe electronics manufacturing, waste minimization, and product stewardship. Recycling laws that impact electronics manufacturers include the following:

- Extended producer responsibilities
- Product declaration
- End-of-life waste electrical/electronic equipment collection and recycling targets
- Substitution of hazardous materials in products
- "Greening" the supply chain
- Decision support guidelines on value reclamation of materials as opposed to energy reclamation

Examples of recycling-related laws for electronics manufacturers in different jurisdictions and regions are provided below, including Europe (European Community member states); North America (the United States and Canada); South America (Brazil); and Asia (China and Japan).

Europe (European Community Member States)

Restriction of the Use of Certain Hazardous Substances in Electrical and Electronic Equipment (RoHS) Directive 2002/95/EC

The objective of the RoHS Directive 2002/95/EC was to encourage environmentally benign recovery, management, and disposal of obsolete electrical and electronic equipments in the European Commission member states. It was also enacted to prevent and mitigate any harmful human health impacts emanating from waste electrical and electronic equipment. The RoHS Directive bans a number of substances that have been found to be hazardous to human health and the environment. Examples of hazardous substances banned in the manufacture of new electrical and electronic equipment (effective July 1, 2006) under the RoHS Directive include lead (Pb), mercury (Hg), cadmium (Cd), hexavalent chromium ($Cr6+$), polybrominated biphenyls (PBB), or polybrominated diphenyl ethers (PBDE). Exemptions were made for analytical medical monitoring and control instrumentations under this directive. Amendments were made under this directive in June 2009, however, by the European Commission to exempt certain specialized applications of lead, cadmium, and mercury in electronics manufacturing.

Waste Electrical and Electronic Equipments (WEEE) Directive 2002/96/EC (2003)

The WEEE Directive 2002/96/EC (2003) was enacted to prevent and/or reduce waste from electrical and electronic equipments in European Union Member States. Directive 2002/96/EC (2003) establishes minimum criteria for the reuse, recycling, and recovery of waste from electrical and electronic equipment devices. It also provides useful guideline information especially to recycling facilities as well as all other stakeholders (i.e., electronic manufacturers, distributors, and consumers) for improved environmental performance along the life cycle and value chains of electrical and electronic products. A collection target of 4 kilograms per person per year is required from member countries in the European Union under this directive. It is, however, suspected that significant amounts (about 67 percent) of waste electrical and electronic equipments are illegally exported out of the European Union, possibly to nonmember states and developing countries where electronic recycling laws are lacking or not stringent. Only 33 percent of electrical and electronic equipment waste is reported to have been processed in treatment facilities in Europe since 2004.

Directive 2005/32/EC on the Eco-Design of Energy-Using Products (EuP)

Under Directive 2005/32/EC, the European Commission stipulated implementing measures for affected EuPs that are widely traded in the European Community and that pose significant environmental impacts in the community. In principle, Directive 2005/32/EC applies to all energy-using products (except vehicles for transport) and covers all energy sources.

Self-regulation by affected industries is an alternative to the implementing measures prepared by the European Commission. Requirements under the self-regulation option for a new entrant industry include the following:

- An initial preparatory study
- Consultation process
- Self-assessments of impacts
- Formation of a regulatory committee
- Evaluation of the entire process of self-regulation by the European Parliament, Brussels

North America (the United States and Canada)

The United States

The United States is the global leader in electronic waste production per annum, with an estimated e-waste mass of 3 million tons in 2010. An estimated 70 percent of all mercury and cadmium in U.S. landfills are believed to be from electronic and electrical waste. Regarding lead in U.S. landfills, 40 percent is derived from obsolete consumer electronic products.

At the federal level in the United States, the laws that apply specifically to electronics and their manufacture include the U.S. Environmental Design of Electrical Equipment Act (EDEE Act) H.R. 2420, and the U.S. Electronic Devices Recycling Research and Development Act (EDRRD Act) H.R. 1580.

The United States Environmental Design of Electrical Equipment Act (EDEE Act) H.R. 2420. The United States Environmental Design of Electrical Equipment Act (EDEE Act) H.R. 2420 is an amendment to the Toxic Substances Control Act (TSCA) of 1976. The goal of the EDEE Act H.R. 2420 is to harmonize restrictions on electrical and electronic equipments. The act provides thresholds for certain hazardous substances effective after July 1, 2010. The EDEE Act affects equipments that have greater than 0.1 percent by weight of the following hazardous substances: polybrominated biphenyls, polybrominated diphenyl ethers lead, hexavalent chromium, and mercury. In regard to cadmium, the restriction under the EDEE Act H.R. 2420 stipulates a limit not to exceed 0.01 percent by weight of the electrical and electronic equipment.

The United States Electronic Devices Recycling Research and Development Act (EDRRD Act) H.R. 1580. The objective of the EDRRD Act H.R. 1580 is to achieve environmentally benign design, manufacture, and recycling of electronics and the minimization of electronic waste over the longer term.

Under the EDRRD Act H.R. 1580, the U.S. Environmental Protection Agency (EPA) could provide competitive grants to research institutions and laboratories to train capacity and to investigate all aspects of electronic design, manufacturing, reuse, and recycling.

State Level Initiatives in the United States. The state of California enacted the Electronic Waste Recycling Act of 2003 (EWRA) in September 2003. The California EWRA stipulates maximum concentration levels of four hazardous substances utilized in cathode ray tubes, liquid crystal display, and plasma display equipments (effective January 1, 2007). The hazardous materials restricted under EWRA include cadmium (0.01 percent by weight maximum); hexavalent chromium (0.1 percent by weight maximum); lead (0.1 percent by weight maximum), and mercury (0.1 percent by weight maximum).

Initiatives in other U.S. states include the following:

- The Illinois Electronic Products Recycling and Reuse Act (2008)
- The Connecticut Electronic Recycling Law (2007)
- The Colorado National Computer Recycling Act (2007)
- The Arkansas Computer and Electronic Solid Waste Management Act (2003)
- The Minnesota Electronics Recycling Act (2008)
- The Missouri Manufacturer Responsibility and Consumer Convenience Equipment Collection and Recovery Act
- The Oklahoma Computer Equipment Recovery Act (2008)
- The Virginia Computer Recovery and Recycling Act (2008)
- The Rhode Island Electronic Waste Prevention, Reuse, and Recycling Act (2008)

Canada

An estimated 180,000 tons of end-of-life electronic waste is produced annually in Canada. At the federal level, Canada has yet to enact a law concerning electronic recycling. However, several provincial governments have enacted laws and regulations on electronic recycling in their various jurisdictions. This includes the following provinces:

- The Alberta Electronics Designation Regulation (Alberta Regulation 94/2004, which went into effect on October 1, 2004)
- The Ontario Waste Electronic and Electronic Equipment Regulation (O.Reg. 393/04)
- The Saskatchewan Waste Electronic Equipment Regulations (R.R.S.c.E-10.21 Reg. 4, which went into effect on February 1, 2006)

In the Canadian provinces of British Columbia and Nova Scotia, requirements for electronic recycling can be found under the British Columbia Recycling Regulation (BC Reg. 449/2004) and the Nova Scotia Solid Waste Resource Management Regulations (N.S. Reg. 25/96), respectively.

Schedule 3, Electronic and Electrical Product Category of the British Columbia Recycling Regulation stipulates the recycling requirements for electronic and electrical products in the Canadian province of British Columbia.

Sections 18J–18Q in the Nova Scotia Solid Waste Resource Management concern solely the recycling requirements for electronic products sold in Nova Scotia.

South America (Brazil)

Brazil's bill (PLS) 173/2009 on restricting the use of hazardous materials in electronics manufacturing was passed in August 2009. This bill (PLS) 173/2009 is likely to be enacted into law by 2011. Bill (PLS) 173/2009 is very similar in content to the European Union's directive on the Restriction of the Use of Certain Hazardous Substances in Electrical and Electronic Equipment (ROHS) Directive 2002/95/EC and subsequent amended versions. Under Bill (PLS) 173/2009, 95 percent recyclability rate of electronic and electrical equipment's parts is required, components and subassemblies should be achieved at the end of product life. The bill also stipulates at least 80 percent energy efficiency in all electronic and electrical equipment sold in Brazil.

Asia

Japan

The principal law that pertains to electronic recycling in Japan is the Home Appliance Recycling System law, which went into effect in 1998. Under this law, it became mandatory to recycle obsolete tube televisions, refrigerators, air conditioners, and washing machines. However, in December 2008, amendments were made to this law, and clothes dryers and liquid-crystal/plasma television sets are now also required to be recycled at the end of their lives. These amendments went into effect on April 1, 2009.

China

An estimated 2.3 million tons of obsolete electronic and electrical equipment waste is generated yearly in China. China is second only to the United States in terms of total electronic and electrical equipment waste generated per annum.

China's Law on the Quarantine, Inspection and Management of Imported Used Electrical and Mechanical Products was enacted in 2003 with the objective of preventing a flood of obsolete electronic goods into the country, in particular from Japan and South Korea.

China's Restriction of the Use of Certain Hazardous Substances in Electrical and Electronic Equipment (China's RoHS) is similar to the European Union RoHS Directive. China's RoHS covers the six hazardous substances used in electronics manufacture. They are lead (Pb), mercury (Hg), cadmium (Cd), hexavalent chromium ($Cr6+$), polybrominated biphenyls (PBB), or polybrominated diphenyl ethers (PBDE). China's RoHS Directive went into force in 2007.

China's Draft Law on Waste Household Electrical Device Collection, Utilization and Management is a draft legal document that has yet to come into effect. The overall objective of the proposed law is to achieve significant reductions in electronic waste. It also aims to obtain higher recycling rates for obsolete electronic products, as well as the ecodesign of new products.

Recycling laws for electronic manufacturers are essential to ensure sustainable stewardship of the Earth's resources. Developed countries are generally ahead of developing countries in regard to enacting recycling laws for electronic products. However, several developing countries have shown a strong interest in following the example of a leader—the European Union—in the formulation and enactment of local recycling laws for their end-of-life electronics. It is worth noting that while trends in recycling law enactments in the European Union follow a classical top-down policy approach from the regional level to the individual member states, the situation in North America (especially in Canada) is exactly the opposite.

See Also: Environmental Law; Environmental Policy; Environmental Policy Act, National.

Further Readings

Ackom, E. K. and J. K. Ertel. "Business of Recycling." In *Green Business: An A-to-Z Guide,* Paul Robbins and Nevin Cohen, eds. Thousand Oaks, CA: Sage, 2011.

Ogunseitan, O. A., et al. "The Electronics Revolution: From E-Wonderland to E-Wasteland." *Science,* 326/5953 (2009).

Emmanuel Kofi Ackom
University of British Columbia

Emerson, Ralph Waldo

Ralph Waldo Emerson (1803–82) was an essayist, a public lecturer, and the central figure of New England transcendentalism. His most widely read works include "The Divinity School Address," "The American Scholar," "Self-Reliance," and the small book *Nature.*

Born and raised in Boston, Massachusetts, he graduated from Harvard College at the age of 19. He took up the study of divinity and was a Unitarian minister for three years, but his increasingly unorthodox religious views and his doubts about church doctrine led him to renounce all forms of organized religion. After leaving the pulpit, Emerson moved to Concord, Massachusetts, and built a successful career as a writer and public lecturer. His importance for green ethics lies primarily in his book *Nature* and in the spiritual dimension he brings to the relationship between humanity and the natural world.

Emerson saw a continuity between nature, humanity, and the Divine. Divinity—God, the Universal Soul, or Spirit—pervades nature. It is the creative force within nature, a process of generation and continual renewal. Nature is "symbolic" of Spirit, and every natural fact is a symbol of some spiritual fact. The same divine Spirit that pervades nature also lives at the very center of our individual being. Every individual soul is part of the Universal Soul, and in every part of nature there is a bit of humanity. To understand the spiritual significance of nature, we must see it whole. The primary question we should put to nature is not about its details, but rather about its unity. Where does it come from and what is its end? What are we to make of the great congruence of humanity and nature, and how is it that we can find parts of ourselves in every bit of nature?

Seeing the underlying unity of nature is difficult because humanity lives in a condition that is alienated from both nature and Spirit. We are also alienated from ourselves, and this is part of the reason we cannot see the unity in nature. Emerson writes in *Nature* that "the reason the world lacks unity, and lies broken and in heaps, is because man is disunited with himself. He cannot be a naturalist until he satisfies the demands of the spirit. Love is as much a demand as perception." The congruity between humanity and nature encourages scientific study, but we are apt to become lost in nature's details and miss its underlying oneness. The unity underlying nature is the same unity that underlies thought. But words break and chop truth, and empirical inquiries can "cloud the sight" and blind us to the whole. When we understand nature only through analysis without unification, we are only halfway human.

To understand nature and become fully human, we must learn to act with what Emerson calls our Reason as well as our understanding. For Emerson, Reason includes perception, insight, and imagination. The best naturalist learns from nature, not through analysis or comparison, but by "untaught sallies of the spirit, by a continual self-recovery, and by entire humility." Human redemption is found in discovering for oneself the soul's unity with nature and the divine. At the same time, the original and eternal beauty of nature is restored through the redemption of the soul. The outward beauty of nature is a sign of inward beauty. Nature's surface beauty is medicinal in the delight that it gives the senses, but more important is nature's spiritual beauty that can be found in combination with human will and virtue. Still more elevated is the relation of nature's beauty to the intellect, which searches out its divine order. Because nature provides the path to spiritual development, there is no need of churches or organized religion, which are actually obstacles to spiritual growth. Nature is moral. Every natural being gives us hints or lessons of right and wrong. "The ethical character so penetrates the bone and marrow of nature, as to seem the end for which it was made."

For Emerson, then, nature is not an end in itself, as it is for many greens. It is an arena for spiritual self-development and the discovery of our original relation to the Divine. This discovery is properly the primary goal of human development and education. One could claim that Emerson's ethical views on nature are highly anthropocentric, and thus run counter to much of green ethics. Nature is valued only as it can benefit the development of the human soul. It is only a "symbol" of a deeper reality and a path through which we can understand that reality. Nature is "meant to serve. It receives the dominion of man as meekly as the ass on which the Savior rode." But this dominion is not one of conquest and submission. All human uses of nature potentially point to Spirit. Because the individual human soul is the same as the Universal Soul that pervades nature, and because there is part of humanity in every bit of nature, a clear distinction between anthropocentric and intrinsic value is not relevant for Emerson's philosophy. To despoil nature is to disrespect the divine and despoil one's innermost self. To neglect the soul is to neglect nature. An ethical relationship with nature requires a unified self, and a unified self is developed through a creative understanding of the unity of Spirit in nature. Untrammeled nature offers the opportunity to experience the world's organic unity most clearly and completely.

See Also: Anthropocentrism; Conservation, Aesthetic Versus Utilitarian; Thoreau, Henry David.

Further Readings

Bishop, Jonathan. *Emerson on the Soul.* Cambridge, MA: Harvard University Press, 1964.
Buell, Lawrence. *Emerson.* Cambridge, MA: Harvard University Press, 2003.
Emerson, Ralph Waldo. *The Collected Works of Ralph Waldo Emerson.* 16 vols. Cambridge, MA: Harvard University Press, 1971–82.
Emerson, Ralph Waldo. *The Complete Works of Ralph Waldo Emerson,* Edward Waldo Emerson, ed. 12 vols. Boston: Houghton Mifflin, 1903–04.
Emerson, Ralph Waldo. *The Journals of Ralph Waldo Emerson,* Edward Waldo Emerson and Waldo Emerson Forbes, eds. 10 vols. Boston: Houghton Mifflin, 1910–14.
Emerson, Ralph Waldo. *Selections From Ralph Waldo Emerson.* Stephen E. Whicher, ed. Boston: Houghton Mifflin, 1957.
Porte, Joel and Saundra Morris. *The Cambridge Companion to Ralph Waldo Emerson.* Cambridge, MA: Cambridge University Press, 1999.
Richardson, Robert D., Jr. *Emerson: The Mind on Fire.* Berkeley: University of California Press, 1995.

Kelvin J. Booth
Thompson Rivers University

Endangered Species Act

The Endangered Species Act (ESA) of 1973, which builds on several laws passed in the 1960s, including the Endangered Species Preservation Act of 1966 and the Endangered Species Conservation Act of 1969, was enacted to prevent species from becoming endangered or extinct due to human impact on natural ecosystems. The ESA bars federal agencies from undertaking actions that harm endangered or threatened species, prohibits government agencies and individual citizens from "taking" endangered animals without a permit, and requires the designation of critical habitat for species once they are listed.

Species are put on the endangered or threatened species list by the Fish and Wildlife Service (FWS) or the National Oceanic and Atmospheric Administration (NOAA). Once a

While the Endangered Species Act has resulted in almost 50 species recovering enough to be removed from the endangered list, some protected species such as this dusky seaside sparrow from Florida have nonetheless gone extinct.

Source: U.S. Fish & Wildlife Service

species is on the endangered list, FWS or NOAA fisheries must create an Endangered Species Recovery Plan, including protocols to protect the species and increase their number to the point where they are no longer endangered, as well as to create a timeline and estimated costs to reach this goal. As part of the recovery plan, the FWS or NOAA are allowed to designate critical habitat since habitat loss is the primary threat to many species on the endangered list.

A case involving the snail darter, a new species of fish discovered in 1973 that was placed on the endangered species list in 1975, was the first challenge for the ESA. In 1966, the Tennessee Valley Authority began construction of the Tellico Dam, located on the Little Tennessee River, for the purposes of flood control, creation of hydroelectric power, and to promote development of the area. However, part of the Little Tennessee River was designated as critical habitat for the snail darter, and a lawsuit was filed in 1976 in the District Court to halt construction of the Tellico Dam until further studies could be conducted. In 1978, the U.S. Supreme Court upheld the decision of the appeals court that the continuation of dam construction would violate the Endangered Species Act. Chief Justice Warren Burger wrote that the language of the ESA was clear—federal agencies were not allowed to carry out actions that threatened the habitat or existence of an endangered species—and that no exceptions would be made for projects already underway when the ESA was enacted.

In response to the Supreme Court decision, two legislators from Tennessee—Congressman John Duncan, Sr., and Senator Howard Baker—introduced an amendment to the ESA, passed in 1978, which allowed exclusion of specific projects from the rule of the ESA. Congress created a mechanism whereby a specific project could be excluded from the Endangered Species Act. The mechanism to achieve this involved forming a so-called God Committee or God Squad (so-called because they could make decisions that would allow the destruction of critical habitat, which could lead to the extinction of an entire species) including at least one member from the affected state that would apply for exemption. However, the committee's application for exemption of the Tellico Dam project from ESA regulation was denied, so in 1979 Baker introduced an amendment that specifically exempted the Tellico Dam Project. It was passed in 1979, and the Dam was completed that year.

The ESA has had mixed success in protecting endangered species. Close to 50 species (including the snail darter, which was successfully moved to other streams) have been removed from the list, the majority due to recovery. Policies attributed to the ESA have enabled the recovery of such charismatic mega fauna as the bald eagle, the peregrine falcon, and the grizzely bear. However, the ESA was not able to save species like the blue walleye (blue pike) of the Great Lakes region, and the dusky seaside sparrow of southern Florida. The ESA has also been criticized for creating perverse incentives in some cases, as exemplified by the red-cockaded woodpecker. After the bird was listed, many landowners cut down forests rather than let them develop into mature forests that could become designated as a critical habitat for the bird.

See Also: Environmental Policy Act, National; Instrumental Value; Intrinsic Value; Land Ethic; *Silent Spring.*

Further Readings

Baur, D. C., W. R. Irvin, and American Bar Association. Section of Environment Energy and Resources. *The Endangered Species Act: Law, Policy, and Perspectives.* Chicago, IL: Section of Environment, Energy, and Resources, American Bar Association, 2001.

Burgess, B. B. *Fate of the Wild: The Endangered Species Act and the Future of Biodiversity.* Athens: University of Georgia Press, 2001.

Czech, B., P. R. Krausman, and Center for American Places. *The Endangered Species Act: History, Conservation Biology, and Public Policy.* Baltimore, MD: Johns Hopkins University Press, 2001.

List, J. A., M. Margolis, D. E. Osgood, and National Bureau of Economic Research. *Is the Endangered Species Act Endangering Species?* Cambridge, MA: National Bureau of Economic Research, 2006.

Murchison, K. M. *The Snail Darter Case: TVA Versus the Endangered Species Act.* Lawrence: University Press of Kansas, 2007.

Stanford Environmental Law Society. *The Endangered Species Act.* Stanford, CA: Stanford University Press, 2001.

Sullins, T. A. *ESA, Endangered Species Act.* Chicago, IL: Section of Environment, Energy, and Resources, American Bar Association, 2001.

Jo Arney
University of Wisconsin–La Crosse

Energy Ethics

Energy ethics is devoted to determining the just distribution of the goods and hazards associated with energy production and use. Given the finite quantity of both energy and the ability to assimilate waste, the distribution of access to energy as well as the wastes associated with its production becomes an object of ethical concern. Two general questions guide the field. First, how are current energy supplies and energy-related wastes to be distributed among the world's current persons and peoples? Since overuse of energy by some exacerbates energy scarcity for others and increases quantities of the hazardous by-products of energy production and use, the waste or excessive use of energy by some can be directly linked to diminished opportunities and heightened hazards for others. Second, how are these energy supplies and wastes to be distributed over time? Finite energy resources and waste assimilation capacities also entail intergenerational harm from overuse by the current generation, as in the future effects of anthropogenic climate change.

These ethical questions arose from the field of ecological economics and its analysis of ecological limits to energy production, applying the First and Second Laws of Thermodynamics to an understanding of the economic process. The first law states that energy can be neither created nor destroyed. The second law states that within a closed system, the totality of all physical processes increases the total quantity of unusable energy, called entropy. Given that Earth is essentially a thermodynamically closed system (defined as allowing heat exchange across the boundary, but no material exchange), the energy available for use is finite, and thus ecological economics concludes that limits also exist for human activity requiring energy.

Within sustainable development literature, energy ethics addresses the allocation of energy resources and the capacity to absorb the resulting waste that provides the third world with the ability to develop without unjustly decreasing the quality of life of the first world. Some suggest that insufficient energy is available for the entire world population to live at the level of the developed world and that Earth is unable to absorb the waste necessary for developing countries to follow the strategy used by the developed world. Energy ethics thus inquires into the ethical balance between energy demands made by developed and developing countries.

However, some deny that there are limits to these resources. This cornucopian response has two main arguments. First, statistical trends suggest that resource availabilities are not decreasing. Second, improvements in technology will increase both resource availability and efficiency of use. Therefore, development is not considered an ethical question. Instead development is regarded as a practical problem of technological innovation. This cornucopian discourse faces two major criticisms. First, the findings used to substantiate the claims of limitless resources are argued to be based on insufficient knowledge. Second, the ability of the human mind to develop the appropriate technological innovations is questionable.

The harmful effects of energy production and use also present a number of ethical concerns. Exposure to by-products of production processes can increase the risk of contracting some diseases. For instance, some carcinogenic chemicals used in the natural gas drilling technique called "hydraulic fracturing" have contaminated drinking water supplies near drilling sites. Energy ethics focuses not only on the allocation of these potential harms but also on the procedure by which they are allocated. Anthropogenic climate change, a result of fossil fuel combustion, presents a second example of an environmental harm caused by energy use. Both energy production and energy use are causes of localized and global harms, all of which are considered to be avoidable to the degree that the energy used is not necessary. To the extent that these harms are avoidable, energy ethics holds that their creation ought to be prohibited.

Moral responsibility to the future is understood both as providing an adequately healthy ecosystem to support the demands of the future and as providing an equitable distribution of available energy for future use. Thus, consideration is given to the effect current energy usage has on the ability for like usage in the future. It is currently uncertain whether sufficient energy is available to allow for equitable future use. Uncertainty also exists regarding the stability of Earth's systems, given the continuation of equal quantities of waste production. This uncertainty generates questions of intergenerational ethics for two reasons. First, though renewable energy sources are not temporally limited, the use of nonrenewable energy resources increases the absolute scarcity of energy for future generations. Second, the negative effects of energy use are not entirely limited to the time of their creation. This is of particular concern when considering the inability of future generations to consent to these harms. Therefore, current energy use beyond that which is necessary creates avoidable harms to future generations.

See Also: Climate Ethics; Club of Rome (and Limits to Growth); Development, Ethical Sustainability and.

Further Readings

Daly, Herman E. *Beyond Growth: The Economics of Sustainable Development.* Boston: Beacon Press, 1997.
Kimmins, James P. *The Ethics of Energy: A Framework for Action.* Geneva: United Nations Educational, Scientific and Cultural Organization, 2001.
MacLean, Douglas and Peter G. Brown, eds. *Energy and the Future.* Totowa, NJ: Rowman & Littlefield, 1983.

Matthew E. Heller
University of Colorado, Boulder

Environmental Ethics Journal

Environmental philosophy emerged as a formal academic discipline in the 1970s, a result of the environmental movement of the 1960s. Despite this growth, publishing of peer-reviewed articles on environmental philosophy was extremely limited in mainstream philosophy journals. Norwegian deep ecologist Arne Naess provided the most common early forum as editor of his philosophy journal *Inquiry.* Dr. Eugene C. Hargrove founded the first journal dedicated entirely to the field, *Environmental Ethics,* in 1978. The full name of the journal is *Environmental Ethics: An Interdisciplinary Journal Dedicated to the Philosophical Aspects of Environmental Problems.* The first volume was published in 1979 out of the University of New Mexico. It moved to the University of Georgia, along with Hargrove, from 1981 to 1990. The journal has since been published out of the Center for Environmental Philosophy, directed by Hargrove at the University of North Texas. It remains the primary journal in the field, although other forums have arisen, including (in order of appearance) *Environmental Values* (1992); *Ethics and the Environment* (1996); *Ethics, Place, and Environment* (1998); and *Environmental Philosophy* (2004).

Early intellectual inspiration for the field came not just from the 1960s environmental movement but from three seminal publications. The first two were articles in the leading mainstream journal *Science.* "The Historical Roots of Our Ecological Crisis" by Lynn White in 1967 suggested that Christian dominion over nature was a root cause of environmental degradation. "The Tragedy of the Commons" by Garrett Hardin in 1968 offered that common-pool resources such as fisheries and public rangelands were more prone to degradation than private property. The third publication was Aldo Leopold's essay "The Land Ethic" in *A Sand County Almanac,* written in the 1940s and popularly republished in the 1960s, which posed that current environmental problems should be solved by a natural progression of ethics over time outward from the human community to the biotic community (nature).

Thus, in its first few years, the journal *Environmental Ethics* included articles addressing such issues as Christian dominion over nature, property rights and related duties to nature, and extension of ethics to the natural world. As one example, the first volume in 1979 published Leopold's 1924 essay "Some Fundamentals of Conservation in the Southwest" for the first time. The essay, located in Leopold's archival papers, was seen as a precursor to "The Land Ethic." It exhibited Leopold's first written inclinations toward moral considerations in natural resource management. The same issue included commentary by Leopold historian Susan Flader and John B. Cobb, Jr.'s article "Christian Existence in a World of Limits." The next issue in 1979 contained Kathleen M. Squadrito's article "Locke's View of Dominion."

An important debate occurred in the journal's early years between proponents of environmental ethics and animal rights. Leopold-based environmental ethics was concerned more with preservation of ecological integrity (species composition), stability (ecosystem process function), and beauty (aesthetics) than it was with rights of individuals within nature. Extension of rights to ecosystems, in the same context as rights are extended to individual humans, was problematic for many philosophers. A key source of debate was J. Baird Callicott's 1980 article "Animal Liberation: A Triangular Affair." After its publication, philosophical studies of animal rights and animal liberation, highlighted by the work of Peter Singer and Tom Regan, took a separate path in other forums, though the topics and work of those authors were still covered to some degree in *Environmental Ethics*.

Another heated, lengthy debate within environmental philosophy—over the existence of objective intrinsic value in nature—occurred within the journal's pages from the 1980s into the 1990s and beyond. Objective intrinsic value is the value of an object in and for itself, independent of human subjective judgment. The notion is problematic because many philosophers agree that for a natural object to have value, there has to be a sentient being to ascribe value to it. The debate went back and forth largely between Callicott's "truncated" nonanthropocentric, subjectivist, intrinsic value of nature, in which value does not exist in nature but instead is created by a valuing agent, and Holmes Rolston III's nonanthropocentric, objectivist, intrinsic value of nature, based on natural entities using nature instrumentally to promote their own good as an end, independent of human valuing. The debate was further enlarged by Bryan Norton's introduction of "weak anthropocentrism," based on the considered preferences of humans and their societal ideals from a pragmatic perspective, eliminating the term *intrinsic value*. Other environmental philosophers such as Eugene C. Hargrove and Ben Minteer weighed in with compromise positions: Hargrove proposing a weak anthropocentric, subjectivist, conception of intrinsic value and Minteer proposing a way in which such intrinsic value could be included within Norton's environmental pragmatism.

Themes of seminal books in environmental philosophy, many of them emerging in the late 1980s, are also represented in the journal through articles by the books' authors, many of them published prior to the books. Examples include Paul Taylor's 1986 book *Respect for Nature,* a foundational text on biocentrism, previewed by his articles "The Ethics of Respect for Nature" (1981) and "In Defense of Biocentrism" (1983). Holmes Rolston III's 1987 book *Environmental Ethics: Duties to and Values in the Natural World,* a key outline of multiple values in nature, is preceded by his journal articles "Values in Nature" (1981), "Are Values in Nature Subjective or Objective?" (1982), and "Valuing Wildlands" (1985).

Additionally, several of Callicott's 12 articles in the journal *Environmental Ethics* reappear in two of his most important books, *In Defense of the Land Ethic* and *Beyond the Land Ethic.*

The content of articles in *Environmental Ethics* has varied between theoretical and applied philosophy. More traditional philosophers find the field too applied, while environmental practitioners can find the field too esoteric. Notable examples of applied studies include Douglas Crawford-Brown and Neil E. Pearce's 1989 article "Sufficient Proof in the Scientific Justification of Environmental Actions." Countering the notion of an environmental policy formed by objective science, the authors found at least five different decisions that involved subjective value judgments within the radon risk reduction policymaking process. Callicott's 1996 article "Do Deconstructive Ecology and Sociobiology Undermine Leopold's Land Ethic?" offered defense of Leopold in the context of newer ecological science that challenges prior notions of ecosystem stability and integrity. In 2002, Irene Klaver and coauthors illustrated the contextual value of restoring an "imperfect" wildness to the Netherlands. Robert Frodeman's 2006 article highlighted the "policy turn" in environmental philosophy. A 2008 special issue on applied environmental ethics in southern Chile, highlighting Ricardo Rozzi's work, was further evidence of "on-the-ground" case studies within the journal.

The range of topics covered in the 30-year history of the journal *Environmental Ethics* illustrates the impressive scope of the field. Issues critically addressed in the context of environmental ethics (along with approximate number of articles in parentheses) show a relatively even distribution: animal rights (30 articles); anthropocentrism, biocentrism, ecocentrism, and intrinsic/universal value (43); deep ecology (19); ecofeminism (20); economics and sustainability (30); endangered species and restoration (25); environmental justice and indigenous environmental ethics (34); environmental policy (28); environmental science (20); Leopold's land ethic (18); miscellaneous (35); monism, pluralism, situational, and postmodern environmental ethics (27); pragmatism (21); religion, metaphysics, and process philosophy (27); and wilderness/wildness (28).

See Also: Callicott, J. Baird; Hargrove, Eugene C.; Intrinsic Value; Naess, Arne.

Further Readings

Callicott, J. Baird. "Animal Liberation: A Triangular Affair." *Environmental Ethics,* 2 (1980).

Callicott, J. Baird. "The Case Against Moral Pluralism." *Environmental Ethics,* 12 (1990).

Callicott, J. Baird. "Do Deconstructive Ecology and Sociobiology Undermine Leopold's Land Ethic?" *Environmental Ethics,* 18 (1996).

Callicott, J. Baird. "Intrinsic Value, Quantum Theory, and Environmental Ethics." *Environmental Ethics,* 7 (1985).

Callicott, J. Baird. "Rolston on Intrinsic Value: A Deconstruction." *Environmental Ethics,* 14 (1992).

Crawford-Brown, Douglas and Neil E. Pearce. "Sufficient Proof in the Scientific Justification of Environmental Actions." *Environmental Ethics,* 11 (1989).

Frodeman, Robert. "The Policy Turn in Environmental Ethics." *Environmental Ethics,* 28 (2006).

Klaver, Irene, Jozef Keulartz, Henk van den Belt, and Bart Gremmen. "Born to Be Wild: A Pluralistic Ethics Concerning Introduced Large Herbivores in the Netherlands." *Environmental Ethics,* 24 (2002).

Minteer, Ben A. "Intrinsic Value for Pragmatists?" *Environmental Ethics*. 23 (2001).
Norton, Bryan G. "Environmental Ethics and Weak Anthropocentrism." *Environmental Ethics*, 6 (1984).
Norton, Bryan G. "Why I Am Not a Nonanthropocentrist: Callicott and the Failure of Monistic Inherentism." *Environmental Ethics*, 17 (1995).
Rolston, Holmes, III. "Are Values in Nature Subjective or Objective?" *Environmental Ethics*, 4 (1982).
Rolston, Holmes, III. "Values in Nature." *Environmental Ethics*, 3 (1981).
Rolston, Holmes, III. "Valuing Wildlands." *Environmental Ethics*, 7 (1985).
Rozzi, Ricardo, Juan J. Armesto, and Robert Frodeman. "Integrating Ecological Sciences and Environmental Ethics Into Biocultural Conservation." *Environmental Ethics*, 30 (2008).
Taylor, Paul W. "The Ethics of Respect for Nature." *Environmental Ethics*, 3 (1981).
Taylor, Paul W. "In Defense of Biocentrism." *Environmental Ethics*, 5 (1983).

William Forbes
Stephen F. Austin State University

ENVIRONMENTAL JUSTICE

Although there is not a standard definition of what constitutes environmental justice, the environmental justice movement grew out of the inequitable distribution of environmental hazards among poor and minority citizens. All people deserve to live in a clean and safe environment free from industrial waste and pollution that can adversely affect their health and well-being. Environmental justice seeks to bring this ideal state of the world into reality. A society that embraced environmental justice would ensure that all of the costs that accompany living in an industrialized nation were equally and fairly distributed among citizens and that a nation's minority and/or underprivileged populations were not facing inequitable environmental burdens. From a policy perspective, practicing environmental justice entails ensuring that all citizens receive the same degree of protection from environmental hazards by the government.

The environmental justice movement combines traditional environmentalism with the conviction that all individuals have the right to live in a safe environment. It started as a grassroots movement during the early 1980s in pockets of the United States where the minority and underprivileged were facing environmental burdens. It has evolved into community initiatives, federal offices, and a presidential executive order. A criticism of the environmental justice movement is sometimes levied by mainstream environmentalists who believe that it is not an environmental movement but a social justice movement.

History

Environmental justice does not have a birthplace or date of conception. It grew out of the many movements by local groups who sought to protect their own health and well-being and that of others in their community. One such grassroots start-up formed in the early 1980s in Warren County, North Carolina, the home of a PCB landfill. The citizens of Warren and surrounding counties had been the victims of midnight dumping as industrial polluters dumped thousands of gallons of PCB-laden oil along their roadways in the middle of the night. The company and contractor accused of dumping explained that the motive for dumping was to avoid paying for recycling costs of the oil. In an effort to deal with the dumping crisis, the state of North Carolina created a toxic landfill in Warren County. The residents of Warren County were mainly African American and it was also one of the poorest counties in North Carolina. Residents felt that the county was picked for the landfill based on its demographic characteristics. Citizens rallied together to try to prevent the dump. They wrote letters, held demonstrations, and blocked trucks. Although they were successful in gaining media attention, the landfill was ultimately developed.

The events in Warren County eventually led to the coining of the term *environmental racism*. This term refers to the targeting of communities for the placement of waste-generating or waste-storing facilities or to discrimination in the enforcement of environmental standards in communities or neighborhoods based on the racial characteristics of the residents. In the book *Dumping in Dixie,* Robert Bullard examines the location of waste disposal facilities in five communities. Using demographic data based on the zip codes in which the facilities are located, Bullard concludes that the prosperity of a community is not as good a predictor of hazardous waste locations as is the race of the residents. He concludes that choosing where to locate a waste facility often includes racism. "Racism" has often been used to describe an individual's decision to discriminate against a person or group of people based on their race. This would be applicable in cases where a company decided to site a hazardous waste-producing facility in a minority neighborhood simply because of the number of minorities living there. Environmental racism is often a form of institutional racism, however. Institutional racism is not based on individual discrimination but rather on discriminatory practices that become part and parcel of the practices, traditions, and policies of organizations or society.

An example of institutional racism may be linked to the "Not In My Backyard" (NIMBY) phenomenon. NIMBY involves groups of citizens engaged in blocking unwanted projects in a given community or neighborhood. NIMBY movements are likely to involve members of a community who have political efficacy and, in turn, believe that they can make a difference. Often disenfranchised individuals in poor and minority neighborhoods do not exhibit much political efficacy. This means that NIMBY groups are less likely to organize. A decision maker looking for a site to place a waste-producing or waste-storing facility would want to avoid areas where there is strong citizen opposition and would be more likely to choose a site based on the path of least resistance. The selection of a minority neighborhood based on these criteria would be an example of institutional discrimination.

Environmental justice is not only about environmental burdens faced by minorities but also about those related to an individual's income or gender. Sociologists disagree on the extent to which poverty and race are separated from each other in reference to discrimination. In both of the Warren County and Bullard examples above, the individuals in the minority neighborhoods and communities facing environmental burdens were also low-income individuals.

Community Response

One shining example of a community response to an environmental injustice is the Dudley Street Neighborhood Initiative in Boston, Massachusetts. Dudley Street Neighborhood is located less than two miles from downtown Boston. In the early 1980s, Dudley Street

Neighborhood had a large amount of vacant land. The neighborhood population was made up of minority and low-income families. The vacant land had become the unofficial city dumping ground as individual citizens and companies discarded their trash, and much of the neighborhood became contaminated by hazardous waste. With the help of volunteers and a small amount of grant money, the citizens were empowered to form the Dudley Street Neighborhood Initiative (DSNI). DSNI involved more than 3,000 residents, businesses, nonprofit organizations, and religious institutions. They became dedicated to restoring the neighborhood, not only into a healthy place to live, but into a thriving community.

The DSNI has been successful in reversing the environmental dangers present to members of the community. More than half of the vacant parcels have been cleaned up and transitioned into affordable housing, a new school, a community center, a community greenhouse, parks, playgrounds, and other public areas.

Government Involvement

President Bill Clinton signed Executive Order 12898 in February 1994. This order required federal agencies to develop strategies to identify and to address health or environmental effects of government programs, policies, and actions on minority and low-income populations. He also created an Office of Environmental Justice. This office is now part of the Environmental Protection Agency (EPA). This office has been scaled down since the Clinton presidency, but the EPA still has a National Environment Advisory Council and heads the Federal Interagency Working Group, which gives out grants and awards for communities trying to tackle environmental injustices.

International Examples

The Green Belt Movement focuses on the environmental hardships borne by women as they attempt to take care of their families in Kenya and is an international example of the principles of environmental justice. The Green Belt Movement was founded by Professor Wangari Maathai. She began this movement in the late 1970s when she organized poor rural women in Kenya to plant trees. Planting trees promised to generate the wood needed for cooking but was also part of a larger effort to combat deforestation in Kenya. Communities in Kenya have organized to prevent further environmental devastation. The movement has also improved both environmental and societal conditions for women in Kenya. Maathai was awarded the Nobel Peace Prize in 2004 for her efforts with this movement.

Criticisms

There is a small group of traditional environmentalists who try to distinguish themselves from the environmental justice movement. They have criticized this movement as an attempt to shift the focus of the environmental movement away from important environmental issues toward more anthropocentric concerns such as racism, classism, and sexism. Other environmentalists embrace environmental justice as a bridge between environmentalism and important social movements.

See Also: Civic Environmentalism; Democracy; Sustainability, Consumer Ethics and; United Nations Millennium Development Goals.

Further Readings

Bryant, B., ed. *Environmental Justice: Issues, Policies, and Solutions.* Washington, DC: Island Press, 1995.

Bullard, R. *Dumping in Dixie: Race, Class, and Environmental Quality.* Boulder, CO: Westview Press, 1990.

Cuesta Camacho, D. E. *Environmental Injustices, Political Struggles: Race, Class, and the Environment.* Durham, NC: Duke University Press, 2002.

Dudley Street Neighborhood Initiative. http://www.dsni.org (Accessed January 2009).

Lester, J., et al. *Environmental Injustice in the United States: Myths and Realities.* Boulder, CO: Westview Press, 2000.

Maathai, Wangari. *Unbowed: A Memoir.* New York: Alfred A. Knopf, 2006.

Shrader-Frechette, K. *Environmental Justice: Creating Equality, Reclaiming Democracy.* Oxford, UK: Oxford University Press, 2002.

U.S. Environmental Protection Agency. "Environmental Justice." http://www.epa.gov/environmentaljustice (Accessed January 2009).

Wenz, P. S. *Environmental Justice.* Albany: State University of New York Press, 1988.

Jo Arney
University of Wisconsin–La Crosse

Environmental Justice Ecology, Radical

Radical ecology is considered a call to action motivated by a nonanthropocentric or nonhuman-centered worldview. Philosophically and as a movement, radical ecology is said to be quite diverse and lacking coherence, even if there is consensus that action is needed to address the current environmental problems, and that nature, including the human and nonhuman, living and nonliving, has intrinsic (as opposed to utilitarian) value.

Radical ecology can be understood in its relation to issues of environmental justice (or a fight against environmental injustice) but is more explicitly connected to an ideology or a set of principles based on particular understandings of ecology and radical praxis. In this article, diverse sets of both theoretical and action-inspired foundations are considered in order to flesh out some of the diversity that exists within radical ecology. As well, critiques from within and outside the field are considered as part of the ongoing radical ecology movement.

One of the often-cited sources of inspiration for the perspective upon which radical ecology is founded is the work of Aldo Leopold. Leopold's "land ethic" argued that there was a moral imperative to perpetuate the integrity and stability of "biotic communities" (ecosystems), and Leopold was keen to draw on ecological science to support his arguments.

According to Carolyn Merchant's *Radical Ecology: The Search for a Livable World,* radical ecology draws from three broad theoretical approaches: spiritual ecology, social ecology, and deep ecology. Spiritual ecology focuses on the relationships or interconnectedness between religions and the environment in order to foster a respect of nature and to address environmental problems. Social ecology, however, is more concerned with the material impacts of environmental problems, and its theoretical foundations include Marxist political economy and anarchism.

Deep ecology, which like Leopold also takes up and builds upon a similar ecocentrism and appreciation for concepts from ecological science, has also been an important theoretical approach connected to radical ecology. Norwegian philosopher Arne Naess first used the term in the early 1970s to distinguish "deep" or holistic as compared to "shallow" or reformist environmentalist approaches. The deep ecology platform was founded on two key values: self-realization, where the self and nature are connected and all members—human and nonhuman—are treated fairly; and ecocentrism, an explicitly ethical argument that all parts of nature possess intrinsic value. Together, such philosophical arguments can be seen as an important part of the foundation upon which radical ecology is built.

In addition to the move toward self-realization sought by supporters of deep ecology, other more explicitly radical environmentalists have employed and advocated direct action against perpetrators of ecological damage. In *Eco-Warriors: Understanding the Radical Environmental Movement,* Rik Scarce presents EarthFirst! as one example among several radical groups whose members, motivated by what they see as the necessity of preserving biological diversity, have employed direct action as a strategic move to garner public and media attention. These mobilizations operate on nonhierarchical models and at small scales, and in some cases are rooted in individual rather than group action.

Radical ecology activists are seen as taking a "no compromise" position as defenders of nature from destructive human activities. They aim to "put the Earth first," following the idea that there is a responsibility to act in defense of nature based on ecocentric or biocentric principles. Similar groups include Sea Shepherd, Animal Liberation Front, and the Earth Liberation Front.

While EarthFirst! and Sea Shepherd have been distinguished from more mainstream environmental organizations such as World Wildlife Fund or the Sierra Club, both were founded by former staff members of such organizations, and they see their work as being complementary to the goals of the mainstream movement. Radical campaigners argue that they can make more far-reaching demands than mainstream groups that must constantly protect their relationships with financial donors and political power brokers; the direct activism of radical organizations may raise the profile of issues and concerns common to both sets of groups, yet at the same time allow the mainstream environmentalists' demands to seem more reasonable.

Within the approaches of radical ecology, both violent and nonviolent measures have been pursued by different activists. Such activities have ranged from tree sit-ins to actions with a greater potential for violence—tree spiking, for example, destroys logging equipment like chainsaw blades or mill equipment and is potentially dangerous to loggers and mill workers.

While these actions have been successful in exposing and drawing attention to some serious environmental problems—the clear-cutting practices in old growth forests in North America, for instance—a variety of critiques have been leveled at these radical approaches. For one, more mainstream environmental groups may view radical approaches as not complementary but detrimental to their own work, particularly with regard to building relationships with government in order to influence public policy.

More serious philosophical critiques, and charges of racism and sexism, have also been made. Founders and other members of EarthFirst! made statements in the 1980s and 1990s that targeted population growth as one of the most important sources of the environmental crisis and that affixed blame for this growth along racial lines. EarthFirst! founder David Foreman, for instance, was accused of making racist statements in advocating strict immigration laws, particularly with regard to Hispanic peoples entering the United States.

Moreover, since at least the late 1980s, ecofeminism, which emerged as a radical approach in its own right, has offered influential critiques of radical ecology. Ecofeminism aims to critique anthropocentrism and can be seen as promoting more of a direct connection with nature; however, many ecofeminists have critiqued other biocentric approaches for ignoring and/or perpetuating mutually reinforcing patriarchal systems of domination over both nature and women.

Over time, such critiques and debates within the movement have served to challenge and push for change within radical ecology. While debate continues, such debate and diversity has also pushed radical ecologists to address new/different/various problems. For example, highlighting both the diversity in the movement and responses to critique, some radical ecology activists, such as leaders within EarthFirst!, have argued that opposing social injustices is just as important as advocating nature protection.

Overall, motivated by a responsibility to act in defense of nature based on principles that recognize its intrinsic value, radical ecology continues both to be the subject of critique and at the front line of pressing environmental issues.

See Also: Biocentrism; Deep Ecology; Ecocentrism; Ecofeminism/Ecological Feminism; Social Ecology.

Further Readings

Drengson, Alan and Yuichi Inoue, eds. *The Deep Ecology Movement.* Berkeley, CA: North Atlantic Books, 1995.

Merchant, Carolyn. *Radical Ecology: The Search for a Livable World.* London: Routledge, 1992.

Scarce, Rik. *Eco-Warriors: Understanding the Radical Environmental Movement.* Chicago, IL: Noble Press, 2005.

Slicer, Deborah. "Is There an Ecofeminism–Deep Ecology 'Debate'?" *Environmental Ethics,* 17 (1995).

Vanessa Lamb
York University, Toronto

Environmental Law

Awareness of the havoc humans can wreck on the environment has increased exponentially over the last century. In response to that awareness, many individuals, corporations, and other entities have voluntarily engaged in actions geared toward creating more sustainable practices. While these efforts have had a tremendous result in improving certain environmental risks, most are unwilling to allow voluntary actions alone to be responsible for creating a greener world. In response, environmental laws seek to regulate certain human behaviors that negatively impact the Earth. Environmental laws may encourage certain sustainable behaviors, such as recycling, through a series of incentives, or prohibit other behaviors, such as particulate emissions, through bans and penalties. Environmental laws focus on two primary goals: the first concerns reducing pollution and remediating its effects, while the second

This photograph documents industrial pollution on the Calumet River near Chicago in the late 20th century. The Clean Water Act of 1972, an example of environmental law, led to the establishment of a permit system to regulate point sources of water pollution.

Source: U.S. Environmental Protection Agency

involves conserving and managing natural resources.

In the United States, environmental laws are the concern of both federal and state governments. Although both federal and state legislative bodies have an interest in these matters, the U.S. Congress at times supersedes state mandates in order to provide consistent and predictable rules. In the United States, laws related to pollution usually deal only with a specific aspect of the environment, such as air, water, or the like. Laws dealing with pollution can emphasize minimizing emissions of pollutants into the environment or establish penalties for exceeding certain maximum standards or liability for cleanup. Resource conservation and management laws tend to focus upon a single natural resource, such as forests, minerals, or specific regions or coastal areas. Other laws that are not specifically environmental in nature, such as zoning, land use, and other such statutes, also affect those interested in sustainability.

Foundations of Environmental Law

Environmental law as a field is influenced by a variety of principles, disciplines, and concepts. Environmentalism is, of course, driven by sometimes conflicting standards of responsibility, sustainability, ecology, conservation, stewardship, management, and preservation. While many agree on the importance of these principles in general, specific disagreements often arise regarding the best ways to protect the natural environment and maintain human health. Environmental law draws upon various disciplines when looking at sustainable practices, including biology, biochemistry, economics, chemistry, physics, environmental studies, engineering, and philosophy. Environmental law ultimately involves maintaining equilibrium between these varying perspectives, as well as balancing individual and group norms with regard to certain benefits and risks.

The balancing of interests affects both laws intended to control pollution and better manage and conserve resources. Pollution control laws focus on protecting the natural environment from waste and contaminants and preventing risks from negatively impacting human health. Laws that concentrate on resource management and conservation balance the economic benefits from utilizing natural resources with the benefits of preservation. Economically, environmental laws focus on preventing present and future externalities, or those costs or benefits not transmitted through current prices. For example, climate change, which is generally attributed to greenhouse gas emissions that result from the burning of coal, gas, or oil, is a negative externality insofar as the costs incurred are borne by the public as well as by future generations who did not agree to it. Environmental laws also seek to prevent the exhaustion of common resources, such as clean water, from individual exhaustion. Environmental laws often impose penalties upon certain groups while benefiting others. As a result, they are often highly controversial, and require an assessment of competing interests. Environmental laws are enacted at the international, national, and local level.

International Environmental Laws

The desire for sustainable development, and to avoid pollution, creates a need for international environmental laws. International environmental law often brings into play questions about sovereignty, as well as issues of comity or legal reciprocity that nations extend to rulings from other jurisdiction's courts. Nations unite in a desire to create environmental laws that have international jurisdiction because pollution does not respect political boundaries. Shared interests in reducing terrestrial, marine, and atmospheric pollution and protecting wildlife and biodiversity have brought together lawmakers from various nations in an attempt to craft solutions that will be binding across borders.

International environmental laws are generally enacted via bilateral or multilateral treaties, which are sometimes known as conventions, agreements, or protocols. Treaties that deal with environmental issues tend to concentrate on specific situations or harms, although occasionally a general treaty might have a clause or two that refers to such matters. Interest in sustainability and the environment have resulted in over 1,000 treaties relating to environmental law being entered into and ratified. No other area of international law has generated such a large degree of interest. The science base undergirding environmental law is constantly growing and changing. As a result, protocols, which are agreements that supplement or supplant portions of a previously agreed upon treaty, are frequently used with international treaties addressing environmental law. Protocols are also useful where it is politically more expeditious to reach agreement on a broad framework while leaving more controversial details to be worked out in the future. International treaties pertaining to environmental law abide by those customs, norms, and rules that bind nations during their normal course of behavior. These behaviors are sometimes referred to as customary international law, and would, for example, require one nation to alert or warn other states about events of an environmental nature or potential damages to which another might be exposed.

Growing interest in sustainability, pollution, and preservation has generated a number of international treaties affecting environmental law. Significant international treaties that have developed environmental law include the United Nations Conference on the Human Environment (1972), the World Commission on Environment and Development (1983), the United Nations Conference on Environment and Development (1992), and the World Summit on Sustainable Development (2002). The United Nations Framework Convention on Climate Change (UNFCCC), for example, developed out of the United Nations Conference on the Human Environment, and resulted in the Kyoto Protocol, which has also proved noteworthy. The UNFCCC's objective is to stabilize the amounts of greenhouse gases in the atmosphere at levels that would prevent anthropogenic interference with the climate. The Kyoto Protocol was adopted as part of a meeting held in 1997 in Kyoto, Japan. The Kyoto Protocol entered into force in early 2005 and has been ratified by 187 countries. A goal of the Kyoto Protocol was for nations to reduce their greenhouse gas emissions below 1990 rates. As of 2010, the United States remains a notable nonparty to the protocol, even though it accounted for over 36 percent of 1990 greenhouse gas emissions. International treaties have great influence over environmental law, in part because of the authority provided by signatories, but also due to the weight of joint cooperation and discussion of environmental topics. International courts, such as the International Court of Justice, the International Tribunal for the Law of the Sea, and the European Court of Justice, although few in number and with limited authority, also play an important role in the development of environmental law. Holdings by such courts often have policy impacts broader than the scope of the individual cases heard.

United States Environmental Law

In the United States, federal, state, and municipal legislative bodies have all passed a variety of laws that address environmental issues. The U.S. Congress has passed landmark environmental legislative programs, but many lesser-known statutes and administrative agency regulations are equally significant. The legislatures of all 50 states have also passed legislation that seeks to reduce pollution and protect and preserve wildlife and other natural resources within state boundaries. Court decisions, from the federal and state judiciaries, also review statutory systems and administrative regulations that address environmental issues.

Federal Environmental Law

The United States' federal government regulates certain activities and policies that affect the environment. Although the United States has seen environmentally significant development for hundreds of years, including industrialization, the clearing of forests, mining, and farming, attention to its effects only became significant during the 1960s. Beginning in the 1960s, interest in the environmental movement surged, and with this came a desire for federal laws regulating activities with an environmental impact. U.S. environmental policy has three primary and complementary goals: the first is to allow future generations to enjoy the environment, the second is to interfere as little as possible with commercial interests or the liberties of the current generation to use and enjoy the environment and its resources, and the third is to provide an equitable way of dealing with environmental costs. While individual environmental laws may not satisfy all constituents, these goals drive environmental policy in the United States.

During the 1960s and 1970s, a number of environmental laws were passed by the U.S. Congress. Chief among these were laws regulating air and water pollution and laws that established the U.S. Environmental Protection Agency (EPA). Laws attempting to control hazardous waste, such as solvents or sludge from petroleum refining, were also passed. Battles between environmentalists and business interests—before, during, and after legislation is passed—are not uncommon, and represent competing interests and different desires with regard to outcomes. These competing interests sometimes slow down environmental legislation, but progress has been made, with increases in air and water quality occurring over the past 40 years.

The Clean Air Act, originally passed in 1963 and amended in 1966, 1970, 1977, and 1990, was one of the first major environmental laws passed by the U.S. Congress. In its current form, the Clean Air Act requires the EPA to develop and enforce regulations protecting the public from airborne contaminants hazardous to human health. The act also requires coordination between state and federal authorities. The EPA issues National Air

Ambient Quality Standards that apply to outdoor air across the United States. There are primary standards, which are designed to protect human health, especially that of the elderly, children, and those with respiratory diseases, and secondary standards that focus on other adverse effects of pollution, such as poor visibility, crop damage, and danger to domestic animals. Since 1997, the National Air Ambient Quality Standards have set stringent criteria regarding permissible levels of ground-level ozone (a major cause of smog) and airborne particulate matter (a major cause of soot). The EPA negotiates with individual states to allow local authorities to take over responsibility for compliance within their borders in exchange for money. To qualify, states must submit a state implementation plan (SIP) to the EPA for approval. SIPs include the regulations a state will use to clean up polluted areas and must be made available to the public. The EPA also provides technical assistance and expertise to the states, often in the form of scientific research, engineering designs, and funding for clean air programs.

The Clean Water Act, based on the same motivating forces that spurred the Clean Air Act, was enacted in 1972, and was later amended in 1977 and 1987. The Clean Water Act sought to eliminate water pollution by 1985, and to ensure that all water would meet standards for sports and recreation by 1983. To that end, the EPA was charged with establishing a permit system regulating point sources of pollution. Point sources are identifiable localized causes of pollution, and include industrial plants such as petroleum refineries, governmental facilities such as military bases, and agricultural facilities such as animal feedlots. Pursuant to EPA policy, point sources are not allowed to discharge pollutants to surface waters without first receiving a permit to do so from the National Pollutant Discharge Elimination System (NPDES), which is a subagency of the EPA. The Clean Water Act initiated the permit system because previous mandates that states establish water quality standards had proven ineffective.

Although certain nonpoint sources of pollution were not originally regulated, such as storm water runoff, evolving understanding of their harm caused these to be regulated after 1987. The Clean Water Act provides states and municipalities funding through federal grants that cover up to 75 percent of a sewage treatment plant's construction costs. The Clean Water Act also provides states with research expertise, engineering consulting, and other services. The Clean Water Act applies to all waters with a significant nexus to navigable waters in the United States and its territories. Some environmental advocates suggested that this definition included intermittent streams, wetlands, and other sensitive areas. In 2006, however, the U.S. Supreme Court ruled in *Rapanos v. United States,* 547 U.S. 715, that the Clean Water Act applied only to streams, oceans, rivers, and lakes. This ruling highlights the interaction between Congress—which passes statutes, administrative agencies—which interpret them, and the judiciary—which is the final arbiter of that interpretation.

Other federal statutes have guided the environmental policy of the United States, including the Toxic Substances Control Act, the Comprehensive Environmental Response, Compensation, and Liability Act, and the Food Quality Protection Act. In 1976, Congress passed the Toxic Substances Control Act (TSCA), which controls the introduction of new chemicals. TSCA prohibits the manufacture or import of any chemicals not on the TSCA inventory, which are sometimes referred to as existing chemicals. Manufacturers of new chemicals, or those not on the TSCA inventory, must notify the EPA prior to the manufacturing process so that the EPA may assess the risk to human health. If the EPA finds an unreasonable risk to human health, it regulates that substance, imposing a rolling set of controls—from limiting a substance's uses, controlling production, or an outright ban.

The Comprehensive Environmental Response, Compensation, and Liability Act of 1980 (CERCLA), sometimes referred to as the Superfund legislation, was intended to provide federal authority to clean up land contaminated with hazardous substances. CERCLA authorizes the EPA to identify releases of hazardous substances, to distinguish the individuals or corporations responsible for the contamination, and compel them to clean up the sites. In the case of contaminated sites where the responsible parties cannot be identified, CERCLA provides a trust fund, initially provided by a tax on the petroleum and chemical industries, which allows the EPA to conduct removal and remediation. Although CERCLA has been responsible for identifying or cleaning up nearly 2,000 sites, it has been controversial insofar as it has sometimes required cleanup from current owners of properties who had nothing to do with the land's contamination.

In 1996, the Food Quality Protection Act (FQPA) was passed. The FQPA requires the EPA to alter the ways in which it evaluates and regulates pesticides. Specifically, the FQPA has mandated that the EPA take a variety of actions to improve public safety, including adopting a new safety standard requiring reasonable certainty of no harm for all pesticides used with food production, to consider risks to infants and children when setting tolerances, to expedite approval of reduced risk pesticides, and to monitor potential harm throughout the food cycle. The FQPA has improved national uniformity of safety standards and improved accountability and communication related to pesticides.

Other Sources of Environmental Law

In addition to international treaties and federal legislation, other sources of environmental law exist, including federal regulations, judicial opinions, and other state and local statutes. Regulations, promulgated by federal agencies such as the EPA, fill in the specifics for broad policies initiated by Congress. Judicial decisions allow the resolution of disputes regarding the meaning of statutes, the appropriateness of regulations, and the interplay between laws and the U.S. Constitution. Holdings in certain sensitive cases, such as *Rapanos,* have a policy impact much broader than the case at hand. Private civil actions also have an effect on environmental law, especially tort cases. Torts are civil actions brought by one party against another for noncontractual, noncriminal wrongdoing. Legal doctrines such as negligence, nuisance, strict liability, and trespass provide those who have allegedly suffered an environmental harm a remedy separate from those contained in environmental statutes or regulations. Plaintiffs are thus able to use common law to seek relief when other sources of law do not support their claims.

State and municipal governments are also a source of environmental law. Although local law is preempted when federal statutes addressing a certain area are in place, there are many areas where state and local governments can play a role in environmental law. Laws pertaining to recycling, land use, and preservation are especially the purview of state, local, and municipal government. These statutes and regulations may control lot size, the types of development permitted, or the creation of nature reserves near and containing parcels of particular beauty or environmental importance.

Future Developments

To date, much environmental law has been promulgated at the national level, as specific needs to deal with contaminants, pesticides, and clean up drove much of the initial legislation on this topic. As national laws have matured, however, many have recognized that a need exists for an international environmental regulatory body to deal with transboundary issues. Attempts to deal with environmental issues in a global manner have been hindered by a lack of mechanisms to enforce environmental policy across national borders. Previous experience with international agreements, such as the Kyoto Protocol, suggests that reaching agreement regarding an international environmental regulatory body may well prove contentious. These difficulties indicate that those supporting such a system to enforce environmental policy across borders should be prepared for a lengthy and arduous struggle before consensus is reached.

As interest in sustainability and the environment increases, environmental law will continue to play a central role in establishing standards for behavior and setting goals for permissible behavior. Although great disagreement may exist about specific actions, consensus regarding the need for governmental regulation exists. Environmental law will continue to evolve to address changes in science and the need for sustainable development.

See Also: Civic Environmentalism; Conservation, Aesthetic Versus Utilitarian; Environmental Justice; Environmental Policy; Forest Preservation Laws; Green Laws and Incentives.

Further Readings

Farber, D. A., J. Chen, R. R. M. Verchick, and L. G. Sun. *Disaster Law and Policy: Katrina and Beyond,* 2nd ed. New York: Aspen Publishers, 2009.

Farber, D. A., J. Freeman, and A. E. Carlson. *Cases and Materials on Environmental Law,* 8th ed. Saint Paul, MN: West Publishing, 2009.

Mutz, K. M., G. C. Bryner, and D. S. Kenney, eds. *Justice and Natural Resources: Concepts, Strategies, and Applications.* Washington, DC: Island Press, 2002.

Rodgers, W. H., Jr. *Environmental Law,* 2nd ed. Saint Paul, MN: West Publishing, 1994.

Stephen T. Schroth
Jason A. Helfer
Jonathan R. Fletcher
Knox College

Environmental Law and Policy Center

The Environmental Law and Policy Center (ELPC) is a public interest, nonprofit organization that provides environmental legal advocacy and green business consulting. Focused on the Midwest region of the United States, ELPC maintains offices in Chicago, Illinois; Columbus, Ohio; Des Moines, Iowa; Jamestown, North Dakota; Madison, Wisconsin; Minneapolis, Minnesota; Sioux Falls, South Dakota; and Washington, D.C. Since its founding in 1993, ELPC has subscribed to an interdisciplinary philosophy in creatively solving environmental problems and in finding innovative ecoentrepreneurial solutions. This philosophy is evident in the diversity of its staff and its board of directors, which are composed of experts in the areas of law, finance, science, business, public policy, government, community outreach, education, communications, and marketing. This perspective is also illustrated by ELPC's creation of a Science Advisory Council, which links the group with academicians and scientists from many different areas of the United States. Based on its mission and its work, ELPC is an important example of an organization that transforms its commitment to green ethics and philosophy into progressive action and implementation.

ELPC's major projects revolve around several key aims, which include making strides for improvement in the areas of extreme global climate change, environmentally friendly transportation, and clean energy; expanding green business; and preserving natural areas, like rivers, forests, and lakes, in the Midwest. In pursuit of these goals, ELPC engages in the promotion of renewable energy, reduced emission transportation systems, and energy-efficient construction. Additionally, ELPC advances innovations in the areas of e-waste recycling, as well as greener architecture, business solutions, and restaurants. Further, ELPC has worked with communities and other nonprofit organizations to engage in substantial conservation efforts for many natural places like the Mississippi River, the Great Lakes, and Wisconsin's Northwoods.

ELPC has worked in legislative, legal, executive, administrative, and private arenas to advance its causes. Legislatively, ELPC has lobbied a variety of state, local, and federal officials to enact and implement a variety of environmental statutes. In 2006, the ELPC consulted with the Governor's Office of Illinois and the Illinois Environmental Protection Agency to promulgate a state mercury pollution reduction rule. An example of ELPC's litigation was a 2001 action before the U.S. Supreme Court, in which ELPC successfully argued for the suspension of a steel company's punitive damages claim against a community group based on a previously filed environmental enforcement action. Finally, as an example of its public-private partnerships with industry, ELPC has helped to raise millions of dollars for the development and enhancement of energy-efficient light rail transportation systems in Illinois, Wisconsin, and several other Midwest states.

In addition to physical connections with a multitude of civic, governmental, and private organizations, ELPC has implemented a comprehensive communications strategy by maintaining a presence on several social networking sites, including Twitter, Facebook, LinkedIn, and on other Internet sites, such as YouTube and Flickr. To extend its reach and influence, ELPC also collaborates with Midwest and national news outlets to contribute to media conversations dealing with the environmental advocacy issues that are at the heart of ELPC's mission. Finally, ELPC publishes a variety of handbooks, studies, presentations, and reports that illustrate the goals and strategies of the organization. This integrated approach to advocacy marketing demonstrates one facet of ELPC's strategic plan to transform its green ethics and philosophy into action.

ELPC has been awarded a variety of accolades and distinctions for its work in environmental legal advocacy and consulting. In 2009, ELPC was recognized by both the American Institute of Architects and by the United States Green Building Council for its work toward the establishment of energy-efficient residential and commercial building codes in Illinois. Additionally, ELPC has attained the National Wind Energy Advocacy Award, and a variety of other associations have honored ELPC's staff and directors for their environmental leadership. These awards are illustrations of ELPC's efficacy in its area of public interest law and policy.

Within the last decade, the concepts of green business and government have taken hold throughout the globe, whereby companies and public entities have actively sought to market themselves as environmentally and socially responsible enterprises. At the root of this movement has been the growth of sustainable green approaches to business, government, and planning. Although many companies and governmental bodies have adopted the mantra of green ethics and philosophy, there remain many legal, regulatory, and economic challenges to creating a more sustainable future. ELPC has sought to address these challenges through its various programs and initiatives, establishing itself as a leading non-profit for the Midwest region of the United States. Based on its extensive record of achievements and awards, it seems that ELPC will continue to serve in a leadership role for the 21st century in the areas of environmental law and policy, clean energy, global warming, eco-business, green entrepreneurship, and conservation.

See Also: Business Ethics, Shades of Green; Conservation; Environmental Law; Green Ethics in the Legal Community; Green Law Trends.

Further Readings

Environmental Law and Policy Center. http://elpc.org (Accessed March 2010).

Finney, Martha I. *In the Face of Uncertainty: 25 Top Leaders Speak Out on Challenge, Change, and the Future of American Business.* New York: AMACOM/American Management Association, 2002.

Learner, Howard. "Emerging Issues in Energy and the Environment Symposium: Introduction: Cleaning, Greening, and Modernizing the Electric Power Sector in the Twenty-First Century." *Tulane Environmental Law Journal,* 14 (2001).

Amanda Harmon Cooley
South Texas College of Law

Environmental Pluralism

It is evident that people have different frameworks for making moral choices. These differences appear within as well as between cultural groups. Western philosophy, for example, has multiple schools of thought: utilitarianism (briefly, the greatest good for the greatest number), the Kantian emphasis on duty, and theories based on concepts of justice or virtue. To observe these differences is merely the descriptive side of pluralism. There has been a movement advocating moral pluralism, not only recognizing that it exists. Christopher D. Stone argued that with regard to environmental ethics, even an individual ought to operate with a moral repertoire. J. Baird Callicott fiercely protested the idea of promoting moral pluralism in environmental ethics or even accepting it as unfortunate but either inevitable or transitional, particularly for individuals. Nevertheless, interest in the idea of moral pluralism in environmental ethics has continued to grow, though it has come in terms, for reasons, and from perspectives that differ considerably from Stone's initial proposal: eco-feminist, constructivist, postcolonialist, religious, and pragmatic. In some ways, ethical pluralism seems especially suited to environmental ethics, as moral diversity can be conceived as part of or at least analogous to Earth's biotic diversity.

First, we need to distinguish between pluralism and other similar ideas. The fact that people make different moral choices or that cultures have different customs and practices does not necessarily mean that moral pluralism is at work. The fact that individuals or societies decide to act in different ways does not mean they are acting from different principles. The principles might be identical, but the contexts or situations in which they are applied might differ. This is more accurately described as "situationalism" or "contextualization" than as pluralism.

Pluralism is often mistaken for, or considered a pretext for, relativism. There is a great degree of resemblance. Relativism, like pluralism, has descriptive and normative sides. Descriptive relativism, like descriptive pluralism, simply acknowledges the fact of moral diversity, and both normative relativism and normative pluralism regard such difference in a positive light. Ethical relativism also shares some key insights with ethical pluralism, including the need for tolerance, recognition of the limits of mutual understanding, and acknowledgment that people of good will may differ on what's acceptable.

Relativism has, however, important differences from pluralism and in its strongest forms has serious limitations. Relativism starts from a premise that values and principles are limited in applicability or comprehension by a particular frame of reference and either need not or cannot be shared beyond that frame. There are implicit or explicit claims of incommensurability and incomprehensibility across difference, with stronger forms of relativism seeing greater incompatibility and weaker forms less. But because some degree of incompatibility is presumed as a starting point, relativism is of least use where environmental ethics is needed most—in the overlap of cultures or competing interests. It can foster a defensive or self-serving posture: I cannot criticize your action because your moral standard is valid relative to your context, so also, you cannot criticize mine. Further, if there is a high degree of incomprehensibility between the philosophical ground of different parties, there is no basis on which we can hold one another to account. Theories could be used merely as rationalizations for decisions actually made on the basis of self-interest or power.

Pluralism differs from relativism in that it does not isolate or insulate different theories or cultures from each other but seeks their interaction at the same time as it upholds difference as desirable. Pluralism also is generally willing to entertain fallibilism, the idea that an ethical theory or principle, including one's own primary principle, might be wrong or at least inadequate or incomplete. At the very least, even in its weakest form, pluralism does not require a synthesis of a set of principles, or a resolution of all principles into an overarching or foundational one, although it may seek mutual understanding. Advocates for normative pluralism will argue that decisions made on the basis of multiple frameworks or on more than one independent principle should have a higher likelihood of being strong decisions. Another common position is that a certain theoretical basis may be more appropriate to one context or situation than another would be, and that humans are capable collectively of making judgments of this sort. Both arguments can be summed up with Mary Midgley's pithy analogy: rather than discuss whether the knife or the fork is superior, it might be better to consider how they work together.

One of the first proposals for pluralism that gained widespread attention in environmental ethics was put forth in 1987 by Christopher D. Stone. Almost simultaneously, other writers such as Peter S. Wenz and Andrew Brennan broached the prospect of moral pluralism in terms of "ethical polymorphism" and the possibility of "pluralist theory," but Stone made the most extensive case for moral pluralism at the time.

Stone was arguing against environmental ethics that relied on "monistic determinate" theories, such as Kantianism and utilitarianism. Monistic theories are those, he said, that are grounded in a single principle or set of principles, and they are determinate insofar as they propose to provide one right answer to every quandary. The problems with monistic theories are twofold, according to Stone. First, morality involves several distinguishable activities, not just choice of conduct, but assessing praise and blame, evaluating institutions, and so on.

Second, morality is concerned with a wide variety of things, particularly as it tries to accommodate environmental issues—persons in proximate community, persons remote in time or space, embryos and fetuses, ecosystems and species. The attempt to force all of these into a single framework forces disregard of moral intuitions. Stone was particularly critical that moral considerability was being conceived as an either-or proposition dependent on a single, salient property, such as sentience. Such attempts distort systems that have been constructed based on human sensibilities and capacities. They do not necessarily serve the objectives of ecological ethics well. Moral pluralism, as Stone saw it, would not make the presumption that all activities regarding all things of concern in all contexts can be determined by the same criteria and subject to the same overarching principles. Stone's argument, however, was framed largely within the modern Western philosophical traditions and relies on those concepts and epistemologies. It was quickly demonstrated how limited a concept of pluralism this was.

Jim Cheney shortly thereafter published a pivotal essay that broke open the conversation not only in terms of whose ethical thought counted but also how we might think pluralistically about ethics. He challenged reliance on abstraction of concepts and theories from their original paradigm setting and exportation to others. Postmodern philosophers had already argued that a theory exported is a useful tool for colonizing the cultures and minds of others, and he was taking that "postmodern turn." He proposed a more thoroughly contextualized ethical discourse employing bioregional narrative. His proposal was as much aimed at grounding postmodernism in "place" (still recognizing the social construction of narratives and their associated ethics) as it was in suggesting an alternative to the abstracted and hegemonic tendencies of environmental ethics up to that point. Whether that proposal was entirely successful or strong is not the point here. What matters is that by drawing on the insights of feminists and epistemologies of indigenous people to make his argument for understanding a subtle dynamic of language and place, as well as by his proposal for contextualized discourse and narrative-oriented method, he showed that moral pluralism had many more dimensions.

It was not long before J. Baird Callicott, in "The Case Against Moral Pluralism," took aim at both Stone and Cheney (or rather, Cheney as a proxy for "deconstructive postModernism" in toto). He seems most incensed by Cheney's acceptance of postmodern premises about totalizing discourses. Callicott wanted to defend the long-standing Western traditions of reason, efforts at constructing comprehensive systems of thought, and the importance of metaphysics. He was not defending necessarily a particular metaphysic, at least not one fully formed. His concern was for maintaining a centrality of metaphysics in moral philosophy and the potential for a more adequate emergent metaphysics that would ground a unified theory, one that could link environmental to human social well-being (a potential he sketched based on development of the land ethic). Tellingly, he wrote wistfully about this as a quest for a "Holy Grail." And while he acknowledged that the monistic theories Stone objected to do not prevent conflict over what actions should be taken even by those who share the same ethical framework, he seemed to genuinely fear that without a shared organizing principle or metaphysics, compounded by the deconstructionist distrust of reason found in Cheney, we could only resolve conflict by resort to "bullets or bucks" as we descended into a new "Dark Ages."

Callicott's criticisms of Stone's proposal include raising the specter of relativism and claiming that moral pluralism would weaken our sense of moral commitment. He takes particular exception to Stone's suggestion that an individual could be pluralist in approaching different decisions based on different ethical systems, illustrated by an example of actions of a hypothetical senator based on utilitarianism, the land ethic, and Kantian duty. This would make an individual's moral framework incoherent, Callicott charged, since the underlying metaphysics of utilitarianism implied a radical individualism and social atomism at odds with ecological relationality of the land ethic. Similarly, to act on the premise of Kantian duty would implicitly presume Kant's anthropocentrism, in which Reason defined human essence, the attribution of moral agency and considerability, and was clearly lacking in animals. How could one hold to that and still claim an ethic based on humans as peer members of an Earth community?

The simple answer is that people do live their moral lives acting out of different frameworks all the time. This does not necessarily mean that metaphysics (or worldview, which Callicott alternately uses for the same purpose) is not important in moral philosophy. Some metaphysics or worldviews are clearly unhelpful as they shape ethical responses to the global, multigenerational ecological dilemmas that we face. But it may be that people are capable of thinking in different metaphysical terms in different contexts, or that an ethical system does not necessarily continue to carry the metaphysical weight of its original framers that Callicott assumes.

It is also true, as William Edelglass has shown with regard to Buddhism, that a unified worldview, even a single text, can still be robustly pluralistic. One clear advantage of a pluralistic approach to environmental ethics is that it provides the opportunity for dialogue with and between long-standing religious traditions, bringing a wealth of wisdom to bear and billions of adherents to engage in ethical deliberation about environmental questions. The decade-long Earth Charter demonstrated that people with different worldviews, metaphysics, and foundational principles could, through conversation, generate working intermediate principles connecting ecological and human social well-being.

See Also: Callicott, J. Baird; Earth Charter; Kantian Philosophy and the Environment; Land Ethic; Utilitarianism.

Further Readings

Bauman, Zygmunt. *Postmodern Ethics.* Oxford, UK: Blackwell, 1993.

Callicott, J. Baird. *Beyond the Land Ethic: More Essays in Environmental Philosophy.* Albany: State University of New York Press, 1999.

Callicott, J. Baird. "The Case Against Moral Pluralism." *Environmental Ethics,* 12 (1990).

Cheney, Jim. "Postmodern Environmental Ethics: Ethics as Bioregional Narrative." *Environmental Ethics,* 11 (1989).

Edelglass, William. "Moral Pluralism, Skillful Means, and Environmental Ethics." *Journal of Environmental Ethics,* 3/2 (2006).

Evanoff, Richard. "Bioregionalism and Cross-Cultural Dialogue on a Land Ethic." *Ethics, Place and Environment,* 10/2 (2007).

Hinman, Lawrence M. *Ethics: A Pluralistic Approach to Moral Theory,* 3rd ed. Belmont, CA: Wadsworth/Thomson Learning, 2003.

Marietta, Don E., Jr. "Pluralism and Environmental Ethics." *Topoi,* 12 (1993).

Midgley, Mary. *Utopias, Dolphins and Computers: Problems of Philosophical Plumbing.* London: Routledge, 1996.

Stone, Christopher D. *Earth and Other Ethics: The Case for Moral Pluralism.* New York: Harper & Row, 1987.

Stone, Christopher D. "Moral Pluralism and the Course of Environmental Ethics." *Environmental Ethics,* 10 (1988).
Wenz, Peter S. "Minimal, Moderate, and Extreme Moral Pluralism." *Environmental Ethics,* 15 (1993).

Nancie Erhard
Saint Mary's University, Halifax

Environmental Policy

Environmental policies are deliberate actions concerned with the prevention or reduction of harmful effects of human activities on our ecosystems. They can be formulated and implemented by public and private organizations.

Environmental policies are needed because environmental values are usually not considered in organizational decision making. There are two main reasons for this omission. First, environmental effects are external: the polluter does not bear the consequences of its actions; the negative effects occur elsewhere or in the future. Second, natural resources are underpriced, assuming infinite availability. Together, this results in a tragedy of the commons. The pool of natural resources can be considered as a commons that everyone can use to their own benefit. From an individual point of view, it is rational to use a common resource without considering its limitations, but this self-interested behavior will lead to the depletion of the shared limited resource—and this is not in anyone's interest. Individuals are nevertheless inclined to do so because they reap the benefits in the short term, while the costs of depletion are experienced by the community in the long term. Since incentives to use the commons sustainably are weak, government has a role in the protection of the commons.

A Brief History of Environmental Policy Making

Public policies aimed at environmental protection have been developed since the late 1800s and early 1900s, when industrialization and urbanization led to the destruction and degradation of nature and natural resources and had severe consequences on human health. This led to rules and regulations for urban hygiene, including sewage and sanitation and housing conditions. Also, nature and wildlife were protected, and this was usually privately financed by rich individuals and private foundations.

In the 1950s and 1960s, people became aware of the harmful effect of emissions and the use of chemicals in industry and pesticides in agriculture. This led to a detailed system of regulation, which is still in place today in many industrialized countries. In these regulations, governments forbid the use of hazardous substances or prescribe maximum emission levels of specific substances needed to ensure a minimum environmental quality. Such regulative systems succeeded in effectively addressing point sources, such as plants, where cause-and-effect relationships between the actors causing the negative environmental effect could be clearly established. Nevertheless, there were some environmental problems that persisted, often because of the many diffuse sources of pollution. Individually, these small sources were not harmful, but the accumulation of the emissions by these sources exceeded the regulative minimum norms for environmental quality. Also, the increasing complexity of chains of cause and effect contributed to the persistence of problems. In the 1980s, the effects of acid rain showed that causes and effects of environmental pollution could be geographically wide apart; it underlined the message that the natural resources on Earth were quickly being depleted.

From the late 1980s, sustainable development became a leading concept in environmental policy making. Instead of emphasizing and limiting the negative aspects of human behavior, this concept emphasized the opportunities of sustainable development. Benefits would be social, economic, and environmental; these values were no longer by definition conflicting and competing. The doomsday scenario of the future was replaced by one of positive action. In about 30 to 40 years' time, in a process of gradual, incremental change, our economy could be reformed from one that was based on the overuse of natural resources and the trespassing of Earth's carrying capacity into one that could be sustained.

For environmental policy making, sustainable development had strong implications. With nature and natural resources being considered as economic drivers, environmental policy making was no longer the exclusive domain of government; instead, it also attributed environmental responsibility to private industry. Also, the concept emphasized that individual people and their communities played a key role in the effective implementation of policies. Participation of those individuals and groups that were affected by policies was therefore considered as an important feature and condition for policy success.

Guiding Concepts for Policy Making

Over the years, a variety of principles have been developed to help policy makers to develop effective policies. Examples of such guiding principles, some of which have acquired a legal basis in countries or regions, are the polluter pays principle, which makes the polluter liable for the costs of environmental damage resulting from his behavior, and the precautionary principle, which states that an activity is not allowed when there is a chance that the consequences are irreversible.

More recently, climate change challenges showed the need to view our Earth as an ecosystem consisting of various subsystems, which, once disrupted, can lead to rapid changes that are beyond human control. In the previous decades, policies have already made a shift from end-of-pipe solutions to prevention and control, in which negative effects were mitigated and, if unavoidable, were compensated, for example, by investing in nature in other places than where the damage was caused. But even if prevention policies were implemented today and carbon emissions were minimized within a short period of time, the odds are high that climate change would continue. Besides policies aimed at the reduction and mitigation of effects, a parallel track of policies is being developed that aim to adapt the living environment to changing conditions and to improve its resilience. Resilience is a concept from ecosystems theory and concerns the capacity of a system to bounce back to its original state after a severe disturbance. For example, in Curitiba, a city in Brazil, some districts flooded each year. The residents of these districts were relocated to higher and dryer places, and their former living areas were transformed into parks that could be flooded without disrupting city life. With these parks, Curitiba became one of the cities with the highest percentages of greenery per citizen in the world.

Environmental Policy Instruments

One of the key questions in environmental policy making is how to change actor behavior. Numerous instruments have been developed to influence actor behavior. A large proportion of environmental policy studies are concerned with developing typologies of policy instruments that aim to identify the conditions under which particular types of instruments work best.

Traditionally, public policy theories focused on regulation, incentives, and information as the tools of government. From the 1980s forward, the effectiveness of traditional public policy instruments for addressing environmental problems declined. These instruments could not effectively deal with the growing responsibilities and ambitions of government and the growing interdependencies between public and private sectors. Many of the goals were beyond the authority and control of governments and required cooperation of and collaboration with private actors. Also within government, responsibility for particular policy goals was shared among many public organizations and authorities. Policy making became a process of interaction and negotiation between multiple actors, public and private, who were entangled in a web of networked, interdependent relationships. Within this setting, conventional public policy instruments had limited effect.

Especially, environmental problems proved to be persistent and provided a breeding ground for policy makers all over the world to experiment with new forms of governance. Governments developed a new range of policy instruments that were congruent with the complex characteristics and multi-actor constellation of policy problems. At the core of this second generation of policy instruments was the understanding that the delivery of public policy goals depended upon a joint effort of multiple actors who were involved in and/or affected by policy problems. These instruments make use of other than the strictly hierarchical mechanisms based on command and control for influencing actor behavior and decision making. The conventional instruments as well as some new instruments that have been adopted in many countries all over the world are discussed below.

Regulation

Regulation is used to impose minimum requirements for environmental quality. Such interventions aim to regulate specific activities and their effects, involving particular emissions, particular input (such as specific hazardous substances), ambient concentrations, risks and damages, and exposure. Often, permits have to be acquired for those activities. The permits have to be renewed periodically. In many cases, local and regional governments are the issuing and/or controlling authority. More specialized activities or potentially hazardous activities with high risks, such as plants treating hazardous chemical substances or nuclear power stations, are more likely to be controlled by a federal or national authority.

Regulation is an effective means to prescribe and control actor behavior. Detailed systems of environmental regulations have resulted in a considerable improvement of environmental quality of air, water, and land compared to the quality several decades ago. The strengths of regulation as an instrument are that regulation is generally binding—it includes all actors that want to undertake an activity described in the regulation—and it treats them in the same way. Regulations are also rigid: they are difficult to change. This can be considered as a strength and a weakness of the instrument. Actors who need to comply with regulations need to rely on the fact that these regulations will not change too suddenly, because compliance may require investments that need to be earned back over a long period of time and that can only be afforded when competitors are also forced to comply. The flip side of this rigidity is that regulations slow down innovation; regulations tend to benefit known incumbent practices and technologies (and thus their owners) over new and unknown ones. Actors who feel that they are aggrieved by regulations will contest these regulations, but it will take time before they are changed. Procedural safeguards ensuring the quality of policy making are one reason why it takes long to change regulations. Another factor that undermines the potential effectiveness of regulation is that actors should be able to comply with regulations. When regulations demand behavior that is difficult or impossible to meet, for example, because the regulated actors lack knowledge, skills, or finances to do so, regulations will not be effective. In practice, regulations are therefore only effective when the majority of the actors to be regulated are able to comply; regulation is not an appropriate instrument to prescribe cutting-edge technology or practices.

One common improvement in environmental regulation in many countries in the past decade is that—when possible—prescriptions of means to meet certain ends have been replaced by performance requirements. The advantage of performance requirements is that actors addressed by the regulation are encouraged to innovate in order to meet the requirements more efficiently. But also, performance requirements cannot prevent actors lacking incentives to achieve more than the minimum requirements. Once actors have lived up to the requirements, they stop or slow down innovation; minimum requirements then equal the maximum achievements.

A final remark about the choice of regulation as a policy instrument concerns the costs and ability to enforce regulation. Regulation requires both that requirements be formulated unambiguously and the ability to monitor and control compliance to these requirements. Many environmental problems are too ambiguous to address in unequivocal requirements and/or monitoring and enforcement costs would be too high.

Financial Incentives

Governments can decide to stimulate behavioral change by giving positive or negative financial incentives, for example, through subsidies, tax discounts, or fines and levies. These incentives can play an important role in boosting innovation and diffusion and adoption of innovations. For example, in Germany, the widespread subsidizing of the use of solar energy by private homeowners has contributed to the large-scale adoption of photovoltaic (PV) panels. Financial incentives or disincentives can also stimulate professional actors to change; for example, subsidizing environmental innovation by small and medium-sized enterprises led to numerous innovations in this sector. Potential drawbacks of financial incentives are that only those actors are reached who are already committed to sustainability and would have changed their behavior anyway, even without the incentive; they distort the market; and, when not used for a limited period, they can make receivers dependent upon the subsidy. A final drawback is that subsidies are expensive instruments, especially when they are open ended, that is, when no maximum amount available for the subsidy is specified or when no end date is specified.

Environmental Reporting: EIA, EMS, and Ecolabeling

There are several instruments that aim to inform decision makers about the environmental effects of their actions. Usually, decisions are based on a cost-benefit analysis of which environmental costs and benefits are not part. Environmental impact assessment (EIA) is an instrument that helps public decision makers to decide on initiatives with a certain environmental impact, such as the construction of roads and industrial plants. On an organizational level, environmental management systems (EMS) have become common instruments to improve environmental performance. On the level of products or services, eco-labels inform customers about the environmental performance of these products or services.

EIA has become a legal requirement in many countries for initiatives that will affect the environment. The instrument requires that environmental effects of the initiative are studied and that opportunities for mitigation and compensation are suggested. Making environmental information available in this way enables decision makers to include the information in the cost-benefit analysis, especially because they have to account for their choice not to select the most environmentally favorable way to realize the initiative. EIA thus cannot prevent initiatives from taking place—sometimes to the disappointment of action groups and citizens—but it can reduce the negative environmental impact of initiatives concerned.

Environmental management systems (EMS) are comprehensive approaches that help organizations to reduce their use of natural resources and to contribute to the closing of loops of natural resources, while reducing costs and—when certified—contributing to a positive image. The most commonly known standard for such systems is the ISO 14000 family of standards for environmental management, first issued by the International Organization for Standardization (ISO) in 1996. These standards can be applied to all sorts of organizations and industries. They help an organization to identify and control its environmental impact, to continually improve its environmental performance, and to systematically formulate and monitor environmental objectives and targets and demonstrate that they have been achieved.

The ISO 14000 family of standards is a typical example of a hybrid policy instrument: it is a so-called third party policy instrument that is negotiated by representative organizations, public and private, from all over the world. It can therefore be considered as a voluntary, agreed-upon standard that has acquired legal status by being included in rules and regulations in many countries. For example, sustainable procurement policies of organizations may require doing business with ISO-certified organizations.

Environmental labels and certificates apply to specific (groups of) products and services and inform consumers about the environmental performance of the product. Sometimes governments require such labels and certificates, such as the "CE" marking in Europe, which certifies that a product has met minimum requirements for consumer safety, health, and environment. To push organizations to develop products and services that perform beyond these minimum requirements, there are also labels that express the environmental friendliness of the product or service. For example, in the United States and in the European Union (EU), labels have been developed that indicate the energy performance level of household appliances. Also, in the food industry ecolabels are often applied, and even for buildings, labels have been developed indicating the energy performance or overall environmental performance.

The underlying assumption of ecolabeling is that informed consumers make environmentally responsible choices that will provide a stimulus for industry to innovate and produce cleaner products. However, the effect of these instruments is often limited. Organizations and individuals do not make decisions as rationally as assumed. For example, they tend to value up-front capital investment costs higher in decision making than long-term benefits. In addition, consumers tend to distrust the quality of products with a good environmental performance compared with the conventional products they are used to. Such images are very persistent and the niche markets for many green products have not been able to penetrate the mainstream markets yet.

Multilateral Negotiations and Agreements

The United Nations (UN) has provided the main forum for international negotiations and agreements on environmental policies and objectives. From the early 1970s, it has hosted a number of conferences aimed to result in policy plans, programs, and targets that could be multilaterally agreed upon and implemented in national regulatory systems. The 1972 Stockholm conference was the first international conference on environmental issues, and was followed by the Earth Summits in Rio de Janeiro in 1992 and in Johannesburg in 2002. The UN also hosted some special conferences on climate change, of which famous ones took place in 1996 in Kyoto and in 2009 in Copenhagen.

These conferences and summits were a response to the growing awareness that some of the most challenging environmental problems were of a global character, requiring international cooperation, capacity building, and the pooling of knowledge and resources. These conferences have been very effective in the past, setting an international agenda for regional and national environmental policy making that resulted in treaties and protocols, also known as "hard law," and in resolutions, statements, and declarations, or "soft law." The Kyoto Protocol was an example of hard law, with clear-cut reduction targets of greenhouse gas emissions for regions and countries. To help nation-states meet the targets, they could make use of three so-called flexibility mechanisms, which should ensure lowering the costs of compliance.

Joint Implementation, the first mechanism, made it possible for countries to invest in lowering emissions in other countries that had ratified the Kyoto Protocol and thus had a reduction target to meet. For industrialized, developed countries that had already invested in emission reduction in their own economy, it was cheaper to invest in emission reduction in economies in transition, where the same investment would lead to greater reductions. The Clean Development Mechanism, the second mechanism, allows industrialized countries that have ratified the protocol to meet their targets in any nation where it is cheapest to invest, that is, in developing countries. This mechanism is not undisputed, since it involves questions of intervention in the economies of developing countries, which may have an impact on the economic development of these countries. To prevent industrialized countries' not reducing emissions within their own borders, the mechanism can only be used in supplement to domestic reductions, but no definition of such supplemental action was given, which led some countries to achieve 50 percent of their reduction target through this mechanism.

Tradable Permits

Trading of carbon emissions is the third mechanism, but the instrument of tradable permits is applied to other emissions and other environmental problems as well. The first emission-trading schemes already date back to 1974, when the United States experimented with emission trading as part of the Clean Air Act. In the EU, there have been experiences with tradable quotas in the fields of agriculture and fisheries, for example, for the production of manure and cow milk and fishing quotas for regulated species.

Emission trading is a market-based instrument and can be applied in the form of voluntary markets or in a mandatory framework, such as the EU Emission Trading Scheme for carbon emissions. Most of the trading schemes are based on a "cap-and-trade" model. A central authority puts a cap on the overall carbon emissions allowed in a country or region. Within this cap, emission rights are allocated to the polluters; emissions produced beyond these rights are penalized. The idea is that polluters choose between investing in emission reduction or in emission permits. By lowering the cap over time, total emission reduction can be achieved. The trade of permits will ensure that reduction is achieved at the lowest costs.

Crucial for the effective functioning of cap-and-trade systems is that the cap creates scarcity; only then will industries be confronted with the choice to invest in permits or to invest in clean technologies. The market will make sure that emission reductions are achieved at the lowest cost. However, designing a perfectly functioning system proves to be difficult. For example, in 2005, the EU established a carbon market, but failed to create a scarcity in the first years of its existence. As a result, trade was limited and incentives to reduce emissions were weak. When the market was created, incumbent industries successfully negotiated that they would freely acquire the number of permits needed for their production. As long as the cap is not lowered, the price for permits remains low. Incumbents maintain a competitive advantage over newly emerging and possibly cleaner industries. To identify and repair such market failures and to make sure that the actors trading permits comply with the rules of the market, the presence of a regulator is required. However, the example also shows that environmental policy making takes place in a political context that cannot be controlled by the policy makers designing the instruments.

Conclusion

The choice of policies and instruments depends on the characteristics of the problem to be addressed and the characteristics of the institutional environment in which they should be implemented. A contingent design of policy instruments—an instrument should fit the environment that it aims to influence—is based on the assumption that such a fit will contribute to the acceptance of the instrument and the instrument will thus be more effective. Today, democratic governments therefore aim to involve the actors concerned in the policy-making process. It is expected that the loss of time resulting from such participation in policy formulation will be compensated during the implementation stage, due to the greater acceptance and effectiveness of policies.

In practice, policies are often implemented by combining instruments. With such a mixture, different groups can be targeted in different ways. Conventional environmental policy instruments are only partly capable of enforcing and stimulating behavioral change. Regulations are long-term instruments that ensure a minimum level of environmental quality, provided that the quality concerned can be put in a regulatory format at all. When there are too many uncertainties about an issue, or when values cannot be formulated in an unambiguous, enforceable way, regulation will be ineffective. Also the effectiveness of financial and communicative instruments is limited, since these target specific groups or only influence specific groups. Market-based instruments are promising, since they rely on the market for making smart and efficient decisions, but require careful design. Voluntary, third-party policy programs can provide an incentive for the private sector to innovate, and sometimes governments support such programs. Multilateral negotiations and agreements form the context within which governments and the private sector can formulate environmental policy goals and objectives.

A mixture of programs and instruments is thus needed to address problems at an appropriate scale and to address the appropriate actors. However, effectiveness of instruments and mixtures of instruments is limited over time. Actors influenced by the instrument learn to comply with the instrument in a way that is most convenient to them. Such forms of strategic learning can render policy instruments ineffective; the goals are no longer achieved. Monitoring and evaluation of effectiveness of policies and instruments is therefore important. This offers information based on which policies can be revised or terminated. Also, the problems addressed by the policy change, as well as the institutional context in which the problem is addressed. Especially, environmental problems are subject to changing information, knowledge, and views on what the problem is and how it can be solved. Policies and instruments need to be adjusted to these changed conditions.

See Also: Adaptive Management; Cost-Benefit Analysis; Ecopolitics; Environmental Law; Environmental Policy Act, National; Environmental Values and Law; Green Laws and Incentives.

Further Readings

Glasbergen, P., F. Biermann, and A. Mol, eds. *Partnerships, Governance and Sustainable Development: Reflections on Theory and Practice.* Cheltenham, UK: Edward Elgar, 2007.
Salamon, Lester E. *The Tools of Government: A Guide to the New Governance.* Oxford, UK: Oxford University Press, 2002.
Stavins, Robert R. "Experience With Market-Based Environmental Policy Instruments." Discussion Paper 01-58. Washington, DC: Resources for the Future, 2001.
Sterner, Thomas. *Policy Instruments for Environmental and National Resource Management.* Washington, DC: Resources for the Future, 2003.
Tews, K., P. Busch, and H. Jörgens. "The Diffusion of New Environmental Policy Instruments." *European Journal of Political Research,* 42 (2003).

Ellen van Bueren
Delft University of Technology

Environmental Policy Act, National

The National Environmental Policy Act (NEPA), the first major U.S. environmental law, was enacted in 1969 and signed into law in 1970. NEPA requires all federal agencies to go through a formal process before taking any action anticipated to have substantial impact on the environment. Part of this process requires the agencies to assess the potential environmental impact of their proposed actions in accordance with NEPA policy goals and to consider reasonable alternatives to those actions. Primary responsibility for overseeing implementation of NEPA rests with the Council for Environmental Quality (CEQ), created by the U.S. Congress as part of NEPA and located in the Executive Office of the President.

The scope of NEPA is limited to agencies of the federal government (i.e., it does not apply to purely state or purely private activities). Because of these exclusions, some states have enacted similar regulations mandating that governmental agencies consider environmental impact as one fact when making state decisions.

Environmental policy initiatives appeared on the government agenda in the late 1960s and early 1970s, in conjunction with growing awareness among the general public of the dangers posed to the environment by many human activities. This environmental concern contrasted with a previous attitude in which growth and industrial development were favored while often little consideration was given to their environmental costs. A number of separate events helped raise public concern about the environment. For instance, public awareness of the dangers of industrial pollution was heightened after the "Donora smog" incident of 1948 (in Donora, Pennsylvania, about 20 miles from Pittsburgh) caused by a temperature inversion coupled with trapped emissions from several industrial plants in the lower atmosphere: the resulting pollution killed 20 people and sickened over 5,000. The publication of Rachel Carson's book *Silent Spring,* in the early 1960s, which documented the damaging effects of pesticides (including DDT) on the environment and the risk they posed to human health, also helped spur public awareness of the need for regulations to protect the environment, as did a widely publicized oil spill in the ocean near Santa Barbara, California, in 1969. The first Earth Day celebration was held in 1970, and President Richard Nixon declared the 1970s the "Environmental Decade," a valid label reflected in the passage of considerable legislation, beginning with NEPA, aimed at protecting the environment.

Any federal agency taking an action that is anticipated to have an effect on the environment is required by NEPA to first create a relatively brief document (the environmental assessment, or EA) that describes the anticipated environmental effects of the action and alternatives to it. The EA is reviewed by Office of Federal Activities within the Environmental Protection Agency, and, if it is ruled that the action will result in significant environment impact, a more detailed evaluation called the EIS, or environmental impact statement, must be filed. The EIS describes the expected environmental effects of the action including adverse impacts, reasonable alternatives, an assessment of short-term and long-term gain and impact, and any irreversible changes that would be caused by the action. Notices of EAs and EISs are published in the *Federal Register,* allowing the general public and any interested organizations the chance to identify issues of interest that they wish to see addressed. The EIS process involves multiple stages, and individuals have the opportunities to comment on draft stages of the document, either in person or in writing. In addition, if members of the public feel their concerns have not been adequately addressed in the EIS, they may appeal to the head of the agency involved or may file suit against the agency in Federal court.

NEPA has certainly increased consideration of the environmental consequences of actions taken by federal agencies, and allows the public to take an active role in the governmental process. However, the EIS system is not a perfect guardian of the environment, the most obvious reason being that it does not prohibit actions that are anticipated to harm the environment, but only requires that they, and alternatives, be considered. In addition, most citizens (including scientists) have limited time and expertise to invest in the political process, so the process is stacked in favor of the governmental agency. In addition, its scope is limited to agencies of the federal government. Finally, the ultimate environmental effects of many actions are unknown.

See Also: Carson, Rachel; Endangered Species Act; Environmental Policy; Instrumental Value; Intrinsic Value; Land Ethic; *Silent Spring.*

Further Readings

Bregman, J. I. and K. M. Mackenthun. *Environmental Impact Statements.* Boca Raton, FL: Lewis Publishers, 1992.

Caldwell, L. K. *The National Environmental Policy Act: An Agenda for the Future.* Bloomington: Indiana University Press, 1998.

Clark, R. and L. W. Canter. *Environmental Policy and NEPA: Past, Present, and Future.* Boca Raton, FL: St. Lucie Press, 1997.

National Environmental Policy Act. http://ceq.hss.doe.gov (Accessed August 2010).

Sarah Boslaugh
Washington University in St. Louis

ENVIRONMENTAL VALUES AND LAW

A NOAA veterinarian assesses a heavily oiled Kemp's Ridley turtle after the BP oil spill in the Gulf of Mexico in April 2010. One of the challenges of environmental law is determining the "value" of such things as a species of turtle or the Gulf of Mexico itself.

Source: National Oceanic and Atmospheric Administration and Georgia Department of Natural Resources

Environmental values have a significant and ongoing impact on the law, both in the United States and elsewhere. In some cases, environmentalists and green activists have been the primary force behind representing a value in law, by petitioning U.S. Congress or other bodies. In other cases, environmental concerns overlap with other concerns, as in the case of environmental justice and human rights.

When speaking of law and the environment, "value" and "values" are important terms used in two related but distinct ways. In economics and business, we speak of the "value" of a thing in the sense of its worth. Is the natural beauty of the environment valuable? What is the value of a given species, or the value lost to us as a result of overfishing? Even when it sounds abstract in environmental contexts, this usage can have practical legal importance: what was the value of the Gulf of Mexico before the Deepwater Horizon disaster? A frequent, critical, and difficult to quantify question in environmental matters is "what is the value of biodiversity?" The importance of this sense of "value" to environmental law is also denoted by the fact that much of the body of law has traditionally been termed *natural resources law.*

"Value" also means one of the constituent motives, moral principles, or preferences of a belief system, as in "environmental values include a high esteem for biodiversity," or "is science free of values?" a question broached for a number of reasons, including investigating the underlying motives of a branch or study of science when evaluating its conclusions. At any given time, the body of environmental law reflects those values that have been sufficiently recognized; the present body reflects degrees of all the core environmental principles, as well as the value we attach to the beauty of nature, to the resources provided by our ecosystem, and to every species of life. Even the ability to patent a genetically engineered life form reflects a value attached to that life.

The field of environmental law is not just complex, it is a hodgepodge. Law professors have complained of the difficulty in teaching it because so much of it is ad hoc rather than being organized or developed around overarching themes. There are separate statutes and regulatory schemes for air pollution, water pollution, soil pollution and contamination, and still further bodies of law related to the pollution of groundwater, the pollution of surface water by oil, and other narrow scenarios. The emergence of the conservation movement in the late 19th and early 20th century and the later development of environmentalism as a social and political force has been the primary catalyst for the development of environmental law. Environmentalism is not without internal conflicts, and many laws and regulations have been adopted as compromises between environmentalist aims and those of others. Modern environmental law quite often takes the shape of an accommodation between private sector rights and the public interest. Different laws in different jurisdictions may stem from different values: is the purpose of a regulation to prevent human harm, to protect an ecosystem or endangered species, to safeguard against unintended consequences, to constrain anthropogenic climate change, or to preserve the natural beauty of the landscape? The same soil pollution regulation may have any of these as a motive, for instance, but its articulation, implementation, and enforcement may vary accordingly.

Furthermore, the field of environmental law is in constant flux, in part because many of its concerns and the area it covers are so new; because it is so closely connected with technologies that are themselves constantly developing and with science, which provides a constantly changing body of relevant knowledge; and in part because even without legislative action on the part of Congress, the regulatory atmosphere is subject to constant change because of new regulations set by government agencies, which may themselves be subject to political forces and trends. The body of law consists of precedents dating as far back as the beginnings of our common legal tradition: even legal hallmarks of the 1970s environmental movement like the Clean Water Act and the public trust doctrine have antecedents in the Rivers and Harbors Act of 1899 and the laws of Roman antiquity, respectively, and cases dealing with natural resource law played perhaps the most dominant role in the first century and a half of the American courts. As early as 1998, seven years before the Kyoto Protocol was even in force and its flexibility mechanisms put into place, Yale environmental law professor Donald Elliott called environmental law "the most complicated and detailed body of law the world has ever known . . . the field has simply gotten too large and complex for anyone to master it all."

Core Environmental and Social Principles

According to Sharon Beder's *Environmental Principles and Policies,* there are six core environmental/social principles, all of which have significant bearing on environmental law. The principles she identifies are the sustainability principle, the polluter pays principle, the precautionary principle, the equity principle, the human rights principle, and the participation principle.

Sustainability Principle

Sustainability may not seem represented in law to a great degree, but as federal funding increases for renewable energy sources and regulations develop to favor the sustainability of wetlands and other ecosystems, it is becoming a more important area. The Willamette College of Law, which offered the first sustainability law class in the United States in 1993 and has pioneered legal education in the area of sustainability law, offers a certificate program in Sustainable Environmental, Energy, and Resources Law for its juris doctor (J.D.) students, and other law schools are initiating similar offerings. Environmental justice, land use planning, dispute resolution, energy law, ocean resources law, and water law are cornerstones of such programs.

Polluter Pays Principle

The "polluter pays" principle has become a significant element of environmental law in much of the world, and has received especially strong support in the European Union (EU). The principle states simply that the party responsible for producing pollution is responsible for paying for the damage done to the environment. Principle 16 of the Rio Declaration on Environment and Development affirms the polluter pays principle. The principle doesn't just inform punitive measures against polluters; it is the underlying principle of ecotaxes or green taxes, a Pigovian tax that balances the negative externalities of environmentally harmful activity with government revenue that can be used to remedy those externalities. The benefit of green taxes is that, assuming they are properly structured, they are more elastic in their response to pollution than regulatory fines are, while providing the same monetary disincentive to polluting behaviors. Carbon taxes, taxes on hazardous wastes, and severance taxes on the extraction of natural resources are examples of green taxes; license fees for fishing and hunting, while conceived long before the polluter pays principle was articulated, nevertheless suit its needs.

The polluter pays principle has not been fully implemented in American law. In many cases there are caps on the amount an offending company can be made to pay to clean up its pollution, particularly if negligence cannot be demonstrated; in other cases, services like drinking water or sewage treatment are subsidized and the liability of polluters limited.

Precautionary Principle

The precautionary principle states that if an action has a suspected risk of harm but there is no scientific consensus that it *is* harmful, the burden of proof that it is *not* harmful falls to the actor. In environmental law this is especially key: a company or other entity must prove the environmental safety of its planned actions, rather than go through the environmental crises of the past when certain activities were permitted simply because they had not yet been proven unsafe. This is a statutory requirement in the laws of the EU, for example. In the United States, it has not been adopted at the federal level, but it is a principle underlying the calls for the U.S. Environmental Protection Agency (EPA) to strictly regulate nanoparticles, a new category of substance that policy advocates say have not been sufficiently investigated to consider safe by default.

There are two forms of the precautionary principle: strong and weak. The strong version calls for regulatory action even when the evidence for harm is speculative or the matter is not well understood, regardless of the costs of regulation (or of avoiding the action entirely). For instance, the 1982 United Nations World Charter for Nature declared that an activity should not proceed if its "potential adverse effects are not fully understood." The strong version of the principle amounts to a refusal to give the benefit of the doubt to an action with consequences that aren't yet fully understood; its advocates support it by pointing to the many cases in the past, from dangerous medical practices like the use of mercury fumes to the widespread use of harmful materials like lead-lined plumbing and asbestos insulation, in which a lack of full understanding led to long-term environmental, public health, and economic consequences.

The weak form of the principle is the least restrictive, and allows for balancing the benefits of an action against the possible costs of its harm. A medical analogy would be making the decision to perform a dangerous surgery when it is the only remedy available, a decision informed by weighing the risks of the surgery against the (nonmonetary) costs to the patient of continuing without treatment. An environmental example might weigh the jobs and economic benefits generated for a town by a new factory against the pollution caused by that factory, and determine that economic vitality made the risk of further pollution worthwhile.

Equity Principle

The equity principle calls for fair and equal access to resources and education, self-determination, and full participation in political life for all people. It is compatible with the foundations of American political and legal philosophy, though there are always new attempts to be made to help guarantee that access.

Human Rights Principle

Human rights, too, are of course of prime importance to the American legal tradition. Where they intersect with environmental law, the law and legal activists are concerned primarily with environmental justice, and with protecting innocents from the unintended consequences of others' environmental harm.

Participation Principle

Similar to the equity principle, the participation principle calls for participation of all affected parties in the decision-making process of a proposed action or policy. The decision to build a polluting factory in a given town, for instance, should involve more parties than simply the owners of the land and the owners of the factory—and more even than the regulatory bodies and governments with relevant jurisdiction. It should involve the people of the town, and of adjacent, downwind, or downriver towns that would or could be affected by that pollution. The participation principle, calling for active local participation in policy making, has taken a central role in approaches to government, regulation, and law since the 1970s, and is invoked by occasional calls for returns to the town meetings that used to be a more prominent part of American political life. It is a reaction, as well, against the many population migrations of the 19th and 20th centuries, which displaced American families from their roots and in many cases made them less likely to participate in local politics, perhaps because of a reduced emotional investment in the local community. The participation principle emphasizes not the civic duty of participation—though it does not challenge that argument—but the pragmatic benefit of participation. Throughout the United States, many proposed policies are subject to public comment, from the ballot issues of many state and local jurisdictions to the town meetings held throughout the country in anticipation of federal healthcare reform.

The participation principle also calls for a greater degree of transparency. True participation on the part of the public requires more than just delegating to other parties the task of informing the public. Transparency means that the public can see for themselves the cogs and wheels of what is going on to form a policy or plan an action. At the U.S. federal level, the cable network C-SPAN, televising sessions of Congress, is a significant early boon for transparency; at the local level, local government or council meetings may be broadcast on local-access cable channels. The Internet is also of great help in making relevant documents and minutes available to the public and the media. The Rio Declaration gives consideration to the participation principle, which is considered paramount to pursuing sustainable development.

See Also: Animal Ethics; Business Ethics, Shades of Green; Carbon Offsets; Democracy; Ecopolitics; Environmental Justice; Environmental Justice Ecology, Radical; Environmental Law; Environmental Law and Policy Center; Forest Preservation Laws; Green Ethics in the Legal Community.

Further Readings

Beder, Sharon. *Environmental Principles and Policies: An Interdisciplinary Approach.* London: Earthscan, 2006.

Finkmoore, Richard J. *Environment Law and the Values of Nature.* Durham, NC: Carolina Academic Press, 2010.

Heazle, Michael. *Uncertainty in Policy Making: Values and Evidence in Complex Decisions.* London: Earthscan, 2010.

Torgerson, Douglas. *The Promise of Green Politics: Environmentalism and the Public Sphere.* Durham, NC: Duke University Press, 1999.

Yeager, Peter Cleary. *The Limits of Law: The Public Regulation of Private Pollution.* New York: Cambridge University Press, 1993.

Bill Kte'pi
Independent Scholar

Ethical Vegetarianism

Vegetarians follow a plant-based diet; they do not eat meat, poultry, fish, or crustaceans. There are many different reasons why a person might choose to become a vegetarian. Some feel a diet without meat is the healthiest option; others practice a religion that does not allow them to eat meat. For many people worldwide, meat is unavailable for consumption because it is prohibitively expensive. The most common rationale for becoming a vegetarian, however, is because of ethical commitments. Ethical reasons for a vegetarian diet fall broadly within two categories: animal rights arguments and environmental impact arguments.

These pigs were being raised in 2000 in a large animal confinement factory farm in southwest Wisconsin, where a manure management system was in use. Ethical vegetarians object to the living conditions for animals on factory farms and to the farms' severe negative impact on the environment.

Source: U.S. Department of Agriculture, Natural Resources Conservation Service

Most ethical vegetarians have chosen to exclude meat from their diet at least in part because they feel the living conditions and amount of suffering that animals go through in the factory farming system is inhumane. In most of the developed world, factory farming practices have replaced the family farm and individual animal husbandry model that was in place until the 1950s. Most factory farms confine thousands of animals indoors, close together, and exposed to high levels of waste and bacteria. Some animals are restrained so that they cannot move or turn around, while others are disfigured to prevent animals from hurting themselves or fighting. For example, chickens' wings and beaks are clipped to prevent the birds from flying away or trying to peck one another. Similarly, pigs' tails are cut off to prevent tail biting. However, these violent behaviors such as pecking one another or biting tails have only been observed on industrial farms. When animals are allowed outside and allowed to engage in natural stress-relieving behaviors, like being with their mothers or playing, aggression toward one another is virtually nonexistent. Ethical vegetarians argue that these conditions are unnecessarily cruel and result in substandard and possibly unsafe meat. It is because of these unsafe and horrible living conditions that livestock are so susceptible to disease, which in turn causes farmers to feed animals growth hormones and antibiotics to help grow muscle in animals that have never walked and to prevent animals from contracting diseases from the manure that is mixed with their food and covering their bodies. However, views differ about the degree of comfort and freedom that farm animals deserve. Some feel that animals raised and slaughtered on small family farms have been treated well, and it is an ethically sound practice to eat meat produced in such a way. Others, especially animal rights groups such as People for the Ethical Treatment of Animals (or PETA), feel it is never right to eat other creatures, especially when other food sources are so plentiful.

People who choose to become vegetarians for environmental reasons believe the modern livestock industry is an inefficient and unsustainable mode of food production that severely and negatively impacts the environment. In 2006, the United Nations announced that the meat industry was the largest contributor to global warming, citing factory farming as responsible for deforestation, water and air pollution, loss of biodiversity, stripping of topsoil, and the overuse of nonrenewable resources such as fossil fuels and water.

As Frances Moore Lappé argued in her canonical work, *Diet for a Small Planet,* hunger around the world isn't caused by a lack of food. Rather, it is caused by politics and a misallocation of resources. Livestock consume an extremely large amount of grain before they are slaughtered to feed humans and factory farming requires a very large energy input in the form of fossil fuels. Ethical vegetarians feel it is irresponsible to grow a large crop of corn or grain to feed livestock, which will in turn feed a very few people. This requires a lot of input energy for very little output energy. Instead, they argue that plant-based diets would feed more people and use far fewer nonrenewable resources.

It used to be that most farms produced many different kinds of crops. However, industrial farms largely practice monoculture, or the cultivation of only one crop. The most common monoculture crops in the United States are soy and corn, which are mostly used for animal feed. While monoculture produces greater yields, it has negative consequences. First, since productivity has grown tremendously, prices have plummeted as demand cannot meet supply, making it difficult for small farmers to stay in business. Second, monoculture decreases biodiversity because all of the plants are genetically almost identical. Third, because crops are not rotated in the monoculture systems, soil nutrients are never replenished and the land is never left to rest. As a result, petroleum-based fertilizers must be introduced into the soil in order for crops to continue to grow year after year. These are the crops that are then consumed by livestock, which are then consumed by humans, even though chemical fertilizers have been linked to cancer and other diseases.

Farms are dependent on oil and gasoline to power their machinery. However, the bulk of fossil fuels used in industrial farming goes toward the production of artificial fertilizers and pesticides. Additionally, food travels hundreds, sometimes thousands, of miles from growth to packaging to table, which uses even more fossil fuels. Factory farms also use a disproportionately large amount of water for plant irrigation and to hydrate the livestock. Unfortunately, industrial farming is also responsible for polluting up to 48 percent of all water supplies, according to a study conducted by the U.S. Environmental Protection Agency.

Perhaps surprisingly, the largest pollutant on factory farms is manure. While a sustainable farm can reuse animal waste as a fertilizer resource, factory farming fits an unnatural amount of animals into one space, resulting in an unmanageable amount of waste. Manure is generally dumped into pits of water and later spread over crop fields. However, manure is often overapplied to fields, which causes it to run off and pollute the water supply. Large numbers of livestock packed together in small spaces are also responsible for the production of massive amounts of methane gas.

There are many who feel that ethical vegetarianism is a misguided choice. They argue that eating meat and supporting agriculture in the United States is not only our right but our duty. Some feel that animals are provided as a food source and that they do not have the same kind of consciousness and capacity for pain or fear that animal rights activists think they do. Additionally, there are arguments that refusal to eat meat is not as effective in decreasing environmental degradation as composing one's diet of local foods. For example, an avocado grown in South America and shipped to Colorado still wastes a large amount of fossil fuels and represents an inefficient use of energy. Still others argue that any personal choices, such as shifts in diet or recycling, will not affect climate change at all until changes are made at the policy level. An ethical vegetarian might counter by saying that personal choices are the only course of action that most people have, eating meat is ethically wrong, and when enough people exercise their consumer rights, policy change tends to follow.

See Also: Animal Ethics; Anthropocentrism; Biocentrism; Local Food Movement.

Further Readings

Lappé, Frances Moore. *Diet for a Small Planet: Twentieth Anniversary Edition.* New York: Ballantine, 1991.

People for the Ethical Treatment of Animals (PETA). http://www.peta.org (Accessed February 2010).

Preece, Rod. *Sins of the Flesh: A History of Ethical Vegetarian Thought.* Vancover, Canada: UBC Press, 2009.

Walters, Kerry and Lisa Portmess. *Ethical Vegetarianism: From Pythagoras to Peter Singer.* Albany: State University of New York Press, 1999.

Katherine M. Cruger
University of Colorado, Boulder

Ethics and Science

What is the relationship between ethics and science? How can and should this relationship bear on environmental philosophy, politics, or policy? At the risk of oversimplification, there are at least three senses in which these two domains intertwine. The social sciences (psychology, cognitive science) may inform us about patterns of moral behavior and decision making in different contexts; values may play a role in the practice of science itself; and scientific inquiry may inform environmental ethics and policy. An elaboration of these three senses will follow, along with a discussion of two examples of the interrelationship between science and ethics in the work of Rachel Carson and Aldo Leopold.

First, psychologists, cognitive scientists, and others have done interesting research on how people reason, make decisions, and generate the moral judgments that they do. For instance, cognitive psychologists have argued that people use certain "heuristics" and often exhibit systematic biases in making choices involving risky options. It turns out that how decisions are "framed" makes a great deal of difference to the choices people make. For instance, whether we perceive a choice as a gain or loss determines how we choose; we are, generally speaking, loss averse. Moreover, in general, people take risks to be more serious if they can be more easily called to mind; and, they are susceptible to group interactions that reinforce views about what is more risky and why. Thus, many people fear terrorism far more than car accidents, even though car accidents are much more common, and, if everyone in one's community fears terrorists, one is more likely to fear terrorists. People also make systematic errors in reasoning about probability. For example, they ignore base rates—or, the rate at which an event might ordinarily occur. So, for example, that the risk of shark attack has recently risen 10-fold may sound terrifying, until you realize that the risk of shark attacks is, on average, quite low. That rationality is "bounded" in these and other ways could usefully shape environmental policy. For instance, knowing how and why we procrastinate might lead us to institute policies that require us to set and meet intermediate deadlines, so that we face serious consequences for failing to plan for the long term before it is too late.

Moral psychologists draw upon the social sciences to describe and explain moral behavior. Arguably, the better we may understand the psychology of moral judgment, the better position we may be able to articulate a workable environmental philosophy. Some environmental pragmatists, for example, are drawing upon this research to generate more effective arguments and tools for persuading people to value the environment. Arguably, rhetoric and communication, the study of how and why we can effectively convince people to change their behavior through imagery, language, and other media, is yet another "science" that could usefully inform both ethical and policy arguments.

In the context of environmental policy, some have drawn upon the evidence from social sciences, and particularly, behavioral economics, to argue for "technocratic" environmental policy. According to one school of thought, in order to avoid heuristics and biases, experts, as opposed to laypeople, should ultimately shape environmental policy. Others contend that laypeople have a "richer rationality" than experts, and that the purported "biases" found by psychologists express moral preferences. Moreover, experts are as subject to some of the same patterns of influence and bias as are "ordinary" folk. Drawing upon this "populist" account, some argue that environmental regulation or law should be grounded in popular sentiment. Arguably, there's a middle ground to be found in deliberative democracies between technocracy and populism, which draws upon both expert information and public interest.

A second way in which ethics and science are related has to do with whether science itself is "value free." This has been a subject of long-standing debate among philosophers, historians, and sociologists of science going back at least to the 1960s—the debate may (at the risk of oversimplification) be broken down into two camps. "Externalists" argue that while the "internal structure" of science is "value free," values do influence "external" features of science. For example, insofar as scientists are beholden to public institutions such as the National Science Foundation (NSF) for funding, they must articulate and defend the social value of their research, and so, values shape what research questions are investigated. Thus, a Democratic administration may be more likely to support stem cell research than a Republican one. Or, scientists might choose a project that may be cheaper, quicker, or easier to complete because of requirements for tenure or other institutional constraints. That there are pragmatic constraints on science is not contentious. Thus, that science is influenced by "external values" is not terribly controversial.

"Internalists," on the other hand, argue that science is shaped by values through and through. They argue that normative values influence, for instance, how a research question might be posed, or, even how such a question is investigated and what count as adequate standards of evidence. For instance, the case of the choice to avoid a type I versus type II error may seem "value neutral," but norms might guide such a choice. Type I errors occur when one claims to observe some effect when there is none; type II errors occur when one claims that there is no such effect when there is one. While the choice to avoid one type of error over another may seem a matter of indifference, the choice may have important implications. It would seem that the culture of science privileges avoiding false positives over false negatives; that is, one is "safer" in saying that we have (not yet) observed any effect, than claiming an effect when there is none. However, for science that has import for policy, precaution runs the other direction. We might rather avoid false negatives, for instance, for the purposes of regulating climate change or toxins in the environment. Another example of a case where choice of a "null hypothesis" has import is for dose response curves. Should one assume as one's null hypothesis, for instance, that below a certain undetectable level, some potential toxin has no effect on human health, a neutral effect, or a positive effect? Philosophy of science investigates the role of values both internal and external to science; this may have real import for environmental law, health, and policy.

Finally, a third way in which ethics and science are related has to do with the use of science in informing ethical decisions. Philosophers have argued that one cannot, at least not deductively, derive an "ought" from an "is." Called the "naturalistic fallacy," the claim is that one cannot infer a claim about what is good from a descriptive claim. Normative or evaluative conclusions, the argument goes, require at least one normative premise. Thus, one cannot infer that "adultery is good" from the premise that "adultery is pleasurable." (For the argument to be demonstrative, one further requires the contentious assumption that whatever is pleasurable is, in fact, good.)

This line of argument has been used against some environmental ethicists who claim to derive their moral theory from ecology. The objection runs that it does not follow, for instance, from the claim that more diverse ecosystems are more resilient (setting aside for the moment whether this is true) and that species diversity is a good in general. Apart from the logical problem with deriving "ought" from "is," there is the problem that there are many empirical generalizations about species and ecosystems that may or may not be features of the world we wish to further. For instance, most species in the history of life have gone extinct; but no environmentalist would wish to infer from "Extinction is common" to "Extinction is good."

While it may not be possible to derive an environmental ethic demonstratively from the science of ecology, there are certainly good reasons for environmental ethicists to draw upon science in reasoning about how and why we ought to value the environment. Arguably, some of the most compelling arguments in favor of environmental protection come from thinkers such as Aldo Leopold and Rachel Carson. What made these arguments compelling was, at least in part, the authors' ability to communicate to a wide audience the importance of human dependence on natural ecosystems. Their arguments drew upon aesthetic and rhetorical strategies, to be sure, but they also demonstrated a careful articulation of the significance of otherwise invisible ecological interactions to human health and well-being. Microbial life had no greater champion than Rachel Carson.

To summarize, there are at least three ways in which science and ethics bear on one another. First, the sciences of human behavior, psychology, and cognition might inform our understanding of how and why we make the ethical choices we do, and thus help shape policy. Second, the sciences might themselves be shaped by values, and how and why they are could have important implications for which environmental regulations we might put in place. Third, and finally, science may substantively inform policy end points. How it does so, however, is not self-evident; the sciences are often operating at some point remote from "practical" consequences.

Carson and Leopold are exemplary in that they exhibit the use of psychology (though not self-consciously) in their arguments, they interleave science and value, and they draw upon science in setting policy endpoints. Each used rhetorical tools that drew upon "folk" psychology—for example, appeals to *pathos*. Both also "read values" into nature, as well as drew evaluative conclusions from empirical claims. The "balance of nature" idea is just one example of the complicated relationship between science and ethics, and environmental ethics in particular.

Rachel Carson

Carson is rightly hailed as one of the founders of the environmental movement; her book inspired environmental legislation of lasting influence. Carson's book, *Silent Spring,* also reinforced what has become a pervasive assumption in the environmental movement.

The idea that nature is "in balance," in some sense, is quite old. In ecology, one can trace versions of this idea back at least as far as the 18th century: nature, in particular, nature absent human intervention, is "in balance." Whatever the history of the idea of the balance of nature in ecology, such an idea is, arguably, as old as the Old Testament. The biblical narrative of the fall from grace is echoed in Carson's opening chapter of *Silent Spring,* a "Fable for Tomorrow." The fable is emblematic of a genre of ominous forewarning that is drawn upon by many environmental writers. Different environmental thinkers have dated the "fall" to the rise of agriculture, civilization, or industry. For Carson, the "fall" is a sort of "fallout" due to pesticide use; not accidentally, there are many parallels between her descriptions of the consequences of pesticide fallout and nuclear fallout. The comparison would be particularly vivid for her readers in the cold war era. Clearly, the themes of a fall from grace, the dangers of technology, and mourning for a greener, more harmonious past struck a strong psychological chord, even if some have argued that these themes have lost rhetorical force of late. (The authors of *Break Through: Why We Can't Leave Saving the Environment to Environmentalists,* Michael Shellenberger and Ted Nordhaus, argue that environmentalists need to give the movement a less "mournful" aspect.)

Carson's "balance of nature" message is an interesting example of the interleaving of empirical with normative claims. On the one hand, she's making the uncontroversial point that humans are evolved, like all other living creatures, and evolution involves coadaptation between species. Communities of organisms are made up of "webs of life"; thus, microbial life in the soil and human life is linked. What we do affects the air, the water, and the soil, and thus all other living things, and our own food supply. This much is uncontroversial. However, her use of sympathetic and shocking examples and evaluative and metaphorical language led many to question her scientific credentials, and much of what she says was considered controversial. Carson argues that the fact that pesticides and herbicides remain in the environment and concentrate in food chains puts the highest levels of the food pyramid at risk (birds, mammals); that pesticides are not effective, because resistant breeds of agricultural pests lead to a self-defeating cycle; that chemical pesticides, because of their biological persistence, may lead to chronic toxicity in humans, aside from their acute toxic effects in some cases in humans and other animals; and finally that there are effective alternatives to synthetic chemicals: namely, biological control. While critics of Carson acknowledged that pesticide use has effects, they disagreed about how and how much. As a result, Carson faced public hearings in front of Congress, where she was called upon to defend her claims about the extent and nature of pesticides and herbicides' impact on the environment and human health.

Carson anticipated many of these objections. In one particularly rhetorically savvy passage, she wrote: "In some quarters nowadays, it is fashionable to dismiss the balance of nature as a state that prevailed in an earlier, simpler world—a state that has now been so thoroughly upset that we may as well forget it. Some find this a convenient assumption, but as a chart for a course of action it is highly dangerous. The balance of nature is not the same as it was in Pleistocene times, but it is still there: a complex, precise, highly integrated system of relationships between living things that cannot safely be ignored any more than the law of gravity. . . . The balance of nature is not a status quo; it is fluid, ever shifting, in a constant state of adjustment." Carson both acknowledges and subtly critiques her opponents in the above passage. While she seems to grant that at least over the long term, community composition, or, species number, relationship, and even population sizes may be fluctuating, she claims that they embody a "precise," and "integrated" state of affairs. This state of affairs is presumably described by ecological "laws," no less inviolable than the laws of gravity; we disturb these very delicate balances at our peril. Carson's analogy of physics with ecology both suggests that ecology has the same "status" as the physical sciences, and reinforces her own value judgments about the "balance" of ecological systems; that is, she is not simply communicating scientific "facts" but reading normative conclusions into the science, as she understands it. While she is surely correct that our actions have consequences, ecological laws are nothing like the laws of gravity; they are far more complex, and there is a great deal of uncertainty about both the most basic parameters and the dynamics of ecological systems. In sum, Carson not only used science to inform her ethical worldview, but also to "read" ethical conclusions into her science.

Aldo Leopold

Likewise, Aldo Leopold deployed his understanding of ecology, as well as metaphor and analogy, in advancing an ethical worldview. Leopold was trained in forestry and game management, but came to view the relationship between species as more or less "in balance," much like Carson. In *A Sand County Almanac,* he articulates his "Land Ethic," a summary of the ethical views he arrived at after a lifetime of traveling in the desert Southwest, working in game management, and as a farmer, fisherman, and hunter on his farm in Wisconsin.

Leopold's "Ethical Sequence" traces what he believes to be the history of the extension of moral consideration from immediate families to the human community at large. Leopold is drawing upon reasoning akin to Darwin's in the *Descent of Man,* where Darwin argued that ethical sensibility, or, the "moral sentiments," evolved by natural selection, or, more precisely, group selection. That is, according to Darwin, altruism evolved in higher primates because groups of individuals that acted altruistically toward one another could better defend and provide for themselves than less altruistic groups. Leopold drew upon this reasoning in support of his view that we inevitably develop moral consideration for beings around us, particularly those we depend on for our survival. The "land ethic" is an extension of this perspective to nonhuman entities such as the rivers, plants, trees, and animals upon which we depend for our survival. Sometimes Leopold speaks of the land ethic as not so much a moral recommendation as a consequence, and possibly an "inevitable" consequence, of human evolution.

This view was a very popular one in early-20th-century ecology. Several early community ecologists argued for "altruism" as a necessary consequence of evolution; however, these claims were by no means met with universal agreement. A very common objection to this line of reasoning is the so-called free-rider problem. That is, altruistic groups are always subject to "invasion" by "free-riders"—individuals who benefit from, but do not contribute to, the group's success. There is a long history of debate in evolutionary biology about the likelihood of altruistic behavior evolving on different assumptions. Needless to say, there's no agreement that altruism is "inevitable," even if it is probable on some assumptions.

Leopold used a combination of evolutionary and ecological "laws" to argue that the "land ethic" is an ethic that emerges from the evolutionary process. The solution to environmental problems, he argued, must rest on sentiments of moral concern toward both the biotic and abiotic community. Enlightened self-interest, he thought, is insufficient for genuine conservation ("a state of harmony between men and the land"). He stated that we need to encourage more than conservation that is profitable to the individual in order to prevent erosion of soils, pollution of streams. The federal government, he argued, cannot be held solely responsible for policing conservation; eventually the policing system will become so large as to be ineffective.

Leopold deployed powerful rhetorical skill in conveying the sense of connection, historical interdependence, and aesthetic appreciation he felt with the land. He also deployed metaphors to convey the significant dependence of human health and well-being on the land. Ecological communities are "a pyramid of trophic levels through which energy flows." Yet, the pyramid is "a tangle of chains so complex as to seem disorderly, yet the stability of the system proves it to be of a highly organized structure. Its functioning depends on the cooperation and competition of its parts." Rapid human intervention in the natural ecosystem "inhibits functioning" of these complex webs. "The process of altering the pyramid for human occupation releases stored energy, and this often gives rise . . . to a deceptive exuberance of plant and animal life. These releases of biotic capital tend to becloud or postpone the penalties of violence." In other words, the elaborate chains of predator and prey, consumer and consumed, through which energy is transferred, are being altered by human use at a rate of speed much faster than evolutionary time. Rapid use and release of all this stored "biotic capital," that is, energy, can give the appearance of indefinite resources, but the long-range consequences of such use, he thought, are potentially dire.

Of course, ecology has evolved a great deal since Leopold's (and Carson's) time; theories of the interactions between human action and ecological systems are much more elaborate, and there is a great deal of uncertainty and disagreement. For example, Leopold's assumption that there was a direct causal relationship between the "stability" of ecological systems and "diversity" is contested, at least in part, because there are at least four different definitions of stability, and almost as many measures of diversity. Likewise, long-term impacts of pesticides on both human health and natural systems are still being investigated. Needless to say, the facts by themselves do not tell us what we ought to do, but a careful consideration of these facts may change our understanding, and (hopefully) our behavior.

See Also: Carson, Rachel; Environmental Policy; Instrumental Value; Instrinsic Value; Leopold, Aldo; *Silent Spring*.

Further Readings

Anderson, Elizabeth. "Uses of Value Judgments in Science." *Hypatia,* 19 (2004).
Carson, Rachel. *Silent Spring*. Boston: Houghton Mifflin, 1962.
Cranor, C. F. "The Normative Nature of Risk Assessment: Features and Possibilities." *Risk: Health, Safety & Environment,* 8 (1997).
Douglas, Heather. "Inductive Risk and Values in Science." *Philosophy of Science,* 67/4 (2000).
Jasanoff, Sheila. *Designs on Nature: Science and Democracy in Europe and the United States.* Princeton, NJ: Princeton University Press, 2005.
Lewens, T., ed. *Risk: Philosophical Perspectives*. New York: Routledge, 2007.
Sinnot-Armstrong, W., ed. *The Biology and Psychology of Morality.* New York: Oxford University Press, forthcoming.
Sunstein, C. "Moral Heuristics." *Behavioral and Brain Sciences,* 28 (2005).
Tversky, A. and D. Kahneman. "The Framing of Decisions and the Psychology of Choice." *Science,* 211 (1981).

Anya Plutynski,
University of Utah

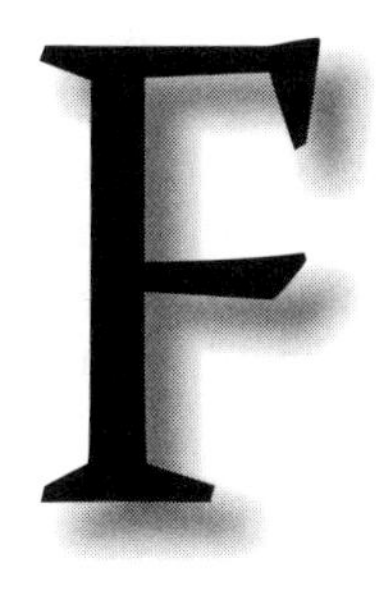

Forest Preservation Laws

A 2008 photograph of a lead mining operation located within the Mark Twain National Forest in Viburnum, Missouri. Unlike U.S. national parks, the country's more than 150 national forests often allow mining, logging, livestock feeding, and other commercial uses.

Source: U.S. Bureau of Land Management

Interest in forest preservation began during the 19th century at the beginning of the environmental movement. Some of the earliest laws dealing with environmental issues were concerned with forest preservation. A combination of private action, federal and state laws, and a change in industrial practices has led to great change in the way Americans think about and interact with forests. As a result of legislation, national and state forests exist across the United States. Unlike land managed by the National Park Service, national and state forests may be utilized for commercial purposes, such as logging, livestock feeding, and the like. Over 150 national forests controlled by the U.S. government total over 190 million acres. This amounts to approximately 8.5 percent of land in the United States. Forest preservation laws seek to balance the interests of preservation and sustainable use of natural resources. Debates regarding these issues are ongoing and occasionally contentious, because the interests of various parties need to be considered.

Origins of Forest Preservation Laws

The 19th-century writer, naturalist, philosopher, and transcendentalist Henry David Thoreau popularized a love of the outdoors and caused many contemporaries to examine how the United States' stewardship of its natural resources was being conducted. As a result of this, concern grew about the disappearance of the forests across North America as well as a perceived future need for timber. An initial attempt to address these concerns resulted in the Timber Culture Act of 1873, whereby the U.S. Congress allowed homesteaders, who had received 160 acres of land, to receive an additional 160 acres if they would plant trees on one-quarter of their original holdings. The Timber Culture Act was seen as a way of ensuring that adequate timber would be available for the Great Plains and could be obtained only after the settler had spent five or more years on the homestead.

While the Timber Culture Act increased the amount of timber available, it did nothing to protect forests. As a result, in 1891, Congress passed the Land Revision Act, which gave the president the authority to set aside any federally owned land covered with timber or undergrowth. This legislation was spurred by Los Angeles–based environmentalists, including Abbot Kinney and Theodore Lukens, who were concerned about harm being done by miners and ranchers to the Los Angeles basin watershed. The Land Revision Act allowed the creation of the national forest, and has resulted in millions of acres of land coming under the purview of the U.S. Forest Service.

Management of forest reserves stems from the Civil Appropriations Act of 1897, which is commonly known as the Organic Act. The Organic Act outlined the management, protection, and responsibilities for caring for the national forests, including the following:

- Specifying the purpose for establishing reserves, as well as the protection and administration necessary
- Granting the secretary of the U.S. Department of the Interior authority to administer the national forests, including making rules and regulations
- Authorizing the General Land Office (GLO) to hire employees for administrative purposes and to open the forests for public use
- Establishing the criteria for designating new forest reserves, including producing timber and protecting watersheds and forests
- Designating the U.S. Geological Survey (USGS) as responsible for mapping the reserves and promulgating these maps

In 1905, the Forest Transfer Act moved responsibility for the national forests from the Department of the Interior to the Agriculture Department, where the Division of Foresters already existed. Statutes regulating forest preservation have been modified since then, most notably by the National Forest Management Act (NFMA) of 1976. The U.S. Forest Service is the agency charged with administration of the nation's forest reserves.

Implementation of Forest Preservation Laws

The U.S. Forest Service employs over 28,000 permanent and 4,000 seasonal workers and has an annual budget of approximately $6 billion. The Forest Service maintains major divisions including the National Forest System, Law Enforcement and Investigations, State and Private Forestry, International Programs, and Research and Development. The Forest Service balances recreational access, resource protection, and resource extraction from the national forests.

Disputes regarding the management of the national forests have sometimes arisen and led to lawsuits to settle competing views regarding appropriate stewardship of specific forest preserves. For example, the Izaak Walton League, a conservation group, has had a long history of disputes with the U.S. Forest Service. During the 1930s, the Izaak Walton League petitioned the Forest Service to abolish all grazing of livestock in forest preserves, and during the 1970s brought suit in federal court claiming that clear-cut logging in the Monongahela National Forest violated the terms of the Organic Act. In 1975, in *West Virginia Division of the Izaak Walton League v. Butz,* 522 F. 2d 945 (the *Monongahela* case), the U.S. Court of Appeals for the Fourth Circuit affirmed a 1973 district court order for the Forest Service to stop such practices. The court so held as the Organic Act only permitted removal of dead, physically mature, and large-growth trees individually marked for cutting.

In response to the *Monongahela* case, the U.S. Congress passed the National Forest Management Act, which controls the management of renewable resources in national forest preserves. Specifically, the NFMA provides the U.S. Forest Service detailed guidance regarding forest plans, including where, when, and how much timber can be harvested. The NFMA requires a great deal of public participation in the preparation and revision of such plans. Despite Congress's attempt at clarity, lawsuits regarding interpretation of the NFMA are not uncommon.

Thirty-eight states have established state forests, including Alabama, Alaska, Arkansas, California, Colorado, Connecticut, Delaware, Florida, Georgia, Idaho, Illinois, Indiana, Iowa, Kentucky, Louisiana, Maine, Maryland, Massachusetts, Michigan, Minnesota, Missouri, Montana, New Hampshire, New Jersey, New York, North Carolina, Ohio, Oregon, Pennsylvania, Rhode Island, South Carolina, Tennessee, Texas, Vermont, Virginia, Washington, West Virginia, and Wisconsin. Similar to the national forests, these preserves balance both recreational and commercial interests. As interest in the use and preservation of forests continues to grow, further debates are expected.

See Also: Adaptive Management; Biodiversity; Environmental Law; Sierra Club; Thoreau, Henry David.

Further Readings

Dale, L. "Wildlife Policy and Fire Use on Public Lands in the United States." *Society & Natural Resources,* 19/3 (2006).

Farber, D. A., J. Freeman, and A. E. Carlson. *Cases and Materials on Environmental Law,* 8th ed. Saint Paul, MN: West Publishing, 2009.

Nie, M. and M. Fiebig. "Managing the National Forests Through Place-based Legislation." *Ecology Law Quarterly,* 37/1 (2010).

Young, R. A. and R. L. Giese. *Introduction to Forest Ecosystem Science and Management,* 3rd ed. Hoboken, NJ: John Wiley & Sons, 2003.

Stephen T. Schroth
Jason A. Helfer
Edward F. Davis
Knox College

Fuller, Buckminster

Richard Buckminster Fuller, known by his friends as "Bucky," is one of the key innovators of the 20th century. The grandson of a Unitarian minister (Arthur Buckminster Fuller), Fuller was also Unitarian. Buckminster Fuller was an early environmental activist and is

Over 500,000 geodesic domes have been built worldwide. Many of the domes, such as this one built for the 1967 International and Universal Exposition in Montreal, Canada, have remained in use.

Source: Flickr/Paula Moya

also well known as teacher, philosopher, thinker, visionary, inventor, architect, engineer, mathematician, poet, cosmologist, environmentalist, humanitarian, and inventor. There are few men who can justly claim to have revolutionized their discipline and, accordingly, Fuller revolutionized many. Fuller was probably one of the first futurists and global thinkers, coining the term *Spaceship Earth,* and his work inspired and paved the way for many who came after him. Fuller published more than 30 books and was awarded 28 U.S. patents. Driven by the belief that humanity's major problems were hunger and homelessness, he dedicated his life to solving those problems through inexpensive and efficient design.

Fuller was born on July 12, 1895, in Milton, Massachusetts, the son of Richard Buckminster Fuller and Caroline Wolcott Andrews and also the grandnephew of the American transcendentalist Margaret Fuller. He attended Froebelian Kindergarten, which was new at this time to American education. He spent much of his youth on Bear Island in Penobscot Bay off the coast of Maine. As a young student, he had trouble with geometry, being unable to understand the abstraction necessary to imagine that a chalk dot on the blackboard represented a mathematical point or that an imperfectly drawn line with an arrow on the end meant to stretch off to infinity. He often made items from materials he brought home from the woods and oftentimes made his own tools. Fuller was sent to Milton Academy in Massachusetts and after that began studying at Harvard University. He was expelled from Harvard twice. By his own admission, he was a nonconforming misfit in the fraternity of education. Years later, he received a Sc.D. from Bates College in Lewiston, Maine, as well as many awarded doctorates.

Between his sessions at Harvard, Fuller worked in Canada as a mechanic in a textile mill and later as a laborer for the meatpacking industry. He also served in the U.S. Navy in World War I as a shipboard radio operator, as an editor of a publication, and as a crash boat commander. After he was discharged, he again worked for the meatpacking industry, acquiring management experience. Fuller married Anne Hewlett in 1917 and went into the construction business with her father. Fuller and his father-in-law developed the Stockade Building System for producing lightweight, weatherproof, and fireproof housing, although the company ultimately failed.

By the time Fuller was 32, he was bankrupt and jobless, living in public low-income housing in Chicago, Illinois. His young daughter Alexandra died from complications from polio and spinal meningitis. Despondent over these failures and family problems, he resolved to focus his energies on a search for socially responsible answers to the major design problems of his time. His intent was to explore what a single individual could contribute to changing the world and benefiting all humanity.

By 1928 Fuller was living in Greenwich Village, decorating the interior of the popular Romany Marie's café in exchange for meals and giving informal lectures several times a week.

Here he met Isamu Noguchi and they collaborated on several projects including the modeling of the Dymaxion car, the car's name derived from the abbreviation of dynamic maximum tension. Fuller designed and built prototypes of what he hoped would be a safer, aerodynamic car. Despite its length and due to its three-wheel design, the car turned on a small radius and could be parked in a tight space. The prototypes were efficient in fuel consumption for their day, getting 30 miles to the gallon. The car never gained popularity and investors backed out; research ended after one of the prototypes was involved in a collision that resulted in a fatality.

In the mid-1940s, Fuller designed an energy-efficient and inexpensive house, the prototype Dymaxion House. It was a round structure shaped like a flattened bell and had innovative features including revolving dresser drawers and a fine-mist shower that reduced water consumption. It was designed to be lightweight and adapted to windy climates, inexpensive to produce and purchase, and easily assembled. It was to be produced using factories, workers, and technologies that had produced World War II aircraft, with the basic model about 1,000 square feet and sheathed in polished aluminum. Unfortunately, the company that Fuller and others had formed to produce the housed failed due to management problems.

Fuller taught at Black Mountain in North Carolina during the summers of 1948 and 1949. There, with the support of a group of professors and students, he began reinventing a project that would make him famous: the lattice-shell structure—the geodesic dome. The geodesic dome had been created 30 years earlier by Walter Bauersfeld, but Fuller expanded upon this design. One of his early models was first constructed in 1945 at Bennington College in Vermont, where he frequently lectured. The geodesic construction is based on extending some basic principles to build simple structures (tetrahedron, octahedron), making them lightweight and stable. They have been used as parts of military radar stations, civic buildings, and exhibit attractions. The patent for geodesic domes was awarded to Fuller in 1954 for his exploration of nature's constructing principles to find design solutions. More than 500,000 geodesic domes have been built around the world and many are in use. One notable dome is Spaceship Earth at Disney World's Epcot Center in Florida—80.8 meters (265 feet) wide. (Spaceship Earth is actually a self-supporting geodesic sphere, the only one currently in existence.)

Fuller spent the final 15 years of his life traveling around the world and lecturing on ways to better use the world's resources. A favorite of the radical youth of the 1960s, Fuller worked to expand social activism to an international scope. One of his well-known quotes is "Making the world's available resources serve 100 percent of an exploding population can only be accomplished by a boldly accelerated design revolution." He was awarded countless awards including the Presidential Medal of Freedom in 1983, 28 patent awards, and many honorary doctorates. Fuller died July 1, 1983, only 11 days before his 88th birthday. He had become a guru of the design and architecture of many alternative communities. His wife of 66 years died 39 hours later, at 86. They are buried in Mount Auburn Cemetery in Cambridge, Massachusetts.

See Also: Conservation, Aesthetic Versus Utilitarian; Consumption, Consumer Ethics and; Human Values and Sustainability.

Further Readings

Fuller, R. Buckminster. *Critical Path*. New York: St. Martin's Griffin, 1982.

Fuller, R. Buckminster and Jaime Snyder. *Operating Manual for Spaceship Earth*. Baden, Germany: Lars Müller Publishers, 2008.

Pawley, Martin. *Buckminster Fuller.* New York: Taplinger, 1991.
Sieden, Lloyd. *Buckminster Fuller's Universe: His Life and Work*. New York: Basic Books, 2000.

Carl A. Salsedo
University of Connecticut

Future Generations

The long-term nature of many environmental problems has forced moral philosophy to pay closer attention to relations between generations. Many effects of present-day greenhouse gas emissions, for example, materialize only after decades or centuries. Intergenerational ethics differs from ethics among contemporaries due to a number of specific features of intergenerational relations, such as the asymmetrical and particular kind of influence the present generation has over future generations. Many issues in intergenerational ethics are debated under the heading of sustainability. Sustainability, however, is a richer concept that is more closely tied to the context of ecological constraints and includes broader concerns such as justice among contemporaries.

Some doubt whether intergenerational relations can be evaluated in moral terms at all. This fundamental doubt pertains particularly to actions that affect persons in the far future, for example, through the disposal of radioactive waste that remains hazardous for millennia. This doubt is mitigated for actions that affect future generations that overlap with contemporaries (thus turning part of present and future generations into contemporaries) and for actions that not only have negative consequences later on but also presently (thus turning the ethical problem partly into a problem of self-interest). Less fundamental doubts about intergenerational ethics pertain to the belief that continued progress of humanity is both possible and likely. On the basis of such a belief, the problems of future generations may seem insignificant compared to today's challenges like poverty. Still less fundamental doubts claim that while the present generation is indeed under a duty to take future generations into account, it is free to give the concerns of future generations less weight than its own concerns. The practice of discounting future utility in economics can be based on this view. Nonetheless, despite these doubts of varying degrees, a morally appropriate relation to future generations is a serious topic for most theorists. Whether we have a duty to bequeath, for instance, equally or only sufficiently much to future generations, what type of value we must bequeath (the general good of well-being or, more specifically, certain environmental goods), and whether there are not only duties on our part but also rights on the part of future generations are all examples of prominent questions discussed by theorists today.

Intergenerational relations differ in significant ways from relations between contemporaries. First, there is a power asymmetry and only limited cooperation between different generations (sometimes hinted at by the question: "What has posterity ever done for me?"). This is a challenge for theories that base the justification of duties on reciprocity or mutual advantage. Such theories will have to rely on, for example, indirect reciprocity (duties to the future are owed in response to what one has received from the past) or on a chain of obligation where the present generation has direct duties only toward those descendants that overlap with itself (and because the same holds true for the descendants and the descendants' descendants, etc., the present generation indirectly cares for the further future, too).

The lack of face-to-face interaction can also be seen as a challenge to moral theories that link duties closely to community ties. It should be noted, however, that many people do have strong relational ties with at least a subset of future generations: their own offspring.

The lack of interaction among generations is, of course, no problem for moral theories that justify duties independently of cooperation and community, such as utilitarianism and many kinds of human rights or religious theories. These theories extend moral concern in a universal manner to all humans, including humans in the indefinitely distant future. Such theories, however, face difficult questions as to where the motivation to comply with such moral demands comes from and questions about how the moral demands will actually be implemented in a democratic process in which future generations themselves have no voice. In response, suggestions for safeguarding the interests of future generations include constitutional provisions or an ombudsman to speak on behalf of future generations.

The second way in which intergenerational relations can differ from relations between contemporaries is that presently living persons can influence future generations in ways that are not common among contemporaries. The present generation can affect in unique ways the preferences and values of future generations or at least the cultural, technological, and political context within which these are formed. The present generation can also influence the population size of future generations. Population size is not only an important issue in terms of its effect on the environment but also as a moral issue in its own right. Given that life is deemed to be something valuable to those who lead it, many questions crop up, for example, whether there is a duty to create ever-more such valuable human lives, how to balance the value of more lives against the average quality of those lives, and what weight to give to the duty to keep the probability of humanity's extinction as low as possible. Further, the present generation not only influences the preferences and the size of future generations but also the identity of the persons that comprise future generations.

This leads to the so-called nonidentity problem. To illustrate this problem, imagine a person (call her Laura) suffering from climate change in 2100 and lamenting the fact that people at the beginning of the 21st century did not pursue radical mitigation policies. There is something puzzling about Laura's complaint: if radical mitigation policies had been pursued, this would not only have prevented climate change but it would also have changed the course of history in infinitely many small and large ways, and this makes it doubtful that Laura's parents would have met and would have conceived a child with the exact same ovum and sperm that led to the child being born actually being Laura. Thus, with the mitigation policy in place, Laura would not be better off than she is now. Rather, she would not have been born at all. There is a large body of literature on the implications of and solutions to the nonidentity problem.

Finally, a third way in which intergenerational relations differ from relations between contemporaries is that intergenerational ethics has to overcome particularly serious epistemic problems. So, in response to these hurdles of uncertainty, precautionary principles have been designed to guide policy choice.

See Also: Conservation; Consumption, Consumer Ethics and; Intergenerational Justice; Sustainability, Seventh Generation.

Further Readings

De-Shalit, Avner. *Why Posterity Matters*. London: Routledge, 1995.

Gosseries, Axel. "Theories of Intergenerational Justice: A Synopsis." *S.A.P.I.EN.S*, 1/1 (2008).

Roberts, Melinda. "The Nonidentity Problem." http://plato.stanford.edu/entries/nonidentity-problem (Accessed April 2010).

Dominic Roser
University of Zurich

Gaia Hypothesis

The Gaia hypothesis, proposed by James Lovelock in the 1960s, purports that the Earth functions as one large single living superorganism. Lovelock developed this hypothesis while working for the U.S. National Aeronautics and Space Administration (NASA) as part of a mission to look for life on Mars. According to Lovelock, the Earth maintains and regulates its biosphere in what he calls a "preferred homeostasis." The hypothesis itself is described in Lovelock's book *Gaia: A New Look at Life on Earth*. Lovelock treats the hypothesis as a scientific theory and not simply an environmental philosophy.

The name *Gaia* is a reference to the Greek goddess of Earth. According to Lovelock, scientists have to invent many different theories to explain how the many processes in nature result in a natural equilibrium on Earth. Lovelock deals with this phenomenon holistically by attributing purpose to a terrestrial self-regulating, living system. This includes Earth's biosphere, atmosphere, oceans, and soil. Lovelock argues that life on Earth provides a homeostatic feedback system operated automatically and unconsciously by the planet itself.

Lovelock's work, and the criticisms of it, has resulted in two forms of the Gaia hypothesis. The first is called the "weak Gaia hypothesis" and it focuses on how the organisms on the Earth have altered its composition. The weak Gaia hypothesis suggests that Gaia is coevolutive, meaning that just as the Earth has an effect on biological evolution of organisms, so does biological evolution affect the Earth. On the other hand, the "strong Gaia hypothesis" suggests that the biological processes of Earth's biosphere, atmosphere, oceans, and soil and organisms manipulate their physical environment for the purpose of creating biologically favorable conditions. The strong Gaia hypothesis focuses on the stability of the oxygen and carbon dioxide on Earth, Earth's temperature, and the salinity of the oceans.

From an environmental philosophy perspective, Gaia can be seen as the ultimate holistic biocentric theory by treating the Earth as one giant organism.

See Also: Anthropocentrism; Attfield, Robin; Biocentrism; Deep Ecology; Instrumental Value; Intrinsic Value; Land Ethic; Lovelock, James Ephraim; Rolston, III, Holmes.

Further Readings

Attfield, R. *Environmental Philosophy: Principles and Prospects.* Aldershot, UK: Avebury, 1994.

Attfield, R., A. Belsey, and Royal Institute of Philosophy. *Philosophy and the Natural Environment.* Cambridge, UK: Cambridge University Press, 1994.

Belshaw, C. *Environmental Philosophy.* Montreal, Canada: McGill-Queen's University Press, 2001.

Chappell, T. D. J. *The Philosophy of the Environment.* Edinburgh, UK: Edinburgh University Press. 1997.

Preston, C. J. *Grounding Knowledge: Environmental Philosophy, Epistemology, and Place.* Athens: University of Georgia Press, 2003.

Jo Arney
University of Wisconsin–La Crosse

Gandhi, Mohandas

Mohandas K. Gandhi (1869–1948) is generally known in the West as Mahatma Gandhi, Mahatma being an honorific meaning "great soul" in Sanskrit. Often considered the greatest statesman in modern Indian history, he was the political and spiritual leader of the movement that culminated in India's independence. He did not live to enjoy the fruits of his work, however, as he was assassinated by an extremist Hindu nationalist. Gandhi is primarily known as a political leader and social reformer, and influential figures including Martin Luther King, Nelson Mandela, and Barack Obama have acknowledged a debt to Gandhi. Unlike them, he was never awarded the Nobel Peace Prize, although he was nominated several times. Gandhi's significance is that he developed a philosophy that encompasses both social and economic justice and care for the environment.

Gandhi was born in Porbandar, India, and studied law in England. After qualifying, he practiced law for two years, and in 1893 accepted a one-year contract in South Africa, which at that time was under British control. He was appalled at the way that Indians (including himself) were treated in South Africa and worked to unify the Indian community into a political force, including the method of *satyagraha*. This Sanskrit word, which literally means devotion to the truth, was his term for the theory of nonviolent protest and mass civil disobedience that he was developing. These activities brought him into conflict with the authorities, and he and many of his followers were often jailed and mistreated by the police. On his return to India, Gandhi began a campaign of resistance with the goal of achieving independence; his strategy included protests and strikes by tenant farmers, a boycott of foreign-made goods and of colonial institutions, and the famous Salt March in protest against the salt tax imposed by the British, as a result of which some 60,000 Indians were imprisoned. In addition to the political gains, including the radicalization of many Indians, the march taught an environmental lesson: that nature provides for the needs of all who are willing to work.

Gandhi believed all of the world's ills including political oppression, economic inequalities, crime, and environmental destruction stem from a single cause: violence in thought and action. The solution is *ahimsa*. This Sanskrit word is usually translated as "nonviolence," but this is somewhat misleading. The prefix "non-" usually means the mere absence of something, as in "nonexistence," but Gandhi argued that a relationship or regime may be violent even if no physical violence is used because the dominant party is able to control by creating an atmosphere of fear—what is commonly referred to today as "structural violence." Moreover, the principle of *ahimsa* requires one not to be complicit in violence. Thus most Hindus and all Jains abstain from eating meat, even though people who eat meat usually have not personally harmed any animal. He also saw *ahimsa* as requiring that one actively help to assist the victims of violence. There is a duty to speak out against injustice and also to employ *satyagraha* against injustice.

Gandhi actively campaigned against all forms of exploitation, including poverty, the unequal status of women, the caste system, and meat eating. He believed that a simple vegetarian diet was sufficient for health and that the production of meat was a wasteful luxury and, indeed, beyond the means of most Indians. Moreover, he believed that because humans have no power to create life, they therefore have no right to destroy life. He stated that vegetarianism was the basis for his practice of *brahmacharya*, spiritual and practical purity. This requires not just that one live a simple life but that one cease to desire all physical pleasures, including those of taste and sex, which he taught exists only for procreation. His saying, "Earth provides enough to satisfy every man's *need*, but *not* every man's *greed*," is widely quoted, and this is how he lived his own life, owning few possessions and dressed only in a simple garment made from cotton that he spun himself.

Gandhi is often considered as a forerunner of today's sustainability movement. He criticized consumerism, referring to the Western lifestyle as a *Bhasmasur* (destructive monster) and taught that one of the main evils of British imperialism was that it sucked up resources from countries such as India so that the British could live in luxury. He was aware that natural resources are limited and that India's vast population (estimated at 345 million at the time of his death; 1.15 billion in 2009) could never aspire to live a Western lifestyle. He criticized air and water pollution, both because of their effects on human health and because they show lack of respect and compassion for nature. He believed in the value of physical labor but criticized working conditions in urban factories. Indeed, he opposed urbanization, arguing that the best form of human settlement is the village. Unlike large centralized, impersonal states, villages are on a scale where the needs of all can be met and everyone can participate equally in social and economic life. Moreover, democratic institutions at a village level date back thousands of years in India. Finally, each village is in control of its own natural resources, growing its own food and cotton crops, maintaining a clean water supply and sanitation, none of which is possible in a city. Thus both human and environmental health will be protected. In 1947, approximately 80 percent of Indians lived in villages, and in 2010 the proportion is still over 70 percent.

Gandhi is famous for taking the long view and he did not expect transformation to happen overnight. Gandhi often counseled that "endless patience" was required for transformation. A famous quote from Gandhi reminds us that we have obligations to future generations. He states, "The earth, the air, the land and the water are not an inheritance from our forefathers but on loan from our children. So we have to hand over to them at least as it was handed over to us."

See Also: Animal Ethics; Civic Environmentalism; Democracy; Development, Ethical Sustainability and; Ethical Vegetarianism; Future Generations; Social Ecology.

Further Readings

Gandhi's Philosophy. "Gandhi's Views on Environment." http://www.gandhi-manibhavan.org/gandhiphilosophy/philosophy_environment.htm (Accessed February 2010).
Khoshu, T. N. *Mahatma Gandhi and the Environment.* New Delhi: Teri Press, 2009.
Narayan, Shrian, ed. *Selected Works of Mahatma Gandhi.* Ahmedabad, India: Navajivan, 1968.
National Gandhi Museum. http://www.gandhimuseum.org/sarvodaya/articles/environment.html (Accessed February 2010).

Alastair S. Gunn
University of Waikato

Genetic Engineering

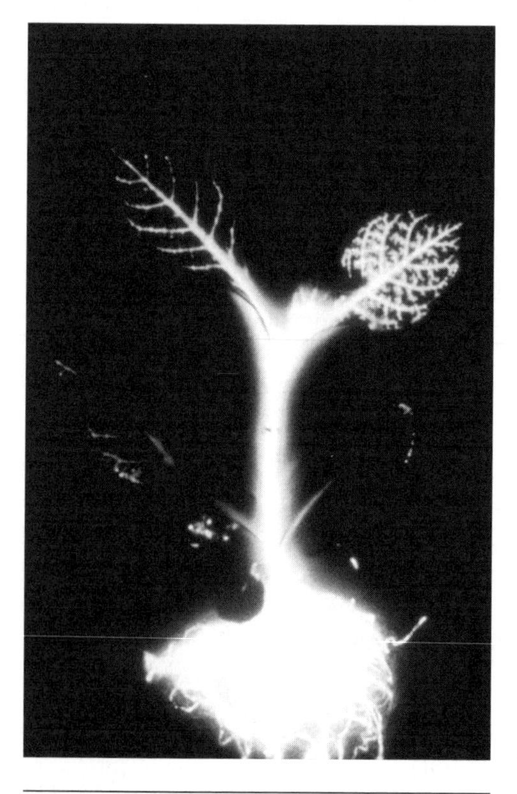

In a study meant to help genetic engineers breed disease-resistant plants, in 2005 researchers at the University of California, San Diego, modified a tobacco plant by adding a gene from a firefly to its DNA. The resulting plant, shown here, glowed like a firefly.
Source: National Science Foundation

Humans have been practicing genetic modification, in a broad sense, for thousands of years, traditionally through selective breeding, and this is not generally controversial. More recently, new varieties of plants, animals, and microorganisms have been created using technology that acts directly at a genetic level. Much of the debate on this subject has been from the perspective of utilitarianism; as well, there are broader ecological and cultural concerns.

The terms *genetic engineering* (GE) and *genetic modification* (GM) are often used interchangeably and neither term has a clear and widely accepted definition. In this article, GE is treated as a subset of GM. A third term, *genetic enhancement,* is also explained.

There is a very broad sense of GM, used here, that includes all techniques to develop a plant or animal for a particular purpose by producing a new combination of genes. This includes techniques ranging from the first attempts at selective breeding of plants by Neolithic farmers 10,000 years ago to the transfer of genetic material in laboratories via gene splicing between plant and animal organisms that are separated by well over a billion years of evolution.

The term *GE* is used in this article to refer to any procedure that changes the genetic sequence of an organism in ways that could not have happened, or at least were extremely unlikely to have happened, without human intervention, or more narrowly, without the techniques of sophisticated biotechnology. Unlike traditional forms of GM, which mimic natural processes, GE bypasses them. This is most obvious in the case of transgenic organisms, where genetic material from one species is introduced into the DNA of an unrelated species.

Genetic enhancement is the transfer of genetic material intended to modify human traits. While the term is sometimes employed to include techniques to rectify disabilities and susceptibility to diseases, it usually refers to enhancing attributes and capacities beyond the normal, commonly, though inaccurately, styled "designer babies." This topic is controversial but is not discussed here.

Traditional Genetic Modification

Traditional GM via selective breeding is not as such controversial; indeed, without it, the world's population would probably be at most a few million people. However, there is increasing concern about the loss of genetic variation and concentration of potential lethal genes due to extreme selection, which will be examined later.

Today, 15 staple crops provide 80 percent of the world's food energy needs; just three grasses (rice, maize/corn, and wheat) provide 60 percent. The wild ancestors of these crops produced relatively low yields of small, hard-to-digest seeds. Neolithic farmers used selective breeding to increase the size of seeds, remove bitter tastes, and increase the length of time for which seeds could be stored for later consumption by humans and livestock and for replanting the following season. Crop yield is the ratio between the number of seeds planted and the number that are available for use. A minimum yield for survival is a ratio of 1:3, with two seeds available for consumption and one for next season's harvest. For most of human history, in most of the world, this ratio was rarely exceeded over any extended period because of floods, droughts, diseases, and pest infestations beyond the control of the farmers. However, if the higher estimates of the population of the indigenous peoples of the Americas are accepted, their yields must have been high. The people of the Southwest grew hundreds of varieties of corn, squash, peppers, beans, and other crops, all developed through selective breeding. Elsewhere, yields also gradually increased, particularly since the 18th century in the West, partly through selective breeding but also through techniques such as irrigation and crop rotation. More recently, due to the development of new, fast-growing, high-yielding crops (the "green revolution"), the mechanization of agriculture, and increasing use of fertilizers and pesticides, yields have increased exponentially. According to the United Nations Food and Agriculture Organization (FAO), between 1961 and 2000, wheat yields per hectare increased by 300 percent or more in the European Union and Asia, including a massive increase of 681 percent in China, while rice yields increased by close to 200 percent or more in the Americas and Asia.

Selective animal breeding has also produced substantial increases in yields. The wild ancestor of the domestic chicken, the red jungle fowl, lays up to 30 eggs per year, but layer hens today, under intensive conditions, lay around 300 eggs per year. Broiler chickens, which are far larger than their wild ancestors, reach maturity in 6–7 weeks. The wild ancestor of today's dairy cows produced around two liters of milk per day, enough to feed one calf, but today's Holsteins and Jerseys produce 20 liters per day. Again, these increased yields are not solely due to GM via selective breeding but they would not have been possible without it.

In passing, it should also be noted that people, especially members of elite groups such as royal families, have long attempted to practice selective breeding among humans, notably by marrying off their offspring to the offspring of fellow members of the elite.

Ethical Issues

Ethical concern about selective breeding is relatively recent because its benefits have been clear and there has been no evidence of any risks. However, there is increasing concern about the vulnerability of animal and plant varieties that have a narrow genetic base. The first genetically induced disaster was the potato blight that led to *an Gorta Mor*, the great famine in Ireland in the period 1845–52 when around a million people died of starvation and related disease while another million emigrated. The same disease struck the Scottish Highlands and, although there were few fatalities, 1.7 million Scots left Scotland during the period. Although English control of Ireland and Scotland systematically denied the local people access to resources—throughout the Irish famine, Irish wheat and beef were exported to England—the main reason for the famine was the narrow genetic base that exposed all crops of the variety to the same disease.

Of current concern is Panama disease (also known as fusarium wilt), which is killing off entire banana plantations around the world. The disease, for which there is no cure, affects only the Cavendish variety, but this type dominates world banana production. Americans eat more bananas (almost entirely Cavendish) than any other fruit—11.9 kilograms (26.2 pounds) annually per capita, while globally, before the introduction of the disease, 47 percent of the 100 million bananas produced annually were of this variety. All Cavendish plants are susceptible to the disease because they are genetically identical. The Cavendish variety was bred to replace an earlier variety, Gros Michel, which had itself succumbed to Panama disease; it remains to be seen whether conventional selection techniques will be able to develop a variety sufficiently similar to the Cavendish; currently, several breeders are experimenting with GE to create a resistant strain of Cavendish.

Animal welfare is an issue for selective breeding. The highly productive hens and cows described above are commonly unhealthy and live miserable lives. This is largely because of the conditions in which they are kept, but chickens and turkeys also suffer from inherited defects, notably disablement from leg abnormalities. Most pedigree dog breeds are at risk of abnormalities due to inbreeding. Hip dysplasia, a painful and eventually crippling disease, is frequent in larger breeds such as German shepherds, Labrador retrievers, mastiffs, and rottweilers. Bulldogs, perhaps the most inbred variety, have a hereditary disposition to numerous skin and eye diseases, malformation of the mouth and teeth, bone disorders, and abnormalities of the respiratory system and the heart; they give birth with difficulty and Caesarean deliveries are common.

Perhaps the most extraordinary example of inbreeding is that of thoroughbred horses. Officially, all thoroughbred horses are descended from three sires: the Darley Arabian, the Godolphin Arabian, and the Byerley Turk. Current research suggests that 95 percent of thoroughbreds can be traced to the Darley Arabian, almost all via Eclipse (1764–89), the outstanding racehorse and sire of his period. Unsurprisingly, thoroughbreds are subject to a vast array of disorders and only a small proportion of horses in training ever win a race.

Genetic Engineering

Conventional modification is a slow and unreliable process. Crossing two organisms with highly desirable characteristics may or may not be successful because most characteristics are determined by a combination of genes or sequence and because the recessive genes may not be expressed for several generations. As horse breeders have learned, to their cost, the progeny of a champion sire and dam may turn out to be useless for racing. And as the examples above show, a gene sequence that produces desired characteristics may produce undesirable characteristics as well. As successive generations have an ever-narrowing genetic base, genes for hardiness, resistance to disease, or even normal development may be lost forever. Genetic engineering seems to offer the possibility of relatively rapid and reliable modification without the risks.

Genetic engineering began in the 1970s when researchers discovered how to transmit genetic material at the bacterial level. In the late 1970s, scientists used recombinant DNA to produce insulin and interferon, though not in usable volumes. In 1982, the first successful gene transfers between species of fruit fly and between species of plants were carried out. The first field tests on plants were conducted on tobacco in Belgium in 1986. The earliest consumer product, the Flavr Savr tomato went into commercial production in 1992, though it was unpopular with consumers and was soon withdrawn from the market. In 1988, Harvard University received the first patent for a variety of mammal, the "oncomouse," so called because of its genetic propensity to develop cancer and therefore its utility in cancer research. In 1996, scientists at the Roslin Institute in Scotland carried out the first cloning of a mammal using the cells of an adult—Dolly the sheep.

Today, GE ingredients are found in many foodstuffs, especially in the United States. Over half of all cheese produced in the United States contains a GE enzyme, chymosin, which has the same function as rennet. Most U.S. dairy cows are injected with GE-derived bovine growth hormone (BGH), a drug that increases milk production. But the main source of GE foods is through soy and maize, most of which is grown from seeds supplied by the Monsanto Company, which produces 90 percent of the world's GE seeds, as well as BGH. Monsanto, which produces the glyphosate-based herbicide Roundup, has developed soy, corn/maize, and cotton varieties that are resistant to Roundup so that fields can be sprayed with herbicide that will kill weeds but not the crop. Currently, 91 percent of all soy, 87 percent of all cotton, and 73 percent of all maize grown in the United States are from GE seeds. Around the world, some 13 million farmers grow GE crops on 125 million hectares of land, which is equivalent in area to the United Kingdom, Ireland, France, Belgium, and Germany combined.

Ethical Issues

The issues discussed earlier in relation to traditional GM, notably the shrinking gene pool, apply equally to GE. Public debate on GE has mostly focused on perceived risks and benefits to agriculture and consumers through widespread cultivation and consumption of GE products. Thus, the debate has largely been conducted at a pragmatic, utilitarian level. In most jurisdictions, though not the United States, it is accepted that government agencies rather than individuals and corporations should determine which, if any, GE crops should be grown and, if so, under what conditions.

Promoters of GE products promise improved food quality, wider availability of food to undernourished people in poor countries, and economic benefits to farmers. They claim that

GE can provide higher yields, pest and disease resistance (thus reducing the need for pesticides and herbicides), drought and frost resistance, more uniform product quality, longer shelf life, and enhanced nutritional properties. Opponents are skeptical of these claims, for which there is little published evidence, and warn of potential (albeit largely unspecified) risks to human health. There are also possible risks to conventional and (especially) organic agriculture via mechanisms such as horizontal gene transfer (the transmission of genetic material from GE plants to non-GE plants, notably via cross-pollination) and the creation of "superweeds" that are resistant to currently available herbicides. Some also fear that GE may threaten biodiversity, where GE crops are grown in the same area as related native species, and there are similar concerns about the risk to native animal species. More generally, there is a risk that the biotechnologies promoted by multinational corporations such as Monsanto will end up being the only biotechnologies that are available.

Assessment of risks and benefits are, of course, necessary for rational decision making, but for many people, a shortcoming of utilitarian cost-benefit analysis is its inability to account for anything that cannot be given a monetary value such as human life, culture, spirituality, and nonutilitarian ethical considerations. The danger is that these aspects of humanity are likely to be undervalued, treated in a token manner, or simply omitted from the decision-making process, effectively bolstering the dominant cultural paradigm at the expense of others. The most obvious risk is to indigenous cultures, which tend to have a more holistic and ecological worldview than their Western counterpart.

Some critics of GE charge that GE organisms, especially those produced by transgenics, are at odds with the natural processes according to which life has evolved. Ecofeminists often write of the need to eliminate an attitude of "arrogance" toward nature. GE is also seen as even more exploitative than conventional agriculture, because the latter merely mimics nature in rearranging existing gene sequences to select for qualities that, under suitable conditions, could have arisen by chance. Thus, it works with nature rather than seeking to conquer nature. Ecofeminsts advocate a relational and sympathetic attitude toward nature, which is incompatible with the level of control over natural processes (to the point of obliteration in the case of transgenics) that is central to GE.

See Also: Agriculture; Biodiversity; Cost-Benefit Analysis; Ecofeminism/Ecological Feminism; Utiiltarianism.

Further Readings

Cuomo, Chris J. *Feminism and Ecological Communities: An Ethic of Flourishing.* New York: Routledge, 1998.

Department of Internal Affairs, New Zealand. "Report of the Royal Commission on Genetic Modification, Wellington: Royal Commission on Genetic Modification, 2001." http://www.mfe.govt.nz/publications/organisms/royal-commission-gm (Accessed January 2010).

Food and Agriculture Organization (FAO) of the United Nations. *Ethical Issues in Food and Agriculture.* Rome: FAO, 2001.

Food and Agriculture Organization (FAO) of the United Nations. *Genetically Modified Organisms, Consumers, Food Safety and the Environment.* Rome: FAO, 2001.

Alastair S. Gunn
University of Waikato

Global Greens Charter

The Global Greens Charter was signed by 800 delegates of global green parties from 72 countries in April 2001, in Canberra, Australia, at the founding of the Global Greens. An expansion of earlier joint green party statements drafted at the 1992 Earth Summit in Rio de Janeiro and among regional affiliations of green parties, the charter emphasized the commonalities of dozens of political parties.

Unlike most political parties, greens—in any country—are by definition concerned with global issues, not simply domestic matters or the relationship of their country with the rest of the world. More than concerns over banking reform, income inequalities, or even trade relations, for example, there is a limit on the extent to which green concerns can be pursued effectively in only one country; certainly more than those concerns, green goals in each country become closer when they are met elsewhere. While climate change can and does have local effects and such effects are not evenly distributed geographically, it is still a truly global concern, as must be any remedies. In deriving the commonalities of the world's green parties, Australian Louise Crossley drew upon the charters of her own party and that of New Zealand, the guiding principles of the European greens and the Brazilian and Mexican green parties, the Ten Key Values of the Green Committees of Correspondence in the United States, and early documents like the 1975 Values Party manifesto.

The six principles so distilled, in addition to the greens' commitment to global cooperation, are the following:

1. *Ecological wisdom.* "We acknowledge that human beings are part of the natural world and we respect the specific values of all forms of life. . . . This requires that we learn to live within the ecological and resource levels of the planet."
2. *Social justice.* "A just organization of the world and a stable world economy . . . the eradication of poverty . . . a new vision of citizenship built on equal rights."
3. *Participatory democracy.* "Individual empowerment through access to all the relevant information . . . all electoral systems are transparent and democratic."
4. *Nonviolence.* "Security should rest not mainly on military strength but on cooperation, sound economic and social development, environmental safety, and respect for human rights."
5. *Sustainability.* "We recognize . . . the need to maintain biodiversity through sustainable use of renewable responses and responsible use of nonrenewable resources."
6. *Respect for diversity.* "We defend the right of all persons, without discrimination, to an environment supportive of their dignity, bodily health, and spiritual well-being."

A little more than half of the charter is devoted to political action, outlining the greens' goals and beliefs in the political sphere, under the headings of democracy, equity, climate change and energy, biodiversity, governing economic globalization by sustainability principles, human rights, food and water, sustainable planning, peace and security, and acting globally. Under each heading is a series of declarative statements meant to reflect the goals of all the signatory parties. These statements vary from the broad ("The greens will work to improve the rights, status, education and political participation of women.") to the specific ("The greens adopt the target of limiting carbon dioxide levels in the atmosphere to 450 ppm in the shortest possible period as requested by the Intergovernmental Panel on Climate Change.") and encompass all areas indicated by the six core principles.

See Also: Deep Ecology; Deep Green Theory; Democracy; Ecopolitics; Green Law Trends; Green Liberalism; Green Party, German; Green Party Ethical System, Four Pillars of the; Ten Key Values of the Greens.

Further Readings

Clapp, Jennifer and Peter Dauvergne. *Paths to a Green World: The Political Economy of the Global Environment.* Cambridge, MA: MIT Press, 2005.
Dobson, Andrew. *Green Political Thought.* London: T&F Books, 2009.
Global Greens. "Green Federations, Networks and National Green Parties." http://www.globalgreens.org/parties (Accessed August 2010).
Radcliffe, James. *Green Politics: Dictatorship or Democracy?* New York: Palgrave Macmillan, 2002.

Bill Kte'pi
Independent Scholar

Globalization

Globalization is commonly understood as the process by which capital, goods, services, and labor cross national borders and develop a transnational character. It is often accompanied by the flow of related taste, ideas, and even values across boundaries, thus helping to reshape local political institutions, social relationships, and cultural pattern. Globalization is an emerging vision of the world and its resources as centrally organized by the global managers (such as the World Bank, International Monetary Fund [IMF], G-8 countries, World Trade Organization [WTO], and other transnational corporations).

Although use of the term *globalization* dates back only a few decades, many references can be found in 19th-century literature and social commentary to the changes in economic and social organization brought about by improved systems of transportation and communication. The most famous may be Karl Marx's observation that the imperatives of capitalist expansion could overrule traditional barriers, such as national boundaries. However we can find many novel trajectories of capital movement in today's globalized world, such as the "electronic herd," a concept used by Thomas L. Friedman in his classic book *The Lexus and the Olive Tree* to describe professionals who move money around the world on a short-term basis (e.g., currency traders) and multinational corporations (such as General Electric) that make long-term investments in foreign countries. Friedman's point is that those with the ability to move large amounts of capital across national borders hold so much power in modern economic life that governmental forces may be unable to control them. Although Friedman's book is one of the most prominent to support and propagate the United States–led current globalization, it does not escape itself from the contradictions and paradoxes inherent in the process.

Philosophical Routes to and Roots of Globalization

Globalization is understood differently from different theoretical perspectives. According to structural functionalists, human society is a comprehensive whole in which different parts are working together in order to create stability and order. Scholars of functionalism may not have any concrete theories of globalization, but their conceptual underpinnings of society from simple to complex speak to what we mean by globalization today.

August Comte, the founding father of sociology, for example, presented three stages of human society: theological, metaphysical, and scientific/positive stage. The positive stage, which corresponds most closely to the kind of society developing under globalization, is largely based on science and its methods, rational calculation, scientific administration, and the whole of humanity as a single unit. The process of globalization has brought the world closer to what Émile Durkheim refers to as "organic solidarity," in which society is characterized by interdependence and specialization rather than the sameness or "mechanical solidarity" that is characteristic of primitive societies.

Neil Smelser, a sociologist from the University of California, Berkeley, showed that there are various forms of shifts from traditional to modern (globalized) society: a move from simple to complex technology, subsistence farming to cash crops, animal/human power to machine power, and finally rural settlement to urban settlement. Talcott Parsons's AGIL model of the core functions of society can be used to describe modern global society: Adaptation to capitalist economy; Goal attainment (liberal); Integration to institutions; and Latency of values.

Modern globalized society corresponds in many ways to a society ruled by formal rationality as described by Max Weber, that is, a society in which decisions were guided by impersonal calculations such as risk. These include (1) appropriation of all physical means of production as disposable property, (2) market freedom or free market in which capital can flow freely, (3) rational technology or computation methods to maximize profits, (4) calculable laws, which are the forms of adjudication and administration for predictable outcomes, (5) free labor market in which capital can move and exploit cheap labor, and (6) commercialization of economic life, which includes the widespread capitalist methods such as bonds, shares, finance, banking, stock market, and so on, to allow capital to be more mobile for profit maximization.

In the modernization theory of W. W. Rostow, the last three stages of economic growth correspond to modern globalization: the drive to maturity; mass consumption; and beyond consumption. In the drive to maturity stage, the economy becomes globalized, technology becomes more complex, and the goal for production becomes not just satisfying basic needs but also maximizing profit to survive in the competitive capital market. In the stage of mass consumption, people take part in a culture of consumption rather than simply satisfying their basic needs, and there will be more societal attention to issues of social welfare and security. In the final stage, beyond consumption, people will seek new forms of satisfaction through exploration and novel experiences such as travel.

The conflict perspective, as propounded by Karl Marx and Friedrich Engels, maintains that society is a complex system in which conflict and tension are the result of each class struggling to satisfy, perpetuate, and enhance their own interests. This perspective focuses on the relations of production, that is, the social, legal, and political arrangement by which a class has access to and ownership of the means of production (technology, capital, labor, tools, knowledge, management, etc.). In Marx and Engel's formulation, the bourgeoisie controls the means of production and holds most of the power, while the proletariats have no option but to sell their labor in the market. The perspective believes that capitalism is driven by naked self-interest, and that colonization was a necessary condition for the development of capitalism. Therefore globalization is not a new social structure, but a new wave of capitalism, relying on the relationships of colonialism. As globalization is driven by the maximization of profit, the conflict perspective argues that it has generated an enormous gap between the rich and the poor that may be very conspicuous both within a country and between countries.

The Inherent Dynamics and Effects of Globalization

Globalization, according to Thomas L. Friedman, is a dream for sale: the middle-class American lifestyle. Friedman's assumption is that the whole world can achieve this dream by sharing the global economic space dominated by the United States. The logic he presents is that in economies based on manufacturing, interdependence was necessary, while the rapid development of electronic technology has produced a qualitative change in the global economy. As many economies shift from manufacturing to services and electronic communications reduce or obliterate the importance of national boundaries, many services can be supplied from anywhere in the world and businesses can relocate to where labor is cheap and laws are amenable.

Friedman further argues that a country's wealth no longer rests on the availability of natural resources so long as there are resources in the minds of its people (i.e., human capital), meaning that mental resources (e.g., ideas, education) are the most important factors in development. Theoretically these mental resources can be obtained by anyone from any country, greatly increasing the opportunities of individuals and countries to rise out of poverty.

New technology in Friedman's view has also increased the power of investors. Capitalists today are more integrated and electronic herds (people who control technology and the flow of capital) can exercise greater influence on countries' economic and social sectors, marking a new phase of international relations (which differs from the cold war period that preceded it). Now international relations are characterized not by international rivalry, but by global integration, and not by military technology, but by economic technology.

Friedman's arguments and characterizations emphasize the positive side of globalization, but his work has been criticized for ignoring many inherent paradoxes and disharmonies of globalization. For instance, while technology may be an important component of social change, how it is used is ultimately determined by the people who control the technology. Friedman's assertions regarding labor and capital also ignore some complexities: for instance, while capital may move to different places to exploit cheap labor, it may not remain there for long time. This ability to easily move manufacturing around the world has had major effects on labor, often not to the benefit of the worker: for instance, people may find their employment opportunities are no longer in stable, wage-based employment, but in contract and project work.

Migrant and guest workers are also a crucial factor in the modern globalized economy and they are frequently victims of abuse and violence. Philip McMichael notes that of 20 million refugees and 100 million migrants around the world, 42 million are official guest workers. Over 80 percent of these refugee guest workers are in Asia and Africa rather than in the developed countries that are the main beneficiaries of globalization. More than 3 million illegal workers are paid less than legal wages, are often paid late, and have no legal recourse. Migrant workers are also vulnerable to changing economic conditions: for instance, in 2002, facing economic downturn, the Malaysian government expelled 400,000 Indonesian workers, who returned home to a land of high unemployment.

One of the troubled legacies of globalization is human trafficking. About 2 million women and children are trafficked annually (for the sex trade, domestic servitude, and other forced employment), and this illegal trade's annual profit is $6 billion. Since 1990, about 30 million women and children have been trafficked for prostitution and sweatshop labor, a rise directly related to the global feminization of poverty and the use of the Internet as a sex forum. Evidence suggests that sex tourism to Thailand contributed to the demand for Thai women overseas. By 1993, there were almost 100,000 Thai women working in the Japanese sex industries, and in Europe, there were over 5,000 in the city of Berlin alone.

Although globalization holds the promise to increase economic growth, alleviate poverty, and to elevate life standards and human dignity, in reality, it does so for only a handful of people while the vast majority experience greater poverty, insecurity, vulnerability, and marginalization. More than 900 million of the world's poorest are living today on less than $1 a day, and living in areas vulnerable to soil erosion, droughts, desertification, and floods. About 200 million people may face rising sea levels due to climate change, and 50 million people are in famine-vulnerable areas subject to climate change. More than 550 million people are already suffering from chronic water shortage. Globalization has contributed to these problems as agricultural modernization and debt restructuring force peasants and rural labor away from secured rural livelihoods. Displaced workers migrate to cities seeking employment. Many end up in urban slums without proper sanitation, drinking water, or healthcare, which may place unmanageable burdens on the family structure.

Conclusion

Apart from some positive aspects, globalization has generated enormous crises worldwide that particularly affected third world countries. The crises include the problems of displacement, food insecurity, and the marginalization of the labor force. Globalization is a powerful force in the modern world, but has often failed to deliver what it promised, mainly helping a few elite while marginalizing others. However, resistance to globalization continues as different social movements have emerged that focus on laying bare the inherent ideology of globalization, which is the "maximization of profit before people and ecology," and to strive for an alternate future where people and ecology will be more important than profit.

See Also: Ecology; Philosophy and Environmental Crisis; Urbanization.

Further Readings

Collins, Randall and Michael Makowsky. *The Discovery of Society.* New York: McGraw-Hill, 2005.

Friedman, Thomas L. *The Lexus and the Olive Tree.* New York: Anchor Books, 2000.

Hoogvent, Ankie. *Globalization and the Post Colonial World: The New Political Economy of Development.* Baltimore, MD: Johns Hopkins University Press, 2001.

Islam, M. Saidul. "Paradoxes of Globalization." *Daily Star,* 5/1074 (June 9, 2007). http://www.thedailystar.net/2007/06/09/d706091503100.htm (Accessed January 2010).

McMichael, P. *Development and Social Change: A Global Perspective.* Thousand Oaks, CA: Pine Forge Press, 2008.
McMichael, P. "Globalization: Myths and Realities." *Rural Sociology,* 61/1 (1996).
Ritzer, George and Douglas J. Goodman. *Classical Sociological Theory.* Toronto: McGraw-Hill, 2004.
Rostow, Walt W. *The Stages of Economic Growth: A Non-Communist Manifesto.* Cambridge, UK: Cambridge University Press, 2005.
Skrobanek, S., N. Boonpadki, and C. Janthakeero. *The Human of Traffic in International Women.* London: Zed Books, 1997.

Md Saidul Islam
Nanyang Technological University

Gore, Jr., Al

Albert Arnold "Al" Gore, Jr., was born on March 31, 1948, in Washington, D.C. He is the son of Albert Gore, Sr., a U.S. representative (1939–44, 1945–53) and senator (1953–71) from Tennessee and Pauline LaFon Gore. Al Jr. spent his early years on the family farm in Carthage, Tennessee, and in Washington, D.C., where he attended St. Albans high school. There, at his senior prom, he met Mary Elisabeth "Tipper" Aitcheson, whom he married five years later. In 1965, he enrolled at Harvard, where he planned to be an English major and worked on a novel. In 1967, he took a climate science course from Professor Roger Revelle, which drew his attention to environmental concerns. Later, he changed his major to government and graduated from Harvard cum laude in 1969.

Al Gore, the former U.S. vice president and current chairman of Generation Investment Management, a sustainable investment firm, addressing a business conference on the topic of sustainability in May 2010.

Source: Flickr/Tom Raftery

Although both he and his father were opposed to the war in Vietnam, Al Gore enlisted in the army, in part so as to not negatively affect his father's difficult 1970 Senate election campaign. Al Jr. completed his basic training and was assigned to be a journalist at Fort Rucker, Alabama. His orders to go to Vietnam were delayed (it is believed that the Nixon administration held them up until his father's Senate election was decided; his father lost his Senate seat). In Vietnam he was stationed with the 20th Engineer Brigade in Bien Hoa and worked as a journalist for the *Castle Courier.* Upon his discharge in 1971, he attended Vanderbilt University Divinity

School on a year-long scholarship from the Rockefeller Foundation for people planning to follow a secular career. After leaving the divinity school, he got a job as a journalist at the *Tennessean* (in Nashville, Tennessee), and in 1973, his first child, Karenna Aitcheson Gore, was born. He attended the Vanderbilt University Law School 1974–76 but took away no degrees, dropping his studies unexpectedly when he decided to run for his father's former seat in the U.S. House of Representatives.

Political Career

Al Gore, Jr., was 28 when he was elected to the House of Representatives and would serve in the House and Senate for a total of 17 years. During his legislative career, he was considered a moderate Democrat, or as he liked to call himself, a "raging moderate." He served on the House Committee on Energy and Commerce and the House Committee for Science and Technology, the committee he chaired for four years. During his time in Congress, Gore was one of the "Atari Democrats" later known as "Democrats' Greens."

He ran for president in the primaries in 1988 and campaigned as "a Southern centrist." However, running against Jesse Jackson, he split the Southern vote and lost the primaries. Although he was initially reluctant to join Bill Clinton's 1992 ticket, after he clashed several times with the George H. W. Bush administration over climate change, he accepted the invitation to be Clinton's running mate. While serving as vice president, he focused on information technology and environmental policies. His initiatives are credited with having fueled the dot.com revolution and, in turn, the constant growth of the U.S. economy in the 1990s. He ran for president in 2000 and lost to Republican George W. Bush following the controversial Florida recount.

Environmental Agenda

Al Gore's activities related to environment and sustainability started during his years in the House of Representatives and were very much connected to his interest in new technologies. As an Atari Democrat on the House Science and Technology Committee "before computers were comprehensible, let alone sexy, the poker-faced Gore struggled to explain artificial intelligence and fiber-optic networks to sleepy colleagues." While in the House of Representatives, Gore held the "first congressional hearings on climate change, and co-sponsored hearings on toxic waste and global warming." Throughout the 1980s, he continued to be one of the most outspoken proponents of environmental policies in the U.S. legislature. In 1990, Al Gore organized and presided over a three-day conference attended by legislators from 42 countries, aimed at creating a Global Marshall Plan "under which industrial nations would help less developed countries grow economically while still protecting the environment."

Following his son's tragic 1989 car accident, Al Gore announced in 1991 that he would not run in the upcoming elections in order to spend more time with his family. In this period, he wrote *Earth in the Balance: Ecology and the Human Spirit,* published in 1992. The book details the Earth's environmental predicament and the policies needed to address the associated issues; it draws upon the conclusions of the 1990 Global Marshall Plan conference. As vice president, he launched the GLOBE Program, which used the Internet to increase student environmental awareness. He strongly supported passage of the Kyoto Protocol, advocating a reduction in greenhouse gas emissions, but the Senate voted unanimously in favor of the much more conservative Byrd-Hagel Resolution.

In 2004, Al Gore colaunched the investment management firm Generation Investment Management, which takes sustainability factors into account when making financial decisions. In 2006, he launched the Alliance for Climate Change in order to "persuade people of the importance, urgency and feasibility of adopting and implementing effective and comprehensive solutions for the climate crisis." In 2006, Al Gore launched the documentary film *An Inconvenient Truth,* which was later accompanied by a book of the same title. He organized a series of conferences and meetings at which he presented the movie and the book. The meetings and extensive media coverage helped promote the sustainability agenda, although they have also drawn considerable criticism, for example, from the High Court of Justice (UK), which claims there are "nine significant errors" in the movie.

Awards

The film *An Inconvenient Truth* won an Academy Award for Best Documentary in 2007. Al Gore won a Primetime Emmy Award for Current TV, which he founded, and the Prince of Asturias Award for International Cooperation. In 2007, Al Gore, Jr., was awarded the Nobel Peace Prize together with the Intergovernmental Panel on Climate Change.

See Also: Conservation, Aesthetic Versus Utilitarian; Democracy; Ecology; Ecopolitics; Environmental Policy Act, National.

Further Readings

Aldred, Jessica. "Timeline: Al Gore." *The Guardian* (October 12, 2007). http://www.guardian.co.uk/environment/2007/oct/12/climatechange1 (Accessed February 2010).

Budd, Leslie and Lisa Harris. *E-Economy: Rhetoric or Business Reality.* London: Routledge, 2004.

CNN. "Albert Gore Jr.: Son of a Senator." (2000). http://www.cnn.com/SPECIALS/2000/democracy/gore/stories/gore (Accessed January 2010).

CNN. "The First Presidential Run." (2000). http://www.cnn.com/SPECIALS/2000/democracy/gore/stories/gore/index2.html (Accessed January 2010).

Howd, Aimee. "Gore With Freshman Dorm Mates at Harvard, Including Actor Tommy Lee Jones." *Washington Post.* http://www.washingtonpost.com/wp-sv/politics/campaigns/galleries/lifeofgore/photo8.htm (Accessed February 2010).

Howd, Aimee. "Staff of the Castle Courier." *Washington Post.* http://www.washingtonpost.com/wp-srv/politics/campaigns/galleries/lifeofgore/photo10.htm (Accessed February 2010).

Ireland, Corydon. "Gore: Universities Have Important Role in Sustainability." *Harvard Gazette.* http://www.news.harvard.edu/gazette/2008/10.23/99-gore.html (Accessed January 2010).

Miles, Sarah. "A Man, a Plan, a Challenge." *Wired* (January 30, 1998). http://www.wired.com/politics/law/news/1998/01/9939 (Accessed February 2010).

Noon, Chris. "Gore Really Does Get the We." *Forbes* (September 21, 2006). http://www.forbes.com/facesinthenews/2006/09/21/gore-google-yahoo-face-cx_cn_0920autofacescan06.html (Accessed February 2010).

Shabecoff, Philip. "World's Legislators Urge 'Marshall Plan' for the Environment." *New York Times* (May 3, 1990). http://query.nytimes.com/gst/fullpage.html?res=9C0CEED7173FF930A35756C0A966958260 (Accessed February 2010).
Stengel, Richard. "Profiles in Caution." *Time* (March 21, 1988). *The Times.* "Al Gore's Inconvenient Judgment." (October 11, 2009). http://business.timesonline.co.uk/tol/business/law/article2633838.ece (Accessed February 2010).
U.S. Senate. "Albert A. Gore, Jr., 45th Vice President (1993–2001)." http://www.senate.gov/artandhistory/history/common/generic/VP_Albert_Gore.htm (Accessed January 2010).
Weaver, Warren, Jr. "Gore as Candidate: Traveler Between Two Worlds." *New York Times* (January 21, 1988). http://query.nytimes.com/gst/fullpage.html?res=940DEFDF133CF932A15752C0A96E948260&sec=&spon=&pagewanted=print (Accessed January 2010).
Wood, Thomas. "Nashville Now and Then: Young Al's Big Decision." *NashvillePost.com* (February 29, 2008). http://www.nashvillepost.com/news/2008/2/29/nashville_now_and_then_young_als_big_decision (Accessed January 2010).

Daniel Tomozeiu
University of Westminster

Green Altruism

Green altruism is perhaps best summed up by the popular bumper sticker: "Live simply so that others may simply live." Many, but not all, environmentally friendly activities require some sort of sacrifice. On the personal level, recycling, using environmentally friendly products, biking to work, purchasing organic food, and supporting environmental institutions all require expending time and/or money. Many argue that this is altruistic behavior, as it distances the actor from the capitalist ideal of unrestrained materialism. However, these actions are often conspicuously carried out in order to achieve some sort of prestige that is associated with environmental altruism.

The concept of altruism has been central, although rarely stated, to the U.S. environmental movement from its inception, from Henry David Thoreau's "roughing it" in his log cabin to better connect to the natural soil to Rachel Carson's plea to use less environmentally damaging pesticides to its most recent manifestation with the popularity of purchasing carbon credits. "Green" cleaning products and organic foods are increasingly making inroads at grocery stores and the price difference between these and conventional goods is slowly decreasing, but the commitment to purchasing these goods still requires a sacrifice on some level.

Or does it? Ecophilosopher Arne Naess has argued that this form of altruism will facilitate the emergence of an awareness of the "ecological self" and that these actions will no longer be seen as altruistic but rather as an act of enlightened self-interest. Other lines of reasoning also point toward self-interest, but perhaps of a less enlightened kind. People who purchase hybrid vehicles, while possibly saving money from lower gasoline prices in the long run, also gain quite a bit of status, or what some researchers have labeled "green capital." Green capital is a sort of social credit that people gain through these acts of altruism that increases others' perception of them as a good person; this has been labeled "competitive altruism," where being more altruistic equates to "winning." Green altruism is a form of prestige that is somewhat different than normal, as green prestige does not always have to come from altruistic behaviors; for example, bicycling to work instead of driving doubles as a form of exercise and might not be considered to be a sacrifice. Green altruism, then, is a subset of ways to increase one's prestige in the eyes of others.

Michael Edelstein, Maria Tysiachniouk, and Lyudmila Smirnova (2007) argue that green altruism is correlated with three main traits: an awareness of pollution at different scales, a sense of responsibility to enact change, and a perception that the costs of doing so are not prohibitive. This last point is important because it maintains the notion of prestige and altruism associated with certain actions such as purchasing higher-priced ecofriendly products. If these products were the same price as conventional goods, then the associated prestige would disappear. This line of reasoning has led, in some instances, to corporate greenwashing, where a company advertises its products as being more environmentally friendly in order to raise prices and capture a different market. This is not to say that a company's sole motivation for green altruism has to be marketing; companies like Patagonia have made attempts at being environmentally friendly from inception and are motivated by this notion of altruism.

Other research (University of Toronto, 2009) has pointed to the fact that green altruism does not lead to other forms of altruism and, in fact, is correlated to higher levels of theft as well as to lying about purchases of conventional products, perhaps further demonstrating that the desire for prestige outweighs a desire to make truly altruistic sacrifices. The study also demonstrated that people are more likely to purchase environmentally friendly products in public than in private, again demonstrating the social influences on green altruism.

Green altruism has also been proposed and enacted with varying degrees of success on the national and international scales, most prominently in conceptions of sustainable development and climate change. Both of these forms of sacrifice are primarily proposed so that there will be a healthy environment for future generations, thus extending the purpose of altruistic behavior temporally. This notion of sacrifice for the environment has led to many arguments in the United States that environmental quality comes at the expense of economic development. There are, however, numerous examples of how the two can complement, instead of detract from, one another. This perception is one of the legacies of the concept of altruism in the U.S. environmental movement.

See Also: Biocentric Egalitarianism; Carbon Offsets; Consumption, Consumer Ethics and; Ethical Vegetarianism; Naess, Arne; Sustainability and Spiritual Values.

Further Readings

Cloud, John. "Competitive Altruism: Being Green in Public." *Time* (June 3, 2009).

Edelstein, Michael R., Maria Tysiachniouk, and Lyudmila V. Smirnova. *Cultures of Contamination*. Bingley, UK: Emerald Group Publishing, 2007.

Seed, John, Joanna Macy, Arne Naess, and Pat Fleming. "Thinking Like a Mountain: Towards a Council of All Beings." Gabriola Island, British Columbia, Canada: New Society Press, 1988.

University of Toronto, Rotman School of Management. "Buying Green Can Be License for Bad Behavior, Study Finds." *ScienceDaily* (October 9, 2009).

Matthew Branch
Pennsylvania State University

Green Ethics in the Legal Community

Since about the 1990s, green ethics have become increasingly more prominent in the legal community. In 1993, for example, the year after the Earth Summit in Rio de Janeiro, the first American law school course was offered on sustainability, at Willamette. Although law schools have lagged behind liberal arts colleges and European institutions in adopting green practices, they have responded to the needs of the incoming student body and the changing legal community and have offered a number of programs on sustainability law, in addition to the older traditional programs in natural resources law. The rise of new technologies, such as biofuels and nanoparticles, ecological disasters like the Deepwater Horizon oil disaster, and new international treaty frameworks like the Kyoto Protocol have all contributed to changing and expanding the field of environmental law, but even attorneys who work in other fields may have an interest in practicing green ethics.

One of the key differences in the rise of "sustainability law" programs and practices is the emphasis of the role of the sustainability lawyer as a protector of natural resources. While the traditional field of natural resources law certainly doesn't position the lawyer as an exploiter, it is generally taught from a value-neutral standpoint: that is, a natural resources attorney is as likely to be hired to find loopholes around environmental regulation or to defend a corporation accused of violating a regulation or sued for negligence as he or she is to be the one suing or prosecuting. Willamette's sustainability law certificate, on the other hand, is described as placing "special emphasis on the role of the lawyer in formulating environmental and natural resources law and policy to sustain and protect our local and global resources and to ensure social justice for all beings . . . to prepare the next generation of lawyer-advocates to lead their communities, the nation and the world toward a more sustainable future." There are valid comparisons to be made to other intersections of legal expertise and advocacy, from the civil rights attorney-activists of the 1950s and 1960s to the human rights and privacy advocates of the 21st century.

A growing number of attorneys concerned with green ethics are not actually working within the environmental law field. More and more attorneys may work with agricultural businesses and family farms to help them navigate the process of achieving organic certification. The growing carbon offsets field has a need for business lawyers, and the many companies associated with green technologies, from solar panel producers to environmentally friendly baby product vendors to farmers growing corn for use as biofuel, require periodic or in-house representation. Even patent attorneys can make use of specializing in green products. This new category of green law is poised to grow extensively in the coming decades.

The Green Guide for Lawyers

The Meritas Leadership Institute, operated by a nonprofit alliance of 170 commercial law firms in 60 different countries, produced its *Green Guide for Lawyers* in 2008—a best practices handbook advocating and delineating sustainable business practices for law firms. The handbook followed closely behind one produced jointly by the American Bar Association and the EPA, and reflected the growing 21st-century concern with the amount of waste produced by law offices. In surveying firms in preparation for composing the guide, lawyers working for Meritas were surprised at the unequal distribution of sustainability awareness throughout the legal community. While a surprising number of firms were only vaguely aware of the term and had given no thought to sustainable development or energy efficiency (or thought no further than saving on the electric bill), there were many firms who were far more advanced in their sustainability awareness than Meritas expected to encounter, firms that in some cases had been having internal discussions about sustainable offices and practices for years. In some cases, the sustainability discussion had been introduced by younger associates; in other cases, it had come about as a result of working with clients who were sustainably minded or ran green businesses.

Those concerned with sustainability fell into two broad categories: the true believers and the bottom-liners. The true believers, such as younger lawyers who may have taken sustainability coursework in law school or in their undergraduate career, saw sustainability as a good in and of itself, though they may not have taken the initiative to pursue it as far as they could, or may have needed advice on ways to adopt sustainable practices. The bottom-liners were primarily interested in sustainability either because of the savings it offered or because it would appeal to a particular clientele. Of course, there is always some overlap between these groups. The *Green Guide* focuses very specifically on practices rather than principles—specific steps that can be taken to achieve sustainability, presented in a way that is equally useful and appealing to either of these groups.

In the long run, the most difficult obstacle for law firms to achieving sustainable practices is their paper use. Those outside the legal community are not always aware of the sheer enormity of paper that is produced, and rarely recycled, by legal work: it is estimated at about 50–100 sheets of paper per work hour. Because certain strategies for paper reduction are not possible in some contexts—for instance, most courts will not accept double-sided documents and require hard copies, the reduction strategies are that much more critical in every area where paper use can be reduced.

The guide defines several "tiers" of firms, based on their commitment to sustainability.

EPA Sustainability Programs for Law Offices

The ABA-EPA Law Office Climate Challenge is a joint effort through the American Bar Association and the U.S. Environmental Protection Agency (EPA). The challenge is designed to encourage law offices to conserve energy and resources and to reduce greenhouse gas emissions and pollution. Much of the challenge focuses on managing office paper more efficiently. The challenge also encourages participation in WasteWise and the Green Power Partnership, and use of Energy Star–rated equipment and participation in the Energy Star program's recommendation to reduce energy consumption in law offices by at least 10 percent. The challenge began in 2007 and is expected to continue through March 2013.

Office paper management, as recommended by the challenge, calls for assuring that at least 90 percent of all paper used in the office—including envelopes—consist of at least 30 percent postconsumer recycled content. Recycling bins should be easily accessible, and at least 90 percent of the paper discarded by the office can be recycled. Double-sided printing and copying should be used for all drafts and internal documents.

WasteWise is an EPA program designed to help companies reduce and recycle solid waste and industrial waste. Many of WasteWise's requirements are met by the challenge's paper management suggestions. Awards are issued to companies that achieve particular goals of waste reduction. In addition to paper waste, the program has initiatives to recycle organic materials like food waste and yard trimmings, packaging waste like cartons, wood, polymer wraps, and shipping containers, and finding beneficial reuses for foundry sand, construction and demolition debris, and coal combustion products, which would be applicable to law firms moving into a newly constructed office. The WasteWise Electronics Challenge calls for donating unneeded electronic equipment to schools and nonprofit organizations, recycling obsolete or nonworking equipment (much of which contains valuable or toxic metals that can be reclaimed rather than allowed to leech into landfills), and buying remanufactured equipment when possible. WasteWise is also available to advise in sustainable building design.

The Green Power Partnership (GPP) program encourages partners to purchase a portion of their electricity from green power products, which is not only better for the environment, but increases profitability for green power producers who may not yet be able to price their products competitively. Participation in the GPP program requires that from 2 to 10 percent of the firm's total energy consumption every year be from green sources: the percentage is based on the size of the firm and its energy consumption, with especially high energy consumers using the 2 percent figure, and small companies using the 10 percent figure. There is a separate algorithm used for law firms that lease their office space rather than own it. The EPA maintains a list of available green power producers and helps law firms and other companies to find them in their area.

See Also: Animal Ethics; Business Ethics, Shades of Green; Carbon Offsets; Climate Ethics; Ecopolitics; Environmental Justice; Environmental Law; Environmental Law and Policy Center; Environmental Values and Law; Green Laws and Incentives; Green Law Trends; Sustainability, Business Ethics and; Sustainability and Distributive Justice.

Further Readings

Beder, Sharon. *Environmental Principles and Policies: An Interdisciplinary Approach.* London: Earthscan, 2006.

Meritas Leadership Institute. *Green Guide for Lawyers.* Minneapolis, MN: Meritas, 2008.

Yeager, Peter Cleary. *The Limits of Law: The Public Regulation of Private Pollution.* New York: Cambridge University Press, 1993.

Bill Kte'pi
Independent Scholar

Green Laws and Incentives

In recent decades, a number of environmental policy makers have been moving away from "command-and-control" paradigms of regulation in favor of policies that operate by changing the incentives that people face when making decisions. In the traditional view, the job of the regulator was to tell people what to do and when in order to bring about desired environmental goals. But today, many regulators strive to avoid such rigid demands, instead seeking to build institutional structures in which individuals are encouraged to make better choices and to share the burden of environmental protection in socially beneficial ways. In this article, we will first review the development of ideas in the field of economics that define the role of incentives in green policy and then explore how some of those ideas have been put into practice. Finally, we will briefly consider some potential objections to incentive-based approaches to environmental policy making.

Theories of Market Failure

In *The Wealth of Nations,* Adam Smith famously observed that even though people typically pursue their own self-interest, they nevertheless help to bring about positive social outcomes as if by an "invisible hand." This happens because in a market economy, individuals can best promote their own interests by producing for other people. When we sell our products to people who value them, we make them better off, and we are also made better off by the profits on the sale. By placing our self-interest in the service of others, the market system encourages us to act in ways that benefit both ourselves and society at large.

In *The Economics of Welfare,* however, Arthur Pigou pointed out that sometimes the benefits of an action to an individual and the benefits of that action to society can diverge. If individuals do not receive all of the costs and benefits from their actions, then they may choose things that are good for them but not ideal from a social standpoint. Or they may avoid doing things that are personally costly but highly beneficial to society. In these situations, the market fails to effectively coordinate the interests of individuals to the benefit of society. The operators of a polluting factory, for example, might profit from the fact that others have to bear the burdens of their activities while they keep all of the proceeds from their products. Or a landowner might neglect to create a refuge for an endangered species on her property because she would have to pay the full cost but would not realize much personal benefit from doing so. Pigou suggested that sometimes the best way to deal with these sorts of scenarios would be for the government to step in and create legislation to make people do the right thing.

Since Pigou's era, thinkers like Ronald Coase and Garrett Hardin have provided the foundation for alternative forms of intervention besides direct regulation. Coase pointed out that sometimes socially inefficient outcomes arise from the fact that people have trouble negotiating mutually beneficial solutions. And Hardin observed that undesirable consequences can result when resources are left for common use and people are unable to preserve them for personal gain—in situations like these, people may use the resources irresponsibly, to the detriment of society as a whole. These insights emphasized the notion that instead of using legislation to tackle problems directly, regulators can help bring about better outcomes by empowering people to negotiate with each other and by creating more functional institutional structures to guide people toward better choices.

From Theory to Practice

Most environmental legislation to date has been administered through centralized commands. Typically, environmentally harmful activities are restricted, forcing individuals to obtain permits for the right to engage in those activities at acceptable levels. Such policies have been criticized for being rigid and vulnerable to the shortcomings of the regulators charged with implementing them. It is difficult for regulators to decide how much of the activity to allow and how permits should be distributed, and the fairest distribution may not be the most cost efficient. Regulators may also be subject to political pressure that could taint the process. But when successful, these policies can address incentive problems by prohibiting the harmful activities in which individuals might otherwise engage.

Policies based more explicitly on adjusting incentives can help environmental policy makers to avoid some of the difficulties associated with centralized, command-based regulations. There are a number of different ways to craft these policies, each with its own strengths and weaknesses. We will focus on three types of examples: privatization and community ownership schemes, taxation and subsidies, and the use of marketable permits.

Privatization takes resources that were formerly owned by the government (or not owned at all) and grants exclusive title over them to private individuals. Community ownership schemes involve the transfer of titles to community groups. Such policies encourage the new owners to be responsible stewards because destroying the resources will make them less valuable in the long run. Ownership-based policies face important challenges in deciding how to distribute resources and how to allocate decision-making authority to the new owners. If owners are not likely to be effective in safeguarding resources (or if they are not inclined to do so), then ownership-based policies may not be appropriate for achieving environmental goals. But in some situations, they can help to overcome adverse incentive problems by allowing owners to capture more of the benefits of socially responsible behavior. In parts of South America, ownership of forests is being transferred to communities with apparently positive results, and in a number of African nations the privatization of wildlife refuges has been instrumental in encouraging conservation.

Regulators can also discourage environmentally harmful actions by imposing taxes on them; similarly, they can promote beneficial actions by subsidizing them. These policies work by bringing the costs and benefits faced by individual decision makers into line with the corresponding costs and benefits for society. Limits on regulators' knowledge can compromise these policies, especially when it is difficult to know how large an effect will be produced by a particular rate of tax or subsidy. And while subsidies are typically not met with too much political resistance, taxes are generally politically unpopular. Accordingly, examples of environmentally motivated subsidies are ubiquitous, while examples of similar taxes are harder to come by. Nevertheless, several European countries have introduced taxes to discourage the use of polluting fuels and waste-generating products.

An alternative scheme, pioneered by John Dales, involves the issuing of permits for environmentally harmful activities as in traditional command-and-control policies, but establishes a mechanism for parties to trade those permits with each other. Marketable permits still force regulators to find fair ways to allocate permits initially, and it is still necessary to decide just how much of the permitted activity will be allowed. But permit-trading schemes have the advantage of enabling parties who would have difficulty reducing their impacts to effectively pay others to take on the reductions in their stead. This means that impact-reduction goals can be achieved as cheaply as possible, and it also means that individuals gain a positive incentive to find inexpensive ways to reduce pollution—if they succeed, then they can sell their excess permits for a profit (or avoid having to buy permits). The European Union has adopted a carbon dioxide emissions trading scheme as part of its climate policy, and the United States has a similar program in place to combat acid rain.

Criticisms

Incentive-based policies have been criticized by environmental advocates for allowing people to pay for the right to engage in morally objectionable actions. For example, if it is morally wrong to emit dangerous pollution, then providing an incentive to stop polluting will be inappropriate: we should prohibit the pollution. Defenders of the policies argue, however, that stopping all pollution would be infeasible—the goal of environmental policy should be to find ways to keep pollution at tolerable levels. A related criticism, which can be traced to Aldo Leopold, is that the constraints and incentives that can be produced by public policies are no substitute for the self-control that only comes with real moral change.

A different kind of criticism can be traced back to ideas articulated by Friedrich Hayek and James Buchanan. On this view, the extreme complexity of the market system and the wide range of different values held by citizens make it impossible for regulators to adjust incentives to reflect what is truly best for society. Defenders of incentive-based policies argue that even if they will never be perfect, regulatory regimes can nevertheless be better than the systematically imperfect market.

See Also: Cost-Benefit Analysis; Ecological Footprint; Economism; Environmental Policy; Utilitarianism.

Further Readings

Dales, John Harkness. *Pollution, Property & Prices: An Essay in Policy-Making and Economics*. Cheltenham, UK: Edward Elgar, 2002.

Pennington, Mark. "Liberty, Markets, and Environmental Values: A Hayekian Defense of Free Market Environmentalism." *The Independent Review*, 10/1 (2005).

Rose, Carol. "Liberty, Property, Environmentalism." *Social Philosophy & Policy*, 26/2 (2009).

Schmidtz, David. "When Preservation Doesn't Preserve." *Environmental Values*, 6/3 (1997).

Dan C. Shahar
University of Arizona

Green Law Trends

Green laws inherently reflect an understanding that good community design will allow nature to perform a number of environmental services, many of them essential to sustaining and fulfilling human life. Often referred to as ecosystem services, they include the provision of clean water, maintenance of livable climates, pollination of vegetation, and maintenance of biodiversity, among many others. Green laws now increasingly require that local planning decisions do not lead to the obliteration of such services. A unique feature of most of the services (as opposed to goods) generated by nature is that although widely acknowledged, they are unaccounted and unpriced and, therefore, have largely remained outside the domain of the market.

A traditional response has been to turn to governments—national and local—for continued supply of ecosystem services through regulation (such as green laws), cost sharing, and other mechanisms. However, the use of market-based instruments to encourage preservation and restoration of ecosystems has recently achieved some attention. The implicit understanding is that communities would derive net benefit from increased production of ecosystem services and that market mechanisms would be the most appropriate means of achieving this. However, key challenges remain regarding the methodologies by which services are valued, the availability of sufficient scientific information, and the development of appropriate mechanisms and policies for market creation.

Urban and local environmental laws promoting public well-being and safety by protecting and enhancing nature in cities and municipalities are also known as green laws. Although

AmeriCorps members planting a tree in an urban environment. Modern tree management ordinances have increased their scope to include tree protection plans for developments, standards for land clearing, parking lot shading, and other new approaches.

Source: U.S. Department of Agriculture, Natural Resources Conservation Service

much legislation concerning the environment has been generated at federal and state levels, local green law has few national or regional precedents and each municipality must create its own. Consequently, each community has its own limitations, opportunities, and policies shaping local regulation.

Commonly, green laws intend to establish urban tree management programs, establish new plantings and landscaping following construction, and preserve existing natural amenities including historic trees, forest lands, wetlands, and unique habitats. They also include site-specific laws that address storm water and all other environmental concerns in built or urbanized settings. While early green laws usually addressed aesthetic concerns, more recently such codes have become more comprehensive and explicitly address the environmental benefits to humans of having nature and ecosystems in cities and other densely populated areas, becoming responsive also to the latest scientific findings.

The Pennsylvania Shade Tree Law of 1700 was, perhaps, the first green law in America, although green laws became more widespread only in the 1930s and 1940s, following the threat of Dutch elm disease blight. Finding themselves overwhelmed by dead and dying trees, cities rushed to put into effect municipal tree programs to deal with the problem. Contemporary green laws date to the late 1960s and 1970s; however, the term *green law* itself emerged only in the late 1980s, when the importance of nature to urban living increasingly found formal recognition.

Tree Management

Trees are the central concern of public tree management ordinances. Until recently, tree management referred simply to publicly owned trees. However, some of the more progressive green laws today also extend municipal control to trees on private land.

Early tree ordinances were primarily designed to establish city forestry programs and to regulate the practice of arboriculture (the cultivation and management of trees within landscapes). Typically, they provided definitions, outlined licensing and insurance requirements, and included basic standards for planting, maintenance, removal, and protection of trees. Contemporary tree management ordinances do more than organize municipal forestry programs and the licensing of arborists. They explore new approaches to urban forestry, require tree protection plans to be submitted with development applications, recommend specific species for planting, and include standards for land clearing, revegetation, parking lot shading, density, and replacement values. The advent of climate change has led to an understanding of urban trees also as contributors to greenhouse gas reduction and as temperature regulators, although climate-related events such as wildfires are beginning to alter public opinion about the utility of trees in densely populated neighborhoods in some areas, raising awareness of hazards potentially inherent in urban trees, which could alter the content of ordinances in some contexts.

Post-Construction Landscaping

Regulations that require landscaping of sites following construction are rapidly becoming the most widespread form of green laws. Commonly called "landscape codes" or "landscape ordinances," they require the revegetation of developed land following construction, reflecting a desire to improve the appearance and quality of both public and private areas. Usually, such regulations set minimal standards for design as well as technical components. Increasingly remarkable for the detail of regulation put in place, they can require landscaping to street edges, enhancement of parking lots (i.e., by using variable parking geometry, pervious parking, and biofiltration), screening of unsightly views, screening of trash areas, street yard planting and street wall planting, and irrigation plans. Concerns about changing climate patterns have led to the revision of landscape ordinances in some municipalities, calling for greater water efficiency in landscape design and drought-tolerant planting, aiming at a low water-loss factor.

Land Alteration

Land alteration is any activity that removes vegetation from, or changes the topography of, the land by tree removal, clearing, filling, excavating, or any other activity that changes the essential character of the natural landscape. Green laws covering this type of activity call for minimal disturbance to the natural environment. Although often including the protection and replacement of trees, land alteration ordinances have also come to include issues such as the protection of soil and water resources (i.e., stormwater management), of surface water flow, of environmentally sensitive areas, of uplands providing significant and important wildlife habitats, and of naturally occurring vegetation not covered under tree management ordinances. With the significance of land use changes to greenhouse gas emissions becoming more recognized, and understandings that land alteration may hold more direct implications for regional climate phenomena, green laws in this category may, gradually, prioritize climate management as impacted by land surface drivers.

Conclusion

Ultimately, whether municipal regulations (green laws), market mechanisms, or a combination of both best protect urban ecosystems remains to be seen. A sustained criticism of both approaches remains their essential anthropocentric (human-focused) character, which supports the protection of the natural environment for human benefit only and not for its intrinsic worth.

See Also: Conservation; Environmental Law; Green Laws and Incentives; Post-Construction Landscape Laws; Urban Tree Management Ordinances.

Further Readings

Abbey, Buck. "Green Laws, Landscape Codes for the Twenty-First Century." www.greenlaws.lsu.edu/greenlaws.htm (Accessed January 2010).

Abbey, Buck. *U.S. Landscape Ordinances: An Annotated Reference Handbook*. New York: John Wiley & Sons, 1998.

Kumar, Pushpam. "Market for Ecosystem Services." Winnipeg, Canada: International Institute for Sustainable Development, 2005.

Whitten, Stuart, et al. "Markets for Ecosystem Services: Applying the Concepts." Canberra: Commonwealth Scientific and Industrial Research Organisation, 2003.

Wolf, Kathleen L. "Trees, Parking and Green Laws: Strategies for Sustainability." http://faculty.washington.edu/kwolf/Australia/Wolf_ParkiNg/Wolf_TreesParkingprt.pdf (Accessed January 2010).

Fanny Thornton
Australian National University

Green Liberalism

Green liberalism is the marriage between ecologically friendly principles and the political philosophy of liberalism. This is hardly a harmonious union, as the core values of what it means to be "green" or "liberal" often contradict one another. Until recently, green thinkers have pointed to the liberal political tradition as the major culprit in our ecological problems. However, there is a growing opinion, especially within environmentalism, that liberalism can be compatible with green principles.

Liberalism as a political philosophy assumes that humans are born free and equal. It emphasizes the rights and well-being of individuals, making human agency the organizing virtue of any liberal society. Liberalism commitment to neutrality ensures that each individual is free to pursue whatever she or he defines as the "good life." Liberalism (or classical liberalism) is also characterized by extending individualism to economic liberty. By emphasizing that economic exchanges between individuals ought to be private transactions, liberalism's laissez-faire attitude asks the government to stay small and to stay out of private interactions. As a result, liberalism also possesses a commitment to neutrality. No one, or no one government, should decide what the "good life" is for any citizen.

Liberalism's emphasis on individualism and human well-being leads to the attitude that humans are apart from nature. This entrenched anthropocentric bias is difficult to reconcile. As a political philosophy, liberalism is concerned with the interests of humans, and nature only makes an appearance in liberal theories as a resource to be used (and even abused) by humans for their gain.

In general, green liberalism tames the fervent individualism and anthropocentrism of the liberal philosophy. Nature, therefore, ought to be an entity that receives heightened consideration. While the degree varies, there is an implicit recognition that human well-being is dependent upon a sound natural environment. Although the anthropocentric angle is still present, green liberalism relaxes it somewhat by adding an ecological component.

There are numerous philosophical and pragmatic approaches to "being green." Generally, however, to be green means a rejection of the belief that nature has only instrumental value of benefit to humans. How, precisely, green liberalism incorporates this rejection and cornerstone belief of green theory is an issue of continued debate.

Two distinct strands of green liberalism deserve attention. The first focuses on the economic, free-market aspect found in classical liberalism. Rather than pointing to liberalism as the cause of environmental problems, green liberals argue that the market can provide the cure. The market allows self-interested individuals to make a profit by providing services and/or products that are directed toward helping the environment. Ecologically friendly values and behavior are best cultivated by incentivizing said attitudes and actions. By privatizing ownership of a forest, for example, the owner is directly responsible for protecting its value. Critics of this approach argue that value is confused with money. The profitability of harvesting the trees for the timber or land development may outweigh the value of protecting the forest for society at large and for future generations. Economic freedom of the forest's owner trumps whatever value (intrinsic or otherwise) it may provide.

The second strand of green liberalism suggests that liberalism can be greened if we think of liberalism as a broad and varied tradition, not one that solely envelops the free market. That is, the role of the state should not be to remain neutral in the pursuit of the good life. In fact, those who support this approach suggest that it is misleading to suggest that liberalism ever supported a metaneutrality principle. Any liberal theory presumes an approach to what is good, or what the good should be. If the environment is conceived as a good to be protected—even preserved—then the government ought not to remain complacently neutral. It is entirely within the realm of liberalism's view of neutrality for the government to limit (not prescribe) the range of lifestyles deemed incompatible with ecological well-being. Green liberals certainly favor more government involvement than their classical liberal predecessors did. They see the benefit of government intervention in crafting environmental policies such as caps on carbon emissions.

This strand of green liberalism also recognizes the strength of liberal democracy. It sees an engaged citizenry as the bridge between mounting ecological problems and an atomistic population. This strain stresses citizen responsibilities to the polity, which include future generations as well as care for the environment, rather than rights, which liberalism sees as a private matter. Ecological citizenship can craft both private and public virtues that are important to improving ecological conditions. Values can be reinterpreted to something that members of a community share as citizens rather than preferences they hold as private consumers.

See Also: Anthropocentrism; Democracy; Economism; Ecopolitics; Instrumental Value.

Further Readings

Dobson, Andrew. *Citizenship and the Environment.* Oxford, UK: Oxford University Press, 2003.

Hailwood, Simon. *How to Be a Green Liberal: Nature, Value and Liberal Philosophy.* Montreal, Canada: McGill-Queen's University Press, 2004.

Wissenburg, Marcel. *Green Liberalism: The Free and the Green Society.* London: University College London Press, 1998.

Wissenburg, Marcel. "Liberalism." In *Political Theory and the Ecological Challenge,* Andrew Dobson and Robyn Eckersley, eds. Cambridge, UK: Cambridge University Press, 2006.

Austin Elizabeth Scott
University of Florida

Green Party, German

Renate Künast addressing a 2005 German Green Party convention in Oldenburg, Germany. A coalition of the German Greens and the Social Democrats was able to introduce a number of popular environmental reforms, including an ecotax, agricultural innovations, and ambitious renewable energy legislation by the early 2000s.

Source: Wikipedia/Till Westermayer

In liberal democratic settings it is common that political parties are related to social movements. If articulate agents represent them, movements will likely attempt to affect the exercise of power by the state by working through existing political parties or forming new ones. The distinctive feature of "green" parties is their postindustrial and global orientation in the progressive tradition, the corresponding social profile of most activists, and an electorate encompassing middle-class professionals. Measured by election results and policy impact, the German Greens are an instance of a particularly successful green party.

It did not start out that way: "green" and "brown" have been intimately connected in German history. The essential background of Germany's Greens is the trauma of fascism, war, and defeat, themselves a result of the country's ultimately tragic experience with nation building. The Nazi (National Socialist) movement inherited and developed elements of 19th-century Romantic thought, social Darwinism with a pronounced organicist conception of society, and historicism emphasizing the uniqueness of German identity. It is indicative of the radically different political choices available to the emergent middle class across the industrializing world that America's dominant social movement at the time (1890–1920) was Progressivism, a middle-class movement with radical strains, which, in contrast to the situation in Europe, moderated polarizing class conflict.

Thanks to the Faustian bargain fascism struck with industry and significant elements of the middle class, the political landscape in postwar Germany was dominated by Social Democrats and Christian Democrats. Liberal ideas migrated into and were selectively absorbed by these two parties, which by the mid-1960s, accounted for more than 90 percent of the electorate. Meanwhile, the cold war setting disallowed an indigenous Communist party in the West German state. While political fascism was banned, legacies of that movement persisted in German society and opposition to it eventually bequeathed West Germany with a special form of youth rebellion in the 1960s.

Birth of the German Greens

Remarkable economic success in the paradoxically protective geopolitical structures of the cold war ushered in the gestative period of the German Greens in the 1960s. Demonstrations against German rearmament in the 1950s augured the emergence of extraparliamentary issue politics that ballooned beyond the array of left-sectarian grouplets that were placeholders in the political space that elsewhere in Europe was occupied by left-socialist and/or communist parties. Some West Germans born into the middle class after the war, coming into political consciousness in the 1960s, were distressed by the integrative strategies of the dominant parties. Out of this came an extraparliamentary opposition called Außerparlamentarische Opposition (APO), a loose collection of groups that rejected compromises with the fascist past and desired as complete a break with its cultural and social legacies as possible.

Then, too, the *Lagerpolitik* of the Weimar era was breaking down under the transformation of settlement patterns and the increasing significance of the service sector. An increasing portion of younger-age cohorts entered institutions of higher education. A new middle class emerged, a critical element of which was oriented to a thorough reform of the Bonn republic. Instead of directing their engagement through the established parties, they organized locally around the "failures of success" of Germany's vaunted export machine, focusing on new issues—pollution, megaprojects, solidarity with the third world, authoritarian structures in all spheres of life, and better distribution of the social product. By the end of the 1960s, more Germans had affiliated with *Bürgerinitiativen* (citizens' initiatives) than were members of political parties, a clear signal that Germany had been transformed by the so-called successor generation.

The German Greens developed in three overlapping stages—from a protest party to a party with a distinct programmatic orientation, and then to a party of government.

The Social Democrats under the popular Willy Brandt reincorporated much of the movement sector political energy by addressing the concerns of a nontraditional (i.e., non–working class) clientele—expanding the welfare state and engaging in the détente policy, his governing coalition's *Ostpolitik* that attempted to move beyond the cold war. These elements felt let down by Brandt's successor, Helmut Schmidt, whose government mandated a radical increase in nuclear power plant construction after the October 1973 Yom Kippur War to diversify energy supply in light of the initial successes of the Organization of the Petroleum Exporting Countries (OPEC) and in 1977 seemed to move decisively against Brandt's détente policies by advocating deployment of a new generation of intermediate-range nuclear missiles in Europe (including Germany) to counter the perceived threat of Soviet SS-20 rockets. Against this background, Green parties cropped up around Germany, progressing from the local to the regional level.

Symptomatically, it was the first direct election to the European Parliament in 1979 that provided the first nationwide electoral test of political potential for the Greens. SPV-Die Grünen was launched, with the German-American Petra Kelly, a former Brandt enthusiast now working for the European Communities in Brussels, heading an umbrella organization of environmental nongovernmental organizations (NGOs) heading the list. The Greens garnered 3.5 percent of the vote, not enough to win any seats but enough to convince the organizers that the potential existed for a national presence. This eventuated, mostly at the expense of the Social Democrats, in the 1983 Bundestag elections, held in the wake of U.S. Pershing and cruise missile deployments, with just above the minimally required 5 percent of the vote.

Die Grünen attracted support and activists from across the political spectrum. It took more than a decade to resolve their political identity. In a first internal clarification, nationalist elements (some with Nazi pasts) were made unwelcome. The liberal vocation of the party won out over socialist elements. Similarly, the so-called Realo wing of the party emphasizing tenacious parliamentary reformism vanquished the Fundamentalo wing, whose adherents were more willing to eschew participation in government for the sake of broadcasting a clearer message of change.

Germany's Political System

All political parties are shaped by the type of political system, the prevailing population of collective (political) identities, and the rules governing elections. West Germany has a competitive multiparty system in which the distribution of seats occurs on the basis of proportional representation from multimember districts. A highly engaged electorate, with actual turnout in national elections averaging over 78 percent of eligible voters, translates into great leverage for small parties, since (with a single exception) all postwar German national governments have been based on parliamentary coalitions of at least two parties.

German unification sorely tested the party. Its antinational, dual-Germany stance contributed to wiping out its Bundestag presence in December 1990, with the exception of a few members from the newly acquired eastern German Länder. With the introduction of the eastern-based Party of Democratic Socialism (PDS) into the national mix, the national political arithmetic of the evolving Berlin republic grew still more complicated, reflecting a fundamentally different national experience and social profile than the West German Bonn republic that incubated Die Grünen.

By 1993, the Greens had merged with the eastern Bündnis-90 (Alliance 90), the same year the party convention in Aachen adopted an explicitly European Union–friendly orientation. In 1995, prodded by parliamentary leader and Realo Joschka Fischer, the party crossed the Rubicon and admitted that the use of force was sometimes the lesser evil in international relations—for example, the support of armed intervention in the Balkans to prevent genocide. The path was thereby cleared for a coalition with the Social Democrats.

With a thin parliamentary majority, the Greens and the Social Democrats formed a governing coalition following the September 1998 Bundestag elections, and were able to renew that combination in 2002 on the basis of popular reforms—an ecotax, a new citizenship law, ambitious renewable energy legislation, a new approach to agriculture—and expanded the Greens' electorate. In the meantime, the Greens proved successful at establishing and increasing their presence in the European Parliament, usually as the largest and leading element of the Greens Group. The synergy between the two levels of governance magnified their impact and arguably produced a much greener European policy profile than would otherwise have been the case. Certainly, the perception of Europe eclipsing the United States in green policy development over the course of the 1990s was at least partially attributable to their presence.

The single most vexing problem for the Greens since 2002 has been the waning appeal of the Social Democrats, whose electoral weakness led first to a "grand coalition" with the Christian Democrats under Chancellor Angela Merkel (former environment minister and midwife of the Kyoto Protocol to the UN Framework Convention on Climate Change) in 2005, and then in 2009, to their worst-ever national electoral showing. It remains to be seen whether or not the Greens have other options for governing, either joined with the Christian Democrats, or within a three-party format.

See Also: Ecopolitics; Green Party Ethical System, Four Pillars of the; Ten Key Values of the Greens; United Tasmania Group.

Further Readings

Baker, Kendall, Russell Dalton, and Kai Hildebrandt. *Germany Transformed: Political Culture and the New Politics*. Cambridge, MA: Harvard University Press, 1981.

Bramwell, Anna. *Blood and Soil: Walter Darré and Hitler's "Green Party."* London: Kensal, 1985.

Corbett, Richard, Francis Jacobs, and Michael Shackleton. *The European Parliament*, 4th rev. ed. London: Longman, 2000.

Kitschelt, Herbert. *The Logics of Party Formation. Ecological Politics in Belgium and West Germany.* Ithaca, NY: Cornell University Press, 1989.

Lipset, Seymour M. and Stein Rokkan, eds. *Party Systems and Voter Alignments: Cross-National Perspectives*. New York: Free Press, 1967.

McGerr, Michael. *A Fierce Discontent. The Rise and Fall of the Progressive Movement in America, 1870–1920*. New York: Oxford University Press, 2003.

Olsen, Jonathan. *Nature and Nationalism: Right-Wing Ecology and the Politics of Identity in Contemporary Germany.* New York: Palgrave MacMillan, 1999.

Reutter, Werner, ed. *Germany on the Road to "Normalcy": Policies and Politics of the Red-Green Federal Government (1998–2002)*. New York: Palgrave Macmillan, 2004.

Rifkin, Jeremy. *The European Dream. How Europe's Vision of the Future Is Quietly Eclipsing the American Dream*. New York: J. P. Tarcher, 2004.

Carl Lankowski
U.S. Department of State

Green Party Ethical System, Four Pillars of the

The four pillars of the Green Party, shared by almost all the world's green parties (approximately 90 worldwide), are Ecology, Social Justice, Nonviolence, and Grassroots Democracy. The principles endorsed go deeper than a mere superficial change in policy, suggesting a qualitative shift in ethical norms, cultural practices, and prevalent paradigms, and a structural change in politics/democracy/governance and economic organization.

Ecology

Ecology or ecological wisdom encapsulates the diverse teachings and philosophies oriented to reducing the negative impact of human civilization on the environment, the biosphere, and the planet, and to finding new ways to cohabitate harmoniously with the living and nonliving entities of the nonhuman world on the planet. While many aspects of this ecological wisdom are drawn from spiritual, indigenous, and non-Western forms of thinking, contemporary science is also an important source. The essence of this pillar is to focus attention on finding ways in which human beings can live sustainably, in harmony with the nonhuman world in a manner in which it is using but not abusing the nonhuman world. In short, living according to ecological wisdom and knowledge means living and interacting with the nonhuman world in a symbiotic and a nonparasitic manner.

Social Justice

Social justice—sometimes Social Equality and Economic Justice or Socioeconomic Justice—reflects the general rejection of discrimination based on distinctions between class, gender, ethnicity, sexual orientation, or culture. Green parties are almost universally egalitarian in their outlook, seeing that great disparities in wealth or influence are caused by the way in which societies are structured politically and economically, which produces inequality and perpetuates social injustice, both within countries and globally between countries.

For example, the common green party critique of conventional economic growth is advanced on the grounds that the idea of exponential economic growth is simply unsustainable in the sense of not being compatible with "one planet living" and living within the environmental and resource constraints of the planet. Another equally powerful and important reason why greens critique economic growth and seek to replace it with a steady state economy and a focus on quality of life and human flourishing is that economic growth under capitalism both requires and reproduces socioeconomic inequality within society. In short, if you are egalitarian and seek to create a more equal and less unjust society, then economic growth is not the way to do this.

Nonviolence

Nonviolence reflects the green movement's policy of rejecting violence as a means to overcoming its opponents or bringing about social change. It expresses the peace, antinuclear, and antiwar origins of green partiers in the 1960s and 1970s. Green philosophy draws heavily on Gandhian and the Quaker tradition, as well as the history and lessons learned from the nonviolent civil rights strategy of Martin Luther King. All of these advocate measures by which violence can be avoided, while not cooperating with those who commit violence. However, green party support for nonviolence should not be confused with pacifism since many greens, following Gandhi, see a vital role for nonviolent direct action (NVDA). The green commitment to nonviolence can also be interpreted in terms of the general commitment by greens to "live lightly" on the planet and also extended in moral terms beyond the human species. For example, since it is a general, nonnegotiable aspect of life on this planet that "life predates on life," living lightly and nonviolently for many greens moves us in the direction of a vegetarian or vegan diet, given the well-documented pain, death, suffering, and violence inflicted on animals reared and killed for the purposes of humans' consuming their flesh.

Grassroots Democracy

For greens, all decisions should be made at the lowest possible level, committing green parties to policies and strategies of decentralization and localization. However, depending on the issue, this can mean that the most appropriate and lowest level is the nation-state or regional bodies—think of climate change, for example. Grassroots or participatory democracy is espoused by greens as the only reliable governance model for achieving radical social change. Green parties favor decision-making procedures whereby local communities are empowered to make decisions that affect their welfare. Hence, green parties would generally support proposals ranging from greater decision making and policy making at local municipality and local authority levels to innovative experiments such as the "participatory budgeting" in Porto Alegre in Brazil where local citizens decide how the municipal budget is spent, and the burgeoning "transition towns" grassroots movement of communities creatively responding to peak oil and climate change.

Many green parties have rejected or constrained the traditional role of leaders and many green party constitutions are configured to prevent the party bureaucracy from accumulating too much power in the organization in order to promote more decentralized and member-driven processes. Green parties internally are the most democratically organized and constituted political parties, standing in stark contrast to the rigid, top-down hierarchical organizations favored by most other political parties, whether they are within representative democracies or not.

See Also: Animal Ethics; Democracy; Global Greens Charter; Green Party, German; "Should Trees Have Standing?"; Ten Key Values of the Greens; United Tasmania Group.

Further Readings

Barry, J. "Choose Life, Not Economic Growth: Critical Social Theory for People, Planet and Flourishing in the 'Age of Nature.'" *Current Perspectives in Social Theory,* 26 (2009).

Barry, J. *Rethinking Green Politics: Nature, Virtue and Progress.* London: Sage, 1999.

Barry, J. and S. Quilley. "The Transition to Sustainability: Transition Towns and Sustainable Communities." In *The Transition to Sustainable Living and Practice,* L. Leonard and J. Barry, eds. London: Emerald, 2009.

Jackson, T. *Progress Without Prosperity.* London: Earthscan, 2009.

Wilkinson, R. and K. Pickett. *The Spirit Level: Why More Equal Societies Almost Always Do Better.* London: Allen Lane, 2009.

John Barry
Queen's University Belfast

Greenwashing

The term *greenwashing* (a portmanteau of "green" and "whitewash") was coined in the 1980s by Jay Westerveld. Westerveld was an endangered species biologist and environmental activist who disapproved of hotel chains that framed the reuse of towels and sheets as an environmentally influenced choice, when really it was an economic decision. Greenwashing was originally used to describe the practice of overselling a product's "green" characteristics or misleading consumers through green advertising in order to capitalize on the environmentalist trend. However, as the green movement gained momentum in our culture and more corporations tried to frame themselves as ecofriendly, the range of greenwashing transgressions has widened. Today, charges of greenwashing can be issued for a broad range of unethical behaviors, such as deceiving marketing practices, untruthful environmental reporting, and fraudulent environmental activism. "Green sheen" and "green screen" are two other neologisms that have been used to describe the same phenomenon.

In the wake of environmental disasters like the Chernobyl accident and the *Exxon Valdez* oil spill, the green movement gained credibility with the American people. Ethical

While the FIJI Water Company planned to become a "carbon negative" company in 2010 by purchasing carbon offset credits, its product, shown here in New York City, still requires the use of more natural resources than any other bottled water, as it must travel more than 5,000 miles to reach the closest U.S. city.

Source: Wikipedia/Verne Equinox

consumerism acquired ground, and citizens wanted to buy "green" products and support companies with forward-thinking environmental policies. Transnational corporations responded with the dawn of corporate environmentalism. However, not all appeals to consumers' environmental sympathies were truthful or reliable.

In December 2007, an environmental marketing company called Terrachoice published a document titled "The Six Sins of Greenwashing," wherein they outlined the most common deceitful marketing practices in North American "green" advertising. First is the "Sin of the Hidden Trade-Off," committed when marketers suggest a product is green because of one attribute, while ignoring other less-green qualities, for example energy-efficient appliances that contain materials that are hazardous to human health and unrecyclable. Second is the "Sin of No Proof," where nothing backs the claims made in advertising, such as beauty products professing to be natural or organic without any certification. Third, we have the "Sin of Vagueness," when companies use ambiguous terms such as *natural* or *green* without qualifying them. Fourth, the "Sin of Irrelevance," frames products as green using immaterial facts, like hairsprays pronouncing they are CFC-free when chlorofluorocarbons (CFCs) were banned more than 20 years ago. Fifth is the "Sin of Fibbing," where inaccurate claims are made, such as builders insisting all the appliances in a home are Energy Star or that all power for the home comes from a sustainable source when this is not the case. Finally, the sixth sin is the "Sin of Lesser of Two Evils," or assertions that are true within a product category but distract the consumer from the greater environmental impact of the category as a whole, such as the marketing of "natural" cigarettes with no additives as the environmentally responsible choice, or advertising that a particular brand of sports utility vehicle (SUV) is the most fuel efficient of its type.

Industry Examples of Greenwashing

One common and very public example of greenwashing is advertising in the oil industry. BP's most recent series of commercials rebrand the company as "Beyond Petroleum," rather than "British Petroleum." The advertisements highlighted BP's investment in alternative energy sources, and predominantly featured different shades of green for its color palette. However, as many critics pointed out, BP spent millions more on the advertising campaign than on any actual alternative energy investment. In the wake of the 2010 oil leak in the Gulf coast, BP's claim to be a green company appears even more ludicrous than it did previously. Similarly, ExxonMobil released a series of commercials touting its newfound commitment to alternative energy while still profiting from fossil fuels and after donating millions of dollars to the climate change denial industry for decades. Shell has also faced critique for sponsoring a wildlife photography exhibit while profiting from nonrenewable resources.

While the oil industry is perhaps the most obvious and visible example of greenwashing, there are countless others. For instance, FIJI Water, a product that has to travel more than 5,000 miles to reach even the closest U.S. city, requires the use of more natural resources than any other bottled water (an industry already widely criticized for being unsustainable). To combat this negative press, FIJI Water Company launched a Website in 2007 using the slogan, "every drop is green." However, the Website is extremely vague about how or why FIJI's water is "green." The company promises to become a "carbon negative" company by 2010 by purchasing carbon offset credits for the transportation of their water by sea, and they advocate transparency about the carbon footprint of their product. However, as of 2010, there was still no information posted about the company's carbon footprint or "sustainable" practices. FIJI Water Company also claims to support recycling "legislation," but this does not change the fact that the recycling rate for plastic bottles in the United States is still extremely low (less than 15 percent).

Other consumer products are guilty of similar greenwashing strategies. "Natural" disposable diapers, for example, are perhaps less toxic for children than the traditional variety, but they still require petroleum for manufacturing and transportation, and they take more than a lifetime to decompose in a landfill.

So why is greenwashing a problem? Obviously, it is ethically questionable to deceive consumers when marketing a product or service. Businesses should be held accountable to an informed citizenry. However, there is more to it than that. Consumers who try to make ethical purchases only to discover that they have been duped may become cynical and feel their efforts were futile. It may cause people to stop aiming for environmental responsibility and settle for complacency if they feel they cannot make a difference. Furthermore, companies that are truly progressive may be considered together with others that are trying to capitalize on the green trend and may not thrive as they should. In fact, many businesses are finding that becoming more environmentally responsible actually does lower costs and increase profits, while businesses accused of greenwashing often see a fall in profits, so the gamble is not worth the risk.

See Also: Business Ethics, Shades of Green; Consumption, Consumer Ethics and.

Further Readings

Enviromedia Social Marketing. "Greenwashing Index" (2009). http://www.greenwashing index.com (Accessed January 2010).

Federal Trade Commission. *Guides for the Use of Environmental Marketing Claims.* http://ecfr.gpoaccess.gov/cgi/t/text/text-idx?c=ecfr&sid=b2333ddf96abf25788ef3037ffcfb40a&tpl=/ecfrbrowse/Title16/16cfr260_main_02.tpl (Accessed January 2010).

Greer, Jed and K. Bruno. *Greenwash: The Reality Behind Corporate Environmentalism.* New York: Apex Press, 1997.

Karliner, Joshua. "A Brief History of Greenwash." *Corpwatch* (March 22, 2001). http://www.corpwatch.org/article.php?id=243 (Accessed January 2010).

Terrachoice Environmental Marketing, Inc. *The Seven Sins of Greenwashing.* http://sinsofgreenwashing.org/findings/greenwashing-report-2009 (Accessed May 2011).

Terrachoice Environmental Marketing, Inc. "The Six Sins of Greenwashing," (2007). http://www.terrachoice.com/files/6_sins.pdf (Accessed May 2011).

Katherine M. Cruger
University of Colorado, Boulder

HAECKEL, ERNST

Impavidi progrediamur—Courageously let us go forward. With this Latin expression, Ernst Haeckel ended his book *The Freedom of Science and the Freedom to Teach* (1879), in which he not only defended his belief in the evolutionary theory but also advocated that it should be taught in schools. It is an expression that might be a good description of his life.

Ernst Heinrich Philipp August Haeckel (1834–1919) was born in Potsdam (then Prussia), spent his first years in Merseburg, and lived his adult life in Jena. His love for natural sciences was determined by his readings of Schiller, Goethe, and Humboldt's and Darwin's travel literature. He studied medicine but never practiced it, instead devoting his life to teaching, researching, and writing. Haeckel had an enormous creative power and was a prolific writer. He produced not only many scientific publications (on zoology, morphology, embryology, evolution, and monism) but also some of the most successful popular works on science, selling millions of copies.

When Haeckel read *The Origin of Species* (1859), he felt he had found his main axiom in life. At that time he was researching *radiolaria*—single-celled marine organisms with silica skeletons. That is when he realized that his own research supported Charles Darwin's theory: the visual similarities of several radiolarian species constituted evidence of a common origin. From then on, Haeckel became Germany's leading public advocate of the evolutionary theory, and he also devoted most of his research to further develop it and prove it. When his work *Monograph on Radiolarians* (1862) was published, he sent it to Darwin, who answered very positively. This started a relationship that would last throughout Darwin's life. Haeckel visited Darwin in England in 1866 and they kept on corresponding, sharing, and praising each other's work.

They had opposing characters. While Darwin was gentle, Haeckel had a combative personality and "courageously went forward" on several controversial issues concerning evolution, religion, and monism. He even had to face charges of scientific fraud. Haeckel became famous for his research on the biogenetic law, postulating that ontogeny (the embryonic development of the individual) recapitulates phylogeny (the evolutionary history of the species). In the first edition of *Naturliche Schopfungsgeschichte* (1868), he included three drawings from early stages of the embryos of a dog, a chicken, and a turtle, challenging the readers to find the differences. Critics found that the three drawings were only one, consequently charging him with scientific fraud. He had indeed used the same woodcut and immediately realized he had gone too far in his wish to communicate these similarities. In the second and following editions, he revised the drawings, admitting his error. But the epitaph of fraud never quite vanished and was used throughout his life up to the present (mainly by creationists), undermining not only the fact of the similarities but also much of his overall research.

Haeckel believed that one could not study organisms independently of their context. He opposed most biologists of his day by stating in the *Generelle Morphologie der Organismen* (1866) that "one must compare these life-forms in their collective entirety, in their general and continuous mutual interrelationships." In the text, he also coined the term *ecology—Oecology*—defining it as "the science of the mutual relationships of organisms to one another." The term *ecology* was, however, ignored for a long period and became widely used only much later, even though Haeckel used the concept throughout his research. He developed the contextual and holistic view of nature, which is still the philosophy behind ecology. In that sense, he anticipated modern ecology seeing the universe as a unified and balanced organism. He was a great defender of monism, arguing that matter and spirit are attributes of a single substance.

Haeckel's great love was Anna Sethe (1835–64). They became engaged in 1858 and married in 1862 after Haeckel's nomination as professor at the University of Jena. But Ernst was widowed only 18 months later. According to R. J. Richards (2008), losing Anna marked Haeckel's life in all senses. He lost faith, became bitter, immune to critics and evermore strong-minded. He later married Agnes Huschke (1842–1915) and they had three children. He was, however, never happy. At 65 years of age, he met Frida von Uslar-Gleichen, 30 years younger, and they fell in love. It was an intense but impossible love affair, during which they exchanged over 900 letters in six years. It ended with Frida's death in 1903.

Throughout his life, Haeckel traveled extensively, making expeditions to several parts of the world. He discovered and researched thousands of invertebrate marine species that he described and illustrated in many publications. These exquisite illustrations (mostly of radiolarian and medusa) even became a source of inspiration to many art nouveau artists. Haeckel himself was an accomplished artist.

After his death in 1919, the Nazis misused some of his ideas about the differentiated evolution of human races. His tree depictions (otherwise famous for being an eloquent graphic way of communicating complex ideas) were efforts to understand evolution, but they damaged his righteous place in history. R. J. Richards (2008) restored the many accomplishments Haeckel achieved in his long and creative life, claiming he was a genius and a contributor to science and to evolutionary thought as very few in his time. *Impavidi progrediamur.*

See Also: Biocentric Egalitarianism; Ecology; Future Generations.

Further Readings

Bowler, Peter J. and Iwan R. Morus. *Making Modern Science: A Historical Survey.* Chicago, IL: University of Chicago Press, 2005.

Richards, R. J. *The Romantic Conception of Life: Science and Philosophy in the Age of Goethe.* Chicago, IL: University of Chicago Press, 2004.

Richards, R. J. *The Tragic Sense of Life: Ernst Haeckel and the Struggle Over Evolutionary Thought.* Chicago, IL: University of Chicago Press, 2008.

Sofia Azevedo Guedes Vaz
New University of Lisbon

Hannover Principles

The Hannover Principles are a self-described living document about building and object design, originally formulated in 1992 by William McDonough and Michael Braungart while planning Expo 2000 (held in Hannover, Germany).

About Expo 2000

Expo 2000 was a World's Fair, held from June 1 to October 31, 2000, which despite 10 years of planning was not financially successful. While previous World's Fairs had typically focused on the cutting edge of technology, Expo 2000 focused more on presenting possible solutions for the future. Only about 25 million of the expected 40 million visitors came to the Expo, which was never fully developed, in part because of the last-minute decision of the United States to participate after originally declining. The Expo's theme and focus was never fully articulated: corporate sponsorship was limited because of the high cost charged to sponsors, which enhanced the prominence of McDonald's sponsorship and their 10 highly visible on-site restaurants, which seemed at odds with the synthetic voice or vocoded theme song commissioned from the German industrial band Kraftwerk. Advertising for the Expo focused on its cerebral design elements and grandiose ambitions, leaving both sponsors and the public unsure about what exactly they were attending and what entertainment the World's Fair would provide. Themed pavilions were set aside for the themes of Knowledge, Mobility, Work, Energy, Nutrition, Environment, Basic Needs, the Future of Health, and the nation of Chile. National exhibits varied from the 360-degree circular movie theater of the United Arab Emirates, built of recycled materials and screening a history of the country, to the Buddhist temple transported from Bhutan to the Expo in 16,000 parts. Most pavilions were made in whole or in part of recycled or recyclable materials, in keeping with the vague forward-looking theme of the fair.

The Principles and the Authors

The Hannover Principles proved to be the most important legacy of Expo 2000. German chemist Braungart, a former activist and the former head of chemistry at Greenpeace, completed his Ph.D. at the University of Hannover in 1985 and was engaged in encouraging environmentally sound design and production in the chemicals industry. McDonough, a Tokyo-born American architect, designed the Environmental Defense Fund's headquarters in 1984, his first major commission, the challenges of which inculcated him with a deep interest in sustainable development. The McDonough Braungart Design Chemistry firm received the commission for Expo 2000 in 1992, and McDonough became the only individual recipient to date of the Presidential Award for Sustainable Development in 1996, while simultaneously working at his own practice, at McDonough Braungart, and as the dean of the School of Architecture at the University of Virginia.

View of the Hannover Principles

A 59-page document, the Hannover Principles that McDonough and Braungart articulated set out development guidelines for the Expo and, more broadly, for building and object design that is conscious of sustainability and environmental impact. About 20 pages are devoted to issues specific to Expo 2000; the rest present a discussion of sustainability and the philosophical roots of the nine Hannover Principles that the document sums up as follows:

1. Insist on rights of humanity and nature to coexist.
2. Recognize interdependence.
3. Respect relationships between spirit and matter.
4. Accept responsibility for the consequences of design.
5. Create safe objects of long-term value.
6. Eliminate the concept of waste.
7. Rely on natural energy flows.
8. Understand the limitations of design.
9. Seek constant improvement by the sharing of knowledge.

McDonough and Braungart were very conscious of the life their document would have outside and beyond the World's Fair. "It is hoped," the introduction states, "that the Hannover Principles will inspire an approach to design which may meet the needs and aspirations of the present without compromising the ability of the planet to sustain an equally supportive future."

McDonough and Braungart went on to coauthor *Cradle to Cradle,* a 2002 book that takes the "life-cycle development" that Braungart helped to articulate at the Environmental Protection Encouragement Agency in the 1990s and expands it into a manifesto calling for a new corrective industrial revolution, one that is geared toward sustainable development, a fundamental act of which is the simple act of considering a product's entire lifespan when approaching its design. Instead of emphasizing efficiency and "doing more with less," as many environmentalist texts of the previous decades had done, McDonough and Braungart call for a new approach altogether, one that emulates natural systems.

See Also: Adaptive Management; Carbon Offsets; Consumption, Business Ethics and; Development, Ethical Sustainability and; Strong and Weak Sustainability; Sustainability, Business Ethics and.

Further Readings

Braungart, Michael and William McDonough. *Cradle to Cradle: Remaking the Way We Make Things*. New York: North Point Press, 2002.

Braungart, Michael and William McDonough. *The Hannover Principles: Design for Sustainability.* Charlottesville, VA: William McDonough Architects, 1992. http://www.mcdonough.com/principles.pdf (Accessed August 2010).

Carter, Alan. *A Radical Green Political Theory.* London: Routledge, 1999.

Dobson, Andrew. *Green Political Thought*. London: T&F Books, 2009.
Radcliffe, James. *Green Politics: Dictatorship or Democracy?* New York: Palgrave Macmillan, 2002.
Torgerson, Douglas. *The Promise of Green Politics: Environmentalism and the Public Sphere*. Durham, NC: Duke University Press, 1999.

Bill Kte'pi
Independent Scholar

Hargrove, Eugene C.

Environmental philosophy emerged as a formal academic discipline in the 1970s. Eugene C. Hargrove was one of the most important founders of the field, having initiated its first journal, *Environmental Ethics,* in 1979. Professor Hargrove also founded the best-known academic program in environmental philosophy, located at the University of North Texas (UNT), beginning in 1990. The UNT Department of Philosophy and Religion Studies, chaired by Hargrove from 1990 to 2004, hosted the world's first master's program in environmental philosophy beginning in 1992 and more recently added a doctoral program, primarily through his efforts. Hargrove also founded and continues to serve as director of the Center for Environmental Philosophy.

Hargrove also authored key publications in environmental philosophy, most notably his 1989 book, *Foundations of Environmental Ethics,* which has been translated into Korean, Italian, and Chinese, and will soon be available in Spanish. The book highlights Hargrove's research interest in the history of ideas, setting the background of today's environmental movement and policy within a chronology of the thinking of early naturalists and the aesthetics of nature. One of the aesthetics examples is the role of 19th-century Hudson River School of landscape painters in generating public concern over preservation of western national parks such as Yellowstone. This provides one of the foundations of the largely upper- and middle-class environmental movement in the United States and to a lesser degree in Canada, Europe, and other English-speaking countries such as Australia and New Zealand.

Notable journal articles or book chapters authored by Hargrove (cited below) that illustrate his wide range of study include the role of environmental ethics in education, use of rules in decision making, ethical role of zoos in conservation, and defining weak anthropocentric intrinsic value. Hargrove found public resistance to insertion of environmental ethics into standard school curricula; thus he suggested it be packaged in the more palatable concept of environmental citizenship. Interest in his views on environmental education has largely come from outside the United States, in Canada, Brazil, Korea, China, and Chile. His study of rules in decision making, published in deep ecologist Arne Naess's journal *Inquiry,* uses reasoning in chess as a model. Hargrove suggests an approach to reasoning in environmental ethics between deontological ethics and virtue ethics, recognizing the importance of the manner in which humans actually reason. Within a lengthy debate in the field over intrinsic value in nature, Hargrove's thought lies between the nonanthropocentrism of J. Baird Callicott's "truncated" intrinsic value of nature and the pragmatic instrumentalism of Bryan Norton's "weak" anthropocentrism. Thus Hargrove defines "weak anthropocentric intrinsic value" as the valuing of living and nonliving entities and systems for their own sake from a human (or subjective) perspective.

Hargrove also helped broaden the field of environmental ethics as an editor of three books and publisher of four (cited below). His edited anthologies address animal rights, religion and environment, and the ethics of space exploration. Through his Center for Environmental Philosophy (Environmental Ethics Books), Hargrove has reprinted books on theology and ecology, environmental aesthetics, process philosophy applied to biological individuals and communities, and his own on historical foundations of environmental ethics.

The University of North Texas, where Hargrove moved in 1990, provided a unique setting for interdisciplinary collaboration with the environmental science program. Hargrove worked with environmental scientist Ken Dickson to build critical thinking into the environmental science curriculum through required courses in environmental philosophy. Other internal collaborators were fellow environmental philosophers Pete Gunter, a leader in establishment of the Big Thicket National Preserve in East Texas in the 1970s, and Max Oelschlager, author of major books on wilderness and religion and environment. As chair, Hargrove added notable environmental philosophers such as Baird Callicott. More recent additions have included Robert Frodeman, with interests in environmental policy, and international faculty members Irene Klaver (Netherlands) and Ricardo Rozzi (Chile). All 12 permanent faculty members have published in environmental philosophy.

Key external collaborators for Hargrove include noted environmental philosophers such as Holmes Rolston III, Bryan Norton, and Mark Sagoff. Professor Hargrove is remembered by many graduate students for generously hosting numerous social functions at his home, providing a forum for students to mingle with visiting environmental philosophers after their formal talks. He is currently writing a book on environmental ethics and the culture war with a focus on problems in environmental education created by that ongoing political conflict beginning with the presidency of Ronald Reagan. Hargrove will leave a legacy as a unique and important founder of the field of environmental philosophy.

See Also: *Environmental Ethics* Journal; Green Altruism; Green Liberalism.

Further Readings

Birch, Charles and John B. Cobb, Jr. *The Liberation of Life: From the Cell to the Community.* Denton, TX: Environmental Ethics Books, 1990.

Cobb, John B., Jr. *Is It Too Late? A Theology of Ecology,* rev. ed. Denton, TX: Environmental Ethics Books, 1995.

Hargrove, Eugene C., ed. *The Animal Rights/Environmental Ethics Debate: The Environmental Perspective.* Albany: State University of New York Press, 1992.

Hargrove, Eugene C., ed. *Beyond Spaceship Earth: Environmental Ethics and the Solar System.* San Francisco, CA: Sierra Club Books, 1986.

Hargrove, Eugene C. *Foundations of Environmental Ethics.* Englewood Cliffs, NJ: Prentice-Hall, 1989; reprint ed., Denton, TX: Environmental Ethics Books, 1996.

Hargrove, Eugene C., ed. *Religion and Environmental Crisis.* Athens: University of Georgia Press, 1986.

Hargrove, Eugene C. "The Role of Environmental Ethics in Environmental Education." In *Environmental Education: Progress Toward a Sustainable Future,* J. F. Disinger and J. Opie, eds. Troy, OH: North American Association for Environmental Education, 1985.

Hargrove, Eugene C. "The Role of Rules in Ethical Decision Making." *Inquiry,* 28/1 (1985).
Hargrove, Eugene C. "The Role of Zoos in the Twenty-First Century." In *Ethics on the Arc: Zoos, Animal Welfare, and Wildlife Conservation,* B. G. Norton, M. Hutchins, E. F. Stevens, and T. L. Maple, eds. Washington and London: Smithsonian Institution Press, 1995.
Hargrove, Eugene C. "Weak Anthropocentric Intrinsic Value." *The Monist,* 75/2 (1992).
Sepanmaa, Y. *The Beauty of Environment: A General Model for Environmental Aesthetics,* 2d ed. Denton, TX: Environmental Ethics Books, 1993.

William Forbes
Stephen F. Austin State University

Hayes, Denis

Denis Hayes (1944–) is the director of the Bullitt Foundation, established to protect, restore, and maintain the natural environment of the Pacific Northwest, and an environmental activist and policy worker whose first claim to fame was as the coordinator of the first Earth Day in 1970.

Raised in rural Washington State, Hayes attended college at Stanford University (B.A. in history) and Harvard's Kennedy School of Government and was an avid backpacker as a student, as well as a an activist involved in the antiwar effort. When Senator Gaylord Nelson (D-WI), a vocal environmentalist, reformer, and small business advocate, organized Earth Day as a teach-in modeled after the Vietnam War teach-ins of the era, he chose Hayes as the event's coordinator. Nelson's principal innovation, a change from the Earth Day proposals that had been made in the previous decade, was to decentralize the event and put the focus at the local level, rather than centering the day around speeches and ceremonies in Washington, D.C. When Hayes read about the planned event in a front-page *New York Times* article in the fall of 1969, which compared Earth Day to the mass demonstrations against the Vietnam War, he traveled to Washington, D.C., to volunteer his services and experience; after making the acquaintance of Nelson and Republican Representative Pete McCloskey, Hayes was asked to drop out of Harvard and direct the event at the national level.

Hayes recruited former students he had known as an activist as well as others to work as his staff in the capital and spend the next six months preparing the event. The movement grew on its own, as thousands of schools and local communities organized their own events, eventually involving more than 20 million Americans, far more than could have been organized under a centralized plan. The event brought together many groups and activists who had been focused on single environmental issues such as pesticides, freeway encroachment, and industrial waste and spills.

In Earth Day's aftermath, Hayes formed the Earth Day Network with other organizers in order to promote progressive and environmental awareness and activism on a year-round basis. The network grew to include nongovernmental organizations, local governments, activists, and nonprofit groups and focused on policy initiatives, environmental awareness campaigns, and the organization of local and regional Earth Day events. In 1990, in the years leading up to the 1992 Earth Summit at Rio de Janeiro, Hayes organized Earth Day 1990 to celebrate the 20th anniversary of the first Earth Day. His was one of two separate and dissenting groups formed to celebrate the anniversary, the other being the Earth Day 20 Foundation founded by 1970 Earth Week project director Edward Furia. Earth Day 20 was the more grassroots focused of the two, primarily bringing together hundreds of local groups; Earth Day 1990, drawing on Hayes's years of experience, was more professional, involving focus groups, e-mail, and diligent fund-raising efforts. Earth Day 1990 also came under considerable criticism by the rival group for its corporate affiliations, which included letting Hewlett-Packard, a major emitter of chlorofluorocarbons (CFCs), onto its board of directors.

In the 21st century, the Earth Day Network, of which Hayes is still chair, is active in 192 countries, with a U.S. program involving more than 10,000 groups in environmental awareness activities throughout the year.

Hayes also headed the Solar Energy Research Institute (now the National Renewable Energy Laboratory) during the Carter administration, before the Reagan administration cut the program's funding, and returned to school to complete his law degree at Stanford Law. Since 1992, he has headed the Bullitt Foundation, based in Seattle, Washington; he has served on a number of governing boards, notably those of Greenpeace, Ceres, and Stanford University; and he has been a visiting scholar at the Woodrow Wilson Center.

See Also: Civic Environmentalism; Conservation; Earth Day 1970.

Further Readings

Carter, Alan. *A Radical Green Political Theory.* London: Routledge, 1999.

Dobson, Andrew. *Green Political Thought.* London: T&F Books, 2009.

Radcliffe, James. *Green Politics: Dictatorship or Democracy?* New York: Palgrave Macmillan, 2002.

Torgerson, Douglas. *The Promise of Green Politics: Environmentalism and the Public Sphere.* Durham, NC: Duke University Press, 1999.

Bill Kte'pi
Independent Scholar

Human Values and Sustainability

Sustainability is a complex, evolving set of human values based on a growing understanding of the negative impacts of our activities on the Earth. These activities affect not only the planet's ecosystems that support our life, but also the quality of our human relationships.

As we are coming to understand it, sustainability, while radical, springs from deep sociocultural roots. It is a holistic approach to our human relationships with each other and the environment, a marked shift in goals, values, and ethical focus from mainstream elements of conservation, environmental, and sustainable development movements of the 20th century. As a result, sustainability is a complex, multilayered response to global and local ecosystem damage brought on by large-scale overdevelopment. Sustainability's ethical underpinnings move beyond its mainly pragmatic antecedents: First, sustainability is envirocentric (biocentric, ecocentric), placing the highest priority on healthy local and global ecosystems. Second, it recognizes the deepest interconnections among humans and our fundamental biological ties with all life on the Earth.

This hillside in the Gallatin National Forest in Montana was damaged by strip-mining for minerals, an example of valuing natural resources for the sake of economic growth. The 2006 photograph shows early attempts at restoring the area's landscape and wildlife habitats after the mine was closed.

Source: U.S. Fish & Wildlife Service/Ryan Hagerty

The sustainability literature specifies social and economic justice as core human values that are central to ecosystem health. Sustainability demands a significant redirection of complex human activities and relationships in the overlapping natural, economic, political, social, and cultural spheres of our life on the planet. As an ethical framework, sustainability demands that we make environmentally responsible and rational decisions to protect our species. We do this by living out our obligation to protect the Earth's often fragile and irreplaceable ecosystems. This responsibility entails envirocentric individual human and group agency within community, national, and global institutions, such as political, financial, and business organizations.

In practice, the sustainability's human values engage us in creating a good quality of individual and collective life, including the environment. The concept of sustainability flows from Aldo Leopold's land ethic, which defines the community as being not only made up of people, but also of the soil, water, plants, and animals. This biological imperative means that humans, whatever our technological prowess, are part of nature and dependent on it for our existence. The land ethic demands that we treat nature—and each other—with dignity and respect. Sustainability's human values directly link the fate of the planet and the human race; as a process that values human-ecological relationships, the successful pursuit of sustainability is more likely to occur in smaller, human-friendly communities.

The Complexities of Sustainability

Sustainability is played out in multifaceted intersections of personal beliefs; personal social interactions, such as those with friends, family, neighbors, community members, and fellow workers; and the widening sociocultural milieu of patterned human interactions at the national and global levels. As a sociocultural artifact that is created and re-created by human and group agency over time, sustainability is not only a goal, it is a values-based worldview and process to enhance all life on the planet. Different approaches to sustainability suggest slightly different perspectives that offer challenges and opportunities, but all can be interpreted to link sustainability with human values intended to improve social relationships and protect the environment, such as the following:

- People, planet, and profits (the three Ps)
- Bearable social and environmental conditions; viable environment and economy; and equitable social and economic life
- Healthy ecosystem, vital economy, social well-being, and healthy people
- Ecology, economy, and equity (the three Es); education and ethics can be added to create the five Es

Each approach is intended to express core human values in theory and practice. For example, the interrelationships of healthy ecosystems, vital economies, social well-being, and healthy people can be woven around human values that support efforts to protect individuals, communities, and the planet. Historically, universal ethical obligations demand that we take care of each other. Sustainability moves to another plane, obliging us to protect nature to ensure our health and survival now and in the future.

The broad and possibly universal theme that unites sustainability and human values is the call for individuals and groups of all sizes to care for each other and the Earth. At its heart, sustainability as a values-laden process is based on radical alternative patterns of social and ecological interactions at all levels of our existence. The human values of sustainability are at once personal, local, national, and global in a world of infinitely complex and interrelated sociocultural processes and ecosystems. But in a dynamic, complex, and fragmented fossil-fuel-oriented global sociocultural system—with wide gaps between the rich and poor and a planet whose ability to support life is already overstretched—the process of moving toward sustainability is daunting. The complexity of sustainability is problematic depending on who is applying the concept and how it is applied in any given place.

Challenges of Applying Sustainability

The complexity of sustainability presents significant ethical and practical problems related to the diversity of human values and agency within the broader global capitalist sociocultural system and its environmental interactions. Individuals, for example, might understand sustainability differently, which creates potential conflicts in the direct sphere of relationships with families, neighborhoods, and communities. Sustainability becomes even more complicated in national and global interactions. First, the actions of individuals singly and aggregated have global implications in a consumption-oriented world. Second, conflicts emerge over the interpretation of human values because of existing inequalities—including natural resource distribution and control and economic and political power relationships—as well as different cultural perspectives. The tension between human sustainability and the local and global land community creates a significant ethical friction point in defining and practicing sustainability values.

The urgency of global environmental problems challenges sustainability as a dynamic sociocultural construct and process. Widespread environmental degradation requires significant behavioral changes for diverse human populations with widely different living conditions and worldviews. Sustainability, with its recognition of a finite planet, offers a framework to express our values by building human-scale social processes and technological approaches to meet environmental crises locally and globally while addressing essential social justice issues. As a radical process, sustainability seeks to level differences in global power and wealth. Without a collapse of global capitalist sociocultural system or its exploited ecosystems, hegemony built on global control of natural resources that create wealth for a relative few presents significant barriers to widespread and rapid changes in overall patterns of human behavior.

Sustainability as a radical worldview, ethical expression, and sociocultural process abandons the hegemonic American and Western models of constant development and growth that have left a widespread legacy of environmental and social damage. Sustainability is not like well-intentioned sustainable development, which still focuses on economic growth while accounting for environmental impacts of development. Sustainability is a new paradigm that puts environmental considerations on at least an equal footing with human needs, and in some interpretations gives the environment primacy. Development and growth are not congruent. Instead, development decisions recognize biological limits based on fulfilling human potential within the local environment's carrying capacity, which may or may not allow for growth.

The persistence of the capitalist development model presents an obstacle to sustainability efforts, which can be overshadowed by the notion of economic prosperity, couched in terms of growing income and wealth for individuals, communities, states, the nation, and the world. Perhaps without fully understanding the longer-run human and ecological implications of their actions, the wealth-seekers who actualized the widely emulated, mythical development model fractured long-standing social and ecological relationships between the First Peoples and the land. The newcomers permanently altered natural ecosystems, shaping the land and its inhabitants to meet their perceived need for security through creation of money and built wealth. In the interest of building a wealthy nation, a narrow form of sustainability, they wrested "free" natural resources from the Earth at an increasing rate to sustain the country's rapid growth. The longer-run implications of these activities are staggering, gradually triggering a recognition of the environmental downside of intensive development that exploited people and the Earth. The hegemony of growth-driven development that rapidly built material and financial wealth persists as a stumbling block to sustainability.

The legacy of the growth and development paradigm as a material and sociocultural phenomenon appears at first glance to be successful, given the country's abundant environment. It took decades to recognize the implications of environmental degradation that ultimately threatens the planet. The threat still seems invisible or, at least, distant to many: in the seemingly prosperous developed world, values of growth and progress are bolstered by material wealth and immediate satisfactions of high living standards. In addition, aspirations for material wealth based on adoption of the development model around the world increase demands on global resources. The growth and development model tends to frame discussions of social justice based on acquiring material wealth. The human values proffered by sustainability are counterhegemonic, questioning how much material wealth is too much and what is needed for a standard of living that provides a good quality of human life in an ecological (land) community.

In a constantly changing global sociocultural system driven by capitalist development, humans alter ecosystems daily, but the scale and intensity of development varies from place to place and over time. Development is shaped by, among other things, different laws and policies, levels and kinds of income and other economic factors, local customs and values, land topography and use, population pressures and migration, and technology. The wide spectrum of a dynamic, exploitive capitalist sociocultural system constantly shapes the human values and ecosystems that are part and parcel of a move toward sustainability, making it a moving target, even at the local level.

A Matter of Scale

The interconnections of humans and environmental decline make global sustainability an imperative, an issue for national governments, nongovernmental organizations, and world bodies such as the United Nations. But global sustainability may seem overwhelming. Local and regional sustainability seem more plausible and manageable. The biological and physical realities of human perception suggest that sustainability is best pursued at a community level as part of recognizing global responsibilities.

Smaller-scale human–nature relationships seem to be a common theme among thinkers who shaped the sociocultural heritage of sustainability. For example, Thomas Jefferson's agrarianism, with its emphasis on farmers as stewards of the land, contributes to ideas about community sustainability. Nineteenth-century transcendentalists, utopians, and naturalists offer some of the earliest thinking in the United States about simplicity, community, and sustainable economies. Henry David Thoreau, Ralph Waldo Emerson, and John Muir contribute different understandings of the place of humans in nature. Succeeding generations of writers as diverse as Edwin Way Teale and Annie Dillard seek to bring readers closer to the natural world with its scientific and affective elements. Early pragmatic conservationists such as Gifford Pinchot and Theodore Roosevelt suggest thrift, efficiency, and scientific management of natural resources for the good of people and communities in the commonwealth. Wendell Berry and Wes Jackson advocate appropriate adaptations by humans to the land community in their writings about sustainable agriculture as a more fully human endeavor.

In literature about humans and the land, human community as part of the local ecology is said to offer deeper meaning for people. Small-scale, community-based sustainability promises a simpler, more meaningful life, nurtured by a sense of place. Place implies a sense of belonging to a community landscape with complementary natural and built elements. It empowers us to engage sustainably in our local ecology, Aldo Leopold's land community. While it may not always be so in reality, the essence of community as place is a sense of shared experiences and purpose. Even though constant global changes challenge the bedrock concept of community, sense of place helps link fulfilling human relationships—cultural, social, political, financial, and built—to the natural landscape. Humans exist in space and reside in places, despite larger forces that cause local change and keep people on the move. The literature offers sometimes conflicting insights into how communities deal with change because of the predominant focus on growth and development. Whether communities can become sustainable places depends in part on leadership and a local culture that nurtures the dignity of humans in their ecology. Smaller scale, with healthy personal contacts and closeness to nature, makes the process of envisioning and moving toward sustainability possible.

Critical perspectives of modern global capitalism highlight sustainability's smaller, human and land community-oriented scale and values. According to its critics, capitalism tends to alienate people from each other, their communities and workplaces, and nature, while leaving people in large parts of the world impoverished. Dissatisfaction with the unfulfilled promises of growth, progress, and consumer goods accompanied the evolution of capitalism, but were never predominant. Widespread acceptance of hegemonic capitalist social relations and values has continued because of relative material abundance, political stability, and the power of ideology fueled partly by consumerism. Material wealth fuels the aspirations of less-developed countries in their pursuit of traditional development.

E. F. Schumacher's *Small Is Beautiful: Economics as If People Mattered* offers an alternative to global capitalism. As a foundational work for sustainability, it moved away from the pragmatism of conservation and early environmentalism. First published in the 1970s, Schumacher's book takes an ethical approach to economics that links people to the land and to each other, where education and analysis have a sense of metaphysical roots. Workings and values of a human-scaled economy help individuals reach their highest potential, offering a deeper meaning for happiness. In his epilogue, Schumacher challenges materialism and the ends of the system that promotes it, while destroying people and nature. Schumacher's alternative vision supports an economic, community-based approach to sustainability and human values.

Community sustainability is based on an understanding that local activities have wider impacts downstream in the watershed and downwind in the air. Communities are not isolated. Moving toward sustainability starts in the community's backyard, but recognizes wider impacts. Community offers a starting place for practicing sustainability and human values on a global scale.

Conclusion

The power of sustainability and human values flows from its diverse origins and philosophical concerns about quality of life on Earth. The core philosophy is that humans thrive and reach their full potential because of direct connections with nature and with each other. These interactions help form a deeper sense of community as a place on a natural landscape, whether rural or urban. Alienation from nature can be mitigated through built infrastructure that creates intimate social and civic settings for people within nature. Whether urban or rural, sustainability proponents focus on human–nature connections based on the scale and intimacy of these interactions. Quality matters more than quantity.

Sustainability's human values are found in individuals' interactions with each other in community, their built environment, and the natural ecology. These values reside in individuals who work together to promote and build sustainable practices; they are reinforced as part of the wider community culture that supports social justice and healthy human relationships. With support from others, individuals choose whether to act on sustainability values each day. Lived sustainability represents a choice that is rational in both the short and long runs. Choosing to act on these values expresses individuals' rights and responsibilities toward each other and the environment based on an understanding of the impacts of our actions on the community and around the world. Responsibility grows out of concepts stressed in *Our Common Future,* published by the United Nations' Brundtland Commission in 1987.

In respecting human dignity, human values of sustainability are progressive, rooted in the importance of people and the ecosystems of their places. The human values of sustainability are also conservative because they demand an approach to the land community that is deliberate, planned, and protective of the ecology locally and globally. Humans are not separate from nature; they act on it, and it acts on them. Human sustainability values are cultural tools that pose an alternative to commonly accepted patterns of thought and action. They are ideals for human agency, guides and goals for behavior that can help individuals and communities move toward sustainability in their place on the planet.

See Also: Biocentric Egalitarianism; Biocentrism; Brundtland Report; Ecocentrism; Future Generations.

Further Readings

Adams, W. M. "The Future of Sustainability: Rethinking Environment and Development in the Twenty-First Century." Report of the IUCN Renowned Thinkers Meeting, January 29–31, 2006. http://cmsdata.iucn.org/downloads/iucn_future_of_sustanability.pdf (Accessed November 2009).

Berry, W. "Conserving Communities." In *Rooted in the Land: Essays on Community and Place,* W. Vitek and W. Jackson, eds. New Haven, CT: Yale University Press, 1996.

Bradshaw, T. K. "The Post-Place Community: Contributions to the Debate About the Definition of Community." Paper presented at the Annual Meeting of the Community Development Society, 2006.

Carnoy, M. *The State and Political Theory.* Princeton, NJ: Princeton University Press, 1984.

Collins, T. "Community Capitals and Land Use: Values for Assessing Sustainability." North Central Regional Center for Rural Development. Community Capitals Framework Workshop, 2007.

Domhoff, G. W. *Who Rules America Now? A View for the '80s.* New York: Simon and Schuster, 1983.

Dresner, S. *The Principles of Sustainability.* London: Earthscan, 2002.

Edwards, A. R. *The Sustainability Revolution: Portrait of a Paradigm Shift.* Gabriola Island, British Columbia, Canada: New Society Publishers, 2005.

Emery, M. and C. Flora. "Spiraling-Up: Mapping Community Transformation With Community Capitals Framework." *Journal of the Community Development Society,* 37/1 (2006).

Hay, P. *Main Currents in Western Environmental Thought.* Bloomington: Indiana University Press, 2002.

Hays, S. P. *Conservation and the Gospel of Efficiency: The Progressive Conservation Movement, 1890–1920.* New York: Atheneum, 1980.

Lasch, Christopher. *The True and Only Heaven: Progress and Its Critics.* New York: W. W. Norton, 1991.

Leopold, A. *A Sand County Almanac, and Sketches Here and There.* Oxford, UK: Oxford University Press, 1968.

McDonough, W. and M. Braungart. "Design for the Triple Top Line: New Tools for Sustainable Commerce" (2002). http://www.mcdonough.com/writings/design_for_triple.htm (Accessed November 2009).

O'Connor, James. *Accumulation Crisis.* Oxford, UK: Basil Blackwell, 1984.

Our Common Future: The Bruntland Report (1987). http://www.worldinbalance.net/intagreements/1987-brundtland.php (Accessed November 2009).

Samuelson, R. J. *The Good Life and Its Discontents: The American Dream in the Age of Entitlement, 1945–1995.* New York: Times Books, 1995.

Savitz, Andrew. *The Triple Bottom Line: How Today's Best-Run Companies Are Achieving Economic, Social, and Environmental Success—And How You Can Too.* San Francisco, CA: Jossey-Bass, 2006.

Scherer, D. and T. Attig, eds. *Ethics and the Environment.* Englewood Cliffs, NJ: Prentice Hall, 1983.

Schumacher, E. F. *Small Is Beautiful: Economics as If People Mattered.* New York: Harper & Row, 1975.

Wilkinson, K. P. *The Community in Rural America.* Middleton, WI: Social Ecology Press, 1999.

Timothy Collins
Western Illinois University

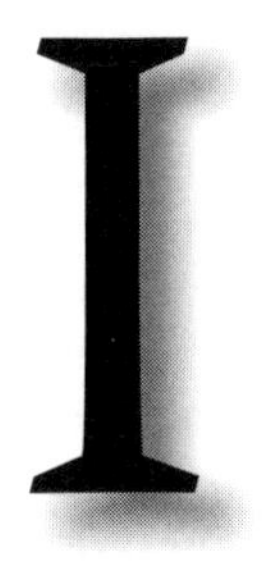

Instrumental Value

Instrumental value considers the value of an object in terms of what can be done with it or what further value can be gained from it. This is often expressed in common speech as considering something as a means to an end and is contrasted with intrinsic value, which means "value in and of itself," or "value for its own sake." A simplistic example of something most individuals associate with instrumental value is an object or tool that they use, such as a bike. A human may value his or her bike because it gets them to work or across town but it is unlikely, however, that a human would value a bike intrinsically—for its own sake and independent of its usefulness to them. On the other hand, there are other things, usually other people, which humans tend to value in and of themselves. We value our family members for their own sake, not because they buy us presents or cook us dinner.

The concept of instrumental value is important to environmental philosophy as it contrasts with intrinsic value. Central to many of the ethical positions that a person can hold toward the environment, such as biocentrism or ecocentrism, is the proposition that life and/or nature can have intrinsic value. This contrasts with a Western/industrial philosophy that imparts only instrumental value to nature objects and views them as a means to an end, such as the production of crops or the provision of timber to build a house or to sell. From this point of view nature is looked on as something to be conquered, overcome, or used, so humans value land because it can be farmed, animals because they can be eaten, trees because they can be used to heat our homes or turned into objects, and so on.

Many Western religions even suggest that the world, or nature, was created specifically for human use, and that humans are considered to exist separately from this natural world. For instance, passages from the Judeo-Christian Bible have been cited to justify the exploitation of the natural world by man. An example is Genesis 1:28: "And God said to them 'Be fruitful and multiply, and fill the earth and subdue it; and have dominion over the fish of the sea and over the birds of the air and over every living thing that moves upon the earth.'" However, such views are not common to all Christian or Jewish theology, and there are alternative interpretations that emphasize, for instance, the responsibility of humans to act as good stewards toward all of God's creations, and thus value the natural world for its own sake. Many Eastern religions such as Buddhism and Jainism also stress the interdependency of life and the intrinsic value of all living things. This notion of intrinsic value challenges the notion of human superiority associated with instrumental value:

regarding all living beings as possessing their own intrinsic value suggests a leveling of the playing field so that man is no longer apart from and superior to nature.

Placing a strictly instrumental value on the natural environment is most often associated with those who hold an anthropocentrist or human-centered position, meaning that they regard humans as the most important creatures in the universe. Such a position is not necessarily incompatible with concern for the environment; however, a person holding anthropocentrist and instrumental views of nature may still be in favor of environmental protections because they believe that it will benefit mankind to do so. For instance, a person could argue that factories should not be allowed to pollute the waterways because humans require a supply of clean water in order to live or because the fish that would be killed by the pollution have an economic value to humans that should be preserved.

Economic valuation methods are increasingly being used to argue for environmentally friendly practices, such as sustainable forestry, because of the long-term gains in productivity and safeguarding of human health through sustainable practices. This approach requires assessing not only immediate gains (such as employment or profits) but also long-term effects on the environment that may impact quality of life, health, or future productivity. These approaches, which may persuasively make the case for sustainable development, do not rely on positing an intrinsic value to nature, but may be defended simply through consideration of the instrumental value of natural resources.

See Also: Anthropocentrism; Biocentrism; Callicott, J. Baird; Deep Ecology; Intrinsic Value; Land Ethic; Leopold, Aldo; Rolston, III, Holmes.

Further Readings

Agar, N. *Life's Intrinsic Value; Science, Ethics, and Nature.* New York: Columbia University Press, 2001.

Lemos, N. M. *Intrinsic Value: Concept and Warrant.* Cambridge, UK: Cambridge University Press, 1994.

Moore, G. E. *Philosophical Studies.* London: Routledge and Kegan Paul, 1951.

O'Hear, A. *Philosophy: The Good, the True and the Beautiful.* Cambridge, UK: Cambridge University Press, 2000.

Ouderkirk, W. and J. Hill. *Land, Value, Community: Callicott and Environmental Philosophy.* Albany: State University of New York Press, 2002.

Rolston, H. *Conserving Natural Value.* New York: Columbia University Press, 1994.

Rolston, H. *Environmental Ethics: Duties to and Values in the Natural World.* Philadelphia, PA: Temple University Press, 1988.

Rolston, H. *Philosophy Gone Wild: Essays in Environmental Ethics.* Amherst, NY: Prometheus Books, 1986.

World Health Organization. The Health and Environment Linkages Initiative. "Using Economic Valuation Methods for Environment and Health Assessment." (2010). http://www.who.int/heli/economics/valuemethods/en/index.html (Accessed August 2010).

Zimmerman, M. J. *The Nature of Intrinsic Value.* Lanham, MD: Rowman & Littlefield, 2001.

Jo Arney
University of Wisconsin–La Crosse

INTERGENERATIONAL JUSTICE

Systematic discussions of intergenerational justice are a relatively recent feature of Western moral philosophy, although documents from other cultures contain codified considerations of what is due to future generations—as in the Iroquois constitution, which dates from the 12th century, and its "seven generation thinking." The increasing attention paid by industrialized societies to the problem of determining what we owe to the future derives largely from increasing interest in ecological concerns, resource depletion, and technological risk. In the 1960s, the relationship between economic and population growth, the exploitation of natural resources, and pollution energized debates that have contributed to a growing body of scholarship. Books from authors such as Robert Jungk (1952), Rachel Carson (2002, 1962), and Paul Ehrlich (1995, 1968) led to increased public discussion of environmental issues, eventually resulting in the emergence within public discourse of the concept of sustainable development, defined by the Brundtland Commission in 1987 as "development that meets the needs of the present without compromising the ability of future generations to meet their own needs."

Discussions of intergenerational justice present specific conceptual difficulties. These are often thought to concern exactly how a conflict of interests is possible between actual and "merely potential" people. Many thinkers have viewed this as a problem to be settled in terms of distributional justice. John Rawls's 1972 discussion sets a pattern by discussing the claims on rationality of a "just savings" principle, based on the assumption that anyone in Rawls's hypothetical "original position" will care for future well-being by ensuring that a "social minimum" of welfare can be provided for. Skepticism about concerns that obligations to future generations can best be fulfilled by conserving utility has served as the basis for much subsequent discussion. Can, for example, future generations be said to have the same interests as people living in the present, and if so, can these be harmed? Should these interests give rise to specific rights for future people, perhaps promoting them to a moral status equal to that of present people?

Aside from the perhaps irresolvable uncertainty surrounding judgments about future outcomes of present actions, two problems arise, both related to how relationships between present and future differ conceptually from those between contemporaries. First, some, like T. Schwartz, deny that future people can have rights at all. Only particular individuals can be rights holders and therefore the beneficiaries of duties, and nonexistent future people, are not considered individuals in this sense. Others like G. K. Pletcher reply that there are general obligations to future people (such as not subjecting them to obvious dangers) that, to be binding, do not need to be derived from rights possessed by particular individuals. A second and more serious objection is also the problem of nonreciprocity, which is at the heart of both D. Hume and J. Addison's remarks: while existing people can harm or benefit each other and do the same to future people, future people can neither harm nor benefit present people. Going further, however, some have suggested that this asymmetry of power has paradoxical consequences: that present people cannot in any coherent sense *harm* future people.

Derek Parfit (1986) argues that actions undertaken in the present can change which particular individuals will inhabit the future. Different policies do not therefore produce different qualities of life for the same set of future people; they should be thought of as producing different populations that experience different qualities of life. If this is true, then policy P1 producing aggregate utility U1 is not better than policy P2 producing a lower aggregate utility U2. For Parfit, we cannot maintain that P2 is worse than P1 on the basis that it produces a greater negative effect on a given set of people, as our actions under P1 and P2 are the cause of different people being born.

Some, like L. H. Meyer, feel that a minimum level of welfare should be set for future people in order to prevent their crossing a "threshold of harm." This approach recalls Rawls, and the idea that a minimum standard of rights (and not the difference between two levels of welfare or utility) should give us an appropriate idea of future-related harm. Another response is to argue that the distributional paradigm itself is inappropriate to relations between present and future people. Rather than focusing on individual utility and harm, it may be advisable to concentrate on preserving the enabling conditions of a good human life. This yields a concept of intergenerational justice based more on positive duties to provide goods that constitute conditions under which capacities may be fulfilled, rather than providing utility to be consumed. An intergenerational community is posited by J. Passmore in which the responsibility of present people to future people is not to prevent harm to them but to "anticipate their care" for the richness and meaningfulness of their world, moving us toward concepts of sustainability. Others like J. O'Neill have suggested that this view contains an answer to Addison's acquaintance's remark: what posterity does for us is take care of the fate of the specific values and projects we care about. This recalls Aristotle's view in the *Nichomachean Ethics* that "both evil and good are thought to exist for a dead man, as much as for one who is alive but not aware of them."

See Also: Energy Ethics; Passmore, John; Precautionary Principle; Strong and Weak Sustainability; Sustainability, Seventh Generation.

Further Readings

Aristotle. *Nicomachean Ethics.* Oxford, UK: Clarendon Press, 1908.

Hume, D. *A Treatise of Human Nature,* 2nd ed. Oxford, UK: Clarendon Press, 1978.

Jonas, H. *The Imperative of Responsibility: In Search of an Ethics for the Technological Age.* Chicago, IL: University of Chicago Press, 1984.

Meyer, L. H. "Past and Future: The Case for a Threshold Notion of Harm." In *Rights, Culture and the Law,* L. H. Meyer, S. L. Paulson, and T. W. Pogge, eds. Oxford, UK: Oxford University Press, 2003.

O'Neill, J. *Ecology, Policy and Politics.* London: Routledge, 1993.

Parfit, D. *Reasons and Persons.* Oxford, UK: Oxford University Press, 1986.

Passmore, J. *Man's Responsibility for Nature.* London: Duckworth, 1974.

Pletcher, G. K. "The Rights of Future Generations." In *Responsibilities to Future Generations,* E. Partridge, ed. New York: Prometheus Books, 1981.

Rawls, J. *A Theory of Justice.* Oxford, UK: Clarendon Press, 1972.

Schwartz, T. "Obligations to Posterity." In *Obligations to Future Generations,* R. I. Sikora and B. Barry, eds. Philadelphia, PA: Temple University Press, 1978.

Chris Groves
Cardiff University

International Association of Environmental Philosophy

The International Association of Environmental Philosophy (IAEP) was founded in 1996–97 by Professors Bruce Foltz (philosophy) and Robert Frodeman (geology), who also served as president and vice president, respectively (Frodeman remains as an executive officer). IAEP has served as a premier global academic/professional not-for-profit organization alongside the older International Society for Environmental Ethics (ISEE). According to its bylaws (Article II: Purpose):

> The purpose of the International Association of Environmental Philosophy is to provide a forum for wide-ranging philosophical discussion of nature and the human relation to the natural environment. We define the Association's breadth of scope to be environmental philosophy in the most comprehensive sense, i.e., not only environmental ethics, but environmental aesthetics, environmental ontology, environmental theology, the philosophy of science, environmental political philosophy, philosophy of technology, ecofeminism, and other areas. We define its breadth in methodology by the diverse approaches we bring to these issues, including many schools of Continental Philosophy, studies in the history of philosophy, and the tradition of American philosophy.

Or, as explained in the introduction to *Rethinking Nature: Essays in Environmental Philosophy* (2004), edited by Foltz and Frodeman, "the IAEP was founded in order to address needs that existed within both Continental and environmental philosophy—in the former case, to expand the conceptual space of European philosophy to include a focus upon the natural environment, and in the latter, to bring the distinctive approaches of the Continental tradition to our concerns with the natural world."

The first—and all subsequent—annual meetings of the IAEP have been held in conjunction with those of the Society for Phenomenology and Existential Philosophy (SPEP), which sheds light upon how it differs from the ISEE, which is perhaps more concerned with issues in environmental ethics. The IAEP is known for providing space for various schools and concentrations of philosophy, as well as encouraging interdisciplinary (and interorganizational) communication in research. Keynote speakers have included Alphonso Lingis, John Sallis, Edward Casey, Mark Sagoff, David Seamon, Kristin Schrader-Frechette, Bruce Foltz, David Wood, Joan Maloof, Stephanie Mills, Leonard Lawler, and Karen Warren.

Since 2004, the IAEP and the ISEE have held the Annual Joint Conference in Environmental Philosophy, a summer retreat held outside Rocky Mountain National Park in Allenspark, Colorado. Featured guests have included Holmes Rolston III, J. Baird Callicott, Bryan Norton, Charles Wilkinson, and Emily Brady. In addition, the IAEP has occasionally held sessions at American Philosophical Society (APA) meetings, both national and regional.

Also since 2004, the peer-reviewed biannual periodical *Environmental Philosophy* has been sponsored and published as the official journal of the IAEP. Cofounding editors were Professors Kenn Maly (philosophy and environmental studies) and Ingrid Stefanovic (philosophy), the latter also serving as director of the Centre for Environment at the University of Toronto, both of whom remain on the editorial board alongside cofounders Foltz and Frodeman and Professors Bob Mugerauer (urban design and planning) and David Seamon (architecture). In 2007, Professor Ted Toadvine of the University of Oregon took over as managing editor of the journal, which is now sponsored by the University of Oregon's Department of Philosophy and Environmental Studies Program as well as the IAEP. Special issues have focused on "Environmental Aesthetics and Ecological Restoration" (Spring and Fall 2007, 4/1–2) and "Species of Thought: In the Approach of a More-Than-Human World" (Fall 2008, 5/2).

Recently, attention has been directed toward the fusion of ecological and phenomenological approaches as examined in *Eco-Phenomenology: Back to the Earth Itself* (2003), edited by Charles S. Brown and Ted Toadvine, although the focus has remained on a holistic environmental philosophy in general.

Membership in the IAEP is open to all, and it includes a subscription to *Environmental Philosophy;* students and retirees receive a discounted rate.

See Also: Green Altruism; International Society for Environmental Ethics; Philosophy and Environmental Crisis; Preservation.

Further Readings

Cataldi, Suzanne L. and William S. Hamrick. *Merleau-Ponty and Environmental Philosophy: Dwelling on the Landscapes of Thought.* Albany: State University of New York Press, 2007.

Higgs, Robert and Carl P. Close. *Re-Thinking Green: Alternatives to Environmental Bureaucracy.* Washington, DC: Independent Institute, 2005.

International Association of Environmental Philosophy. http://www.environmentalphilosophy.org (Accessed May 2010).

Journal of Environmental Philosophy. http://ephilosophy.uoregon.edu (Accessed May 2010).

Brandon B. Rowley
University of Idaho

International Society for Environmental Ethics

Founded in 1990 at the Eastern Division meeting of the American Philosophical Association, the International Society for Environmental Ethics (ISEE) is one of the main scholarly associations devoted to promoting research in environmental ethics. The founding of ISEE at that time stemmed from the need for an organization dedicated specifically to issues of environmental ethics, given the interest of philosophers and other scholars in environmental ethics that had been increasing since the 1960s. The first president of ISEE was the eminent environmental philosopher Holmes Rolston III; a number of other prominent environmental philosophers including J. Baird Callicott, Dale Jamieson, and Mark Sagoff have also served as president of the society. Although philosophers have typically constituted the majority of the membership of ISEE, from its inception the society has often met in conjunction with other philosophical groups. ISEE is interdisciplinary in nature and includes members from a number of different academic disciplines including, for example, religious studies, geography, and environmental studies. The society is truly international in its makeup, and as of 2010, the society has over 350 members representing over 20 different nations; ISEE has regional representatives from areas of Africa, Asia, Australia, Europe, Central America, South America, and North America.

The constitution of the International Society for Environmental Ethics was first adopted in 1990 and revised in 2007. In addition to specifying the various officer positions of the organization and determining other procedural issues, the constitution formulates the basic purpose of the society. In the broadest terms, the constitution affirms that the purpose of ISEE is to advance research and education in environmental ethics and to promote the understanding of and respect for the natural world. The constitution delineates a number of activities that the society should pursue in order to accomplish this purpose. The activities noted include such things as sponsoring meetings, conferences, sessions, and workshops on environmental issues; providing members with information on materials of use to the teaching of environmental ethics; promoting undergraduate and graduate education in environmental ethics; and working to relate issues in environmental philosophy and ethics to other disciplines. While not exhaustive, the items listed give a general sense of the kinds of activities specified in the ISEE constitution. The full text of the constitution can be found in the fall 2007 edition of the society newsletter, which is available online at the ISEE Website.

Throughout its history, the society has supported scholarship in environmental ethics, as well as the teaching of environmental ethics at colleges and universities, in a number of ways. One of the primary means by which ISEE has encouraged scholarly work in environmental ethics is through its sponsorship of scholarly gatherings devoted to issues in environmental ethics. From its inception, the society has sponsored group sessions at the various annual divisional meetings of the American Philosophical Association. Together with the International Association for Environmental Philosophy, ISEE also cosponsors an annual meeting on environmental philosophy. In addition, the society has supported sessions on environmental ethics at the national meetings of a number of other academic organizations in cognate disciplines, such as the American Academy for the Advancement of Science, the American Institute of Biological Sciences, and the Society of American Foresters. At the international level, ISEE has held sessions in conjunction with the group meetings of organizations such as the Canadian Association of Learned Societies and the Australasian Association for Philosophy.

In addition to promoting scholarly discussion of environmental ethics at various academic meetings, the society has also developed a number of significant resources for its members and other scholars working in environmental ethics. The society newsletter, issued three times a year, provides members with updates on society activities as well as a wealth of topical information of interest to those working in environmental activities, including calls for papers, announcements, environmental news, and notes on Websites, films, and books of interest to environmental ethicists. While the newsletter provides an ongoing record of recent works of relevance to those working on environmental issues, ISEE has also afforded researchers in environmental ethics with an extremely significant resource by compiling the largest bibliography in the world on environmental ethics, containing over 7,000 entries. The bibliography is available at the society's Website.

In order to aid in the teaching of environmental ethics, the society Website hosts a syllabus project that contains links to numerous syllabi for courses in environmental ethics provided by members as well as information on textbooks in environmental ethics. The ISEE Website also contains links to other topics of interest in environmental ethics, including information on graduate programs in environmental ethics, funding opportunities, and other scholarly associations. Further, the ISEE hosts a listserv that affords a general forum for discussions among students, researchers, and others interested in issues in environmental ethics and for the dissemination of information of interest to those working in the field. In addition to these academic activities, the society has also pursued its mission of promoting awareness of environmental issues by acting as an official observer and sending delegates to a number of conferences on environmental issues sponsored by the United Nations.

See Also: *Environmental Ethics* Journal; International Association of Environmental Philosophy; Philosophy and Environmental Crisis; Rolston, III, Holmes.

Further Readings

International Society for Environmental Ethics. http://www.cep.unt.edu/ISEE.html (Accessed March 2010).

Light, Andrew and Holmes Rolston. *Environmental Ethics: An Anthology.* Hoboken, NJ: Wiley-Blackwell, 2002.

Varner, Gary E. *In Nature's Interests?: Interests, Animal Rights, and Environmental Ethics.* New York: Oxford University Press, 2002.

Daniel E. Palmer
Kent State University, Trumbull

Intrinsic Value

In ethics, ascribing intrinsic value to something means that it is valuable for its own sake rather than as the means to some end. This is a crucial concept in many moral judgments: for instance, in many cultures, human life is considered to have an intrinsic value that does not need to be justified by, for instance, the economic value of the person's labor. Similarly, preservation of the natural world may be conceived of as having an intrinsic value without regard to the economic or other advantages that may be gained from it. Intrinsic value is contrasted to extrinsic value, which refers to assigning value to a thing according to how it may be used: if intrinsic value considers the value a thing may have "in its own right," then extrinsic value looks at the value a thing may have as "a means to an end."

The concept of intrinsic value is important to environmental philosophy for many reasons. Central to many of the ethical positions that a person can hold toward the environment, such as biocentrism or ecocentrism, is the proposition that life and/or nature can have intrinsic value. A biocentrist, for example, would argue that all life is intrinsically valuable, where an ecocentrist would argue that it is the system (or ecosystem) that is intrinsically valuable, that is, that the environment should be valued for its own sake. This stands in contradiction with the Western conception of wild nature and the

The photograph shows a rare remnant of original oak savanna habitat, which has been preserved in the William L. Finley National Wildlife Refuge in Oregon. The concept of intrinsic value may lead to the recognition that humans have a duty to protect natural resources even when they do not offer any economic or other benefits to humanity.

Source: U.S. Fish & Wildlife Service/George Gentry

often concomitant attitude that nature is something for humans to conquer or overcome.

Intrinsic value is also important to environmental philosophy because of the association between value and duty. In the tradition of axiology (the study of value), the word *value* has a similar meaning to the words *goal* or *purpose*. That is to say, values can generate duty. If humans recognize that something has value, then they have a duty to protect it, promote it, or bring it about.

Intrinsic value is central to Aldo Leopold's "land ethic," which rejects the view that humans should be dominant over nature. Instead, Leopold asserts that humans ought to value the land for its own sake. Leopold contends that "a thing is right when it tends to preserve the integrity, stability, and beauty of the biotic community." The call for respect focuses in on the concept of intrinsic value. Likewise, much of the work of Holmes Rolston III is dependent of the concept of intrinsic value. In his book *Philosophy Gone Wild,* Rolston argues that once humans discover that nature has intrinsic value, they are met with a duty to protect it.

Believing that nature has intrinsic value does not preclude the recognition that there are many instrumental uses to which the natural environment can be put. Individuals who believe nature has intrinsic value do not necessarily support refraining from using the natural environment to meet human needs. Rather, valuing something intrinsically requires humans to adopt a respectful and reciprocal stance toward the natural environment so that instead of "conquering" nature, humans should try to "take care of" nature.

The belief that nature has intrinsic value is opposed to the view of anthropocentrism, which argues that humans are the central and most valuable form of life. As such, an anthropocentrist would believe that the environment exists to be exploited for human purposes and that its value is therefore instrumental in nature.

See Also: Anthropocentrism; Biocentrism; Callicott, J. Baird; Deep Ecology; Instrumental Value; Land Ethic; Leopold, Aldo; Rolston, III, Holmes.

Further Readings

Agar, N. *Life's Intrinsic Value; Science, Ethics, and Nature.* New York: Columbia University Press, 2001.

Lemos, N. M. *Intrinsic Value: Concept and Warrant.* New York: Cambridge University Press, 1994.

Moore, G. E. *Philosophical Studies.* London: Routledge and Kegan Paul, 1951.
O'Hear, A. *Philosophy: The Good, the True and the Beautiful.* New York: Cambridge University Press, 2000.
Ouderkirk, W. and J. Hill. *Land, Value, Community: Callicott and Environmental Philosophy.* Albany: State University of New York Press, 2002.
Rolston, H. *Conserving Natural Value.* New York: Columbia University Press, 1994.
Rolston, H. *Environmental Ethics: Duties to and Values in the Natural World.* Philadelphia, PA: Temple University Press, 1988.
Rolston, H. *Philosophy Gone Wild: Essays in Environmental Ethics.* Amherst, NY: Prometheus Books, 1986.
Zimmerman, M. J. *The Nature of Intrinsic Value.* Lanham, MD: Rowman & Littlefield, 2001.

Jo Arney
University of Wisconsin–La Crosse

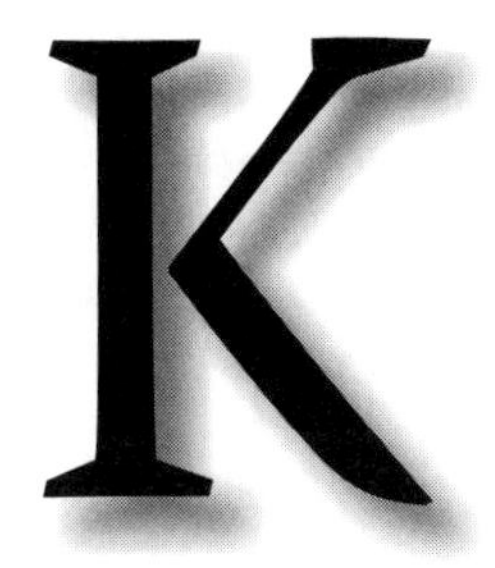

KANTIAN PHILOSOPHY AND THE ENVIRONMENT

Immanuel Kant's ethical theory is one of the most influential theories in Western philosophical history. While this is the case, how to interpret and apply Kant's theory is the subject of lively debate within both theoretical philosophy and applied philosophy.

One of the complicating factors in understanding Kant's theory is that some of the terms he uses do not easily translate into contemporary English. For instance, Kant claims that humanity has a dignity or worth that elevates humanity above all else in nature. This claim is often taken as meaning that humanity has the highest value—that it is more valuable than any and all nonhuman entities. However, Kant does not take "dignity" to imply such a value claim. Instead, "dignity" is to be understood as a status that humanity has in virtue of its capacity for moral agency. The dignity of humanity, then, is to be understood in the sense that those who hold positions of influence are held in high esteem. Those who hold such positions are highly esteemed and considered special because of the role they play in society and the accompanying responsibilities.

Consider, for example, governmental officials such as politicians and police officers. Politicians are special in the sense that they are charged with the responsibility of honestly representing the values of their constituents when shaping and creating laws and public policy. Police officers are special in the sense that they are charged with the power to enforce laws. However, politicians and police officers are not special in the sense that they are inherently more valuable than others who do not hold those positions. Instead, they are special because of their respective responsibilities. Similarly, for Kant, humans are special (i.e., have a dignity) because of the moral responsibilities they have. So, because we are free to act according to any principles we choose, we are autonomous; because we are autonomous, we have the capacity for moral agency; because we have the capacity for moral agency, we have moral responsibilities, and these responsibilities are what give us a dignity.

The dignity of others in itself does not ground our duties to them. Put differently, your dignity is not the reason that a person has moral obligations to you. Instead, it is the person's own unconditional commitment to following the moral law and the dictates of the moral law that generate my moral obligations to you. So, it is not the other who morally obligates an individual, but his or her commitment to following the moral law's dictates for how a person should treat you.

Kant's ethical theory, then, is one of individual moral responsibility: the fundamental focus is the individual (from the first-person perspective); it is about the individual's moral development. The implication of this view is that the primary focus of morality is not the other's well-being, but how an individual responds to the well-being of others.

This understanding is both useful and important when responding to those who criticize Kant's theory with regard to the morally proper treatment of animals and inanimate nature. Kant's explicit view is that moral agents can have no direct duties to nonmoral agents, and that any duties we feel we have to them are actually duties to humanity. While Kant does claim that we are to avoid callousness and indifference toward animals, it is not the animals themselves that matter morally. For when we avoid such treatment or even treat them well, the duty is not to the animal but to humanity. Based on the autonomistic, individualistic interpretation of Kant's theory, we can see that the above criticism loses strength, since the primary focus of morality is not even other humans but one's own moral development.

One of the main points of contention within these debates is the distinction between Kant's personal views and his philosophical defense of his theory. According to Kant scholars (i.e., scholars who devote most or all of their philosophical careers to understanding and interpreting Kant's works), Kant's writing (in German) is difficult to understand because he is not always careful with the terms he uses and the style of his writing is confusing at times. Consequently, translating Kant's work into English is not as easy as directly translating the text—it also requires an understanding of his historical context and the common style of writing during his time. The reason for pointing this out is that many of the critical views of Kant, especially within the field of environmental ethics, do not seem to account for some of the difficulties in not only understanding Kant's theory but also its application to real-world situations. So, the primary aim of this article is to clarify some of the misunderstandings of Kant's theory as it has been applied to the question "How ought we to treat nonhuman nature?"

Even though traditional anthropocentric ethical theories are thought to provide a faulty foundation for a proper environmental ethic, some environmental ethical theories are based on revisions of traditional ethical theories. In particular, Kant's conception of intrinsic value and the role it plays within his ethical theory has significantly influenced many nonanthropocentric environmental ethical theories. While the implications of Kant's ethics for both nonhuman animals and the natural environment were considered "abhorrent" within the charter group of environmental ethicists, Kant's ethics nonetheless provided a framework for how to conceptualize the relationship between intrinsic value and moral obligation. Furthermore, since the capacity for rationality was thought to unreasonably restrict the moral community (i.e., the category of individuals to which direct moral consideration is owed), nonanthropocentrists began proposing various alternatives to rationality—such as Tom Regan's "subject of a life," Peter Singer's "sentience," Aldo Leopold's "ecosystem," and Albert Schweitzer's "will to live"—as appropriate intrinsic value–conferring properties. Thus, by implementing their understanding of Kant's concept of the relationship between intrinsic value and direct moral duties, nonanthropocentrists began developing theories that aimed at extending moral value and moral considerability well beyond humanity.

The major problem with Kant's ethical theory, as understood by the nonanthropocentrists, is that it is ultimately based on a particular conception of value—namely, the intrinsic value of humanity. Since Kant's ethical theory is (allegedly) founded on the absolute or intrinsic value of humanity (i.e., moral agency), moral agents are the only beings to which direct moral duties are owed; all other earthly entities are mere means and moral agents have no binding moral duties concerning their welfare or preservation.

While the nonanthropocentrists' interpretation of Kant's ethical theory is in line with the conventional textbook interpretation of Kant's theory, it is not the only defensible reading available. The alternative reading, which can be called the "autonomistic interpretation," is fundamentally different from the conventional one in that the former does not take the value of humanity as the foundation of Kant's ethical theory; and removing value as the foundation allows Kant's ethics to be interpreted as helping us to understand our morally proper relationship to and with nonhuman existents.

The general implication of this alternate interpretation of Kant's ethical theory for developing an ethic of the environment is that we ought to live up to the dignity of our humanity, which means that we ought to widely develop our feelings of compassion and sympathy for all sentient beings and our appreciation and awareness of natural beauty. By developing such human capacities, we more deeply explore our own humanity.

From this interpretation, it is easy to see the connection between Kant's theory and the American transcendentalists—namely Henry David Thoreau—who also had a significant impact on the development of contemporary environmental ethics.

See Also: Animal Ethics; Anthropocentrism; Emerson, Ralph Waldo; Intrinsic Value; Leopold, Aldo; "Should Trees Have Standing?"; Singer, Peter; Thoreau, Henry David.

Further Readings

Dean, Richard. "Non-Human Animals, Humanity, and the Kingdom of Ends." In *The Value of Humanity in Kant's Moral Theory*, Richard Dean, ed. New York: Oxford University Press, 2006.

Guyer, Paul. "Duties Regarding Nature." In *Kant and the Experience of Freedom,* by Paul Guyer. New York: Cambridge University Press, 1996.

Hayward, Tim. "Kant and the Moral Considerability of Non-Rational Beings." In *Philosophy and the Natural Environment,* Robin Attfield and Andrew Belsey, eds. New York: Cambridge University Press, 1994.

Korsgaard, Christine M. "Fellow Creatures: Kantian Ethics and Our Duties to Animals." The Tanner Lectures on Human Values, delivered at the University of Michigan, February 6, 2004. http://www.tannerlectures.utah.edu/lectures/documents/volume25/korsgaard_2005.pdf (Accessed April 2010).

Lucht, Marc. "Does Kant Have Anything to Teach Us About Environmental Ethics?" *American Journal of Economics and Sociology,* 66/1 (2007).

O'Neill, Onora. "Kant on Duties Regarding Nonrational Nature: Part II." *Aristotelian Society Supplementary Volume,* 72/1 (1998).

Sensen, Oliver. "Kant's Conception of Human Dignity." *Kant-Studien,* 100 (2009).

Timmermann, Jens. "When the Tail Wags the Dog: Animal Welfare and Indirect Duty in Kantian Ethics." *Kantian Review,* 10 (2005).

Wood, Allen W. "Kant on Duties Regarding Nonrational Nature: Part I." *Aristotelian Society Supplementary Volume,* 72/1 (1998).

Keith Bustos
University of St Andrews

Land Ethic

The land ethic invites humans to include land—including soil, water, plants, and animals—in their moral calculations as they consider different courses of action. The land ethic was developed and argued for by Aldo Leopold in his book *The Sand County Almanac,* first published in 1949.

Leopold rejects the traditional view of human dominance over nature. In its place he proposes what he titles a "land triangle." As Leopold explains, this triangle is made up of many different layers. Each of the layers shares one single characteristic. Members of any given layer are alike in the type of food or energy that they consume. Each layer of the triangle is also dependent on the layer below it for survival. The bottom layer is made up of the soil and in the layers above it there are plants, insects, birds, reptiles, and mammals in ascending order. Humans, since they are not true carnivores, are not at the apex of the triangle. Leopold suggests that if one of the layers of the triangle were to collapse, all of the layers above it would also collapse. Thinking of nature in the form of the land triangle suggests that humans are dependent on the land and not dominant over it. Leopold calls this a community of independent parts.

Recognizing the human relationship with the land is not Leopold's end goal, however. The land ethic requires individuals to directly consider this land community as an independent consideration in their ethical dilemmas and debates. Traditionally, humans have only considered other humans in their moral calculations. A utilitarian approach asks humans to choose an option that provides the greatest amount of good for the greatest number of people. Other approaches to solving ethical dilemmas ask people to consider whether their actions would hurt anyone. Over time, humans have extended this type of ethical consideration to animals. Leopold's land ethic asks humans to consider the land community. In other words, a person should ask himself or herself whether an action would produce the greatest good for the land community or whether the action would directly hurt the land community. This approach focuses on the ecosystem.

Leopold calls this the land ethic because it expands the circle of ethics. There is a passage in the land ethic in which he invites readers to consider whether or not an action will affect the "integrity, stability, and beauty of the biotic community." The action is right, Leopold claims, if it upholds these values and wrong if it does not.

See Also: Biodiversity; Intrinsic Value; Leopold, Aldo.

Further Readings

Leopold, A. *Round River: From the Journals of Aldo Leopold.* New York: Oxford University Press, 1953.

Leopold, A. *A Sand County Almanac, and Sketches Here and There.* New York: Oxford University Press, 1949.

Leopold, A., et al. *Aldo Leopold's Wilderness: Selected Early Writings by the Author of* A Sand County Almanac. Harrisburg, PA: Stackpole Books, 1990.

Leopold, A., et al. *For the Health of the Land: Previously Unpublished Essays and Other Writings.* Washington, DC: Island Press, 1999.

Jo Arney
University of Wisconsin–La Crosse

Leopold, Aldo

Aldo Leopold (1887–1948) was one of America's preeminent environmentalists and a pioneer in developing the fields of wildlife management and environmental ethics. Born in Burlington, Iowa, Leopold studied forestry at Yale University and worked for the U.S. Forest Service before becoming a professor in the Department of Agricultural Economics at the University of Wisconsin in 1933. Leopold is well known among the general public for his book *A Sand County Almanac* (1949), which has sold over 2 million copies and has been translated into nine languages.

Leopold was a prolific author of articles in scientific and technical journals, but wrote *A Sand County Almanac* with the deliberate intent of communicating the importance of conservation to the general public. It consists of a collection of essays (some previously published) that combine philosophy and natural history with closely observed descriptions of the natural environment. The first section consists of essays arranged into chapters corresponding to the months of the year, and describes changes over the course of the year to the environment of his farm, located near Baraboo, Wisconsin. The second section describes various experiences in different locations around North America. The final section consists of philosophical essays about conservation, and concludes with an essay titled "A Land Ethic," which states his belief that man should exist in a state of harmony with nature.

In "A Land Ethic," Leopold expressed many ideas that have become key to the modern conservation movement. He objected to the then-common belief that land use was entirely an economic issue, and believed that the well-being of human beings and of the land was closely intertwined. He argued that it was an ethical imperative for people to work together for the benefit of the community, but that the concept of "community" referred not only to human beings, but also to "the land" collectively considered, and including animals, plants, soil, water, and so on. He also argued that it was important for people to maintain a personal relationship with the land in order to be able to value it properly.

In place of the philosophical view that man should dominate nature, Leopold proposed what he called "a land triangle," made up of many different layers, each of which shared one single characteristic. Members of any given layer are alike in the type of food or energy that they consume and are dependent on the layer below it for survival. The bottom layer is made up of the soil, with the layers above it being plants, insects, birds, reptiles, and mammals, in ascending order. However, the upper layers are also dependent on those below, and if one of the layers of the triangle were to collapse, all of the layers above it would also collapse. In this view, humans are dependent on the land rather than dominant over it. Leopold calls this model a community of independent parts.

Leopold argues that the land, as he defines it, has an intrinsic value, and must be considered alongside the needs and desires of humans in moral and ethical calculations. This is in contrast to, for instance, the utilitarian approach, which argues that the best option is one that provides the greatest amount of good for the greatest number of people, or simple moral approaches that ask whether a particular action would harm any person. As some philosophers have extended ethical consideration to animals, so Leopold's land ethic argues that similar considerations should be extended to the land community. So, in considering the ethics of any action, it must be asked if it would produce the greatest good for the land community or whether their actions would directly hurt the land community.

See Also: Callicott, J. Baird; Instrumental Value; Intrinsic Value; Land Ethic.

Further Readings

Anderson, P. *Aldo Leopold: American Ecologist.* New York: Watts, 1995.

Callicott, J. B. *Companion to* A Sand County Almanac: *Interpretive and Critical Essays.* Madison: University of Wisconsin Press, 1987.

Callicott, J. B. *In Defense of the Land Ethic: Essays in Environmental Philosophy.* Albany: State University of New York Press, 1989.

Costanza, R., B. G. Norton, and B. D. Haskell. *Ecosystem Health: New Goals for Environmental Management.* Washington, DC: Island Press, 1992.

Leopold, A. *Round River: From the Journals of Aldo Leopold.* New York: Oxford University Press, 1953.

Leopold, A. *A Sand County Almanac, and Sketches Here and There.* New York: Oxford University Press, 1949.

Leopold, A., J. B. Callicott, and E. T. Freyfogle. *For the Health of the Land: Previously Unpublished Essays and Other Writings.* Washington, DC: Island Press, 1999.

Lorbiecki, M. *Aldo Leopold: A Fierce Green Fire.* New York: Oxford University Press, 1999.

Meine, C. *Aldo Leopold: His Life and Work.* Madison: University of Wisconsin Press, 1988.

Meine, C. and R. L. Knight. *The Essential Aldo Leopold: Quotations and Commentaries.* Madison: University of Wisconsin Press, 1999.

Yannuzzi, D. A. *Aldo Leopold: Protector of the Wild.* Brookfield, CT: Millbrook Press, 2002.

Jo Arney
University of Wisconsin–La Crosse

Local Food Movement

With the rise of industrial agriculture and advances in transportation and food preservation, it is common to find foods from all corners of the world in the neighborhood grocery store. Foods have become global commodities traded over long distances between and within nations. In the United States, food travels an average of 1,500 miles from field to plate. In the last few decades, increasing concern over the detrimental environmental, economic, and social consequences of food globalization has given rise to the local food movement. Local food advocates argue that a shorter supply chain promotes better environmental stewardship, strengthens local economies, empowers consumers with greater knowledge of food origins and food choices, and offers a more nutritious and better-tasting product than its long-distance alternative.

The Obama administration established a new vegetable garden at the White House in order to use and promote local food. Here, First Lady Michelle Obama and White House Horticulturist Dale Haney join students from Washington's Bancroft Elementary School in breaking ground for the garden in March 2009.

Source: White House/Joyce N. Boghosian

While agribusinesses dispute the veracity of these claims, rising demand for local food has increased the number of farmers markets, agricultural cooperatives, and small farms in the United States and other countries as more consumers seek alternatives to fast food and large chain supermarkets. Regional food policy councils and Buy Fresh, Buy Local campaigns have formed to improve access to local food in the United States while popular books and films have raised awareness about the agricultural industry, the obesity epidemic, and food issues. Notable public figures who have promoted local food include Alice Waters, owner of Chez Panisse restaurant in Berkeley, California, and a proponent of nutrition education in schools; Wendell Berry, author of *The Unsettling of America* (1977), a critique of factory farming; Barbara Kingsolver, Alisa Smith, and J. B. MacKinnon, authors who chronicled their adventures in farming and local eating; and Michael Pollan, author of *The Omnivore's Dilemma* (2006) and *In Defense of Food* (2008). Local food has become so popular in the last decade that the *New Oxford American Dictionary* declared *locavore*—a person whose diet consists only or principally of locally grown or produced food—as the Word of the Year in 2007, and even the Barack Obama White House has established a vegetable garden on its premises to publicize its efforts to source local food for the kitchens.

What defines "local" is not a standard distance but rather the boundaries of the community as the consumer sees it. For example, in 2002, the U.S. state of Utah began a promotional program called Utah's Own that framed local food in the context of the entire state (an area of almost 85,000 square miles). The campaign created labels and advertisements to encourage consumers to buy food from Utah-based producers and boost the state economy. As in this case, buying local can be an expression of pride in one's state. It can also be extended to pride in one's home region or country. Ideally, the consumer will know the producer and the farmer will be a trusted member of the community, as recognizable as a teacher or shop owner.

This type of relationship is cultivated at farmers markets and other examples of "direct-to-consumer" exchanges, including community-supported agriculture (CSA) arrangements and informal barter networks. CSAs are farms that sell produce, typically fruits and vegetables, to buyers who are located close enough for pickup or delivery. In exchange for a weekly, monthly, or seasonal flat fee, the customer receives regular deliveries of the in-season harvest. The farm gains the security of guaranteed income that is not tied to the success of any one crop variety. Informal barter networks offer a way for backyard and hobby producers to exchange their goods with consumers, sometimes facilitated by bartering Websites, and are perhaps the most local distribution schemes of all. Each of these arrangements promotes interaction between local businesses and community members and each is flexible enough to accommodate smaller harvests or batches of unique and short-season foods in a way that large orders from grocery stores cannot. All of these close relationships allow consumers to understand and monitor the way their food is raised.

As the local food movement has grown, the large transnational corporations that benefit the most from the global food trade have disputed the benefits claimed by the local food movement. The most common counterclaim is that while local food may have a lower carbon footprint because it requires less transport, it may be possible to produce food more sustainably in distant regions, even to the extent that the comparative advantage in process outweighs the disadvantages of the extra miles. An often-cited example is that of greenhouse tomatoes raised in regions with short growing seasons. It is argued that tomatoes shipped from regions with longer growing seasons come out ahead on the carbon emissions balance sheet because they do not need to be heated and lighted; therefore, the local produce is not the most environmentally friendly choice. However, local food advocates counter by pointing out that nonindustrial food producers are extremely unlikely to raise food in fossil fuel–powered greenhouses. They are usually too small to afford such infrastructure and cannot sell to a global food transporter. While consumers should not rely simply on the designation of "local" as an indication of sustainability, they should be able to trust food from nearby farms that focus on raising diversified crops in season with the local climate as the environmentally sound choice.

Small-scale operations that lack the capacity to ship long distances (whose products are local by default) also lack the resources to implement heavy industrial farming methods. Without large amounts of land, equipment, and capital, they may not be able to invest in pesticides, petroleum-based fertilizer, or tractors that can plant and harvest large monocropped fields to any great extent. Of course, only the U.S. Department of Agriculture "Certified Organic" label imposes concrete restrictions on farming practices, but the economics of farm management limit the artificial inputs that small producers can afford. Consumers must also be aware that not all farmers who practice sustainable management can officially advertise as such because they may not be able to pay the high fees associated with organic certification. Additional barriers exist for people who want to sell sustainable, local meat. Because of food safety laws in the United States, livestock can only be sent to a few large slaughterhouses, which may be far away or prohibitively expensive for many small producers.

As the local food movement becomes mainstream, consumers will have more options to choose from when considering which food to purchase and where to find it. Though the movement began to promote the small-scale production of healthy, traditional regional food in opposition to the dominance of food corporations, fast-food companies, and big agribusiness, the latter have significant incentives to incorporate local food into their stores and restaurants. A number of grocery stores have begun to promote local food to capitalize on demand, and all of them, per the 2009 Country of Origin Labeling law, must provide shoppers with information about where their food comes from. Whether this law and the wider local food movement will change food culture in the United States remains to be seen.

See Also: Agriculture; Berry, Wendell; Globalization; Organic Trend.

Further Readings

MacKinnon, J. B. and Alisa Smith. *The 100-Mile Diet: A Year of Local Eating*. Toronto, Canada: Random House Canada, 2007.

Pollan, Michael. *In Defense of Food: An Eater's Manifesto*. New York: Penguin, 2008.

Sustainable Table. http://www.sustainabletable.org (Accessed August 2010).

Sophie Turrell
Independent Scholar

Love Canal, Carter Administration and the

Love Canal was a little-known suburb of Buffalo, New York, before its clash with the U.S. government made international news in 1978. The residents' battle for protection against the toxic waste that plagued their community played a pivotal role in shaping America's environmental policy and philosophy. The struggles of Love Canal led the Carter administration to redefine what is considered a disaster, pass Superfund legislation, and might have cost Jimmy Carter the presidency. Its effect on policy was matched only by its effect on American environmentalism. Prior to Love Canal, the American environmental movement was defined primarily in terms of wilderness and wildlife protection. Love Canal expanded this definition to include issues of hazardous wastes. Investigations into U.S. toxic waste made the public aware that there were an "unknown number of Love Canals" throughout the country, reshaping people's perception of risk as seen in the development of the "Not In My Backyard" (NIMBY) phenomenon. The story of this community has left an indelible mark on the collective psyche of America and its environmental policy.

From 1942 to 1952, the Hooker Chemical Company buried roughly 21,000 tons of toxic chemicals on 16 acres of land. In 1953, Hooker's executives grew concerned that the encroaching suburbs were making their land a liability and they sold it to the local school board for $1. Provisions in the deed indemnified Hooker against all future liability. In 1954, a school was built on the grounds to service the growing suburban area of Love Canal. By 1978, this community directly abutted the site. It housed 2,618 residents, primarily blue-collar workers, none of whom held deeds that carried disclaimers about nearby chemicals.

The contamination of Love Canal set a precedent for future disaster relief and led to the start of the Superfund program in 1980. This U.S. Army Corps of Engineers official was inspecting a polluted lagoon during a 1982–84 Superfund cleanup in Bridgeport, New Jersey.

Source: U.S. Army Corps of Engineers

In the 1950s, the level of environmental concern among the U.S. population was not yet well developed. It was not until the 1962 publication of Rachel Carson's book *Silent Spring* that the broader public became aware of the possibility that chemical companies hide the harmful effects of their products. Carson is often credited with initiating the U.S. environmental movement. Despite Carson's concern about environmental toxins, this early environmental movement has been characterized as putting a higher priority on protection of wildlife and nature than on toxic waste issues. Love Canal residents gave the newly formed Environmental Protection Agency (EPA) this same characterization, arguing their prime objective was to protect nature, not people.

In 1978, only a year after Carter entered into office, the story of Love Canal hit national news. An international team monitoring the Great Lakes found pesticides in Lake Ontario and traced them back to the Love Canal area. This was featured in a local paper, setting off a firestorm among residents who were concerned about their health and plunging property values. Being the first elected president after the resignation of Richard Nixon, Carter entered into office during a time the public wanted accountability. Up until then, U.S. companies were not required to report on their hazardous waste disposal practices. Thus, when Love Canal first entered the public consciousness, there was little information about where toxic waste sites were located across the country and how toxics were previously disposed of. Furthermore, the laws regarding who was responsible for toxic cleanups were ambiguous. When the EPA tried to sue the responsible companies, it was unclear whether they could prosecute those who had passed along ownership of the land or had gone out of business.

In early August 1978, President Carter declared a state of emergency for Love Canal. To its residents this came with the expectation of relief funds. However, this was the first federal emergency in American history to be declared for something other than a "natural disaster." Unlike the sudden-impact natural disasters, which the federal government commonly dealt with, Love Canal offered a situation that was unprecedented. Man-made disasters developed gradually, allowed for blame, and offered those responsible the chance to mount a defense. The federal government had no protocol to follow. The real question was who (if anyone) would pay for remediation of the land and relocation of the residents. Everyone involved knew this would likely set the precedent for future disaster relief and

Carter appointees from the Office of Management and Budget and Federal Disaster Assistance Administration were reluctant to reinterpret existing laws, not knowing what implications this would have for their organizations' future responsibilities.

For over a year, the Love Canal community was trapped by governmental indecision, with no financial aid forthcoming. In March 1979, Congressional hearings began to examine the toxic waste issue. Soon after, Carter submitted a legislative proposal to Congress, which he would later sign into law as the Superfund Act. This bill established a $1.6 billion fund to clean up abandoned hazardous waste sites and allow the government to sue responsible parties for compensation. The bill, however, was criticized for only funding environmental remediation, leaving out provisions suggested by Carter for compensating community members.

The road to these legislative changes was littered with bad publicity for Carter. The protests of Love Canal activists, including those outside the Democratic Convention, were widely seen. These activities fed into the larger dialogue pushed by Carter's political opponents, who portrayed Carter as inept at dealing with crises, possibly contributing to his loss for reelection. This publicity brought a greater awareness of the toxic waste issue to the American public and an increase in risk perception. In doing so, it may have contributed to the increase in NIMBYism that still lingers today.

See Also: Environmental Justice; Precautionary Principle; *Silent Spring*.

Further Readings

"Environmental Policy, 1977–1980 Legislative Overview." In *Congress and the Nation, 1977–1980*. Washington, DC: CQ Press, 1981.

Levine, A. *Love Canal: Science, Politics, and People*. Lexington, MA: Lexington Books, 1982.

"Love Canal, A Special Report to the Governor and Legislature." Albany: New York State Department of Health, 1981.

Magoc, C. J. *Environmental Issues in American History: A Reference Guide With Primary Documents*. Westport, CT: Greenwood Press, 2006.

Kerry Ard
University of Michigan

Lovelock, James Ephraim

James Lovelock was born July 26, 1919, in Hertfordshire, England, and is an independent scientist, inventor, and environmentalist. He invented the electron capture detector for gas chromatography—an instrument whose exquisite sensitivity has subsequently been central to several important environmental breakthroughs. For example, during the 1960s it enabled the documentation of widespread dissemination of harmful and persistent pesticides like DDT, and later the technique was extended to the polychlorinated biphenyls (PCBs). Lovelock himself used the technique to chart the ubiquitous presence of chlorofluorocarbons (CFCs). He also developed instruments for exploring planets other than our own, including those aboard the two Viking spacecraft that went to Mars in 1975.

Lovelock is best known for the Gaia hypothesis, which presents the Earth as a carefully interconnected, self-regulating cybernetic "super organism" in which life creates and maintains the condition for life. This view of the planet and the life that lives on it as a single complex system, in some ways analogous to a homeostatically self-regulating organism, is what has given rise to the field we now know as "Earth System Science." He is one of the intellectual fathers of the modern green/environmental movement, having spent decades promoting the need for resource and environmental preservation and conservation.

In more recent years his status as an environmentalist has become more problematic and controversial. Lovelock has emerged since the mid-2000s as the world's leading climate pessimist and environmental stoic. By his estimation, it is not only too late for climate legislation as currently proposed—it is too late for any legislation, however radical. Cataclysmic climate change will hit in the coming century, he believes. Any efforts to pretend otherwise only delay the necessary work of preparing for the climate apocalypse. An early advocate of the need for concerted and urgent international action on climate change, he has since 2004 advocated nuclear power as a form of low-carbon energy that can help in the battle against climate change, while ensuring "the lights don't go out." As he put it in 2005, "I am a Green, and I entreat my friends in the movement to drop their wrongheaded objection to nuclear energy." He has also joined a group of green/environmental thinkers and activists who could be reasonably described as "hard green" in the sense of being extremely strong-headed and "survivalist" in their orientation about sustainability issues. Along with others such as "peak oil" and climate change authors such as James Howard Kunstler and Richard Heinberg or novelists such as Cormac McCarthy in his novel *The Road,* Lovelock's "ecological realism" has become pronounced in recent years. He is also a prominent supporter and member of the group Environmentalists for Nuclear Energy

In his 2006 book *The Revenge of Gaia,* Lovelock advocates what he terms a "sustainable retreat" in opposition to "sustainable development," which he dismisses as a utopian fantasy we missed out on implementing in the 20th century. His most recent view, in his latest book *The Vanishing Face of Gaia* (2009), claims that there is no point in trying renewable energy, carbon dioxide emissions trading systems, or attempts to negotiate international treaties on reducing carbon dioxide (CO2), recycling, or any of the other usual components of sustainable development. His apocalyptic account sounds remarkably like the earlier "limits to growth" predictions from the 1970s. Lovelock foresees crop failures, drought, death, and massive social disruption on an unprecedented scale around the globe. The population of this hot, barren world could shrink from about 7 billion to 1 billion by 2100 as people compete for ever-scarcer resources. "It will be death on a grand scale from famine and lack of water," Lovelock told Reuters in an interview. "It could be a reduction to a billion (people) or less." According to Lovelock, the human species should be adopting a "survivalist" perspective and investing in efforts to create safe havens in areas that will escape the worst effects of climate change. In *The Vanishing Face of Gaia* he states, "We have to stop pretending that there is any possible way of returning to that lush, comfortable and beautiful Earth we left behind some time in the 20th century." And in a chilling statement: "The Earth, in its but not our interests may be forced to move to a hot epoch, one where it can survive, though in a diminished and less habitable state. If, as is likely, this happens, we will have been the cause."

However, despite his pessimism, Lovelock (2009) does see some positive and redeeming features of what for him is the inevitable transition from a fossil fuel to a low-carbon economy:

> Just as in 1939 we had to give up on a massive scale the comfortable lifestyle of peacetime, so soon we may feel rich with only a quarter of what we consume now. If we do it right and with enthusiasm, it will not seem a depressing phase of denial but instead, as in 1940, a chance to redeem ourselves. For the young, life will be full of opportunities to serve, to create, and they will have a purpose for living.

See Also: Climate Ethics; Club of Rome (and Limits to Growth); Conservation; Gaia Hypothesis; Preservation.

Further Readings

Kristof, N. "Nukes Are Green." *New York Times* (April 9, 2005).
Lovelock, J. *The Revenge of Gaia*. London: Allen Lane, 2006.
Lovelock, J. *The Vanishing Face of Gaia: A Final Warning*. London: Allen Lane, 2009.

John Barry
Queen's University Belfast

M

Marketing, Consumption Ethics and

The CEO of an association of peanut farmers in Malawi celebrating a Fair Trade sales agreement, which included a $110 per ton premium, with supermarket executives from the United Kingdom in 2006.

Source: U.S. Agency for International Development

The relationship of consumption ethics and marketing is an important focus in sustainability. The tendency of businesses to market products that consumers may not need or to market products produced by questionable practices are two common areas of concern. Much of the focus on consumption ethics turns toward individual consumer choices. Frances Moore Lappé's *Diet for a Small Planet* was a popular, early 1970s call to rethink individual consumption habits.

However, focus on the capitalist system and marketing's role in consumption has also been popular. John Kenneth Galbraith, in his 1958 book *The Affluent Society,* highlights (as did earlier Marxist critiques of capitalism) the strong influence of production, marketing, and advertising in determining consumption preferences and behavior. Vance Packard's *The Hidden Persuaders* was another important, critical look at marketing from the 1950s, focusing on psychological research and subliminal messages to induce desire for products.

William Leach also suggests, in his 1993 book *Land of Desire: Merchants, Power, and the Rise of a New American Culture,* that overconsumption is fostered by corporate marketing. He traces the trend back to the late 19th and early 20th century, when corporations started to convince previously frugal and spiritual citizens of the benefits of the "good life." Leach notes that "the average American (now) sees approximately 21,000 commercial messages per year."

Donald Mayer also illustrates, in the anthology *The Business of Consumption,* how corporate marketing and domestic and international law have institutionalized overconsumption. He cites root causes, such as valuing consumption as an "end in itself, rather than a means to some higher human purpose." Mayer points out that the U.S. economy was not always built on consumer demand, but now it consumes two-thirds of our economy, making both citizens and businesses recoil at the prospect of cutting back on consumption.

Despite these critiques that place responsibility on the capitalist system, corporate marketing, or deficient government policies, consumers themselves are still a focus of moral discussions about overconsumption. Some reviewers such as John Peloza are pessimistic, suggesting most consumer response—for example, in the case of smaller cars—is based on gas prices, not social responsibility. He calls for refocusing responsibility on both the consumer and marketing efforts.

Recent studies indicate that green marketing may find a suitable customer base. Studies in wildlife conservation and conservation psychology, respectively, suggest that self-identity formed by peer pressure and early life experiences with nature are primary motivators in sustainable behavior. Thus a status symbol of conspicuously different consumption could evolve that would provide an expanding customer base.

Martha Nussbaum outlines the Aristotelian capability approach in the anthology *Ethics of Consumption: The Good Life, Justice, and Global Stewardship.* She authors a chapter on capability, based on work by herself and Nobel Prize–winner Amartya Sen, which suggests basing most policy decisions on commonly agreed norms for human flourishing. These norms can often be agreed on by surprisingly diverse cultures. Thus green marketing could achieve economies of scale. One example is Fair Trade. The CEO of TransFair, a U.S. labeling organization, suggested the term *fair* is a historical core American value that transcends less important values of organic, sustainable, and so on. Approximately half of citizens recently surveyed have unclear notions of the definition of sustainability.

One of the more recent, broad investigations into consumerism is the Cultures of Consumption research program in the United Kingdom that involved 26 projects over a five-year span. Authors reinforce the importance of green marketing efforts such as Fair Trade, which seems off to a bigger start in the UK than in the United States. They suggest that green marketing campaigns are more effective at the collective level, such as creation of Fair Trade towns. According to these investigations, as individuals, "consumers feel easily overwhelmed by appeals to change their own lifestyle to save the planet." It may be more effective to emphasize "sensual and spiritual pleasures of a different lifestyle, say, one less dependent on cars, noise, and traffic jams."

Ethical Consumption

One of the projects more specific to ethical consumption was conducted by geographers, led by Clive Barnett, and titled "Governing the Subjects and Spaces of Ethical Consumption." Barnett et al. suggest that ethical consumption opens up new spatially complex ways to look at a relationship typically seen as linear and information based—falsely assuming that if an individual consumer has the right information, good decisions and policy will follow to adjust a linear commodity chain. Sometimes Fair Trade links are set up directly between a consuming and a producing town. However, the nonprofit aid group Oxfam switched to purchasing directly from aid villages to buying from a Fair Trade network, deciding instead to focus aid on helping producers gain access to markets.

Alex Nicholls and Charlotte Opal note that Fair Trade involves more than sustainability. One of its primary focuses is to provide fair markets for developing-nation producers, who are often exploited by middlemen and low prices in traditional commodity chains. Aspects of Fair Trade include agreed-on minimum prices; direct purchasing from producers; extra funding (often 10 percent) for development and technical assistance; cooperative, not competitive, dealings; provision of credit and market information; no labor abuses; democratically run cooperatives; and sustainable production. Sustainable production involves mandatory resource management plans, prohibition of certain pesticides, and often organic certification (which also brings higher prices).

Fair Trade started after World War II with nongovernmental aid organizations such as Oxfam. Highly diverse handicrafts and textiles dominated the early market, but since the 1990s the primary products are food, which are more easily certifiable and are marketed as not only responsible but high-quality products. The rise of Fair Trade has been exponential in cases such as coffee in the UK and United States. Fair Trade bananas gained a 50 percent market share in Switzerland. A survey indicated 65 percent of UK consumers consider themselves "green" or "ethical" consumers. Free-range eggs captured 40 percent of the UK egg market.

Nicholls and Opal include examples of success at both ends of the commodity chain, including a before-and-after story at a cacao bean cooperative of 60,000 producers in Ghana; differences between villages with and without Fair Trade programs in Bangladesh; and the remarkable rise of Fair Trade products in the UK supermarket chain Tesco, from 12 percent to 32 percent of the UK market in one year. Fair Trade has been so successful that related marketing efforts are quite competitive, and numerous producers outside the certification process who are less profitable find themselves identified simply as "alternative" trade. Certification systems in forestry also vary, as the more stringent Forest Stewardship Council (FSC) is endorsed by major environmental organizations, while the less-stringent Sustainable Forestry Initiative is endorsed by pulp and paper companies. FSC has increased its scope to 100 million hectares globally since starting in the 1990s. Half of Switzerland's population now recognizes the FSC label on paper and wood products.

The most philosophically oriented recent book on consumption ethics is titled *The Ethical Consumer.* The introductory chapter, by Clive Barnett, Phil Cafaro, and Terry Newholm, addresses consequential and deontological aspects of ethical consumption and finds that virtue ethics may be more applicable for participants. Consequential approaches focus on "good" outcomes, often on what is best for most (this includes but is not restricted to utilitarianism). Teleological approaches, such as having consumption as the end goal (described by Mayer), fall under this category. The authors note that the most prominent philosopher in animal rights, Peter Singer, uses a consequential approach to illustrate the effects of various amounts and types of consumption on remote animals. Information is a key driver in defining what is "good." The authors see this reliance on information (vast amounts of which are often required to make a choice) and a correspondingly vague notion of "good" as problematic.

Deontological approaches focus on what "ought" to be done as the "right" action, often in regard to moral rules (this includes Immanuel Kant's universal moral imperatives and John Rawls's theory of justice). Rawls in particular took exception to consequential/teleological approaches, especially utilitarianism, as they could allow exploitation of some if the results benefited the greatest number. He sought a more pluralistic approach incorporating multiple values into one set of rules. The authors see this approach as also problematic, in that it relies on highly universalized, general assumptions (such as the precautionary principle) being applied to specific, unique cases.

For Barnett, Cafaro, and Newholm (as well as Martha Nussbaum and Amartya Sen), the tendency to simply call for less consumption, as outlined in consequential and deontological approaches, naively ignores the role of goods in shaping people's identities. They also claim that actions (such as boycotts) are not simply altruistic but also help form self-identity. Thus, both consequential and deontological ethics set up a false dichotomy of good/bad or inside the rules/outside the rules.

The authors suggest that virtue ethics provides a more flexible framework to consider cases such as child labor, which can be exploitive; yet in some cases, child labor is adequately policed and greatly benefits poor laborers, while higher prices may affect the purchasers' families if their income is also marginal. The question then would focus on what kind of virtues the purchaser wants to reflect, such as compassion and generosity. A remaining danger in virtue ethics is a tendency to assume one culture has certain virtues that other, more remote cultures may not, or may have less of. Thus a tension to be resolved is finding a combination of spatially closer "intimate caring" and broader "humanitarian caring" within virtue-based consumption ethics.

Barnett, Cafaro, and Newholm also differentiate between the terms *ethical consumption,* seen as a mode to display personal values, and the *ethics of consumption,* more of a critique of modern capitalism. They note that there is also a false dichotomy between ethical consumption and traditional consumption, which can alienate the general public. Traditional consumption may involve more than self-interest, extending concerns to family, friends, and so on. Barnett, Cafaro, and Newholm call for more everyday public participation in debates over the meaning and purpose of ethical consumption.

Categories of Green Consumers

A final source, Joel Makower's *Strategies for a Green Economy,* offers related ideas and cautions on marketing and ethical consumption. The Ecological Roadmap, conducted by Earth Justice's Cara Pike based on the American Values Survey, is summarized in the book. She segments the U.S. population based on environmental concern. The "Greenest Americans," 9 percent of the population, are the only group to highly value ethical consumerism. They are highly skeptical of "greenwashing," the selling of a company's products based on image rather than on environmental performance. The Greenest Americans thus scrutinize advertising that is not endorsed by environmental or Fair Trade groups. Even this group indicates making individual choices in ethical consumerism is a "chore."

The largest group, making up 24 percent of the population, is called Compassionate Caretakers, based largely on the value that "healthy families need healthy environments." This group is seen as the next green market, as ecological concern and ethical consumerism are important values, just below flexible families and group egalitarianism. As in the discussion on virtue ethics, this group places value on "being a good person rather than politics or individual expression." Marketing keys are that green products should be "affordable, of good quality, and available where they already shop."

Other groups described as Postmodern Idealists (3 percent) and Borderline Fatalists (5 percent) also value ethical consumerism for different reasons. The four groups combined make up 41 percent of the population that are a potential existing market for ethical consumerism. The largest group resistant to ethical consumerism is the Proud

Traditionalists (20 percent), although increasing links between organized religions and ecology may have future influence. Similar studies on emerging economies such as China and India are appropriate.

Consumer ethics and marketing is an arena involving complex issues that may best be solved not only through advanced information and technology but through value systems flexible for different scenarios. Consumption will be ingrained into culture in the future. Operating in collectives helps reduce complexity of environmentally friendly choices and problems linking intimacy with remoteness. Markets for green consumerism may link groups of different values and thus make up a larger potential customer base than many realize.

See Also: Consumption, Business Ethics and, Ecological Footprint; Greenwashing; Sustainability, Business Ethics and.

Further Readings

Barnett, Clive, Phil Cafaro, and Terry Newholm. "Philosophy and Ethical Consumption." In *The Ethical Consumer,* Rob Harrison, Terry Newholm, and Deirdre Shaw, eds. London: Sage, 2005.

Barnett, Clive, Paul Cloke, Nick Clarke, and Alice Malpass. "Articulating Ethics and Consumption." Working Paper Series: Cultures of Consumption. School of Geographical Sciences, University of Bristol, UK, 2004.

Clayton, Susan and Gene Myers. *Conservation Psychology: Understanding and Promoting Human Care for Nature.* Oxford, UK: Wiley-Blackwell, 2009.

Crocker, David A. and Toby Linden, eds. *Ethics of Consumption: The Good Life, Justice, and Global Stewardship.* Lanham, MD: Rowman & Littlefield, 1997.

Fair Trade Federation. http://www.fairtradefederation.com (Accessed March 2010).

Fair Trade Labeling Organizations International. http://www.fairtrade.net (Accessed March 2010).

Forest Stewardship Council. http://www.fscus.org (Accessed March 2010).

Galbraith, John Kenneth. *The Affluent Society.* Cambridge, MA: Riverside Press, 1958.

Leach, William. *Land of Desire: Merchants, Power, and the Rise of a New American Culture.* New York: Random House, 1993.

Makower, Joel. *Strategies for the Green Economy: Opportunities and Challenges in the New World of Business.* New York: McGraw-Hill, 2009.

National Science Foundation. "Peer Pressure Plays Major Role in Environmental Behavior." Press Release 09-132. http://www.nsf.gov/news/news_summ.jsp?cntn_id=115049 (Accessed March 2010).

Nicholls, Alex and Charlotte Opal. *Fair Trade: Market-Driven Ethical Consumption.* London: Sage, 2005.

Packard, Vance. *The Hidden Persuaders.* Brooklyn, NY: Ig Publishing, 1957.

Peloza, John. "What About Consumer Responsibility?" Commentary. Canadian Imperial Bank of Commerce Centre for Corporate Governance and Risk Management, 2008. http://business.sfu.ca/cibc-centre/comments (Accessed March 2010).

Trentmann, Frank. "4-1/2 Lessons About Consumption: A Short Overview of the Cultures of Consumption Research Programme." Birkbeck College, University of London: http://www.consume.bbk.ac.uk/researchfindings/overview.pdf (Accessed March 2010).

Westra, Laura and Patricia H. Werhane, eds. *The Business of Consumption: Environmental Ethics and the Global Economy.* Lanham, MD: Rowman & Littlefield, 1998.
World Fair Trade Organization. http://www.wfto.org (Accessed March 2010).

William Forbes
Stephen F. Austin State University

Marsh, George Perkins

George Perkins Marsh is a figure often overlooked by historians of Western conservation movements and overshadowed by the later achievements of individuals such as Gifford Pinchot, John Muir, Aldo Leopold, and Rachel Carson. Nonetheless, through his groundbreaking work *Man and Nature* (first published in 1864), Marsh provided one of the earliest arguments supporting the conservation of natural resources in the United States. His work was highly influential to later conservationists such as Pinchot and Muir and set a high standard for the scientific analysis of human impacts on the natural world. Though historians have shown that many of Marsh's ideas were not entirely unprecedented for his time, he nonetheless deserves acknowledgment as one of the fathers of conservation and scientific resource management in the modern West.

George Perkins Marsh was born in 1801 into a wealthy family in Woodstock, Vermont. He was an exceptionally precocious child, developing a life-long love of languages and the natural world. His multiple intellectual interests created the great breadth of knowledge that, later in his life, made the writing of *Man and Nature* possible. Following graduation from Dartmouth College in 1820, Marsh briefly served as a professor of Greek and Latin at Vermont's Norwich Academy before working as a lawyer in Burlington. Due in part to his upper-class origins and social connections in Vermont and abroad, Marsh rose quickly in local and national politics, eventually becoming a member of the U.S. House of Representatives. In 1849, Marsh was appointed as United States' minister to Turkey, where he served until 1853. This brief tenure in Turkey proved highly influential to Marsh's ecological thought. In the Mediterranean region, Marsh observed desertification, erosion, and soil depletion based on centuries of overly exploitative farming practices. Upon returning to Vermont in 1854, Marsh was surprised to see similar damaging practices among American farmers and loggers. Without careful intervention and planning, he realized, the United States could end up denuded of resources, much like portions of Turkey. Marsh devoted himself intently to surveying scholarly literature on the environmental sciences and resource depletion in an effort to find further connections between human activities and environmental damage. In 1861, President Abraham Lincoln appointed Marsh as the United States' first minister to Italy. Back in the Mediterranean region, Marsh completed his research and, in 1864, published it as *Man and Nature.* He later revised and expanded the work and, in 1874, renamed it *The Earth as Modified by Human Action.*

Man and Nature presented one of the earliest and most thorough analyses of the adverse changes to the natural world caused by human actions. Marsh believed that many of the damages he witnessed in Turkey and Italy could easily be avoided in the United States with the implementation of scientific resource extraction policies. Beyond simply advocating resource conservation and restoration, though, Marsh articulated a kind of enlightened anthropocentrism, influential to later conservationists, and warned of impending ecological disasters if changes were not made. While Marsh made an anthropocentric argument for conservation—it was for human flourishing, not for any inherent values or rights of nature, that resource management was necessary—he questioned the religious anthropocentrism of many of his contemporaries who believed that God specifically gave the Earth to humans for exploitation.

Like many early ecologists, Marsh also believed that the natural world tended toward equilibrium. Humans, with their independent rational abilities, were the creatures most able to significantly unbalance that natural equilibrium. In order to protect their societies and the continued usefulness of the natural world, then, humans needed to scientifically manage their resource use. Despite his confidence in the ability of humans to manage their environments, Marsh did not believe that success was inevitable. One of the most unique elements of *Man and Nature*—one that resonated later with environmentalists in the 20th century—was Marsh's contention that, without proper management, humans could make the Earth unsupportive for life. Moreover, Marsh believed that humans were already on that course and needed to work diligently to turn themselves toward more sustainable ends. Humans were not necessarily headed toward God-given success over nature, he thought, but through their own scientific ingenuity humans could still manage to improve their lives while preserving the resources of nature for future generations.

Marsh died in Florence in 1882. Of his many lifetime achievements, from his political positions to groundbreaking linguistic work, *Man and Nature* perhaps stands as the most significant. As the United States increasingly drove toward expansion and industrial development following the Civil War, Marsh's was one of the few voices calling for restraint. Though others recognized similar problems, *Man and Nature* provided a foundation for later scientists and politicians to continue Marsh's research and develop strategies for resource conservation and wilderness preservation in the United States.

See Also: Conservation; Muir, John; Pinchot, Gifford; Roosevelt, Theodore.

Further Readings

Judd, Richard W. "George Perkins Marsh: The Times and Their Man." *Environment and History*, 10/2 (2004).

Lowenthal, David. *George Perkins Marsh: Prophet of Conservation*, rev. ed. Seattle: University of Washington Press, 2000.

Marsh, George Perkins. *Man and Nature*, David Lowenthal, ed. Seattle: University of Washington Press, 2003 (1864).

Marsh, George Perkins. *So Great a Vision: The Conservation Writings of George Perkins Marsh*, Stephen C. Trombulak, ed. Hanover, NH: University Press of New England, 2001.

Joseph Witt
University of Florida

Marshall, Alan (Libertarian Extension)

Alan Marshall is a significant scholar in the field of environmental ethics and its history. He has identified three broad approaches to environmental ethics over the last 40 years: the Libertarian Extension, the Ecologic Extension, and Conservation Ethics.

The Libertarian Extension includes the work of philosophers Arne Naess (who used the term *deep ecology*) and Peter Singer; it is characterized by its similarity to libertarianism in their shared emphasis on equal rights and civil liberties. But while libertarianism is politically concerned only with humans or the local community, the Libertarian Extension extends civil liberties to nonhumans as well. Not to do so would be to be guilty of *speciesism,* a term coined by British psychologist Richard D. Ryder in his work in the early 1970s to describe "the widespread discrimination that is practiced by man against other species . . . and like all discrimination it overlooks or underestimates the similarities between the discriminator and those discriminated against." Ryder originally used the term in his work opposing animal experimentation, though some commenters in the years since have constructed ethical arguments in favor of (certain types of) animal experimentation that claim not to be speciesist. Others have accepted the speciesist label and described it as a rational, acceptable attitude for humans to take.

Different philosophers whom Marshall would group as part of the Libertarian Extension take different positions on what criteria exist for a species to be granted those equal rights. Singer was one of the earliest prominent writers to defend sentience as sufficient grounds; Ryder originally agreed but in recent years has introduced the awkward neologism *painience* as his criterion. His 2005 opinion piece for the *Guardian,* "All Beings That Feel Pain Deserve Human Rights," summed up the position in its title. It is this extension of the notion of civil rights that has inspired some animal rights activists to compare the human treatment of animals to the treatment of fellow humans during the Holocaust.

Singer, though, took a "biocentric" or "sentience-centric" view of the Libertarian Extension, distancing himself from deep ecology's argument that the environment itself—and thus the nonsentient entities within it, such as plants and mountains—has an intrinsic value as great as that of humans and other animals.

The Ecologic Extension includes much of deep ecology, as it depends on the fundamental interdependence of all living things—the complexity of the living world, which is harmed as a whole whenever harm is done to one of its constituent parts. James Lovelock's Gaia hypothesis, which assigns a value to the planet Earth as a whole and treats it as a single complex organism, is the best-known example of the Ecologic Extension.

Marshall's third category, Conservation Ethics, focuses on the environment purely or primarily in terms of its usefulness to humans. Conservation Ethics is the oldest form of environmental ethics, the rhetoric responsible for the creation of the national parks system, for wildlife preserves, for water conservation efforts, and later for the responsible use of nonrenewable resources. Why conserve use of petroleum fuels? Not because crude oil has an intrinsic worth and the Earth would be lessened without its presence, but because of the energy crises that would be caused in human societies as oil becomes first scarce, then rare, then a thing of the past. Conservation Ethics rhetoric is notable, then, for being used at times even by parties who do not consider themselves environmentalists or concerned with green notions (which is not to say that such ethics are opposed to green thought).

See Also: Anthropocentrism; Biocentrism; Climate Ethics; Deep Ecology; Gaia Hypothesis; Singer, Peter.

Further Readings

Carter, Alan. *A Radical Green Political Theory.* London: Routledge, 1999.

Dobson, Andrew. *Green Political Thought.* London: T&F Books, 2009.

Marshall, Alan. "Ethics and the Extraterrestrial Environment." *Journal of Applied Philosophy,* 10/2 (2008).
Marshall, Alan. *The Unity of Nature.* London: Imperial College Press, 2002.
Radcliffe, James. *Green Politics: Dictatorship or Democracy?* New York: Palgrave Macmillan, 2002.
Ryder, Richard D. *Victims of Science: The Use of Animals in Research.* London: Davis-Poynter, 1975.
Torgerson, Douglas. *The Promise of Green Politics: Environmentalism and the Public Sphere.* Durham, NC: Duke University Press, 1999.

Bill Kte'pi
Independent Scholar

Mathews, Freya

Freya Mathews (1949–) is a well-known Australian ecological philosopher and ecofeminist. She has taught in Australian universities since 1979 and until 2008 was associate professor in the Philosophy Program at La Trobe University, Melbourne, Australia, where she coordinated the Environmental Enquiry major and coedited the journal *PAN* (Philosophy Activism Nature). She is currently a research fellow at La Trobe. Her writings focus on ecological philosophy, metaphysics, epistemology, ethics, and politics as well as a variety of themes, which she identifies as cosmology, place, identity, and indigeneity versus modernity; many of these writings are available on her Website. She cites Baruch Spinoza as her main philosophical influence. Although her work is highly regarded in academic environmental philosophy circles, her approach is wider than that of most academics, especially in its emphasis on spirituality. She is also a published poet, and many of her philosophical writings have a poetic quality. Central to her thinking is the idea of the "mythopoetic," lyrically presented in an article on the Centre for Education and Research in Environmental Strategies (CERES).

Mathews is best known in academic circles for her 1991 book *The Ecological Self,* in which she addresses the question: how can an ecological view of the world change our understanding of nature and the human self? She believes this is fundamentally a metaphysical question that cannot be adequately answered by conventional environmental philosophy, which, she believes, fails to address the spiritual implications of ecology. While other environmental philosophers have recognized the fundamental truth of ecology—that everything is connected—they have focused on the material connections, ignoring the possibility that our relationship with the world can be "dialogic"; in her later works she explores panpsychism, an ancient metaphysical doctrine that everything in the universe is conscious and aware. There is thus a sense in which the universe itself has—is—a self and this is why developing a deeper understanding of the world leads to a deeper understanding of one's own self. Panpsychism was displaced by modernism, with the result that nature came to be seen as an inert background to human activity and thus as legitimately subject to human control and exploitation. We cannot hope to solve the "environmental crisis" unless we reject this view and embrace a truly ecological metaphysic.

The recognition of the interconnectedness of all things, which she refers to as "ecocosm," leads one to recognize that the universe, like an individual self, is "self-realizing" and this is what gives intrinsic value to both systems and individuals. Our lives find meaning not merely in our own self-realization but also in our relationship with the wider self: "Meaningfulness is to be found in our spiritual capacity to keep the ecocosm on course, by teaching our hearts to practice affirmation, and by awakening our faculty of active, outreaching, world-directed love."

Despite her criticism of the limitations of environmental philosophy, Mathews shares many of the views of its leading scholars. In particular, she believes that the massive environmental damage that humans have caused has an ideological basis: that is, we have been able to exploit the Earth because we believe that it is ours to exploit, for our own goals. In this anthropocentric view, only humans have intrinsic value—value as ends; everything else has only instrumental value—value as a means. As philosophers, we need to show the weakness of this theoretical underpinning and develop new, fruitful ways of thinking about the world from which environmentally responsible behavior can flow. Second, she recognizes that our understanding of the world must be based on ecology; hence her emphasis on the interconnectedness of all things. Third, she values wild nature highly and advocates that humans intervene in nature as little as possible rather than trying to mold it to some human ideal; however, she advocates rehabilitation of damaged areas.

Like other ecofeminists, such as Karen Warren, Mathews rejects what is often regarded as a central tenet of deep ecology, namely, its rejection of the distinction between the individual and the world, indeed its rejection of the concept of the individual, insofar as deep ecologists advocate a complete merging of the self with the world. In Mathews's worldview there are many selves, and their relationships with each other occur at many levels and both contribute to self-realization and are sources of intrinsic value. Thus her love poems often interweave her feelings for her beloved with vivid accounts of nature, while nature is addressed as "my beloved." Another value that she shares with other ecofeminists is her emphasis on the importance of place. This is reflected in her belief that ecological philosophy has much to learn from the traditions of people who have a long-standing history of living in an area, and she frequently cites material from indigenous traditions, including those of China and, especially, Australia. While she recognizes that Europeans have been in Australia for only two centuries (whereas the aboriginal people have lived there for at least 40,000 years), she advocates that an ecological philosopher spend lengthy periods in a particular relatively natural location in order to establish a familiarity with and love for a place. Again, like Warren and others, she makes no claim to "objective" knowledge of metaphysics and value and describes her work in terms such as *exploration, suggestion,* and *proposal.*

The most distinctive feature of Mathews's philosophy is its emphasis on spirituality, which permeates her work. While she is a philosopher in the broadly analytic tradition, with no affiliation to organized religion, she believes that the Western tradition has neglected our faculties of intuition and poetic perception and responsiveness, which she believes can be cultivated through spiritual exercises conducted in colonies of friends.

See Also: Anthropocentrism; Deep Ecology; Ecofeminism/Ecological Feminism; Warren, Karen.

Further Readings

Freya Mathews. http://www.wisenet-australia.org/profiles/freyam.htm (Accessed February 2010).

Mathews, Freya. "CERES, Singing Up the City." *PAN,* 1 (2000). http://search.informit.com.au/documentSummary;dn=758422922060153;res=IELHSS (Accessed February 2010).

Mathews, Freya. *The Ecological Self.* New York: Routledge, 1991.
"Panpsychism." *Stanford Encyclopedia of Philosophy.* http://plato.stanford.edu/entries/panpsychism/#2 (Accessed February 2010).

Alastair S. Gunn
University of Waikato

Military Ethics and the Environment

Up to 20 percent of worldwide environmental degradation may be attributable to the world's militaries. These sailors from a U.S. guided missile destroyer in Souda Bay, Greece, in 2008 were carrying out a standard U.S. operating procedure of deploying oil spill containment booms around ships during refueling.

Source: U.S. Navy/Paul Farley

The world's militaries are estimated to be the largest single polluter and consumer on Earth, accounting for up to 20 percent of global environmental degradation. No international treaty drafted to protect the environment has been invoked as a direct response to war. The ecological marginalization of people as well as both environmental scarcities and environmental concentrations of natural resources can contribute to civil violence and armed conflicts, and global climate change is now regarded as a threat to the national security of the United States and other nations. Somewhat surprisingly, there is little discussion of any of this in either environmental ethics or military ethics literature.

Military ethics commonly fall under the rubric of the just war tradition (JWT). The JWT is one of four classic positions on the ethics of war and peace along with militarism, realism, and pacifism. Militarism is characterized by a cultural bias or even lust for war, an intrinsic value placed on war, and oppositional "us versus them" thinking whereby anyone not enthusiastically on one's own side in a conflict is an enemy. To the degree that wars and military activities result in deleterious environmental effects, militaristic nation-states and nonstate political actors can exacerbate these effects.

Realism is defined by the claim that ethics is not relevant in war or that ethics should be set aside in matters pertaining to war. Historically, many political scientists and international relations scholars have claimed theoretical allegiance to realism, and there has been considerable discussion of the relationship between the environment and armed conflicts in terms of "environmental security" where environmental degradation and resource use are seen as contributing—and sometimes major—causes of violent conflicts and wars. From a realist perspective, the impacts of wars and military activities are desirable or undesirable insofar as these impacts help or hinder the vital interests and national security of states and other political actors.

War and the Environment

Pacifism is a moral renunciation or opposition to war. Typical-variety antiwar pacifists argue that all wars should be avoided and indirectly address military impacts on the environment by seeking to prevent the impacts from occurring in the first place. Pacifism can take on a variety of different forms that include the moral condemnation of war and war systems because of their negative environmental impacts—what some call "ecological pacifism."

Military impacts on the environment are commonly grouped together under the label *ecology of war.* A partial list of these impacts includes the following: (1) formation of craters and compaction, erosion, and contamination of soils; (2) other forms of land pollution; (3) defoliation, deforestation, and land degradation; (4) contamination of surface waters and groundwater; (5) atmospheric emissions and air pollution; (6) direct and collateral killing of animals and plants and loss of habitat; (7) degradation and destruction of protected natural areas; (8) noise pollution; (9) damage and destruction of agricultural products, foodstuffs, water storage and distribution systems, waste collection systems, human structures, and power grids; and (10) dislocations of human populations and social and economic structures.

There are three environment-specific international treaties that regulate the conduct of war. Concerns about the U.S. use of herbicides in the Vietnam War motivated the United Nations Convention on the Prohibition of Military or Any Other Hostile Use of Environmental Modification Techniques of 1976—known as the ENMOD Convention. ENMOD prohibits the deliberate modification of natural forces and environmental modifications such that they can be used as weapons of war. The second environment-specific international treaty is the Protocol Additional to the Geneva Conventions of 12 August 1949 and Relating to the Protection of Victims of International Armed Conflicts—known as Protocol I (1977). Article 35(3) of Protocol I prohibits any methods or means of warfare intended or expected "to cause widespread, long-term and severe damage to the natural environment." The third environment-specific international treaty is the Rome Statute of the International Criminal Court, July 17, 1998. Article 8.2(b)(iv) of the Rome Statute defines causing "widespread, long term and severe damage to the natural environment" as a war crime. The United States has never ratified Protocol I or the Rome Statue. No international environmental treaty has ever been enforced.

The JWT classically is separated into *jus ad bellum* criteria and *jus in bello* criteria. The former consists of conditions that must be met for a war to qualify as a just war, and the latter concerns conditions that must be met for a war to be justly fought. *Jus ad bellum* criteria typically consist of legitimate authority, just cause, right intention, macro proportionality, likelihood of success, and last resort. *Jus in bello* criteria typically consist of micro proportionality, discrimination or noncombatant immunity, and military necessity.

Jus in bello might be the natural place to situate environmental considerations within the JWT. There have been various limitations placed on the damage and destruction of buildings, towns and cities, and the economic base of civilian life that go back at least as far as the book of Deuteronomy (20:19–20) that contains an injunction concerning the battlefield destruction of trees. Destruction of the environment can be regarded as a violation of proportionality insofar as environmental consequences typically outlast the duration of wars. Destruction of the environment also can be regarded as a violation of discrimination insofar as the environment is regarded as a noncombatant akin to civilians, regardless of whether the environment intrinsically has value in and of itself or is instrumentally valuable as resources upon which humans depend. Finally, destruction of the environment can be regarded as a violation of military necessity because victory only necessitates destroying enemy (human) forces and the enemy's military and political cohesion.

The Role of Military Necessity

Military necessity, however, cuts both ways. In the heat of battle, military necessity will probably override environmental protection, no matter how well aimed and informed this protection is. Time runs short to make decisions and deliberate over the possible ecological consequences of one's actions, and military commanders might not think twice about blowing up a rock formation that possibly contains a metapopulation of endangered species of *Parnassius charltonius* butterflies in order to protect their own troops. Invoking some reference to a doctrine of double effect that excuses the destructive, secondary, and unintentional effects of one's actions might justify most environmental destruction as regrettable but unavoidable collateral damage.

One possible way out of this military necessity trap is to more closely tether *jus in bello* criteria to *jus ad bellum* criteria such that the foreseen and predictable negative environmental consequences of war—regardless of their unintentional collateral status—are factored into moral deliberation about whether or not to go to war. If an upcoming war might involve significant attacks against human and/or nonhuman environmental targets, and if adherence to the *jus in bello* criteria of proportionality and discrimination is not likely to be forthcoming, this might count as a reason not to go to war. The precautionary principle could be invoked as an application of the *jus ad bellum* criterion of proportionality, building environmental considerations into the macro measure calculation of overall harms and benefits that must be balanced in order to prosecute a just war. Proportionality is related to a likelihood of success criterion, and damage and destruction of environmental targets might count against satisfying this criterion. Finally, if the environment is to be proactively protected, sacrificing it for political and/or military goals is problematic. This might factor into the just cause criterion in that a defense of one's homeland should not include military activities that damage or destroy the land in order to supposedly protect the home. The just cause that informs humanitarian interventions to stop violations of human rights abuses and/or help populations to meet their basic needs for survival might also be extended to include ecological interventions to prevent significant environmental destruction.

Wartime military conduct might be further restrained or prevented by more effective articulation of newly emerging *jus post bellum* criteria. Given that negative environmental impacts usually linger after wars end, just settlements could include ecological restoration and rehabilitation to repair human and nonhuman communities and restorative justice to repair environmental relationships. A better state of peace might require some notion of environmental health and an environmental likelihood of success criterion through which it can be measured.

See Also: Climate Ethics; Environmental Justice; United Nations Conference on the Human Environment.

Further Readings

Austin, Jay E. and Carl E. Bruch, eds. *The Environmental Consequences of War: Legal, Economic, and Scientific Perspectives.* Cambridge, MA: Cambridge University Press, 2000.

Drucker, Merrit P. "The Military Commander's Responsibility for the Environment." *Environmental Ethics,* 11 (1989).

Eckersley, Robyn. "Ecological Intervention: Prospects and Limits." *Ethics and International Affairs,* 21 (2007).

Reichberg, Gregory and Henrik Syse. "Protecting the Natural Environment in Wartime: Ethical Considerations From the Just War Tradition." *Journal of Peace Research,* 37 (2000).

Woods, Mark. "The Nature of War and Peace: Just War Thinking, Environmental Ethics, and Environmental Justice." In *Rethinking the Just War Tradition,* Michael W. Brough, John W. Lango, and Harry van der Linden, eds. Albany: State University of New York Press, 2007.

Mark Woods
University of San Diego

Muir, John

Renowned preservationist and founder of the Sierra Club John Muir is known for his work advocating the preservation of wilderness and natural places in America during the 19th century. His passionate views and writing about nature and the environment have constituted an important part of the American environmental movement.

Born in Dunbar, Scotland, in 1838, Muir moved to America in 1849 at the age of 11, first settling in the state of Wisconsin, where his father acquired land for farming. The land required much effort from the Muir family, including from Muir himself, the third child and eldest son in a family of eight children. Although Muir is better known for his work in the western United States, he retained strong links to Wisconsin throughout his life and explained that his early years there were important in developing his environmental ethic and respect for nature. In his early 20s, Muir spent time studying botany and philosophy at the University of Wisconsin–Madison but left without obtaining a degree. The university ultimately awarded him an honorary degree in 1897.

During the Civil War, Muir left Wisconsin to live and work near Lake Huron in Canada, where he would not be subject to the Union draft. He continued to pursue his interest in botany when time allowed, and eventually returned to the United States after the war to work in a factory in Indianapolis. It was here that many have traced a pivotal change in direction for Muir: a factory accident that temporarily blinded him also inspired him to devote his life more fully to God, through his work and experiences with nature and wilderness. Muir's religious commitments did not suddenly begin in Indianapolis, however. As a child, religious practice was part of daily life for Muir. His father, Daniel—often referred to as a religious zealot—favored strict teaching and practice of religion.

Throughout his life, Muir maintained a deep commitment to the Christian religion. At the same time, he meant to integrate his understanding of God to that of nature. Overall, Muir argued for a move to a more biocentric—rather than a human-centered—preservationist ethic. This differed from the prevailing ideas of the time, which preferred to see humans as controlling or dominating nature; Muir wanted to see nature valued for its inherent qualities.

In 1899, a short biography of Muir appeared in the first edition of *Who's Who in America*. His inclusion at the age of 61 is indicative of the fact that it was not until later in Muir's life that he became a popular environmental figure. Muir spent the last 46 years of his life based in the western United States, mostly in California, where in 1892, he was instrumental in founding the Sierra Club, an environmental organization that aimed to establish national protected areas such as national parks and national wilderness preservation systems. Muir served as the Sierra Club's president from its formation until his death in 1914. He has often been referred to as the "guardian" of Yosemite and the Sierra Nevada mountains because of his work to preserve these natural spaces.

However, before moving to California in 1868, Muir did travel to other spaces across the United States, including a trip from Indiana to Florida that he documented in *A Thousand-Mile Walk to the Gulf*. His writings on travel, nature, and adventure were important and proved to be numerous. During his life, he published 12 books and hundreds of articles; his work appeared in "eastern monthlies" such as *Harper's*. His writing is considered important in inspiring and mobilizing public sentiments around nature as sacred space in need of protection, but also in advocating the perspective that individuals should be able to leave stressful city and civilized life behind and needed a place to go to experience nature or wilderness as recreation. He also saw a role for the government to play in protecting nature, and he was an important part of discussions around the start of the national parks system.

In 1878, after many years of living in and writing about the wild spaces of the American West, Muir was persuaded to return to a more urban life. Through close friends, he was introduced to Louisa Strentzel, the daughter of a well-known physician and avid horticulturist who resided near Oakland, California. Muir and Strentzel married in 1880; however, he still left the city at times to return to the mountains, even into his later years. In 1914, he contracted pneumonia while on one of his trips and returned to the city for treatment. He died at the age of 76, leaving behind his wife and their two daughters.

The first part of Muir's memoirs, *The Story of My Boyhood and Youth,* which focused on his childhood in Scotland through his years at the University of Wisconsin, was published just one year earlier, in 1913.

See Also: Preservation; Religious Ethics and the Environment; Sierra Club.

Further Readings

Fox, Stephen. *John Muir and His Legacy: The American Conservation Movement.* Boston: Little, Brown. 1981.

Lewis, Michael, ed. *American Wilderness: A New History.* New York: Oxford University Press, 2007.

Muir, John. *The Story of My Boyhood and Youth.* Madison: University of Wisconsin Press, 1913.

Muir, John. *A Thousand-Mile Walk to the Gulf.* Boston: Houghton Mifflin, 1916.

Vanessa Lamb
York University, Toronto

Mumford, Lewis

Lewis Mumford (1895–1990) has provided wide-ranging interpretations on human conditions of life, and their particular natural and social environments, throughout his historical, philosophical, and urban works. Commonly recognized for his insights on cities and techniques, he should also be acknowledged for his contributions to green thought, when considering his attention to ecological depletion, urban renewal, organic outlooks, industrial degradation, and mechanization of life. As he recognized natural environments as prime factors in the development of human cultures, he was an advocate for a balance between resource use, technological advancements, and moral and political values of well-being.

Mumford was deeply influenced by Scottish biologist and urban planner Patrick Geddes and his ecological understanding of cities that encompassed the dynamics between biological and social conditions. In a series of essays on regionalism published in the *Sociological Review* in the 1920s, Mumford highlighted the mistakes of American development in ignoring the harmony of local environments, considered as a whole made of means and physical contours, types of community, institutions, and industries. In this respect, he also revived the ecological thinking of conservationists such as George Perkins Marsh, particularly by criticizing misuses and destabilizations of the natural order, as, for example, in deforestation.

To counter such environmental devastation, Mumford promoted a notion of limited and decentralized urban forms that balanced natural circumstances with human needs. This notion was at the heart of the Regional Planning Association of America (RPAA), established in 1923, which included Mumford, conservationist Benton MacKaye, architects Clarence Stein and Charles Whitaker, planner Henry Wright, economist Stuart Chase, and so on. They aimed at cities as an integral part of regions, planned according to natural contexts, and connected by highways and hydroelectric systems. Inspired by Ebenezer Howard's Garden City in England, RPAA's experiments comprised the housing projects of Sunnyside, Queens, and Radburn, New Jersey. This last project was portrayed in the documentary *The City* (1939), commented by Mumford, as an alternative to the disproportionate industrial city. In this environment, the human scaling would encourage new communities, authentic to local values and involved in democratic deliberation.

Regional planning presented for Mumford a more positive task than conservationism, due to the latter's sole attention to protecting wilderness areas while neglecting other areas. In *The Culture of Cities* (1938), he outlined the emergence of the organic viewpoint in regional cities, which could create adequate habitats for its residents. In his perspective, equivalences could be established between natural and cultural environments, since organisms and people, both with their functions and surroundings, were defined as interrelated wholes. Borrowing his terminology from Geddes, Mumford had already envisaged in *Technics and Civilization* (1934) the appeal in the 20th century of a "neotechnic" phase, where human achievements were guided by organic principles, grounded on nonpolluting sources of energy, long-lasting materials, and synthetic chemical compounds. This last phase would restore equilibrium present in the "eotechnic" and shattered in the "paleotechnic." From 1300 to 1700, the first phase had less environmental impact, relying on wood, wind, and water. After 1750, the second was defined by resource depletion, polluted milieus, and industrial insalubrious conditions, under goals of power, profit, and efficiency.

There is an ecological sense in Mumford's writings regarding his emphasis on energy and resources, the degrees of environmental degradation, the organic integration of commodities and activities, or the values of life and renewal, as we see, for example, in *The City in History* (1961 National Book Award). In his *New Yorker* column, for over 30 years, he was also an attentive critic of the growing homogenization and deterioration of urban architecture. In the postwar period, his writings grew more apprehensive about the outcomes and choices made in the name of progress. He questioned the sprawl of new suburbs and the prevalence of private automobiles, and he opposed several urban projects from planners like Robert Moses in New York. He was also against the development of the atomic bomb and skeptical of the benefits of nuclear energy.

In the two-volume *The Myth of the Machine* (1967, 1970), Mumford presented a more comprehensive and somber diagnostic of human civilization, dominated by the model of "megamachines" and its elements of power, regimentation, conformity, and absolutism. This mechanized order—steered by capitalism, science and technology, bureaucracy, and totalitarian government—shaped a uniform setting under the belief that there were no natural limits to production and consumption. But Mumford argued equally for a restoration of human purposes based on an organic sense of unity with the environment, rather than aggressive conquest. Even if his expectations varied throughout his work, he still believed in the human capacity to transform, control, and choose its own direction, in cooperation with others and with nature. Thus, Mumford was a pioneer in establishing the connections between natural conditions and human institutions, in an attempt to articulate ecological efforts with historical contexts. In the end, human societies would be able to respect the environmental, technical, scientific, economic, political, and cultural developments without detriment to one or another.

See Also: Conservation; Social Ecology; Technology; Urbanization.

Further Readings

Guha, Ramachandra. "Lewis Mumford per WC, the Forgotten American Environmentalist: An Essay in Rehabilitation." In *Minding Nature: The Philosophers of Ecology,* David Macauley, ed. New York: Guilford Press, 1996.

Hughes, Thomas P. and Agatha C. Hughes, eds. *Lewis Mumford: Public Intellectual.* New York: Oxford University Press, 1990.

Minteer, Ben A. *The Landscape of Reform: Civic Pragmatism and Environmental Thought in America.* Cambridge, MA: MIT Press, 2006.

Susana Nascimento
ISCTE-IUL, Lisbon University Institute

Naess, Arne

Arne Naess (1912–2009) was a philosopher, social and environmental activist, and mountaineer. Appointed to the only chair of philosophy at the University of Oslo at 27, Naess became the most prominent Norwegian philosopher of his time. He was a prolific scholar who addressed all the major subdisciplines of philosophical research in several hundred publications. Since the 1970s, however, Naess has been most widely known as the foremost exponent of deep ecology as a philosophical view and a justification for action.

Naess coined the term *deep ecology* in "The Shallow and the Deep, Long Range Ecological Movement," published in 1973. In this brief article Naess distinguishes "deep ecology" from the "shallow ecology" movement, which, he argues, is primarily concerned with mitigating pollution and resource depletion to preserve the health and affluence of people living in the industrialized West. The new environmental awareness and activism that Naess calls the deep ecology movement, on the other hand, is motivated by an understanding of interdependence and relationality among all beings and a recognition of the intrinsic value of nonhuman nature.

Naess formulated seven principles to describe the philosophical basis of the deep ecology movement inspired by ecological thinking. Naess begins with a metaphysics of interdependence, in which individual identity is a function of relations with others. First, the others with whom one is in relation to are both human and nonhuman beings, and Naess's relational metaphysics is accompanied by a "value axiom" that every form of life possesses an equal right to flourish, at least in principle; Naess terms this view "biospherical" or "ecological" egalitarianism. Second, the flourishing of all life is formulated by Naess as a valuing of diversity and symbiosis. While Naess derives this principle from the ways in which cooperation leads to a multiplicity of life forms flourishing in complex ecological systems, it is also intended to apply to diversity in human cultures and practices. Third, Naess emphasizes, however, that while deep ecology embraces diversity, it rejects differences, such as class, that are based on exploitation or oppression. Fourth, deep ecologists, in agreement with shallow ecologists, fight pollution overload and resource depletion, but they understand their resistance in the larger context of class and cultural diversity, and their actions are informed by the other principles. Naess's final two principles—complexity, not complication, and local autonomy and decentralization are both grounded in what he regards as an ecological attitude. Ecosystems are characterized by unifying principles; the parts cooperate to form a whole. Naess sees in this complexity both a model for human life—we ought to lead complex lives, in which we pursue a diversity of activities, but not complicated lives that lack order—and a warning not to intervene with the complexities of nature when we do not fully understand them. Naess's affirmations of local autonomy and decentralization are informed by the ways in which stable and complex ecosystems are vulnerable to invasive species, and also based on the observation that self-sufficient communities pollute less and are more likely to sustainably draw on their resources.

While Naess does not believe that his are the only guiding principles for deep ecological views, he does regard them as normative; the principles provide a hierarchy of values that leads to prescriptions for both policy and individual action. The normative aspect of Naess's thought is more explicit in the eight-point deep ecology platform he developed with George Sessions in the 1980s. The first point in the platform is that all life has intrinsic value—the claim of nonanthropocentrism—from which the rest follow, culminating in the point that if one subscribes to the platform, one is obligated to work toward implementing change according to deep ecology principles. By formulating a platform that could embrace a diversity of nonanthropocentric views, Naess expanded the scope of what could be included under the rubric of deep ecology. At the same time, the platform provides a more accurate description of the deep ecology movement, which includes activists who may not agree with Naess's seven principles from the 1970s. This made deep ecology a broader foundation for activism that could be motivated by a number of worldviews grounded in various spiritual and philosophical traditions, so long as there was agreement that nonhuman nature possessed intrinsic value.

Within the broad category of deep ecology, Naess believed one ought to create one's own ecological worldview, one's own ecosophy. Naess termed his worldview "Ecosophy T," after his high mountain cabin, Tvergastein, that provided inspiration throughout his life. Ecosophy T, informed by Spinoza, Gandhi, process philosophy, Gestalt ontology, and Buddhist and Hindu traditions, possesses, Naess writes, one ultimate norm: self-realization. Grounded in the first deep ecological principle that individuals are a function of their relations, Naess develops an account of a social and ecological self that can be in relation with an ever-expanding community and thereby identifies with this wider web. The self is not an autonomous unit; rather, as we mature, according to Naess, the self identifies with widening circles of family, community, surrounding culture, all people, and also the places where we live and their nonhuman creatures, as well as other, more distant ecosystems and their inhabitants. Self-realization, then, is identification with the widest possible circle of human and natural beings, through which we come to recognize that defending the interests of the oppressed, defending the interests of nonhuman natural beings, is, ultimately, in our own self-interest.

With his clear and accessible prose and his own commitment to a life simple in means and rich in ends, Arne Naess continues to inspire students, scholars, and activists seeking to make our thinking and practices more responsive to the needs of all forms of life.

See Also: Deep Ecology; Intrinsic Value; Kantian Philosophy and the Environment.

Further Readings

Naess, Arne. *Ecology, Community and Lifestyle,* David Rothenberg, trans. and ed. Cambridge, MA: Cambridge University Press, 1989.

Naess, Arne. *The Ecology of Wisdom: Writings by Arne Naess,* Alan Drengson and Bill Devall, eds. Berkeley, CA: Counterpoint, 2008.

Sessions, George, ed. *Deep Ecology for the 21st Century: Readings on the Philosophy and Practice of the New Environmentalism.* Boston: Shambhala, 1995.

William Edelglass
Marlboro College

Organic Trend

A horticulturalist consults with organic farmer Phil Foster (left) on his organic farm in San Juan Bautista, California, a state that accounts for 20 percent of U.S. organic farming operations. In 2006, less than 1 percent of all farms within the United States were organic.

Source: U.S. Department of Agriculture Agricultural Research Service/Scott Bauer

As Homo sapiens are biological beings who depend upon plants and animals for sustenance, how humans procure calories becomes an issue of central concern when discussing contemporary issues of sustainability. This becomes more evident given that humans currently appropriate approximately 40 percent of the global net primary productivity of the biomass energy photosynthesized and produced by various planetary systems. Agriculture, or the growth, transport, and distribution of food items that humans consume in order to survive, contributes a significant portion to this percentage. This article examines the development, continued trends, and likely futures of one type of global agriculture that some of its practitioners claim is motivated by concerns about sustainability—organic agriculture. It specifically looks at the history and goals of organic agriculture; its various regulatory bodies; disputes about organics as a method of agriculture and as a marketing regime; present and future growth of organic agriculture and the organic industry; and the present and future target markets of such agricultural products.

Sir Albert Howard (1873–1947) helped develop the concept and practice of organic agriculture in 1940s England. Howard observed agriculture in India—especially the use of animal manures to make compost that was then applied back to agricultural fields—and combined his observations with a growing sensibility in the West about holism and organicism and began advocating an Earth-friendly sustainable agriculture; this agriculture was to be based on building soil fertility by the use of compost and was to recognize the limits of natural systems so that farmers would strive to work in harmony with these systems. Organic agriculture was compared to and was offered as an alternative to the fossil fuel–based agriculture gaining ascendancy at the time (this is still a major impetus behind organic agriculture). This latter type of agriculture, called conventional agriculture, was and still is based upon inputs of pesticides, fungicides, herbicides, and various fertilizers. This fossil fuel–based method of agriculture, coupled with the development of hybrid seeds, a result of the work of Norman Borlaug (1914–2009), came to be known as the green revolution. Numerous criticisms of conventional agriculture exist, beginning in Howard's era; followed in the 1960s/1970s with Rachel Carson's classic *Silent Spring* and the back-to-the-land movement; and continues today with numerous studies and international nongovernmental organization (NGO) advocacy groups criticizing conventional agriculture for ecological, social, and nutritional reasons. Lady Eve Balfour (1899–1990), who was the first president of the Soil Association (founded in 1946, this is one of the major organic advocacy and certifying bodies in England) and J. I. Rodale (1898–1971) (who in 1942 began publishing the magazine known today as *Organic Gardening*) were other early advocates of organic agriculture. Vandana Shiva and Wendell Berry are two of the more globally famous advocates of organic, sustainable agriculture in contemporary times.

The farming practices of early Western organic pioneers became codified in various certifying bodies, including the California Certified Organic Farmers and Northeast Organic Farming Association of the late 1970s/early 1980s, the Soil Association in England, and the Demeter Association, which certifies biodynamic products, especially in Europe. The U.S. government passed a law in 1990, the Organic Food Production Act, which made organic growing methods (but not the philosophy and ethical reasons behind the turn/return to organic agriculture) beholden to legally recognized and defined practices. The National Organic Standards Board was created by legislative fiat as part of this legal process; this board developed a set of organic standards that in 2002 became the basis of national U.S. organic standards. These standards regulate organic production of all agricultural products in the United States, including meats, dairy products, eggs, fruits, seafood, vegetables, manufactured and value-added goods (including grocery items, vitamins and supplements, and body-care products), and alcohol. Certifying agencies have developed in most Western nations, some African, Asian, and Latin American nations, Australia, and Japan. Some governmental bodies have also regulated domestic organic agricultural practices and products, including the U.S. government and the European Union. One future organic trend will be the continued legislation of organic standards in a growing number of countries and the possible development of internationally agreed organic standards that will be recognized under treaty and/or trade bills.

Regarding global organic statistics, Helga Willer and colleagues undertook a global survey of organic agriculture in 2008, compiling statistics available from 138 countries. They found that in 2006, 700,000 organic farms existed globally, making up 0.65 percent of the agricultural lands that this team surveyed. They also found that sales of organic products have skyrocketed in the past decade, especially in North America and Europe, with global sales exceeding US$38 billion in 2006. Also in 2006, the U.S. organic industry accounted for 2 percent of total food sales, 0.3 percent of crop- and pastureland, 2 percent of vegetable acreage, and less than 1 percent of all farms within the United States. The trajectories of these various figures for certified acreage, sales, and number of agricultural products certified and regulated should only continue to grow in the coming decades; however, it is not likely that they will assume any sort of global dominance in acreage or market shares (fully 70 percent of the United States does not purchase any organic items). While the overall rate of sales might decrease, the overall global dollar amount spent on organic products and overall global acreage devoted to organic agriculture will almost certainly increase. Furthermore, once the world passes through peak oil and less fossil fuel is available for agricultural inputs, these numbers will increase by default.

Humans are motivated to purchase, consume, and farm organically for many reasons, be they health, nutrition, epicurean, ethical, environmental/ecological, ideological, and/or religious/spiritual reasons. These reasons for farming and consuming organic products should continue to hold for the foreseeable future. One recent organic trend is a backlash by some organic "purists" who maintain that government intrusion into organic agriculture has watered down and corrupted the original spirit and practice of organic agriculture. These organic advocates are looking to move "beyond organic" by developing new marketing methods, new certifying bodies, and new ways to promote, grow, and consume organic agricultural products. Some even claim that the label "organic" no longer has any meaning and that large-scale organic monoculture farms and large food corporations have bought and assumed dominance over what used to be reputable organic companies and farms.

These critics point to the growth of Whole Foods, or the entrance of Walmart into selling organic products, or the effort of agribusiness lobbyists to water down and corrupt organic standards at a national level as examples of the sullying of organic agriculture. This centralizing trend, and criticisms of it (seen in the recent increase in documentaries and books about the conventional food system in general and large-scale organics in particular), will surely continue into the near future. A contemporary trend that will most likely increase are the creative attempts to keep "organics" pure, as seen in the recent rise of community-supported agriculture, food co-ops, farmers markets, the slow food movement, restaurants featuring locally grown organic products (such as the internationally recognized Chez Panisse in Berkeley, California), Fair Trade products, and biodynamic and other types of "beyond organic" agriculture like urban and guerrilla gardening. Lastly, organic products over the past 20 years have tended to be purchased by affluent members of the countries in which organic products are sold. This target audience will most likely remain the largest consumer bloc of organic products, but some farmers and businesses are attempting to market and make available organic products to citizens with lower incomes and who are not stereotypical environmentalists or "foodies." Community-supported agriculture/box schemes are one example of this attempt, as is the development of food co-ops (whether of producers, especially in the global south, or consumers, for example, the People's Grocery in low-income, African American, and Latino Oakland, California).

See Also: Agriculture; Animal Ethics; Berry, Wendell; Borlaug, Norman; Carson, Rachel; Genetic Engineering; Local Food Movement.

Further Readings

Balfour, Lady Eve. *The Living Soil.* Bristol, UK: Soil Association Ltd., 2006.

Berry, Wendell. *The Unsettling of America: Culture and Agriculture.* New York: Avon Books, 1977.

Fromartz, Samuel. *Organic, Inc.: Natural Foods and How They Grew.* New York: Harcourt, 2006.

Holthaus, Gary. *From the Farm to the Table: What All Americans Need to Know About Agriculture.* Lexington: University Press of Kentucky, 2009.

Howard, Sir Albert. *An Agricultural Testament.* Emmaus, PA: Rodale Press, 1972, 1943.

Jackson, Wes. *New Roots for Agriculture.* Lincoln: University of Nebraska Press, 1985.

Kimbrell, Andrew, ed. *The Fatal Harvest Reader: The Tragedy of Industrial Agriculture.* Washington, DC: Island Press, 2002.

King, Franklin Hiram. *Farmers of Forty Centuries: Organic Farming in China, Korea, and Japan.* Mineola, NY: Dover, 2004.

Obach, Brian. "Theoretical Interpretations of the Growth in Organic Agriculture: Agricultural Modernization or an Organic Treadmill?" *Society and Natural Resources,* 20/3 (2007).

Pawlick, Thomas. *The End of Food: How the Food Industry is Destroying Our Food Supply—and What You Can Do About It.* Fort Lee, NJ: Barricade Books, 2006.

Pollan, Michael. *The Omnivore's Dilemma: A Natural History of Four Meals.* New York: Penguin Press, 2006.

Schlosser, Eric. *Fast Food Nation.* New York: Harper Perennial, 2005.

Shiva, Vandana. *Stolen Harvest: The Hijacking of the Global Food Supply.* Cambridge, MA: South End Press, 2000.

Thompson, Paul. *The Spirit of the Soil: Agriculture and Environmental Ethics.* New York: Routledge, 1995.

U.S. Department of Agriculture Organic Farming Portal. http://www.usda.gov/wps/portal/!ut/p/_s.7_0_A/7_0_1OB?navid=ORGANIC_CERTIFICATIO&navtype=RT&parentnav=AGRICULTURE (Accessed April 2009).

Willer, Helga, Neil Sorensen, and Minou Yussefi-Menzler, eds. *The World of Organic Agriculture: Statistics and Emerging Trends 2008.* London: Earthscan, 2008.

Todd LeVasseur
University of Florida

PASSMORE, JOHN

John Arthur Passmore (1914–2004) was an influential Australian philosopher. He taught at the University of Sydney, Australia; the University of Otago, New Zealand; and was professor of philosophy at the Australian National University from 1965 until 1979. He was primarily known as a historian of ideas, especially in the areas of 20th-century philosophy, ethics, and aesthetics, and as a pioneer in applied ethics. He was not an "environmental philosopher" in the sense of critically examining the philosophical foundations of environmentalism. His book *Man's Responsibility for Nature,* first published in 1974, was the first attempt to examine environmental issues from the perspective of the Western intellectual tradition: he argued that as a matter of urgency we must reduce our demands on the biosphere. His work was, and remains, controversial because of its anthropocentric approach and rejection of deep ecology.

The importance of *Man's Responsibility* is threefold. First, 20th-century Anglo-American-Australasian philosophical ethics (or "moral philosophy" as it was more commonly called) was almost entirely concerned with ethical theory, including analysis of the meaning of ethical terms such as *good* and *right,* and there was little discussion in the literature of ethical issues in practice. His book was therefore the first work in what is today called applied ethics. Second, he was the first writer to provide a systematic analysis of environmental problems and to try to provide an approach to dealing with them. Passmore's position was anthropocentric, but he strongly believed that humans must recognize their dependence on natural processes and he advocated an ethic of stewardship and cooperation with natural systems. Third, the environmental literature of the late 1960s and early 1970s consisted largely of accounts of the "environmental crisis," often written in a pessimistic or even apocalyptic tone, the titles featuring expressions such as "Doomsday" and "The Last Decade." While Passmore was by no means complacent, nor even particularly optimistic, his work was very much focused on the possibility of what would today be called a sustainable future, provided only that we recognize our responsibilities, understand how natural systems work and how we affect them, and limit our demands on them.

The book is remarkably contemporary in identifying the roots of environmental destruction in terms of attitudes such as ignorance of environmental processes, effects and limitations, a short-term approach to environmental management, and greed. He emphasized the fragility of the biosphere and identified four main areas as needing urgent attention: pollution, resource conservation, population growth, and the loss of natural beauty.

Passmore considered that animals and natural features cannot have independent moral standing, let alone rights, because only conscious beings can have such standing and humans are the only beings that are conscious. However, because animals are sentient beings, he rejected the idea that animals are mere resources to be treated however we wish. He rejected the idea that nature has intrinsic value, independent of its utility to humans, and advocated a secular worldview. Thus, he particularly opposed the idea that nature might have any spiritual or religious value, though he emphasized that the Jewish (though not, on the whole, the Christian) tradition of stewardship is a key resource for environmental responsibility in the Western tradition.

Passmore distinguished between "conservation," which is the protection of nature for the sake of human interests such as aesthetics and biodiversity resources, and "preservation," which is the protection of nature for its own sake. However, he emphasized that in many cases conventional human-centered ethics will often lead to the same conclusion as nonanthropocentric ethics. Thus one person may value a redwood tree for its intrinsic value as a 1,000-year-old inhabitant of the forest while another person may seek to preserve it for its ability to provide aesthetic enjoyment for future generations.

Philosophers such as Richard Routley had earlier called for a rejection of anthropocentrism in favor of what he referred to as a new, environmental ethic. Passmore rejected this, arguing that one cannot just decide to accept a new ethic. Rather, an ethic must arise from existing cultural traditions, which in the West is secular and humanistic and whose central values include rationality and a commitment to scientific method.

Passmore was a controversial figure in Australian environmental circles in the 1970s and 1980s because of his opposition to deep ecology, a theory developed by Arne Naess. Deep ecologists argue that we cannot hope to understand, let alone deal with, "the environmental crisis" without reconceptualizing the human–nature relationship. This distinction between "deep" or "radical" and "shallow" or "reformist" has continued to divide environmentalists, despite efforts by philosophers such as Bryan Norton to redirect the focus from theoretical differences to shared goals.

See Also: Anthropocentrism; Deep Ecology; Intrinsic Value; Naess, Arne.

Further Readings

Norton, Bryan. *Towards Unity Among Environmentalists.* New York: Oxford University Press, 1994.

Passmore, John. *Man's Responsibility for Nature.* New York: Charles Scribner's Sons, 1974, 1980.

Sylan, Richard. "Is There a Need for a New, an Environmental Ethic?" *Proceedings of the X11 World Congress of Philosophy,* 1. Varna, Bulgaria, 1973. http://www.uq.edu.au/~pdwgrey/web/res/sylvan.neweth.html (Accessed February 2010).

Alastair S. Gunn
University of Waikato

Philosophy and Environmental Crisis

From the time of Plato and Aristotle, philosophers have sought to answer questions regarding existence, values, and ways of knowing. Philosophers have primarily addressed the human condition. With the onset of the environmental crisis, a new specialization has emerged that considers how we define relationships and responsibilities to nonhuman beings. The environmental crisis can be defined through multiple lenses, including the effects of pollution, destruction of habitat, decimation of plant and animal species, and a declining quality of life. The effect of humans on the environment is a continually debated issue, with those maintaining an anthropocentric view insisting that the resources of the planet are primarily defined in their utilitarian relationship to humans, while others like deep ecologist Arne Naess argue that humans must respect the intrinsic value of all life regardless of its usefulness to humans.

Philosophers have often been blamed for the initial dualisms that mark the separation between mind and body, or humans and nature. One can view this from an empirical, metaphysical, or epistemological perspective. René Descartes, John Locke, Immanuel Kant, and others are critiqued for their role in elevating rationality and human superiority to justify the exploitation of natural resources for the benefit of humans. However, 20th-century Continental philosophers such as Martin Heidegger, Maurice Merleau-Ponty, and Edmund Husserl have influenced a new generation of thinkers to reconsider phenomenology and the role of the body in engagement with the world as well as the integration of knowledge and practice.

Phenomenology applied to environmental thought borrows from Merleau-Ponty and describes a way of assigning meaning modeled on perception and intercorporeality. There is contestation of Husserl's anti-naturalism through the loosening of boundaries between flesh and nature, and between intentionality and causality. Phenomenology advanced as an objective study of the "things themselves" can lapse into the same duality it hoped to collapse. By incorporating the natural world itself into this study, paradox is allowed to surface and subjectivity can be revealed much as it is in the philosophy of science and Hans-Georg Gadamer's *Truth and Method.*

Poststructuralists often set up opposing positions: one that emphasizes a view of nature rooted in origins and urges a return to a pristine state prior to human involvement in the natural world, and a postmodern view that sees nature as "difference" beyond experience. Contemporary continental philosophy presents an alternative that positions "nature" as a social construct, and views humans as an embedded part of the natural world. Postmodern critics demand a contextualizing of social and cultural contingency and prefer to keep the natural world as a human construct that cannot be defined in terms of origins.

Pragmatic Approaches

Pragmatic approaches to the environmental crisis through the views of John Dewey, William James, and George Herbert Mead are concerned with the empirical experiential and scientific approaches to our global condition. The split between natural and moral human worlds is often differentiated by the scientific analysis of the natural world and the difficulty of applying such measures to human irrationality. Pragmatists advocate education as a tool for reform. Some would contend that if our scientific and technological advances have created imbalance in the ecosystems of the world, it is these same tools that must be employed to repair the damage.

Considering other American philosophical approaches to the environment, one cannot neglect the transcendentalists. Ralph Waldo Emerson cultivated a romantic idealism around the value of the natural world's aesthetic beauty, while Henry David Thoreau allowed a scientific humanism to influence his poetic musings.

Environmental ethics has played an important role in the consideration of instrumental versus intrinsic values regarding the more than human world and tends to be the discipline that most scholars refer to when discussing environmental philosophy. Issues of moral standing, rights, risk/benefit analysis, and sustainable behaviors are vigorously debated. The publication of three important essays—Lynn White's "The Historical Roots of Our Ecological Crisis," Garrett Hardin's "Tragedy of the Commons," and Aldo Leopold's "The Land Ethic"—are often considered the foundation of environmental ethics. Some might argue that Rachel Carson's writings on the danger posed by pesticides and DDT in particular to the environment heralded a new era of consideration for the more than human world in philosophical debates. J. Baird Callicott has defended the land ethic in numerous books and articles.

In recent years, the rights of animals have been a focus of debate for ethical philosophers including Peter Singer and Tom Regan. Singer argues that animals deserve moral consideration like ethnic minorities or women. He uses Jeremy Bentham's reasoning that all things that have the capacity to feel pleasure and pain deserve moral standing to back up his argument and claims that even if animals cannot reason, they still suffer. Singer employs the idea of sentience as a measure of whether animals have interests that humans must acknowledge. Regan argues it is wrong to treat animals as means to an end, for they are ends in themselves and possess intrinsic value. He views humans as moral agents in that they are free and rational and can be held responsible for their choices. Animals, on the other hand, can be seen as incompetent or immature, much like young children who cannot act morally on their own but can be acted on morally. Thus, he encourages an egalitarian theory of justice by which animals must be treated with respect.

Political philosophers contend the environmental crisis is the result of a neoliberal capitalistic approach that gave us industrialism, urban sprawl, pollution, globalization, and global warming. Unlimited growth and the tendency of capitalism toward overproduction contribute to the crisis and the system is unsustainable. Neoliberal economists might advocate that if human activities driven by market forces created the problem, it is the same market forces that can solve the problems of environmental degradation. Political ecology is a growing subset of political theory and social philosophy that examines social movements and critiques such as Green party politics, socially responsible investing and business, the "natural capitalism" promoted by Paul Hawken, and the green anarchism and neoprimitivism of theorists like John Zerzan identified by the term *radical ecology.*

Social ecology, a movement founded by Murray Bookchin, posited that environmental problems are the result of social problems. Rather than faulting technology, Bookchin took aim at the market model of society and critiqued competitive growth models, the profit motive, industrial expansion, and corporate self-interest. He equated the domination of the natural world with the hierarchical model of society and the stratification of a class system. He called for a natural spirituality that would restore humanity's role as moral agents fostering an appreciation of diversity, reducing planetary suffering for human and nonhuman life and advancing an ethics of complementarity in which humans play a supporting role in the evolutionary processes of the planet.

Critiques

A strong critique has emerged in the form of ecofascism that suggests that privileging the natural world over humans is reminiscent of the Nazi ideology known as "blood and soil" that linked bioregionalism and the protection of the natural world with a movement toward racial purity. Michael Zimmerman cautions against a romantic retribalization or a centralized governmental solution to the environmental crises, noting that changes in behavior and institutions are imperative to avoid authoritarian policies that would place the protection of nature ahead of human freedom. Zimmerman collaborated with Sean Esbjorn-Hargens to write *Integral Ecology* based on the perspectives of Ken Wilbur to complexify ecological patterns through the lens of the integral framework that maps four quadrants of experience (subjective), behavior (objective), culture (intersubjective), and systems (interobjective).

Ecofeminism is an emerging discipline that seeks to address the interconnections of the continuing domination of women, marginalized humans, and the natural world through liberative theories, strategies, and systems. According to Karen J. Warren, there are four unifying claims that underlie ecofeminist thought including the interconnections of the continuing domination of women, marginalized humans and the natural world, the use of gender as a category of analysis, the importance of considering these insights with regard to environmental philosophy, public policy, and social justice, and the need to replace the dominating structures and practices with those that are nonoppressive, liberating, life-affirming, cooperative, and just. Some of the classic books in this genre are *The Death of Nature* by Carolyn Merchant, *Woman and Nature: The Roaring Inside Her* by Susan Griffin, *New Woman/New Earth* by Rosemary Radford Reuther, *Gyn/Ecology* by Mary Daly, and *Green Paradise Lost* by Elizabeth Dodson Gray.

One approach to bringing these issues into a manageable perspective is to consider the role of indigenous peoples as the first environmental philosophers. In the lore of Native American, Mayan, Hawaiian, African, South American, and many other First Nations, balance and respect are crucial elements in human relationship to the natural world. Humans in these original cultures were taught gratitude for the plants, animals, mountains, lakes, and other nonhuman entities. Reciprocity and interconnection were acknowledged and passed down through oral tradition and reinforced by experiential observation. David Abram attempts to bring these views into conversation with environmental philosophy through the exploration of language, ritual, and cultural context in *The Spell of the Sensuous*. His work applies a more embodied view to these concepts.

Another non-Western philosophical view emerges from Buddhist thought, which posits the understanding of interdependent origination as a way of demonstrating the interconnection between humans and the natural world. Buddhist philosophy encourages being content with what is, rather than falling prey to the human tendencies toward selfish desire, greed, and materialism.

Environmental philosophy continues to evolve and to contend with the consequences of environmental crisis and consider a variety of possible approaches to create a more sustainable way of being in the world.

See Also: Animal Ethics; Bookchin, Murray; Ecofeminism/Ecological Feminism; Ecophenomenology; Warren, Karen.

Further Readings

Zimmerman, Michael E., J. Baird Callicott, Karen J. Warren, Irene Klaver, and John Clark. *Environmental Philosophy: From Animal Rights to Radical Ecology*, 4th ed. Upper Saddle River, NJ: Pearson Prentice Hall, 2005.

Stephanie Yuhas
University of Denver

Pinchot, Gifford

At the beginning of the 20th century, efforts to preserve forested land in the United States were driven by those like John Muir who favored preserving them for aesthetic purposes, and those like Gifford Pinchot, who advocated managing them for utilitarian reasons. One of the first scientifically trained foresters born in North America, Pinchot was the first chief of the United States Forest Service, founder and first president of the Society of American Foresters, and a founder of the Yale School of Forestry. As chief forester, he served as an adviser to President Theodore Roosevelt on the environmental issues of the day and helped shape that administration's conservation agenda. Pinchot made utilitarian conservation, or the regulated use and scientific management of natural resources for the public good, the cornerstone of his long public career, which also included serving as Pennsylvania's commissioner of forestry (1920–22) and two terms as its governor (1923–27, 1931–35). Through his other roles, he helped establish professional forestry in the United States.

In numerous articles and books, but in particular *The Fight for Conservation* (1910) and his autobiography *Breaking New Ground* (1947), Pinchot argued how conservation was the means to an end, with that end being democratic equality for all. Three principles of conservation defined this commitment to democratic equality: conservation was dedicated to the development and use of natural resources, to the elimination of unnecessary waste, and to providing "the greatest good for the greatest number for the long run." Rational development and distribution of natural resources was, he argued, a moral obligation. He strongly believed that conservation as he defined it could serve as the foundation of world peace if adopted by the nations of the world. Though he was a lifelong outdoor enthusiast, hunter, and fisherman who also supported the preservation of land and wildlife for aesthetic reasons, Pinchot's anthropocentric version of conservation left him open to criticism in his day and subject to ridicule by later-day environmentalists.

The oldest child of a wealthy New York merchant, Pinchot graduated from Yale University in 1889 and then briefly studied forestry in Europe at a time when there were no American forests under management and no forestry schools in North America. At the end of 1890, he returned to the United States fully intent on introducing forest management as well as influencing federal forestry policies.

On private land in western North Carolina, Pinchot initiated America's first large-scale systematic forest management plan in 1892. He published *Biltmore Forest* (1893) to document and to promote his work. It was the first of many publications on forestry and conservation. He then spent the next four years consulting and writing articles and promotional books that explained the need for forest management in the United States and what form it should take.

His efforts brought him to the attention of national forest conservation leaders. Appointed to the National Forestry Commission, which was established in 1896 to formulate a national forest policy, and then as a special forest agent for the Interior Department, he helped shape national forest policy. In 1898, Pinchot accepted appointment as chief of the U.S. Department of Agriculture's Division of Forestry (since 1905 called the Forest Service). To train foresters for the fledgling agency, Pinchot and his family donated $150,000 to establish a graduate program in forestry at Yale in 1900. Pinchot also established the Society of American Foresters in order to give the new foresters immediate professional credibility and standing.

Under President Theodore Roosevelt's patronage, Pinchot led the emerging national conservation crusade, with forestry as its focal point. As chief forester, Pinchot introduced conservation policies for national forests. By merging philosopher Jeremy Bentham's adage of "the greatest good for the greatest number" with the time frame of his forestry profession, Pinchot gave the Forest Service its guiding principle of "the greatest good for the greatest number in the long run." Over the next four decades, what this meant to Pinchot evolved in response to changing circumstances. In the 1910s, he began pressing for federal regulation of logging on private lands. By the 1930s, he viewed the issue in social justice and environmental justice terms and became an advocate of "new conservationism," or the idea of protecting and preserving human life through conservation. The United States' entry into World War II moved him to press President Franklin Roosevelt to consider the role of conservation in creating what Pinchot called a "permanent" peace. His idea reached fruition three years after his death with the 1949 United Nations Scientific Conference on the Conservation and Utilization of Resources, which met to discuss in part how conservation measures could improve the standard of living around the world.

After his dismissal from the U.S. Forest Service in 1910, as a result of his accusations against President William Howard Taft's secretary of the interior, Pinchot remained active in national forestry matters through the Society of American Foresters. From 1920 to 1922, as Pennsylvania's commissioner of forestry he established a forest program that in many ways resembled his best work as Forest Service chief. As governor, he carried out Progressive policies such as regulating utilities and creating the Sanitary Water Board, the first antipollution agency in the country.

See Also: Conservation; Roosevelt, Theodore; Utilitarianism.

Further Readings

Miller, Char. *Gifford Pinchot and the Making of Modern Environmentalism.* Washington, DC: Island Press, 2001.

Pinchot, Gifford. *Breaking New Ground.* New York: Harcourt, Brace, 1947.

Taylor, Bob Pepperman. *Our Limits Transgressed: Environmental Political Thought in America.* Lawrence: University of Kansas Press, 1992.

James G. Lewis
Forest History Society

Post-Construction Landscape Laws

Post-construction landscape laws require such work as revegetation and soil restoration, which may help prevent damage such as this severe erosion near a housing development in Ankeny, Iowa.

Source: U.S. Department of Agriculture, Natural Resources Conservation Service/Lynn Betts

A post-construction landscape law is a green law, adopted and implemented by different levels of government, to preserve and to improve the environment in sites with different types of construction—buildings or infrastructures—and different uses. The goal is to advance the overall quality of green areas in new or in redeveloped urban areas or in sites where major infrastructures have been built. It can be a government regulation covering environmental issues, a municipal ordinance, a section in a municipal zoning ordinance, or a section of a wider municipal land development code.

A municipal post-construction regulation may require, for example, landscaping of sites after the end of the construction, including the full recovery of previous soil standards, and revegetation in new developments as well as in redeveloped brownfield sites. This type of municipal regulation establishes standards for the urban forest in newly (re)developed sites (e.g., landscaping of streets, parking lot shading, tree density and diversity, irrigation standards, buffering between different land use areas). The establishment of post-construction follow-up and monitoring protocols is also a possibility to be considered. The constructor remains responsive to the design team and to the owner, for a certain period, in relation to both buildings and surrounding landscape in that particular development.

It may also require post-construction monitoring of the effects of major infrastructures, as is the case for wind farm projects (e.g., on avifauna and on bats), motorways, dams, and other major infrastructures on wildlife. The extent and duration of this and other types of post-construction monitoring are usually based on the level of risk admitted by the pre-construction environmental impact studies. New minimization and mitigation measures may be identified and established during the post-construction period, as well as other compensatory measures.

Post-construction landscape laws may require the preservation or restoration of soil standards, bringing back—as much as possible—the original and undisturbed characteristics of the native soil in order to allow the use and fulfillment of the site, especially when there are hazardous substances, pollutants, or contaminants as a result of previous uses in that particular site. This soil restoration is also necessary to reduce runoff and its negative effects on wildlife and in the built environment.

In order to prevent these costly solutions, post-construction landscape regulations should include preventive measures for the preservation of soil characteristics before the construction itself is done, and additional measures to be applied after the end of the construction work. If promoters fail to keep the soil at its original capacities, they shall be compelled to restore the soil characteristics by adding compost or other measures able to devolve the soil to full capacity.

There are different soil treatment solutions, varying according to the soil's characteristics. The first option, however, shall be to maintain the native soil undisturbed, protecting it from compaction and other negative actions, during the construction phase, and to bring it back to its full use after that. An alternative is to store the native soil in situ and to reapply it again in its original position in the post-construction stage. A second solution is to make improvements in the existing soil after the end of the construction process. Finally, a third possibility may involve the import of soil with a certain proportion of organic matter content and its application in the affected area.

Post-construction landscape laws may also include a post-evaluation review some years after the end of a new construction (building or infrastructure) or major urban regeneration project, or, for example, an industrial waste treatment facility, in order to evaluate the implementation and the performance of the actions taken to treat the initial problem (e.g., soil degradation, level of pollutants and contaminants in the recovered area, air quality, noise, water pollution), especially when hazardous substances or toxic waste remained in the site affected by the construction, or in the nearby area, in order to check whether pollutant levels are confined within the established legal criteria. This may include monitoring air quality, groundwater, water supply, and biodiversity, for example, among other monitoring and evaluation procedures, as well as assessing whether environmental impacts in the post-construction stage fall within what was defined in the approved license.

See Also: Environmental Law; Environmental Policy; Green Laws and Incentives; Preservation; Urbanization.

Further Readings

Bruce, Sarah and Glenn Barnes. *Survey of Local Government Post-Construction BMP Maintenance and Enforcement in North Carolina.* Chapel Hill: University of North Carolina, 2008.

City of Seattle. *Best Management Practice. Construction Site Management in Landscapes,* 2006.

U.S. Environmental Protection Agency. *Comprehensive Five-Year Review Guidance.* Washington, DC: U.S. Environmental Protection Agency, 2001.

Carlos Nunes Silva
University of Lisbon

Pragmatism

Environmental philosophy and environmental policy were both rooted in the same rationale, that is, the growing care and concern about environmental problems from the 1960s on. Environmental philosophy dealt with ethics and concentrated mainly on the intrinsic value of nature and the dichotomy of anthropocentrism versus nonanthropocentrism.

Environmental policy, on the other hand, focused on problem solving and preventing. Thus, both evolved within different premises, the first focusing on a philosophical grounding of what is thought to be at the root of environmental problems, and the latter aiming to solve and prevent those in the short and medium term.

Some environmental philosophers became increasingly worried about this disconnection between environmental philosophy and policy. Inspired by the American pragmatists, they proposed "environmental pragmatism" as a new strategy in environmental thought, realizing that "theoretical debates were hindering the ability of the environmental movement to forge agreement on basic policy imperatives." Environmental pragmatists wanted to demonstrate that philosophers could contribute to the practical resolution of environmental problems. A. Light (2002) argues that environmental ethicists should help "the environmental community to make better ethical arguments in support of the policies on which our views already largely converge." His point is that it is possible to continue the lively philosophical debates and at the same time be more politically proactive.

To understand environmental pragmatism, it is important to go back to American pragmatism. Its founders and main thinkers—Charles Peirce (1839–1914), William James (1842–1910), and John Dewey (1859–1952)—also questioned the troubled relationship between theory and practice. Even though they disagreed on many issues and had different concerns, H. Putman (1994) derived the following four basic and common "pragmatism characteristics" from their work:

- *Antiskepticism:* pragmatists hold that doubt requires justification as much as belief.
- *Fallibilism:* pragmatists hold that there is never a metaphysical guarantee to be had that such-and-such belief will never need revision.
- *Fact–value continuity:* the thesis that there is no fundamental dichotomy between "facts" and "values." Other dichotomies should also be rejected such as thought and experience, mind and body, analytic and synthetic.
- *Primacy of the practice:* the thesis that, in a certain sense, practice is primary in philosophy.

According to these pragmatism characteristics, environmental philosophy should above all look into the meaning, the causes, and the possible solutions for environmental crisis and avoid locking itself into theoretical discussions. Environmental pragmatism has been described as an "open-ended inquiry into specific real-life problems of humanity's relationship with the environment."

Environmental pragmatic philosophers advocate moral pluralism, ensuring, however, that this position does not take them into a dead-end relativism. The idea being that in spite of the fact that some approaches might be theoretically incommensurable, it is still possible to reach consensus on policy positions. For example, one can accept the moral consideration of animals, either using the criterion of sentience, or the criterion of respect for the teleological center of life. These two criteria are theoretically distinct, but their purpose—moral consideration of animals—is the same. The environmental pragmatists see philosophy as a real contribution to finding viable solutions to environmental problems rather than as a discipline that delays possible solutions because of their endless theoretical incompatibilities.

The three central aspects of environmental pragmatism are therefore to accept moral pluralism, to reduce the importance of theoretical debates, and to consider that practical matters allow us to more easily arrive at political consensus.

Building on the notions of classical American pragmatism, environmental pragmatism rejects the idea of absolute knowledge or metaphysics and focuses on the importance of real-life experiences. It highlights the search for values in their multiple, complex, and indeterminate dimensions as the basis for the analysis of environmental philosophy problems, rejecting the monism of the intrinsic value of nature. Environmental pragmatism sees individual organisms as part of their environment and acknowledges the continuity theory between biological creatures and nature. Thus it dismisses the recurring dualisms or dichotomies in environmental ethics, such as anthropocentrism versus nonanthropocentrism, individualism versus holism, intrinsic value of nature versus instrumental value, and so on, which have been highly disruptive for constructive dialogues. In a pragmatic perspective, the anthropocentrism versus nonanthropocentrism controversy is meaningless because it considers that it is impossible to disentangle the human well-being from the well-being of their environment.

Pragmatism also believes that the values of environmental philosophy should coevolve with the new social practices regarding the human relationship to the natural world. As these are characterized by complexity and uncertainty, there is no reason to search for a unique value in environmental philosophy. Thus, according to scholars, we should accept that this is a "time for experimentation in the expression and language of environmental thought," as this will enable the "creation and evolution of new environmental values and a new relationship with the natural world." The urgency of the ecological crisis requires a form of a metatheoretical compatibilism between opposing and conflicting theories, and makes an appeal for tolerance among philosophers and a joint commitment to solving environmental problems.

A distinction should also be made between public and private practice; philosophers should leave difficult questions to their private debates. Some argue that the public is generally human centered and so environmental philosophers should adopt a "strategic anthropocentric" position for its discourse to have communicative power. This has been criticized by some, who view this strategic anthropocentrism as morally repugnant and feel that it encourages disingenuous arguments.

R. Eckersley (2002) also criticizes environmental pragmatism, be it in a sympathetic way. She sees ecocentric theorists and activists as "advocates" and environmental pragmatists as "mediators," acknowledging that all should be given space to coexist. Mediators are good listeners, flexible, and open minded, while advocates seek to inspire, move, and persuade others. Nevertheless, she identifies three main limitations in environmental pragmatism, namely, (1) "its narrow focus on problem-solving makes it insufficiently critical and emancipatory," (2) "it is too instrumentalist in the way it seeks to close off non-instrumental democratic encounters and the opportunity for the parties to engage in dialogue for dialogue's sake. Their method of inquiry is reductionist in the sense that it seeks to filter out arguments that do not address questions of practical necessity," and (3) "there is ultimately nothing especially environmental about the kind of democratic inquiry defended by environmental pragmatists, in the sense that it ultimately rests on a liberal humanist moral premise rather that any explicit environmental values."

Pragmatism attempts above all to agree on different philosophical assumptions for different specific environmental problems. The primacy of practice over theory is of such importance that it justifies these attempts according to the level of assumptions and not to the level of theories. Light defends that "a more fully responsible environmental ethics must abandon the wholesale rejection of anthropocentric reasons for protecting the environment, at least as part of our public philosophical task." His idea is to develop a more public philosophy focusing on arguments "that resonate with the moral intuitions that most people carry around with them on an everyday basis." Light argues that obligations to future generations are a powerful intuitive reason that most people will easily relate to.

Examples of environmental pragmatism concepts include the following:

1. A case of conflict between different water policies, where P. B. Thompson (1996) realized that using foundational ethical theories to derive a correct moral policy did not function. In fact, it hindered the resolution of the conflict as it mainly provided each stakeholder with a sound theoretical argument for their favored policy. Instead, it was more fruitful to use "James's idea of pragmatic necessity and Dewey's notion of the reconstruction of community" to solve the conflict.
2. Another example of using a pragmatic approach in solving environmental problems was about a controversy over the definition of wetlands. The way problems are framed affects and contributes to how solutions are found. Therefore the question as to how to redefine wetlands was a sensitive issue to authorities, scientists, and environmental activists. Since the definition would determine the amount of land to protect, a consensus was very difficult to achieve. The solution was to define wetlands not by their qualities but by their functions. This solution is based on the pragmatic idea that there are no final or absolute definitions and they should be open to revisions whenever needed.
3. E. N. Castle (1996) also suggested the use of a pragmatic and pluralistic methodology in the case of forest resource management. To manage the complex interaction of natural systems with social systems (in which a wide variety of human preferences coexist) requires interdisciplinary cooperation, and constructive discussions are essential. The use of a sole strategy for the formulation of environmental policies, for example, cost-benefit analysis, is clearly inappropriate. The variety of relationships people have with the natural world is not consistent with a monistic environmental philosophy. The pragmatic tradition sees truth as an evolving concept and not an absolute one, the relationship between the subject and the object as interactive and not static. Thus, there is no one single truth but various truths about the natural world and these are always open to reinterpretation, developments, changes, and updates. The complexity and uncertainty associated with both the natural and social systems further support the usefulness of environmental pragmatism approaches in many situations.
4. A final example of the usefulness of environmental pragmatism relates to teaching environmental ethics in order to facilitate conflict management in environmental policy. While knowledge of environmental ethics does not necessarily change the positions of participants, it can help opponents in conflict situations to appreciate each other's points of view, instead of seeing "the other" as extremists. This can help pave the way for a dialogue through which consensus is more likely to be achieved. Mutual respect is gained, once again proving that ethical theories can be useful in practical cases.

Understanding reality from different perspectives is a way of enriching it. Pragmatism requires openness to the idea of possibility and creativity in problem solving. It also emphasizes that the richness of human existence lies in its multiple relationships within the natural world.

Environmental pragmatism should not be regarded as yet another theory of environmental ethics in alternative or opposition to other ones. Its main objective is to bring theory and practice closer together in order to avoid that theoretical discussions become so delinked from reality that they become meaningless. It aims to contribute directly to solving environmental problems by providing a platform for understanding between different theories.

See Also: Animal Ethics; Anthropocentrism; Philosophy and Environmental Crisis.

Further Readings

Castle, E. N. "A Pluralistic, Pragmatic and Evolutionary Approach to Natural Resource Management." In *Environmental Pragmatism,* A. Light and E. Katz, eds. London: Routledge, 1996.

Eckersley, R. "Environmental Pragmatism." In *Democracy and the Claims of Nature,* Ben A. Minteer and Bob Pepperman Taylor, eds. Lanham, MD: Rowman & Littlefield, 2002.

Light, A. "Contemporary Environmental Ethics From Metaethics to Public Philosophy." *Metaphilosophy,* 33/4 (2002).

Light, A. and E. Katz, eds. *Environmental Pragmatism.* London: Routledge, 1996.

Light, A. and E. Katz, eds. "Introduction. Environmental Pragmatism and Environmental Ethics as Contested Terrain." In *Environmental Pragmatism.* London: Routledge, 1996.

Moriarty, P. V. "Pluralism Without Pragmatism." *ISEE IAEP Papers* (International Society for Environmental Ethics and International Association for Environmental Philosophy, 2006). http://www.environmentalphilosophy.org/ISEEIAEPpapers/2006/Moriarty.pdf (Accessed August 2010).

Putman, H. "Pragmatism and Moral Objectivity." In *Words and Life.* Cambridge, MA: Harvard University Press, 1994.

Schiappa, E. "Wetlands and the Politics of Meaning." In *Environmental Pragmatism,* A. Light and E. Katz, eds. London: Routledge, 1996.

Thompson, P. B. "Pragmatism and Policy: The Case of Water." In *Environmental Pragmatism,* A. Light and E. Katz, eds. London: Routledge, 1996.

Varner, G. E., S. J. Gilbertz, and T. R. Peterson. "Teaching Environmental Ethics as a Method of Conflict Management." In *Environmental Pragmatism,* A. Light and E. Katz, eds. London: Routledge, 1996.

Sofia Azevedo Guedes Vaz
New University of Lisbon

Precautionary Principle

The precautionary principle calls for early measures to avoid and mitigate environmental damage and health hazards in the face of uncertainty. It has been invoked in various fields of risk debates and decision making, most commonly in environmental law and policy. Examples include marine and fisheries protection, the conservation of the natural environment and biodiversity, climate change and global warming policies, chemicals regulation and the protection of the ozone layer, the risk governance of modern biotechnology, and the risk debate over nanotechnology.

The essence of taking precautions is captured by some English proverbs such as "better safe than sorry" and "an ounce of prevention is worth a pound of cure." Generally speaking, the precautionary principle says that if in doubt, decide in favor of the environment. A commonly agreed predecessor of the principle can be found in *Vorsorgeprinzip,* which was introduced to German environmental law and policy in the 1970s. The first explicit mention of the precautionary principle in an international environmental treaty was in the ministerial declaration of the Second International Conference on the Protection of the North Sea in 1987.

There are many formulations of the precautionary principle in official documents, and several definitions have been proposed in the related academic literature. Besides the phrase *precautionary principle,* other terms have been used—most often *precautionary approach*—and they may have slightly different meanings and implications. Arguably the most prominent formulation is that adopted at the United Nations Conference on Environment and Development in Rio de Janeiro in 1992. Another well-known formulation, the Wingspread Statement on the Precautionary Principle, was introduced at a conference organized by the Science and Environment Health Network (SEHN) in 1998.

Different formulations of the precautionary principle may be thought as a function of a trigger condition and precautionary response. When a situation fulfills the prerequisites described by the trigger condition, the stated precautionary response should be taken (or taking the precautionary response is justified). The trigger is twofold. It consists of damage and knowledge thresholds that determine the necessary and jointly sufficient preconditions for the application of precaution. A damage threshold specifies the relevant harmful or otherwise undesirable outcomes. They typically include serious and/or irreversible environmental damage and health hazards. A knowledge threshold defines the required level of scientific understanding of an identified threat. According to a common view based on a decision-theoretic classification, the principle can be applied when the (objective) probability of a risk cannot be established (i.e., in the state of scientific uncertainty), or when the magnitude or severity of a risk is uncertain or contested (i.e., in the state of ambiguity).

Precautionary response means taking preemptive measures. These may take the form of outright bans or phaseouts, moratoria, premarket testing, labeling, and requests for extra scientific information before proceeding. Another kind of precautionary response might be establishing new precautionary risk assessment methodologies. The focus is, then, not only on how to deal with the identified threats, but also on the methods to anticipate and assess threats in the first place. (When these methodologies are in use, they may be considered to belong to the trigger side, as they change the conditions for taking precautions.)

It has become common to distinguish between weak and strong interpretations of the precautionary principle. The weak interpretation embodies the requirement for (at least some kind of) scientific evidence as a trigger condition, and for balancing costs and benefits. It does not necessitate taking precautions but merely offers a justification to do so. Furthermore, the precautionary response may consist of a wide variety of measures. The strong interpretation, in its turn, is reflected in strict, unconditional and cost-oblivious prohibitions and bans, in strong emphasis on catastrophic outcomes regardless of their probability, and in total reversal of the burden of proof. While the Rio Formulation is typically thought to represent a paradigm example of the weak form, the Wingspread Statement is the most frequently cited example of the strong form.

The precautionary principle may be justified on the basis of ethical and sociopolitical grounds and/or as a form of rational action. The principle per se does not imply a commitment to any particular ethical theory or tradition, but it is in accordance with many of them, for example, with certain biocentric and anthropocentric justifications. There are several grounds for taking precautions. The European Environment Agency's report "Late Lessons From Early Warnings: The Precautionary Principle 1896–2000" examines 91 case studies on taking no precautions in the state of uncertainty, and the detrimental consequences of these omissions. Yet the precautionary principle, and in particular its implementation, have remained matters of debate. Several criticisms have been leveled against the principle. It has been argued that the principle is too vague to guide actual decision making, that the principle is inherently incoherent, and that its implementation would result in adverse effects. Some scholars also claim that the principle blurs the boundary between science and policy in an unacceptable way.

Certain decision rules, law principles, and general approaches in environmental decision making have connections with the precautionary principle. The maximum rule according to which one should choose the option that maximizes the possible minimum outcome comes close to its basic idea. A typical argument for the difference between the prevention principle and the precautionary principle is that the former is to be applied when the probability of an unacceptable threat can be assigned in risk assessment. If the probability of the risk cannot be assigned, then precaution may be applied. The distinction between these principles is more complicated in legal practice, however. Some authors have also drawn attention to the precautionary principle's relationship to the principle of sustainable development and to participatory decision-making practices.

See Also: Environmental Policy; Environmental Values and Law; Ethics and Science; Human Values and Sustainability; Technology.

Further Readings

European Environment Agency. "Late Lessons From Early Warnings: The Precautionary Principle 1896–2000." http://reports.eea.europa.eu/environmental_issue_report_2001_22/en/Issue_Report_No_22.pdf (Accessed January 2010).

"Ministerial Declaration of the Second International Conference on the Protection of the North Sea." London, November 25, 1987.

O'Riordan, Tim, James Cameron, and Andrew Jordan, eds. *Reinterpreting the Precautionary Principle.* London: Cameron May, 2001.

"Rio Declaration on Environment and Development." United Nations Conference on Environment and Development, Rio de Janeiro, June 3–14, 1992.

Sandin, Per. "Dimensions of the Precautionary Principle." *Human and Ecological Risk Assessment,* 5 (1999).

Sunstein, Cass R. *Laws of Fear: Beyond the Precautionary Principle.* Cambridge, UK: Cambridge University Press, 2005.

Trouwborst, Arie. *Evolution and Status of the Precautionary Principle in International Law.* London: Kluwer Law International, 2002.

"Wingspread Statement on the Precautionary Principle." Wingspread Conference on the Precautionary Principle, Racine, Wisconsin, January 1998.

Marko Ahteensuu
University of Turku

PRESERVATION

A group of snowmobile riders stop to photograph the Old Faithful geyser in Yellowstone National Park. Snowmobile use in the park exemplifies the conflict between preservationists who view parks as preserved, historic landscapes and those who want to retain access for a wide variety of uses.

Source: National Park Service/William S. Keller

Preservation is an argument for setting aside or taking certain spaces out of human use on the basis of their intrinsic value. Preservation is commonly used to refer to the protection of nature or wilderness areas, although a similar set of concepts and practices has also been applied to historic buildings, monuments, or human landscapes. In much of the literature on nature or wilderness preservation, it has been contrasted with or distinguished from conservation, which has been defined as a more utilitarian activity, more inclusive of human and/or economic activities. Yet scholarship in environmental history has shown that the distinction between conservation and preservation is not so clearly delineated. It can be problematic to rely on a division between these two concepts because of their substantial overlap and the common use of these terms interchangeably. Preservation as concept and practice has also changed over time and across different places, leading to multiple histories of preservation. Scholars and activists have also critiqued the preservationist argument for excluding humans and for privileging certain environmental uses over others.

However, most scholars and advocates of environmental preservation trace its roots to late-19th- and early-20th-century campaigns for wilderness preservation and for the establishment of national parks and wilderness areas in the United States. Key figures identified with the early American preservation movement include Henry David Thoreau, Ralph Waldo Emerson, Aldo Leopold, and John Muir.

Muir's work in particular has been held up as a key early example of the preservationist argument. His efforts at wilderness preservation are often contrasted with the conservation-minded work of his contemporary Gifford Pinchot, the first head of the U.S. Forest Service. Pinchot advocated the wise use of the nation's resources from the perspective of a scientific forester, seeing the natural spaces that Muir and others had sought to preserve under park and reserve systems as important sources of natural resources that should continue to be used for human benefit. While Muir and Pinchot as individuals worked together and respected one another's views on nature and wilderness preservation, their views differed substantially, a conflict that was well illustrated in their disagreements over the Hetch Hetchy dam project proposed in Yosemite National Park.

Pinchot saw the Hetch Hetchy project as necessary for supplying water to the growing city of San Francisco. The proposed dam and large reservoir would flood what was to Muir an essential part of Yosemite National Park, and consequently, he fought unsuccessfully against the project along with other members of the Sierra Club, an organization that he had helped establish. The Hetch Hetchy affair shows some of the tensions between approaches to conservation and preservation; this debate is considered to be at the root of current understandings of these concepts. Other early groups and campaigns that shaped the development of the preservationist argument in the United States included the Free Niagara movement, the Appalachian Mountain Club, and the National Audubon Society.

One of the earliest modern examinations of the preservation movement, including the Hetch Hetchy affair, was Roderick Nash's 1967 work *Wilderness and the American Mind.* Widely read by both scholars and activists alike, Nash argued that the American perspective on wilderness had shifted from seeing it as evil or dangerous (as at initial European contact and settlement) to a more romantic view that saw in wilderness the potential to cure the ills of an industrial society; wilderness emerged from this view as something that needed and deserved to be "preserved." This change in perspective was important because it provided a platform from which to rationalize or legitimize political and economic decisions to establish government-protected spaces.

Nash's work appeared soon after the 1964 Wilderness Act was passed. This legislation set aside 9.1 million acres of land for recreational use, free from some of the things that preservationists at the time deemed inappropriate for wilderness areas: roads, vehicles, and logging equipment like chainsaws.

Long before the passage of the 1964 act, however, Aldo Leopold had already been influential in establishing the first national wilderness preservation area. It was set aside in 1924 at Leopold's suggestion and as a result of advocacy from the Wilderness Society, of which he was a member.

Around this time, Leopold has also been credited with bringing ideas from ecological sciences into conversation with preservation. His ecology-based arguments stood in contrast to the economic or utilitarian concerns that had typified debates such as Hetch Hetchy.

The inclusion of science-based, ecological arguments is still important to the environmental movement that emerged in the 1960s and 1970s. This environmentalism retained some of the key concepts and figures of preservation such as Leopold and Muir, but expanded its scope to address a wider range of issues and sites of concern and in so doing, became a more broadly based movement.

Connected to an increased emphasis on environmental issues, the 1980s and 1990s saw systematic critiques of preservation emerge, offering new histories that took preservation in theory and practice to task for an underlying racism and for overlooking issues of class and gender. In the 1980s in particular, criticism of environmental racism revealed the greater burden and impact of pollution as experienced by racialized minority groups, particularly in American cities. Scholars and activists also began to explore the ways that ideas of preservation, and wilderness preservation in particular, had impacted and excluded a diverse spectrum of groups and communities, including the urban poor, rural farmers, and Native Americans. For instance, scholars of environmental justice critiqued the underlying assumptions of "wilderness"—that the vast spaces of the United States that needed to be preserved were uninhabited—and showed how this had served to erase Native American histories of use and settlement in these very spaces.

Also important were William Cronon's 1995 arguments in "The Trouble With Wilderness; or, Getting Back to the Wrong Nature." Cronon, an environmental historian, argued that while wilderness preservation positioned itself against commercialization or industrialization, the treatment of human habitation or use as contamination and the consequential focus on environments "out there" had led advocates to neglect the local and urban environmental issues that sprang directly from industrialization.

Moreover, although preservation is typically associated with American campaigns, it is not only an American idea; similar practices have been widely implemented around the world. Ramachandra Guha has critiqued wilderness preservation in the third world, arguing that its basic tenets are not appropriate to the landscapes of rural production that typify large areas of Asia and Africa. Guha has also questioned U.S. and global environmentalism for not addressing more pressing issues stemming from overconsumption.

Today, while critical issues and questions about preservation are pursued by academics, activists, and preservation practitioners alike, tensions remain within the present preservationist stance. Debates abound about access and use of national parks, for instance, bringing to light some ongoing tensions between preservation and tourism or recreation. One recent example is the unresolved debates about snowmobile use in Yellowstone National Park that have highlighted the conflict in purpose between parks as preserved natural and historic landscapes and wilderness and parks as places for the "benefit and enjoyment of the people" as noted in the 1872 dedication of Yellowstone as the United States' first national park. Such debates highlight contradictions that must continue to be addressed, including not only concerns regarding wildlife, ecological or economic impacts, noise pollution, and air quality concerns but also concerns regarding who has access to these public spaces and for what uses. These conflicts return to the key tenets of preservation as they have emerged historically, in concern about the overuse, degradation, and disappearance of nature.

See Also: Conservation; Conservation, Aesthetic Versus Utilitarian; Leopold, Aldo; Muir, John.

Further Readings

Cronon, William. "The Trouble With Wilderness; or, Getting Back to the Wrong Nature." In *Uncommon Ground: Rethinking the Human Place in Nature,* William Cronon, ed. New York: W. W. Norton, 1995.

Guha, Ramachandra. "Radical American Environmentalism and Wilderness Preservation: A Third World Critique." *Environmental Ethics,* 11 (1989).

Lewis, Michael, ed. *American Wilderness: A New History.* New York: Oxford University Press, 2007.

Nash, Roderick. *Wilderness and the American Mind.* New Haven, CT: Yale University Press, 2001 (1967).

Wellock, Thomas R. *Preserving the Nation: The Conservation and Environmental Movements 1870–2000.* The American History Series. Wheeling, IL: Harlan Davidson, 2007.

Vanessa Lamb
York University, Toronto

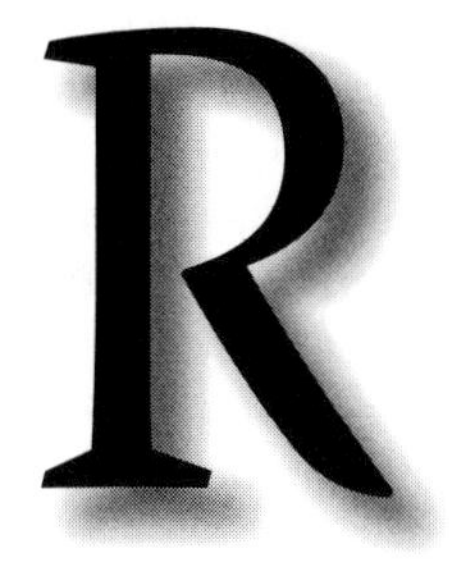

Religious Ethics and the Environment

As environmental concerns confront a planet facing diminishing species, climate change, and finite natural resources, religious leaders have been increasingly aware of their responsibility to both their human adherents and the nonhuman world. Threads of morality, virtue, and ethics inform all the world's religions and, while many religious texts and practices have an anthropocentric lens, the environment has been a focal point for indigenous religions and engaged Buddhism in particular. The Abrahamic faiths of Christianity, Judaism, and Islam have been primarily focused on human relations with God, but as issues of species extinction, resource depletion, climate change, and pollution emerged in the latter half of the 20th century, religious leaders began reconsidering their relationship with the natural world.

Christianity

The Christian understanding that dominates the developed Western world and particularly the United States was reflected in biblical passages such as "Be fruitful and multiply, and fill the earth and subdue it, and have dominion over the fish of the sea and over the birds of the air" (Genesis 1:26). Lynn White's article "The Historical Roots of Our Ecologic Crisis" appeared in a 1967 publication of *Science* and launched a critical reinterpretation of scripture and stewardship in the Christian church that continues to the present time. White critiqued Christian worldviews as exploitive, arrogant, and self-serving, without regard for the intrinsic value of the rest of the natural world with whom humans share the Earth. In addition to noting the scientific and technological achievements that provided man the capacity to devastate the environment at a pace unheard of prior to the 1800s, White associated the idea of progress with Christianity's purging of paganism. White explains that the animistic view was to placate the guardian spirits of a place before cutting a tree, mining minerals, or changing the course of a stream. He noted that ancient understandings of time were more cyclical as opposed to the modern linear understanding of time. In pointing to the ancient ideas of transcendent time without beginning or end, White contrasts this with the teleological idea of progress, which he equates with the biblical story of Creation. White considered the Judeo-Christian understanding to be an anthropocentric view that placed a higher value on humans over the rest of creation. He notes that this human-centered Western Christian view established the dualism between humans and nature and inferred that it was God's will that humans control the natural world in His name. White's claim is that the exploitation that followed was a result of the disregard for relationships with the land supplanted by a focus on the transcendent saints and heaven.

While the article does not examine the development of utilitarianism as an ethical practice, White indicates that as the authority of science and technology began to replace the dominance of the church in Western society, the natural world was seen as more of a resource for human development instead of an interrelated aspect of living systems.

White's alternative view centers around St. Francis, whose approach to the natural world centered on developing the virtue of humility in humans. The Franciscan notion that all creation held equal value challenged the already-accepted notion that nature existed simply to serve humanity. The echoes of this challenge to the dominant paradigm resonate in the developing Christian practices of environmental stewardship and reinterpretations of the theological view of Earth and its creatures as sacred. However, many ecotheologians are still doubtful that humans can easily rise above the inherent paternalism in the notion of stewardship, which assigns humans responsibility for the Earth and its creatures. The image of the shepherd protecting his flock is often evoked as a metaphor for the relationship of reverence and respect that humans should cultivate toward the natural world. Yet while some Christian denominations develop creation-centered theology that focuses on humans as only one part of a common Earth community, others maintain that environmental protection is linked to self-interest.

Various Christian denominations are putting theology into action as they install solar panels on their churches, create biodynamic gardens on their land, and implement zero-waste recycling programs. Many denominations have developed social justice ministries that seek to protect oppressed and disenfranchised peoples from environmental pollution and degradation. These ministries seek to defend those who have the least power in society and are frequently subject to living with polluted air and water and who have limited access to healthy food and farmland. Particularly in developing countries, the consequences of first world consumerism and industrialism are borne by those who suffer from pollution of drinking water from chemical plants or from mining tailings seeping into the farmland or fishing waters of indigenous peoples.

Ecojustice takes several forms, including ecofeminism, liberation theology, and economic justice. Ecofeminists like Rosemary Radford Reuther link the oppression of women with the domination of the natural world. Liberationists like Gustavo Gutierrez and Jose Bonino view the disenfranchisement of oppressed peoples and the environment as stemming from the same political and cultural ethos and urge a redefinition of the church's role in the plight of the poor including their economic and ecological welfare. John Cobb and Herman Daly coauthored *For the Common Good,* which provided a set of criteria for revising the interrelationship of economics, ecology, and human interaction.

Ascetics of all traditions have considered the natural world as a sanctuary to be preserved for its ability to convey spiritual experiences and as a place for retreat and refuge. John Muir described the groves of giant sequoias as cathedrals. Emerson and Thoreau wrote from their secluded forest outposts on the virtues of transcendentalism.

When Father Thomas Berry wrote *The Dream of the Earth* in 1988, the new science of systems theory, quarks, and holons was beginning to question the modernist empirical approach that descended from Bacon, Newton, and Galileo. Rather than the traditional theological focus on redemption, Berry sought to harness the creative energy as a way to reinvigorate humanity's relationship with its surroundings. Berry articulated the reenchantment of the Earth with an interconnected, holistic approach that reunited mind and emotions, faith, and action. Berry insists that technology alone is not the answer and describes a confrontation between the industrial and the ecological worldviews. He encourages a psychic awareness, similar to ancient shamanic techniques that can restore human fulfillment rather than making instrumental calculations about genetic material. His vision was taken up by Brian Swimme and Richard Tarnas, whose goals are to formulate a new Creation story that could be embraced by sacred and secular persons alike.

Judaism

Like Christianity, Judaism is basically concerned with human morality to other humans. However, the Torah does imply a sense of human responsibility for the well-being of the natural world through explicit commandments. During the first three years of growth, the fruits of trees and vineyards are not to be eaten as they are considered God's property. Fruit trees are to be protected and not cut down during wartime. Other prohibitions include pollution of air and water, hunting animals for sport, the destruction of cultivated plants, overgrazing the countryside, overconsumption of any kind, or wasting resources. There is an injunction against breeding different varieties of cattle or sowing fields with different species of seed. The Torah provides strict laws regarding diet and which animals are deemed unclean and unfit for human consumption. Some scholars have reasoned that in forbidding eating various water animals and birds of prey, the rabbis were aware of the need to protect the ecological equilibrium. The just allocation of nature's resources connects "the moral quality of human life and the vitality of God's creation."

As Hava Tirosh-Samuelson explains in an excellent and succinct summary of Jewish environmental ethics, the ancient Israelites were primarily farmers whose agrarian festivals Sukkot, Pessach, and Shavuot were assigned religious meaning in the Bible. Virtues of humility, modesty, moderation, and mercifulness were among the rabbinic teachings that cultivate a Jewish ecological ethic.

Islam

In considering the environment, Muslim scholars and clerics consult the Koran and the Islamic legal code as written in the Shari'a. Textual sources that cite the words of the Prophet Muhammad are taken as the ethical authority. Unfortunately, the texts primarily address issues of social injustice among humans. As one of the world's fastest-growing religions, Islam is practiced with slightly different interpretations based on the cultural context of the region and local societies. Many Muslims follow secular or scientific models of environmental ethical actions since the textual tradition does not specifically address the problems of climate change, pollution, or large-scale development of open lands. Since Islam focuses on the transcendent rather than the immanent earthly concerns, the environment has not received the attention that ecologists hope for.

Persons dissatisfied with their own faith traditions' approach to the environment have sought wisdom from other traditions that maintained what is perceived as more value in the more-than-human world. Indigenous peoples are admired for their appreciation of biodiversity and sense of reverence for nature. While their relationship with the land extends back many generations, the sense of continuity has often been disrupted by colonialism, relocation to less desirable areas, and even the well-intended efforts of relief organizations to provide them with technology to produce higher-yield crops or educate them in the ways of industrial societies. One must use caution to avoid romanticizing indigenous peoples; however, there is much to learn from them if one approaches with respect for their knowledge.

Indigenous Religions

For indigenous peoples, religion is inextricably linked to the land integrated with a local sense of place. Beyond ceremonial practices, the land is interwoven in the oral history of the people and has provided an interdependent source of sustenance on physical and spiritual levels for generations. Indigenous peoples around the globe live by what they consider natural law, although the definition of this principle has much more to do with nature and its intrinsic value than it does with law. Balance, harmony with the natural world, the interrelationship of all beings, and not taking more than you need are key principles. Many Native Americans understand that life is cyclical, following the cycles of the moon, the seasons, and the life of the people. Within this is knowledge that one's actions today affect the quality of life for future generations. There is a notion of reciprocity that one has with the animals hunted and the plants harvested and that offerings are made to the Creator for these gifts of life. Another key tenet is sharing. Individual ownership has not traditionally been a part of the Native spiritual understanding. Traditional ceremonies and individual actions acknowledge gratitude for the gifts they bestow on humans and are punctuated by offerings made to the animals and plants for their gifts of food and medicine. Indigenous stories and rituals convey a sense of interdependence and coexistence between humans and the natural world.

Indigenous ethics arise from a sense of responsibility more than from a set of laws. Their approach to sustainability encompasses whole ecosystems and is informed by observations and practices that have been handed down within families and tribal societies for generations. Their sense of religion is inclusive and penetrates all activities rather than being one among many options for meaning-making. Knowledge is gained through personal experience and is embodied rather than studied. Indigenous wisdom is collective and while individuals may have revelations through dreams and visions, their learning is less for self-gain and more for the good of the community. The epistemology of indigenous peoples varies between locales and is not a static knowledge. Their sense of good evokes what is best for the whole and includes the entire ecosystem in which they reside. The roles and responsibilities held by Native peoples often do not correspond with the worldviews of Western society. It is important not to take specific examples or stories out of context, but to engage with the people themselves if one wishes to gain from the spiritual wisdom they offer.

Buddhism

Buddhism is another of the world's religions that environmentalists look to for ways to relate to the natural world. Key leaders such as the Dalai Lama and Thich Nhat Hanh provide ethical guidelines for interacting with the natural world. Of course, one must keep in mind that the original texts of the Buddha's teachings did not provide explicit instructions regarding the environment. Buddhist principles are consistently being reinterpreted through commentaries, and the modern adherent can find several aspects of the texts that can be applied to environmental concerns.

Respect for animals is evoked throughout the Indian subcontinent and the Tibetan plateau. Buddhists are encouraged to be vegetarians as a way of doing less harm to other sentient beings. The teachings of the Buddha speak of the Wheel of Life and the various realms that a being experiences on the path to enlightenment. Although only humans are believed to have the ability to achieve this state of liberation from the cycle of *samsara,* or earthly suffering, animals are viewed as very close to humans. Certain foods like fish are avoided because of the many eggs they produce, and thus the numerous potential lives that one extinguishes by eating them. Animals are excluded from the Buddhist diet on the principle of *ahimsa,* or nonviolence to other creatures. This principle is also central to Jains and Hindus. Critics recall that the Buddha ate meat when it was offered, and the response from Buddhist teachers is that intentional killing is the most harmful, so monks are instructed to accept meat only when they have not witnessed the slaughter or learn that the killing was done specifically on their behalf.

Hindu cosmology contains an emphasis on the sacredness of animals, the importance of protecting trees and forests, and places value on vegetarianism. These values inspired the Chipko Movement in which activists in the Himalayan region used tactics of nonviolent resistance such as hugging trees to protect them from the lumber companies. However, these values are frequently outweighed by a focus on *moksa,* or spiritual liberation and practices that encourage the acquisition of wealth over simple living and frugality.

Buddhist teachings cultivate a sense of nonattachment that resists the current Western consumerist tendencies. Arising from desire, the idea of hoarding material wealth or clinging to things as a way of establishing one's identity is considered to be a sign of ego or self-clinging. One of the primary goals in Buddhism is to let go of attachments, although this does not mean, as is often misunderstood, to deny oneself all worldly goods and activities. Rather it is about cultivating a sense of nongrasping and not using external objects to prop oneself up. Mahayana Buddhists seek a "middle way" that does not go to either extreme—not denying oneself the nutrition to be healthy but neither gorging oneself on rich delicacies.

Using Buddhist philosophy as her point of reference, Buddhist scholar Rita Gross articulates her ethic in terms of the Four Noble Truths. The First Noble Truth states that conventional lifestyles inevitably result in suffering; the Second Noble Truth states that suffering stems from desire rooted in ignorance. Gross clarifies the translation of the Sanskrit *trishna* as "addiction," "compulsion," "grasping," "clinging," "craving," or "fixation." More than just a fleeting impulse of desire, the Second Noble Truth paints the desire as more of an insatiable lust and Gross contends it is at the root of "both excessive consumption and overpopulation." However, Buddhist thought does not condemn us to eternal suffering. Instead, it suggests cultivating a sense of detachment and equanimity that allows us to rejoice in what is, rather than in what might be. While this may be a good strategy for the wealthy, Gross admits that "those in poverty are often too consumed with survival to develop equanimity and enlightenment." Gross's premise is that consumption and overpopulation are interconnected in a circular dynamic. In order to cut the root of suffering, she asks us to recognize that "more is not better, whether it is more people or more consumables."

Interrelationship with all life on the planet is a basic Buddhist understanding. To return to the Wheel of Existence, it is said that in the cycle of reincarnation all beings have touched our lives at one time or another. An ethical principle derived from this understanding is that one should extend loving-kindness and compassion toward all creatures. There are numerous accounts of Buddhists who go to great lengths to avoid killing insects, worms, or the smallest life forms. The Jataka Tales, which provide accounts of the Buddha's previous lives, describe incarnations as animals and trees that influenced the religion's attitudes toward the environment. The Vinaya, which guided the discipline of the early Theravadin monks, prohibited travel during the rainy season to prevent monks from stepping on worms that surfaced in the wet weather and warned them not to drink unstrained water and to treat plants and wild animals with respect and kindness.

In his recent book on sustainability, Thai monk and activist Sulak Sivaraksa notes the Buddha suggested living by the four immeasurables as a path to ethical action. These approaches include loving-kindness, compassion, altruistic joy, and equanimity. These principles are also thought of as ways of working with interdependent origination, a concept popularized by American scholar Joanna Macy.

Conclusion

In surveying a variety of ethical practices and philosophies among diverse religious traditions, it is apparent that there is no singular approach to the environment. Religious ethics have primarily concentrated on human relationship with a transcendent source, or recommended behaviors that would be favorable to conditions after life on Earth. Issues of environmental degradation have inspired religious leaders to reinterpret scriptures to meet the demands of followers concerned about the decline of species, water quality, and pollution. While the focus of religious ethics has been cast through an anthropocentric lens, the desire for a more sustainable relationship with the ecosystems that support life on Earth has prompted religious leaders to consider the care of animals, trees, and rivers and to adopt practices that value the Earth's resources.

See Also: Animal Ethics; Anthropocentrism; Berry, Reverend Fr. Thomas; Ecofeminism/Ecological Feminism; "Ecological Crisis, The Historical Roots of Our"; Ethical Vegetarianism.

Further Readings

Berry, Thomas. *The Dream of the Earth*. San Francisco, CA: Sierra Club Books, 1990.

Coward, Harold and Daniel C. Maguire, eds. *Visions of a New Earth*. Albany: State University of New York Press, 2000.

Gottlieb, Roger, ed. *A Greener Faith*. New York: Oxford University Press, 2006.

Gottlieb, Roger. *The Oxford Handbook of Religion and Ecology*. New York: Oxford University Press, 2006.

Kearns, Laurel and Catherine Keller, eds. *Ecospirit: Religions and Philosophies for the Earth*. New York: Fordham University Press, 2007.

Schweiker, William, ed. *The Blackwell Companion to Religious Ethics*. Malden, MA: Blackwell Publishing, 2005.

Zimmerman, Michael E., J. Baird Callicott, Karen J. Warren, Irene Klaver, and John Clark. *Environmental Philosophy: From Animal Rights to Radical Ecology*, 4th ed. Upper Saddle River, NJ: Pearson Prentice Hall, 2005.

Stephanie Yuhas
University of Denver

Rolston, III, Holmes

Holmes Rolston, III (1932–) is a Distinguished Professor of Philosophy at Colorado State University. He is the founder of the journal *Environmental Ethics* and has been active in establishing, shaping, and defining the modern discipline of environmental philosophy. Rolston was born in Rockbridge Baths, Virginia, and studied at Davidson College (physics and mathematics), Union Seminary (he is a Presbyterian minister, as was his father and grandfather before him), and the University of Pittsburgh (philosophy) before earning his Ph.D. from the University of Edinburgh. Rolston has published a number of seminal books in environmental philosophy including *Philosophy Gone Wild, Environmental Ethics,* and *Conserving Natural Value.*

Rolston has sought an expansion of a philosophical understanding of the natural world. He argues that the natural world carries intrinsic values, meaning that the environment should be valued for its own sake rather than merely as a means to an end. This is in contradiction to the more traditional Western concept of the relationship between humans and the environment in which nature is something separate from man that should be conquered or overcome, a concept reinforced by some Western religions that have been interpreted as stating that the world, or nature, was created specifically for human use.

Initially, Rolston had difficulty getting his ideas published in philosophy journals: his first major success was the article "Is There an Ecological Ethic," which was published in 1975 in *Ethics.* In this article he argued that in contrast to the typical conception that all values come from human conception, nature contains intrinsic value and should be treated with respect. In 1986, he published *Science and Religion: A Critical Study,* and in 1987, he published *Environmental Ethics,* both of which reject anthropocentrism in ethical analysis. Also in 1986, Rolston published a collection of essays, *Philosophy Gone Wild,* which use the concept of intrinsic value to draw a connection between value and duty. Rolston argues that if humans recognize that something has value, then they have a duty to protect it, promote it, or bring it about. These values exist not only at the level of individual organisms but also in species, ecosystems, and natural processes. By this principle, he argues that humans have a duty to the natural world, and that they need to protect species and ecosystems from destruction.

Rolston was selected to deliver the Gifford Lectures in 1997–98 (a prestigious series of lectures at four Scottish universities delivered over the course of an academic year), and published *Genes, Genesis, and God* in 1999, a book developed from the lectures. In 2003, Rolston was awarded the Templeton Prize for Progress Toward Research or Discoveries About Spiritual Realities, established by Sir John Templeton in 1972 to honor those who advance spiritual values. His works have been translated in 18 languages, and he has lectured widely through the United States, Europe, South America, Australia, and Asia.

See Also: Attfield, Robin; Biocentrism; Deep Ecology; Instrumental Value; Intrinsic Value; Land Ethic; Leopold, Aldo.

Further Readings

Attfield, R., A. Belsey, and Royal Institute of Philosophy. *Philosophy and the Natural Environment.* Cambridge, UK: Cambridge University Press, 1994.

Chappell, T. D. J. *The Philosophy of the Environment.* Edinburgh, UK: Edinburgh University Press, 1997.
Elliot, R. and A. Gare. *Environmental Philosophy: A Collection of Readings.* St. Lucia, Australia: University of Queensland Press, 1983.
Light, A. and H. Rolston. *Environmental Ethics: An Anthology.* Malden, MA: Blackwell, 2003.
Ouderkirk, W. and J. Hill. *Land, Value, Community: Callicott and Environmental Philosophy.* Albany: State University of New York Press, 2002.
Rolston, H. *Conserving Natural Value.* New York: Columbia University Press, 1994.
Rolston, H. *Environmental Ethics: Duties to and Values in the Natural World.* Philadelphia, PA: Temple University Press, 1988.
Rolston, H. *Genes, Genesis, and God: Values and Their Origins in Natural and Human History: The Gifford Lectures, University of Edinburgh, 1997–1998.* Cambridge, UK: Cambridge University Press, 1999.
Rolston, H. *Philosophy Gone Wild: Essays in Environmental Ethics.* Amherst, NY: Prometheus Books, 1986.
Rolston, H. *Science and Religion: A Critical Survey.* New York: Random House, 1987.
Winkler, E. R. and J. R. Coombs. *Applied Ethics: A Reader.* Oxford, UK: Blackwell, 1993.

Jo Arney
University of Wisconsin–La Crosse

Roosevelt, Theodore

Theodore Roosevelt exemplified the youth, enthusiasm, and ambitions of his times. As the 26th president of the United States, Roosevelt guided his nation into the 20th century, establishing an influence on not only his but also future generations. His legacy has allowed biographers and historians to continuously present various and diverse aspects illustrating Roosevelt's strenuous life, which included the outdoors.

The strenuous life that Roosevelt described involved "a life of effort, labor and strife." As he embodied this philosophy throughout his life—as a sickly young boy until his death in October 1919—he developed an appreciation and passion for his natural environment. At Harvard University, the future president studied natural history and natural sciences. Following his academic career, his devotion to nature continued as he rose up the political ranks, becoming New York governor, then U.S. vice president, and in 1901, president after the assassination of William McKinley. As the country's chief executive, Roosevelt endorsed a progressive agenda calling for social, economic, and political reform and stressing a Square Deal and New Nationalism. Roosevelt expanded the role of the U.S. presidency and the federal government in preserving the nation's environment and natural resources.

As president, Theodore Roosevelt intertwined his lifelong passion for nature, his progressive ideas, and his great concern for the welfare as well as the future of the United States. Throughout his career, Roosevelt maintained his interest in ornithology and established lifelong associations with numerous naturalists along with advocates for conserving and preserving the nation's resources. These associations included U.S. chief forester Gifford Pinchot,

In 1903 President Theodore Roosevelt designated Pelican Island in Titusville, Florida, as the first national wildlife refuge in order to protect its pelicans, some of which are shown here, and other birds. There are now over 530 national wildlife refuges in the United States.

Source: U.S. Fish & Wildlife Service/ George Gentry

John Muir, and John Burroughs. Along with his conservation friendships, Theodore Roosevelt demonstrated his passion for nature with his numerous explorations throughout the United States and the world. During his lifetime, Roosevelt's adventures included the Badlands of the Dakotas, the Sierra Nevada, the Adirondacks, Yellowstone, the Blue Ridge Mountains, the Mississippi River, Africa, and the Amazon. These travels not only exhibited his fervor toward nature but also provided a firsthand account on environmental issues such as reclamation, irrigation, forest removal, animal extinction, and land misuse.

During his presidency, Roosevelt strove to make conservation of the nation's natural resources a priority. In the nearly eight years of his administration, his policies toward conservation involved designating land for preservation and establishing or restructuring federal agencies to conserve and to protect the country's forests, rivers, wildlife, and public lands. President Roosevelt designated 150 national forests, the first 51 federal bird reservations, 5 national parks, the first 18 national monuments, the first 4 national game preserves, and the first 21 reclamation projects. Federally protected lands during his term totaled about 230 million acres, an area equivalent to the East Coast states from Maine to Florida. Along with designating conservation a priority, the young president established commissions to observe and evaluate the various aspects of the nation's natural resources and the then-current legislation regarding these issues and to establish long-term guidelines or principles to underpin continuing national efforts toward conservation. These commissions included the Public Lands Committee, the Inland Waterways Commission, the Conference of Governors, the National Conservation Commission, the Country Life Commission, the Joint Conservation Congress, and the North American Conservation Conference.

President Roosevelt's directives toward conservation surpassed any of his predecessors in the White House and set a precedent for his successors. His actions represented his belief that the nation's chief executive, along with the federal government, should serve as steward of the country's natural resources and its environment. However, his conservation policies were just one aspect of his political agenda and progressive platform involving a Square Deal and New Nationalism. As a Progressive, Roosevelt believed that improving one's environment would assist in improving the quality of life for the individual. Along with this belief, Roosevelt emphasized that everyone was entitled to a Square Deal that not only provided equality and justice for all Americans, but that the government should assist those who made an effort to improve their own quality of life.

While emphasizing a Square Deal for all Americans, Roosevelt also believed that this would fulfill the early-20th-century philosophy of New Nationalism that placed national interests above individual, sectional, or regional interests. Roosevelt's urge for a Square Deal and carrying out the doctrine of New Nationalism was a call for responsibility by Americans to improve their own lives, to work to improve their communities, to leave their environment better for their children, and to endeavor to create a better a nation. These principles provide a historic link to understanding sustainability today.

Roosevelt's philosophy on conservation derived from his passion for nature. He continued to demonstrate this devotion as president by advancing the role of this office and the government to serve as stewards to preserve the nation's natural resources for future generations to benefit socially as well as economically. More importantly, his views on conservation evolved around individual as well as collective responsibility to lead to a better nation.

See Also: Conservation; Muir, John; Pinchot, Gifford; Preservation.

Further Readings

Brinkley, Douglas. *The Wilderness Warrior: Theodore Roosevelt and the Crusade for America.* New York: HarperCollins, 2009.

Collins, Timothy and Stephen R. Hicks. *Theodore Roosevelt's Country Life Centennial 1909–2009: A Foundation for Sustainable Rural Community Development.* Illinois Institute for Rural Affairs, 2009. http://www.iira.org/clc (Accessed November 2009).

Croly, Herbert. *The Promise of American Life.* Cambridge, MA: Belknap Press of Harvard University Press, 1965.

Gurney, Scott. "Biographical Portrait: Theodore Roosevelt." *Forest History Today* (Fall 2008). http://www.foresthistory.org/Publications/FHT/FHTFall2008/Gurney.pdf (Accessed November 2009).

Roosevelt, Theodore. "President's Eighth Annual Address to Congress." Speech delivered to Congress December 8, 1908. http://www.presidency.ucsb.edu/ws/index.php?pid=29549 (Accessed November 2009).

Roosevelt, Theodore. "President's Seventh Annual Address to Congress." Speech delivered to Congress December 3, 1907. http://www.presidency.ucsb.edu/ws/index.php?pid=29549 (Accessed November 2009).

Roosevelt, Theodore. "The Strenuous Life." Speech delivered to Chicago's Hamilton Club, Chicago, Illinois, April 10, 1899. http://www.historytools.org/sources/strenuous.html (Accessed November 2009).

Roosevelt, Theodore. *Theodore Roosevelt: An Autobiography.* New York: Macmillan, 1913.

Stephen Hicks
Western Illinois University

S

SAN PEDRO BORDER FENCE

This controversial fence in the San Pedro Riparian National Conservation Area has contributed to the disruption of migrations by some of the 300 animal species in the area and increased the amount of human and vehicle traffic over previously untraveled land.

Source: U.S. Department of Homeland Security

The San Pedro Riparian National Conservation Area (RNCA) consists of approximately 57,000 acres of public land in Arizona's Cochise County, adjacent to the Mexican border and the riparian zone (where land meets river) of the San Pedro River. It was designated as an RNCA by Congress in 1988 in order to protect the local ecosystem, one of the few desert riparian ecosystems to survive the settling of the American Southwest. Administered by the Bureau of Land Management, the RNCA is home to some 300 species of animals, and provides a winter habitat to 250 migrant bird species.

The area has also long been a travel corridor for illegal immigrants from Mexico. As efforts to stop illegal immigration increased, including the statewide Operation Gatekeeper in California, Operation Hold the Line in Texas, and Arizona's Operation Safeguard, an extensive system of border walls was approved by Congress as part of H.R. 6061—the Secure Fence Act of 2006. The Secure Fence Act directed the secretary of homeland security to "provide at least two layers of reinforced fencing, installation of additional physical barriers, roads, lighting, cameras, and sensors" for various stretches of the border including "from 10 miles west of the Calexico, California port of entry to five miles east of the Douglas, Arizona port of entry."

In the RNCA, a pedestrian fence was erected across the length of the area, with a river-crossing vehicle barrier on the river itself. Temporary roads were constructed for the sake of construction crews and border patrols. The immediate effect of the construction was that it noticeably increased erosion in the river, even in the first year. Wildlife populations and migrations have been disrupted, and both illegal immigrants and the border patrols searching for them have strayed into the nearby wild lands, where their presence and vehicles disrupt the rare ecosystems of the Sky Island mountain ranges that provide oases of moisture to unique desert species. Environmentalists are particularly concerned about migratory butterflies that travel through the area, and two wildcats native to the area—the ocelot and the jaguarundi. Nearly half of the area that jaguars travel through, from the Baboquivari Mountains to the Pelloncillo Mountains, has been blocked by fencing.

The plan was criticized as a boondoggle from the start, and among the criticisms leveled against it has been its environmental impact, because, under the George W. Bush administration, the border walls would extend through federal protected lands like the San Pedro RNCA. In March 2010, President Barack Obama halted the expansion of the "virtual fence" (the system of cameras and sensors in the interstices between the walls), and was among the senators voting in favor of the bill at the time of its passage, but as of 2010 the government had not seriously addressed the matter of the fence's presence in federally protected areas.

See Also: Animal Ethics; Civic Environmentalism; Conservation; Ecological Footprint.

Further Readings

Carter, Alan. *A Radical Green Political Theory*. London: Routledge, 1999.

Dobson, Andrew. *Green Political Thought*. London: T&F Books, 2009.

Radcliffe, James. *Green Politics: Dictatorship or Democracy?* New York: Palgrave Macmillan, 2002.

Torgerson, Douglas. *The Promise of Green Politics: Environmentalism and the Public Sphere*. Durham, NC: Duke University Press, 1999.

Bill Kte'pi
Independent Scholar

"Should Trees Have Standing?"

Christopher Stone is (as of 2010) the J. Thomas McCarthy Trustee Chair in Law at the University of Southern California. He is an expert in environmental issues, particularly environmental law and environmental ethics. In 1972, he published the widely influential article "Should Trees Have Standing? Toward Legal Rights for Natural Objects." This article became a foundational piece in the literature of the emerging field of environmental ethics and philosophy. The legal status of natural entities continues to be a widely debated topic, and his work still serves as an important resource for thinking through such issues.

Beginning his article, Stone credits Charles Darwin for defining what is called "moral extensionism." As our species has evolved and progressed, our moral scope has expanded to include more entities under its consideration; people who were once regarded as "objects" or "property" have gradually become incorporated into our interpersonal ethics.

Stone points out a parallel phenomenon in the history of law; the class of people and entities that are possessors of rights has likewise increased. This process is always a difficult transition that at first appears inconceivable. At each step, the extension and inclusion of previously rightless "things" severely challenges and disrupts the status quo, appearing at first ridiculous and wrong. Such extensions are perceived in this manner because they involve a profound shift in the status of an entity from being a thing, which the previous rights-bearers could use as they wish, to being a rights-bearer itself, which other rights-bearers must respect in certain ways. It is often difficult for a society to grant rights to something until they can appreciate the value of the object in question for its own sake; however, this is difficult to achieve for many until we actually grant the object in question rights, which consequently shifts our thoughts about and behaviors toward the object.

Stone proposes that we give rights to natural objects. He attempts to preempt typical responses to such calls by stating that such a proposal need not entail that natural objects have the same rights as humans or even equal rights among all objects in nature, and explains what it might mean for a natural object to be a possessor of rights. As Stone sees it, in typical cases where a legal dispute arises involving natural objects, the resolution typically revolves around the human interests at stake. For example, an industry draining large amounts of water for its operation and reducing the flow may be held responsible to the farmers downstream who cannot irrigate their farms adequately as a result. The business might be forced to pay the farmer directly for his or her lost earning capacity. Any damages, that is, are damages that must be paid to other human beings, and not to the natural object itself. What is at stake is not the right of the river to flow freely or be free from pollution, and so on, but rather competing human interests in the river.

Stone wants to give rights to natural objects such that they would be able to benefit legally from harms or other offenses inflicted upon them and be able to receive rewards to ameliorate such offenses. He suggests a "guardianship" model to accomplish this. A natural object would have a guardian to represent it in court, and any monetary rewards that the object may receive from damages would go to a fund administered by the guardian in the best interest of the natural object. In short, granting legal rights to natural objects would entail that they themselves have legal standing in their own right and can receive relief for their own benefit. The apparent difficulty that natural objects cannot speak for themselves is erased due to the fact that their guardians can perceive damages to them and can speak for them in court—the same way lawyers do for children, corporations, and other rights-bearers who cannot fully represent themselves in court.

Stone questions whether granting natural objects rights is really the best or only way to go. He suggests that it is, because speaking in terms of rights is powerful and can have a significant effect on the way we think about things. Many environmentally concerned thinkers have suggested that we need a shift in consciousness—away from the idea that nature exists for us, toward some notion of nature existing in its own right. Stone agrees; however, he also argues that a shift in consciousness may be too simplistic given the degree to which we live in a world dominated by institutions. Nevertheless, even if we seek institutional reform, we need to alter the way we understand our place within the natural world. Granting rights, Stone suggests, can contribute to ushering in the necessary shift in consciousness to confront the environmental challenges we face as a society.

Stone's work, groundbreaking at the time, still is discussed and influential today.

See Also: Anthropocentrism; Environmental Law; Environmental Policy; Environmental Policy Act, National; Environmental Values and Law.

Further Readings

Stone, Christopher D. "Should Trees Have Standing? Toward Legal Rights for Natural Objects." *Southern California Law Review*, 45 (1972).

Tuhus-Dubrow, R. "Sued by the Forest: Should Nature Be Able to Take You to Court?" *Boston Globe* (July 17, 2009). http://www.boston.com/bostonglobe/ideas/articles/2009/07/19/should_nature_be_able_to_take_you_to_court (Accessed January 2010).

Jonathan Parker
University of North Texas

Sierra Club

The Sierra Club is the oldest grassroots environmental organization in the United States. It was founded in 1892 by the pioneer environmental preservationist John Muir (1838–1914), who was its first president. It is particularly associated with the protection of large areas of wilderness in as natural a state as possible and was instrumental in the creation of several national parks, notably Yosemite. However, today it also has policies on a much wider range of issues.

The Sierra Club was established as a regional California organization with the goal, as its name implies, of preserving mountainous areas, in particular the Sierra Nevada. From the beginning it was a lobbyist organization, targeting both politicians and public opinion. Its successes include the preservation of much of the mountainous areas of California including Sequoia and Kings Canyon National Parks and it was also influential in the establishment of Glacier and Mount Rainier National Parks.

The Sierra Club's first major losing battle was also the first time that it faced organized economic and political opposition. The Hetch Hetchy reservoir is an important case because it illustrates a key issue in U.S. environmental history. The proposal was to dam the Tuolumne River in order to create a reservoir to provide water for San Francisco for both general consumption and firefighting, in the wake of the disastrous fires following the earthquake of 1906. The Sierra Club, and Muir personally, objected to the dam both because of the special character of the beautiful valley that would be inundated and especially because it was (and still is) part of the Yosemite National Park, which was supposed to be a protected area. Hetch Hetchy was also a clash between the conservationist values of Gifford Pinchot (1865–1946), the first chief of the United States Forest Service, and Muir's preservationist views, discussed elsewhere in this volume. Muir referred to the valley as a cathedral and a holy temple of the people. After many years of opposition from the Sierra Club, the dam was approved in 1913 and opened in 1923: today it provides 85 percent of San Francisco's water supply and also generates some 200 MW of electricity. Since 1987, the Sierra Club has been at the forefront of the movement to restore Hetch Hetchy to its original condition.

The Sierra Club was for many years a small organization with relatively few members and almost entirely reliant on volunteers. It had no chapters outside California until 1950, when the Atlantic chapter was established; membership stood at 7,000.

In 1952, the club appointed David Brower (1912–2000) as its first professional executive director. Brower, a famous mountaineer (he was the first to climb Shiprock, New Mexico, in 1939), was one of the most influential figures in conservation in the 20th century: he also founded Friends of the Earth, the League of Conservation Voters, and Earth Island Institute. Brower led a number of campaigns in the early 1960s that rapidly led to the organization's becoming a national force, including successful opposition to the Echo Park Dam on the Green River in Dinosaur National Monument and to two proposed dams in the Grand Canyon. The campaign included full-page advertisements in the *New York Times* and the *Washington Post* with the slogan, "Should we also flood the Sistine Chapel so tourists can get nearer the ceiling?" The Sierra Club also joined with the Wilderness Society to help pass the Wilderness Act of 1964.

Brower was determined to turn the Sierra Club into a professional organization and to establish the Sierra Club Foundation as a funding source for its activities. He also arranged for the club to produce a number of collections of high-quality photographs of wilderness. These volumes were loss makers for the Sierra Club but have been credited with increasing membership, which rose to 30,000 in 1965 and 75,000 in 1969. Today it is a national organization with chapters in every state and some 1.3 million members.

The Sierra Club is governed by a 15-member board of directors, elected by the members. Notable directors have included the following:

- Ansel Adams (1902–84), distinguished environmental photographer who was a director for 37 years.
- William O. Douglas (1890–1980), the longest-serving associate justice of the U.S. Supreme Court. Douglas was an activist justice, especially in areas of civil liberties and the environment, who supported the idea of legal standing for environmental objects in the case of *Sierra Club v. Morton* (1972), commonly known as the Mineral King case.
- Anne Ehrlich (1933–), ecologist and population expert.
- Dave Foreman (1947–), radical environmentalist, founder of the organizations Earth First! and the Wildlands Project.
- Paul Watson (1950–), founder of the Sea Shepherd Conservation Society, a marine protection organization with a focus on the protection of marine mammals.

The Sierra Club continues to be involved in successful campaigns to preserve natural areas and to influence environmental legislation: examples include the defeat of the National Timber Supply Act, defense of the Clean Air Act, promotion of the National Forest Management Act, and support for reauthorization of a strengthened Superfund legislation and the Clean Water Act. Among the natural areas preserved following Sierra Club campaigns are the Columbia Gorge National Scenic Area (1986), Great Basin Natural Park (1986), the Escalante-Grand Staircase National Monument (1996), and the Tongass Natural Forest (2000).

Although the Sierra Club is still concerned with the natural environment, its current focus is increasingly on climate change and, according to its Website, "Most of all, the Sierra Club is about people." It has policies on a number of issues including energy (it generally opposes nuclear and coal electricity generation), waste and pollution, land use including agriculture and urban planning, transportation, and water, including the oceans; it strongly supports the precautionary principle.

Perhaps the most controversial issue the Sierra Club has faced in recent years is immigration. It has long been concerned with population, and indeed, published Paul Ehrlich's *The Population Bomb,* a somewhat apocalyptic book warning of mass global famine unless governments adopt coercive population policies. From the late 1980s until 2004, some members and directors (including Anne Ehrlich and Dave Foreman) tried to persuade the organization to adopt policies of immigration reduction, in order to achieve population stabilization. However, the Sierra Club has maintained a policy of neutrality on this issue.

See Also: Conservation, Aesthetic Versus Utilitarian; Pinchot, Gifford; Precautionary Principle; Preservation; "Should Trees Have Standing?"; Wilderness Act of 1964.

Further Readings

Cohen, M. P. and Sierra Club. *The History of the Sierra Club, 1892–1970.* New York: Random House, 1988.
Ehrlich, Paul. *The Population Bomb.* New York: Sierra Club and Ballantine Books, 1968.
Sierra Club. http://sierraclub.org (Accessed February 2010).
Sierra Club v. Morton, 405 U.S. 727 (1972). http://supreme.justia.com/us/405/727/case.html (Accessed February 2010).

Alastair S. Gunn
University of Waikato

Silent Spring

Silent Spring, written by Rachel Carson, has proved to be one of the most influential books in the modern environmental movement. It was first published in serial form in the *New Yorker* in 1962, and then in book form the same year by Houghton Mifflin. It was influential in providing the impetus for tighter control of pesticides and was widely read by the general public, becoming a *New York Times* best seller as well as a Book-of-the-Month club selection. *Silent Spring* has been honored on many lists of influential books, including being named to *Discover* magazine's list of the 25 greatest science books of all time, and being placed fifth on the Modern Library List of Best 20th Century Nonfiction. The title *Silent Spring* was inspired by a line from the John Keats poem "La Belle Dame sans Merci," and evokes a ruined environment in which "the sedge is wither'd from the lake/ And no birds sing."

Carson was a biologist and science writer who earned a master's degree in zoology from Johns Hopkins University in 1932, and in 1936 became only the second woman to be appointed to a full-time, professional position with the U.S. Bureau of Fisheries. As an adjunct to her scientific work as a biologist, Carson was a prolific writer throughout her career and published successfully in many formats including radio scripts, official brochures, and articles for the popular press (including the *Baltimore Sun* and the *Atlantic Monthly*). She became chief editor of publications at the Bureau of Fisheries in 1949.

Carson's breakthrough as a writer came with her second book, *The Sea Around Us,* which was a major popular and critical success. It was on the *New York Times* best seller list for over a year, was abridged by *Reader's Digest,* adapted into a documentary film, and won the George Westinghouse Science Writing Prize and a National Book Award, among other honors. This financial success allowed Carson to quit her job with the Bureau of Fisheries in 1952 to concentrate on her writing career. Although she had been aware of the use of synthetic pesticides since World War II (when DDT was widely use to control

Children study a museum exhibit on Rachel Carson's influential best seller *Silent Spring* at the U.S. Fish & Wildlife Service National Conservation Training Center in West Virginia in 1998.

Source: U.S. Fish & Wildlife Service

malaria and typhus), she did not concentrate on this topic until 1957 when she was recruited by the Audubon Society to investigate the dangers of the then fairly uncontrolled use of DDT and other pesticides. To this end, she conducted extensive interviews with scientists and physicians as well as reading the scientific literature and attending FDA hearings on the use of chemical pesticides on food crops.

Silent Spring is a radical book that documented the many harmful effects pesticides were already known to have on the environment and argues that they should properly be called "biocides," because in most cases they can and do harm organisms other than the target pest. It noted that DDT is classified as a chemical carcinogen implicated in causing liver tumors in mice, and accused representatives of the chemical industry of spreading disinformation contradicted by scientific research. She also accused government officials of uncritically accepting the chemical industry's claims of safety and, more radically, questioned the then-dominant paradigm of scientific progress and the philosophical belief that man was destined to exert control over nature. She argued that the success of pesticides is necessarily limited because the target pests tend to develop immunity, while risks to humans and the environment will increase because pesticides tend to accumulate in the environment. However, *Silent Spring* did not call for the cessation of all pesticide use, but for greater moderation and care in their use. To provide some context, bear in mind that in the late 1950s the U.S. Department of Agriculture (USDA) authorized aerial spraying of DDT and other chemicals mixed with fuel oil over Long Island, New York, a practice that dispersed the pesticides over broad areas and exposed many species besides the intended target of fire ants.

Upon publication of *Silent Spring,* Carson was attacked as an alarmist and was accused of trying to reverse scientific progress and return mankind to the Dark Ages. The chemical industry mounted a counterattack, and this book is still cited by opponents of chemical regulation as an example of how an overzealous reformer can stir up public opinion and militate for the passage of regulations that ultimately do more harm than good. However, Carson's claims were vindicated in an investigation ordered by U.S. president John F. Kennedy, and they led to an immediate strengthening of regulations regarding the use of chemical pesticides.

Although Rachel Carson died in 1964, *Silent Spring* remained influential far beyond her lifetime: for example, it was influential in campaigns against the use of DDT, which was banned in the United States in 1972, and in 2004 was banned internationally, except when used for the control of malaria-causing mosquitoes. It also provided a model of radical environmental activism that questioned prevailing attitudes about the unquestioned benefits of scientific progress, and the attitude that humans should take toward nature.

See Also: Carson, Rachel; Endangered Species Act; Instrumental Value; Intrinsic Value.

Further Readings

Carson, R. *Silent Spring.* Boston: Houghton Mifflin, 1962.
Graham, F. *Since Silent Spring.* Boston: Houghton Mifflin, 1970.
Hynes, H. P. *The Recurring Silent Spring.* New York: Pergamon Press, 1989.
Levine, E. *Rachel Carson: A Twentieth-Century Life.* New York: Viking, 2007.
MacGillivray, A. *Rachel Carson's Silent Spring.* New York: Barron's, 2004.
Marco, G. J., R. M. Hollingworth, W. Durham, and R. Carson. *Silent Spring Revisited.* Washington, DC: American Chemical Society, 1987.

Sarah Boslaugh
Washington University in St. Louis

Singer, Peter

Peter Singer speaking at MIT on March 14, 2009. While he is known for his work on animal welfare, and especially for his 1975 book *Animal Liberation,* he has explored many other applied ethical issues.

Source: Wikipedia/Joel Travis Sage

Peter Albert David Singer (1946–) is an Australian philosopher. He is the Ira W. DeCamp Professor of Bioethics at Princeton University and laureate professor at the Centre for Applied Philosophy and Public Ethics (CAPPE), University of Melbourne. He specializes in applied ethics, approaching ethical issues from a secular preference utilitarian perspective. He is one of the world's best-known (and controversial) philosophers.

One of the foundational issues for Singer is how to weigh the different interests of different beings (human and animal). His favored principle of equal consideration of interests does not dictate equal treatment of all those with interests, since different interests warrant different treatment. All sentient beings have an interest in avoiding pain, for instance, but relatively few have an interest in cultivating their abilities. Not only does his principle justify different treatment for different interests, but it allows different treatment for the same interest. For Singer, a starving person's interest in food is greater than the same interest of someone who is only slightly hungry.

His focus on and interest in applied ethics has led Singer to examine some of the most pressing and controversial ethical issues. These have ranged from his long-standing concern with world poverty and the obligations of justice of those better off to change their lifestyles to help them. In his well-known article "Famine, Affluence and Morality" (1972), Singer makes the intuitively simple argument that there is an obligation on those who can meet the needs of others or ensure their survival by sacrificing some unnecessary wants. As he puts it, "if it is in our power to prevent something bad from happening, without thereby sacrificing anything of comparable moral importance, we ought, morally, to do it." This concern with eliminating world poverty has continued with his 2009 book *The Life You Can Save: Acting Now to End World Poverty.* Other topics Singer has tackled include abortion, infanticide, and euthanasia. His defense of the latter has brought him into dispute with antiabortion, Christian religious, and "pro-life" and disability activists—especially in the United States.

Singer is perhaps best known for his work on animal welfare and environmental ethics, taking the utilitarian view that much of how we treat and (ab)use animals is morally unjustified. As he puts it in his book *Animal Liberation* (1975), "If a being suffers there can be no moral justification for refusing to take that suffering into consideration. No matter what the nature of the being, the principle of equality requires that its suffering be counted equally with the like suffering—insofar as rough comparisons can be made—of any other being." The term *speciesism* is used by Singer to denote this ethically unjustified treatment, on the same logic of racism and sexism, namely, treating entities as representatives of types rather than as morally considerable individuals. The implications of his utilitarian argument for animal welfare are radical. The society-wide application of it would require the abolition of meat eating based on industrial factory farming.

Singer has always been consistent in stressing the need to move from abstract ethical thinking to practical action, and he himself has initiated and gotten involved in many ethically inspired campaigns and issues. For example, in 1993, he founded the Great Ape Project together with Paola Cavalieri, an Italian philosopher and animal advocate. The objective of this project is to grant some basic rights to the nonhuman great apes: life, liberty, and the prohibition of torture through medical experimentation, for example.

Apart from these applied ethical issues, Singer is also politically active, supporting left-wing and green political causes. He was a founding member of the Green Party in Victoria, Australia, and was a Green Party senate candidate in 1996. He is also an explicitly secular thinker prominent in advocating a humanist approach to political and ethical issues; in 2004 he was Australian humanist of the year.

See Also: Animal Ethics; Ethical Vegetarianism; Green Altruism; Green Liberalism.

Further Readings

Singer, P. *Animal Liberation: A New Ethics for Our Treatment of Animals.* New York: Random House, 1975.

Singer, P. "Famine, Affluence and Morality." *Philosophy and Public Affairs,* 1/1 (1972).

Singer, P. *The Life You Can Save: Acting Now to End World Poverty.* New York: Random House, 2009.

John Barry
Queen's University Belfast

Social Ecology

The concept of social ecology was developed by the radical writer and social philosopher Murray Bookchin (1921–2006) as a critique of contemporary socioeconomic and political practices. Social ecology provides a sustainable, environmentally minded, and ethical framework for social planning. The theory of social ecology highlights the need for radical new perspectives on social and environmental issues, which incorporate ecologically sound, direct democratic, and confederal political structures. From an ideological perspective, social ecology includes an understanding of an ethical approach to political economy that aims for diversity, post-scarcity, and nonhierarchical social relations, alongside coexistence between humans and other species.

Social ecology extends the belief that humanity can take the lead in planning a sustainable model for the planet, with all species and societies free from the human exploitation and ecological degradation that has characterized patriarchal capitalist societies throughout history. In addition, the theory of social ecology highlights the significance of independent, sustainable, and utopian perspectives when planning for future societal and ecological frameworks. Social ecology as a concept is concerned with the development of grassroots politics for all communities, through a platform that recognizes the primary importance of politics born from ethical principles when dealing with human and ecological issues.

Social ecology is based on an understanding that current environmental issues emanate from anthropocentric social problems. Following from that premise, the main tenet of Bookchin's "social ecology" approach holds that environmental issues must be framed in a manner that incorporates human and ecological issues together in order to proscribe sustainable solutions. Therefore, in order to solve any of the current crises facing the planet today—be they socioeconomic or focused on human rights or inequalities of gender or ethnicity, race or religion—the social ecology perspective holds that the primary issue to be addressed first must be the exploitative relationship between humankind and the other species and bioorganisms that share the planet. All other forms of degradation or exploitation can then be dealt with in a genuinely reformative process.

Bookchin spells out this premise with the following view on these challenges of hierarchical systems of development:

> Ecology raises the issue that the very notion of man's dominance of nature stems from man's dominance of man. Feminism reaches even further and reveals that the domination of man by man actually originates in the domination of woman by man. Community movements implicitly assert that in order to replace social domination by self management a new type of civic self . . . must be restored . . . to challenge the all pervasive state apparatus.

Social Ecology and the Environmental Crisis

Social ecology as a concept holds that the roots of the current environmental crisis have emerged as a result of the hierarchical structures of power and dominant authoritarian impulse that has characterized the political structures and cultural mores that have shaped societal governance over centuries. The human impulse to dominate and exploit nature, and other humans, has emanated from these distorted political traditions. These forms of exploitation are the fundamental issue that social ecology seeks to address in order to prepare the way for a more sustainable pattern of living. According to Bookchin in *The Ecology of Freedom,* "The domination of nature by man stems from the very real domination of human by human."

For Bookchin, social ecology provides an alternative to the politics of exploitation; this alternative form of society is based on ecological principles. These principles include a respect for diversity, freedom from exploitative hierarchies alongside and emphasis on the creation of respect for the interdependent relationship between all forms of species and biological systems that cohabit the planet. From a social ecology perspective, the transformation of human society from an exploitative one into a nonexploitative set of relations would then transform humankind's relationship with all other species and forms of life. Bookchin argues that social ecology presents a coherent and broad philosophy that presents a more developed and pragmatic ideology than either deep ecology or ecofeminism, as social ecology addresses all forms of exploitation.

Key Principles of Social Ecology

The core principles of social ecology hold that ecological problems arise from deep-seated social problems based on human exploitation. Therefore, ecological issues cannot be fully understood or addressed without overcoming the political structures that support exploitation of humans, other species, or the planet. Social hierarchy and class inequalities create further exploitation, while capitalism and consumerism cause human alienation and ecological degradation. Bookchin's work highlights the links between concepts of "growth" and the impulse to exploit humankind and nature alike and that all species share a common destiny and that any exploitation on the planet causes suffering to all. He argues that the principles of those who fight against racism, sexism, and inequality are based on the same principles of those who fight against environmental degradation. Bookchin argues that social ecology provides a plan for a holistic "ecological sensibility" and "species ecumenism" that would end the exploitation and degradation of the planet.

According to Bookchin, social ecology can be understood in the following quote, given as part of a speech that made the distinction between social ecology and deep green thinking:

> Social ecology is neither deep, tall, fat, nor thick. It is social. It does not fall back on incantations, sutras, flow diagrams, or spiritual vagaries. It is avowedly rational. It does not try to regale metaphorical forms of spiritual mechanism and crude biologism with Taoist, Buddhist, Christian, or shamanistic Eco-la-la. It is a coherent form of naturalism that looks to evolution and the biosphere, not to deities in the sky or under the earth for quasi-religious and supernaturalistic explanations of natural and social phenomena.

From a philosophical or theoretical perspective, social ecology has emerged from the organismic tradition within Western philosophy. This thinking begins with Aristotle and Hegel, and was developed further through the work of the Frankfurt School's critique of logical positivism later utilized by Arne Naess, and the primitivistic mysticism of Martin Heidegger, both of whom are used in the conceptualization of deep green thought. Social ecology further challenges existing conceptualizations of hierarchical societies, including the Malthusianism that underpins deep green philosophy. For Bookchin, social ecology emerges from the radical anarchistic communitarianism of Peter Kropotkin and the critique of capitalism put forward by Karl Marx, the anarchist tradition and the counterculture and New Left radicalism of the 1960s. Social ecology is aligned with the European "Red-Green" tradition, and the American ecofeminist movement and the movements for human emancipation.

Ethically and philosophically, social ecology maintains the humanistic tradition that emerged from the Renaissance. To social ecology, humanity and nature are part of natural evolution, rather than having distinct destinies on the planet. According to Bookchin:

> Natural evolution is nature in the very real sense that it is composed of atoms, molecules that have evolved into amino acids, proteins, unicellular organisms, genetic codes, invertebrates and vertebrates, amphibians, reptiles, mammals, primates, and human beings—all in a cumulative thrust toward ever-greater complexity, ever-greater subjectivity, and finally ever-greater mind with a capacity for conceptual thought, symbolic communication of the most sophisticated kinds, and self-consciousness in which natural evolution knows itself purposively and willfully.

Bookchin argues that social ecology goes beyond issues that most "environmentalists" focus on, which he sees as only revealing the symptoms of planetary problems rather than proscribing any solutions. Examples of issues that Bookchin thought were the subject of misguided concerns by the environmental movement include concerns about technology and population growth. For Bookchin, population levels are not the key issue for environmentalists, whereas exploitation was, and he argues that social ecology would provide a framework for dealing with other social issues such as the media's focus on consumerism or the threats posed by technologies to rainforests; nothing would change unless the impulse to exploit using such technologies was challenged. Social ecology would provide a framework for profound societal change, rather than providing improvements to existing social processes that were based on exploitation.

Therefore, social ecology is presented as an ideology that goes beyond contemporary understandings of sociocultural reformation, offering instead a platform of complete transformation of human relations and species cohabitation on the planet. Equally, the concept of social ecology is highly critical of trends toward "green capitalism" or "green consumption," with Bookchin arguing that the human impulse to exploit and dominate would soon resurface under such arrangements, and that the desire to seek profit would overcome any ecological aspect devised under any such endeavor. In addition, Bookchin is suspicious of the "New Age mysticism" that pervaded much of deep green ecology. He argues that social ecology would provide a platform for an egalitarian and ecological ethics that would be more socially compelling than the vagaries of deep green self-realization. Social ecology would provide for a form of spiritualism that could incorporate naturalism, rather than focus on the internal requirements of individual humans, as deep green philosophies did. Social ecology demonstrates that environmentalism has areas of ideological overlap in many paradigmatic areas, but the distinction between authentic green radicalism and "shallow" compromises that tolerate high levels of pollution for profit can be clearly identified.

Social Ecology and the Natural World

Social ecology argues that humans are an integral part of a profound process of natural relations, and that humankind and other species are intertwined and interdependent.

Bookchin argues that human consciousness has developed to understand the complexity of our relationship with the natural world, and that human thought reflects this critical awareness of the relationship between humans and nature. From a social ecology perspective, this locates humans within a process of biological evolution, which Bookchin sees as the primary process of "First Nature." Furthermore, the development of human consciousness around ecological issues reflects the growth of a process described as "Second Nature." This second nature is witnessed in humankind's capacity to develop various technologies, sciences, social structures, and institutions, all of which are also linked to the instincts developed from first nature. Nonetheless, Bookchin sees no special position for humanity within the wider set of relations with the natural world.

Social ecology rejects the hierarchical concepts that underpin both anthropocentricity and biocentricity. Anthropocentrism argues that humans are free to dominate nature for their own ends and locates humans at the top of a species hierarchy. Alternatively, biocentricity argues that all species carry an equally intrinsic value. Social ecology argues against both positions and purports to combine both first and second natures, with humanity and nature combining in a process of evolution. Therefore, social ecology argues that human development is not the cause of the ecological crisis; rather, it is the impulse to overexploit and overconsume that creates planetary destruction. Exploitation of the planet originates with human exploitation of other humans. Social ecology challenges this impulse to exploit, while Bookchin claims that social hierarchies create the framework for further exploitation of humanity and nature. He argues that humanity could overcome the impulse to exploit nature only by building a society without hierarchical structures or human inequality.

In order to replace the existing exploitative socioeconomic system, social ecology proposes an egalitarian society based on mutuality and communitarian ethics, with an emphasis on the interests of the collective that are inseparable from those of each individual. The idea of private property would be replaced by collective ownership, with all resources belonging to the community as a whole. Bookchin refers to this communitarianism as a "commune of communes," where property would not belong to private individuals, corporations, or the state. This transformation would be achieved through the ending of all exploitative relations and through challenging domination through radical collective action and societal changes, supported by progressive social movements. Social ecology argues that oppressive hierarchies and societal inequality are at the root of all human and planetary problems and that only a move toward communitarianism could overcome existing human and environmental crises. Ultimately, in social ecology, Bookchin has left a conceptual legacy that challenges any elitist tendencies that arise from deep green thought, one that challenges all environmentalists to strive for an equal and pragmatic form of ecological thought and practice.

See Also: Anthropocentrism; Biocentrism; Bookchin, Murray; Deep Ecology; Deep Green Theory; Ecofeminism/Ecological Feminism; Naess, Arne.

Further Readings

Bookchin, Murray. *The Ecology of Freedom: The Emergence and Dissolution of Hierarchy.* Oakland, CA: AK Press, 2005.

Bookchin, Murray. *Social Ecology and Communalism.* Oakland, CA: AK Press, 2007.

Bookchin, Murray, Dave Foreman, Steve Chase, and David Levine. *Defending the Earth: A Dialogue Between Murray Bookchin and Dave Foreman.* Cambridge, MA: South End Press, 1999.
Light, Andrew. *Social Ecology After Bookchin.* New York: Guilford Press, 1998.

Liam Leonard
Institute of Technology, Sligo

Steffen, Alex

Alex Steffen, writer and cofounder of Worldchanging.com, is a leading champion of solutions-based journalism, pursuing a dream of a "bright green" planetary future. Steffen's mission, in both his own writing and in his leadership of Worldchanging.com, has been to describe a future of "one-planet" prosperity, the dynamism and enticement of which is impossible to resist. Navigating between what Steffen puts forward as a false choice between the asceticism of "deep greens" and the wanton consumption of "light greens," he uses the media platform of Worldchanging.com to broadcast a vision that is "bright green": a world that is designed to dematerialize energy and solid material consumption while expanding the possibilities of human experience and social integration.

For Steffen, the dire state of the planet is not due to a fault in the human soul—its desires for convenience and luxury and peer rewards that are so difficult for all but a few to voluntarily deny—but due to a fault of the human mind. Consequently, his position within the broader green movement is far from the deep ecologists, such as Arne Naess and Thomas Berry, who begin their inquiries with a reevaluation of anthropocentrism and ground their actions on the principles of integrated universal space and the right of all beings—human, faunal, and floral—to self-realization into perpetuity. Steffen's work is not grounded explicitly in philosophy, but in practice; it is not seeking to reorient the world as much as to remake it. There is no return to nature, or language of stewardship and caretaking: the Earth is still to be used by humanity, but more intelligently. For Steffen, the focus of sustainability is not ethical reorientation, but intelligent design. Consequently, his inspirations within the green movement, and those with whom his thought and practice is most complementary, are the designers: John Thackera, Bruce Sterling, and William McDonough.

Steffen was born January 7, 1968, to George and Dolores Steffen in Oakland, California. He grew up the oldest of four brothers in a self-described "hippie" household for which the woods of Humboldt County in northern California were a second home. Steffen recalls that he was taught from an early age that explaining solutions is more effective than yelling ecological values. Growing into adulthood during the height of the Pacific Northwest "forest wars" reinforced Steffen's environmentally aware childhood, leading Steffen to believe that human society was set on an unsustainable path. After graduating from Allegheny College, a small liberal arts college in northwestern Pennsylvania, Steffen began what would become a career of social activism as an environmental reporter for the *Japan Times* in Osaka, Japan. His environmental reporting experience in Japan, including breaking a 1991 story on the international shipping of nuclear waste from Japan to France, demonstrated to Steffen that the conflicts between current structures of economic growth and environmental health that plagued the United States were also worldwide problems.

In the late 1990s, Steffen became known for his work in urban planning politics and in supporting the arts and artists in Seattle, Washington. From 1998 to 2000, he served as president of Allied Arts, a Seattle institution begun in 1954 dedicated to enhancing the cultural livability of the city through creation of a network of citizens concerned about historic preservation, urban design, and the arts. During his tenure, Allied Arts successfully lobbied to increase the city's financial support of the arts and unsuccessfully opposed expansion of the Convention Center. In 2000, Steffen joined the board of directors of Seattle-based Fuse Foundation, whose mission was to provide life-changing financial grants and mentorship support to young artists, or those Steffen has called "cultural innovators."

Building upon his belief that cultural work is the most powerful lever to move public opinion, in September 2003 Steffen and Jamais Cascio turned to their keyboards to take up an active role in building what Jim Moore has called "The Second Superpower": the "emergent democracy" embodied in blog-writing and online citizen activism, with "each individual making sense of events, communicating with others, and deciding whether and how to join in community actions." Making their aspirations explicit, Steffen and Cascio named their blog Worldchanging.com, and Moore's argument merited the first post.

The value of their aggregation and framing of others' work as well as their own long essays on planetary solutions was quickly recognized, and within six years Worldchanging .com had logged 10,000 posts, recruited hundreds of contributors, and been recognized by Pop!Tech, *Utne* Magazine, the Webby Awards, Design Indaba, PicNic, and Ted. In 2006, Steffen edited *Worldchanging: A User's Guide for the 21st Century,* based on his solutions-based journalism approach and the Website's coverage of "bright green" solutions to the planet's structural problems. Al Gore wrote the foreword, and Stefan Sagmeister designed the cover and slipcase.

There are three signatures of Worldchanging and Steffen's contribution to the green movement. First is the focus on networked, collaborative work, both in the production of ideas and in their transformation into action. Formation of the network itself, the coalition of persons involved in writing and reading the Website, is an action that is understood as activist and integral to planetary solutions. From its inception, Worldchanging has offered an open invitation to the public to not only comment on postings but to contribute posts and original essays; the majority of the contributors remain unpaid as they freely share their expertise. As such, it functions more like a public space of dialogue than a business.

Second, neither Steffen nor Worldchanging offer a critique of existing planetary conditions or of solutions-based programs that have not worked. That the situation of human society and the planet is dire is taken as a given, as is the necessity of radical change. Projects that have attempted radical change or claim green or sustainable credentials are not investigated to see if their implementation holds up to their publicity. The focus is resolutely positive, with the goal of awakening a broader public to the conviction that solving the planetary crisis is not only possible but already in the making. The mission is to be inspirational, and the orientation is futurist.

Third, Steffen's bright green planetary future that brings prosperity for all is decidedly urbanist. Access is privileged over ownership, and experience over materiality, both of which are made possible through compact, dense settlements where public space and goods are prioritized over private space. It is a city where there are significantly fewer cars. Power drills are not purchased, but loaned from a tool library. Residents walk to a shared gym, rather than installing one in an extra room in a private house. In this way, Steffen argues that prosperity is decoupled from planetary impact. He advocates a systems approach to urban living, where interpersonal interactions, public encounters, and shared spaces dominate, and individual consumption of goods is significantly reduced. It is a city that prioritizes working more like nature, rather than looking like it. The goal of the bright green city is to flourish within the intelligent limits of the planet. While this city remains a future idea that has not yet been built, Steffen has pointed to the Beddington Zero Energy Development (BedZED) in South London and Dockside Green in Victoria, Canada, as urban developments that are constructing not just environmentally conscious urban spaces, but spaces that create desirable and seductive conditions of human living—spaces that are bright green.

Steffen does not self-identify as an environmentalist as much as a creator and networker of ideas that seek to redesign how humanity lives today, yet still be generous to its progeny. While he speaks and writes as an advocate and strategist of that redesign, his work at Worldchanging to foster emergent, public democracy through collaboration and communal action is also a technique.

See Also: Berry, Reverend Fr. Thomas; Bright Green Environmentalism; Deep Ecology; Gore, Jr., Al; Naess, Arne.

Further Readings

Moore, J. "The Second Superpower Rears Its Beautiful Head." In *Extreme Democracy*, Jon Lebkowsky and Mitch Ratcliffe, eds. Lulu.com, 2005.

Steffen, A. *Worldchanging: A User's Guide for the 21st Century.* New York: Abrams, 2006.

Worldchanging.com. http://www.worldchanging.com (Accessed April 2010).

Shannon May
University of California, Berkeley

Strong and Weak Sustainability

The distinction between "strong" and "weak" sustainability has grown out of the discourse and thinking around "natural capital" and specifically in relation to the substitutability between "human" and "natural" capital and how much "critical natural capital" we need for sustainability. Proponents of weak sustainability maintain that human-made and natural capital can be substituted for one another, while those who support strong sustainability believe they cannot, arguing instead that human and natural capital should be seen as complementary, not interchangeable and substitutable. The debate between weak and strong sustainability raises profound ethical questions about the intrinsic value of the nonhuman world and the limits of "economic" ways to conceptualize the natural world and our relationship to it. And the debate foregrounds the extremely important ethical question of what do we wish to preserve and pass on to future generations? The world as it is now with its current stock of natural capital, landscapes, ecosystems, and biodiversity? Or a changed world in which "losses" of natural capital, biodiversity, and species can be compensated for by passing onto future generations a higher level of human capital?

Weak Sustainability

The essence of weak sustainability is that it conceptualizes the nonhuman world not in biophysical terms but in terms of its economic value. That is, unlike strong sustainability

Karst remains, which are left behind after phosphate mining, in the center of Nauru Island in 2004. The devastation of Nauru, which affected about 80 percent of the 21 sq. km. island nation, is an example of weak sustainability.

Source: U.S. Department of Energy, Atmospheric Radiation Measurement Climate Research Facility

discussed below, it deals with the "value" of the forest—that is, how much monetary value are the timber and the ecosystem services worth and then, using that economic benefit (when the forest is developed or liquidated or sustainably managed) to "substitute" for the loss of the "natural capital." In short, it translates or transforms natural into human capital so that although there is "loss" recorded on the natural capital balance sheet, this is compensated for or perhaps even outweighed by the "gain" on the human capital balance sheet.

An example of dangers of weak sustainability is the small Pacific island nation of Nauru. In 1900, one of the world's richest phosphate deposits was discovered on Nauru and today, as a result of intensive phosphate mining, about 80 percent of the island is devastated. At the same time, the people of Nauru have enjoyed, over the past decades, a high per capita income. Income from phosphate mining enabled the Nauruans to establish a trust fund. Interest from this trust fund should have ensured a substantial and steady income and thus the economic sustainability of the island. Unfortunately, the Asian financial crisis of 1989, among other factors, wiped out most of the trust fund. The people of Nauru now face a bleak future. Their island is biologically impoverished and the money for which Nauruans traded their island home has vanished. The development model of Nauru followed the logic of weak sustainability and shows clearly that weak sustainability may be consistent with a situation of near-complete environmental devastation. This case illustrates a telling argument against weak sustainability.

Another serious problem with weak sustainability is the way in which it seeks to establish an economic valuation or "price" for the nonhuman world. For some, such as the environmental philosopher Alan Holland, the very notion of natural capital is a dangerous ethical oxymoron. "[C]onstrued as a commitment to nature, the commitment to natural *capital* is therefore hollow," he wrote in 1999, in the sense that viewing nature as "capital" is to express a commitment and relationship to it that is instrumental, that is, what it can contribute to human welfare. His argument, and one shared by most environmentalists and greens, is that to prescribe the use of an economic language and economic way to understand the nonhuman world is to commit a category mistake. Such forms of economic valuation (contingent valuation and cost-benefit analysis techniques) "crowd out" noneconomic forms of valuation and therefore misrepresent people's environmental values, according to Barry (1999). Forcing people to put a price on nature in this way is to replace the economist with the cynic in Oscar Wilde's well-known phrase "to know the price of everything and the value of nothing." In other words, there are both normative-theoretical as well as empirical arguments and evidence that weak sustainability relies on the systemic corruption and misrepresentation of peoples' values, beliefs, and preferences about the value and relationship they have with nature. Simply put, weak sustainability relies on the nonhuman world's only having economic or instrumental value. Yet, we have both normative arguments and empirical evidence of both the noneconomic and the intrinsic value people place on nature. People have aesthetic, spiritual, and other noneconomic relations to and therefore valuations of the natural world. We also have evidence that people express these noneconomic values when they take part in such economic valuation exercises. The widespread experience of people expressing "protest bids," that is, astronomically high monetary valuations on landscapes, species, or other aspects of nature, does indicate the existence of these noneconomic valuations of the environment.

Strong Sustainability

Under the strong sustainability criteria, minimum amounts of a number of different types of capital (economic, ecological, and social) should be independently maintained, in real physical/biological terms. The major motivation for this insistence is derived from the recognition that natural resources are essential inputs in economic production, consumption, or welfare that cannot be substituted for by physical or human capital. Another reason is the acknowledgment of the environmental integrity and intrinsic value of and even "rights" of nature. But perhaps the strongest motivation for strong sustainability and its insistence on the preservation of some natural capital from being "developed," that is, liquidated/depleted, is that some critical/minimum level of natural assets is required for human (and nonhuman) life support. There are some natural processes, entities, and services for which there is no technological or other human-created substitute. Think of the hydrological cycle or the carbon cycle, or soil fertility or the pollination work that bees do, or any other number and variety of ecosystem services the nonhuman world provides and for which there is no human substitute. The biggest one of all here is, of course, the maintenance of a stable and life-supporting climate; outside of techno-fantasies of "terraforming" and "earth management," humanity does not (and perhaps never will or ought to try) have the capacity to manage the Earth's climate system in a life-supporting manner.

The United Nations project called The Economics of Ecosystems and Biodiversity (TEEB), which is a global effort to develop more accurate economic valuations of the nonhuman world, can be characterized as a strong sustainability initiative. This project is based on the notion of the nonsubstitutability of ecosystem and biodiversity natural capital and the importance of taking ethical (and not just economic) considerations into account when making decisions. One specific example of strong sustainability—moving us in the direction of avoiding any net reduction in overall natural capital—would be compensatory afforestation schemes, especially if such schemes replace "like with like" (in terms of the type of trees and quantity).

The reality is that substituting financial or other forms of human-made capital for natural resources is incompatible with maintaining a suitable physical environment for the flourishing of the human species (or other nonhuman species). Therefore strong sustainability implies that we must step outside the conventional (but sadly dominant) economistic framework in order to establish the conditions for maintaining human and nonhuman survival, well-being, and flourishing. In short, it implies that sustainability requires preserving actual nature, the "stuff" of biophysical entities and not the "economic value" of the nonhuman world. In short, the preference for strong over weak sustainably comes down to preferring that scientists, not economists, manage and measure what we need to preserve in nature.

Summarizing, one can see that "strong" and "weak" sustainability represent a continuum of how we should use (or not use) and think about the nonhuman world (or rather our relationship—both material/metabolic and moral—with the nonhuman world). The choice between which of them to choose is a mixture of scientific, ethical, prudential, and political choices. The continuum of weak and strong sustainability demonstrates that there is no "objective" or "value-free" way to understand and implement sustainability.

See Also: Cost-Benefit Analysis; Environmental Justice; Human Values and Sustainability; Instrumental Value; Intrinsic Value; "Should Trees Have Standing?"

Further Readings

The Economics of Ecosystems and Biodiversity (TEEB). "The Economics of Ecosystems and Biodiversity" (2010). http://www.teebweb.org (Accessed July 2010).

Gowdy, J. M. and C. McDaniel. "The Physical Destruction of Nauru: An Example of Weak Sustainability." *Land Economics,* 75 (1999).

Holland, A. "Sustainability: Should We Start From Here?" In *Fairness and Futurity: Essays on Environmental Sustainability and Social Justice,* A. Dobson, ed. Oxford, UK: Oxford University Press, 1999.

John Barry
Queen's University Belfast

Sustainability, Business Ethics and

Business ethics are a fundamental concept of sustainability, and sustainability considerations often lead businesses toward more ethical actions. The modern sustainability movement traces its roots back to the United Nations (UN) World Commission on Economic Development. Defining concepts of sustainability, and its application in the business world, were further developed by John Elkington. Today, investors are increasingly demanding that businesses not only act in an ethical and sustainable manner, but that they disclose their actions to the public. Many large companies today are producing sustainability reports in an effort to be unceasingly transparent and accountable to the public.

At the end of the 20th century, the UN became increasingly aware of global pollution issues, such as pesticide bioaccumulation and ozone depletion. The international community was also facing increasing industrialization among developing nations of the world. It was generally acknowledged that the path many nations had taken in their industrial development was not the best for the health of humans or the environment, and a more careful path to development would be prudent.

To answer the global communities' questions on how to best guide development, the UN convened the World Commission on Environment and Development (WCED) in 1983, chaired by Gro Harlem Brundtland, the prime minister of Norway. Due to her pioneering work to address the issues surrounding continuing global development, the WECD came to be known as the Brundtland Commission. The Brundtland Commission was created to address the growing concern "about the accelerating deterioration of the human environment and natural resources and the consequences of that deterioration for economic and social development." The Brundtland Commission's 1987 report, *Our Common Future,* defined sustainable development as the following:

> sustainable development is development that meets the needs of the present without compromising the ability of future generations to meet their own needs.

This definition is considered the original, and probably the most quoted, definition of sustainable development. As a result of the Brundtland Commission's report, many efforts emerged to promote sustainable development within governments. But the integration of sustainability into the world of business is often credited to the work of John Elkington.

In his 1998 book, *Cannibals With Forks: The Triple Bottom Line of 21st Century Business,* John Elkington coined the phrase *triple bottom line* (TBL) as a new means of accounting, expanding the traditional reporting framework to take into account environmental and social performance in addition to the traditional business metric of financial performance. TBL represents the three elements (or pillars) of sustainability as environment, economy, and society, which must all be balanced for a truly sustainable development to occur.

Commitment to this triple bottom line is considered a step beyond environmental awareness of corporations into the realm of corporate social responsibility.

When accounting for social impacts in the business context, the emphasis is often on the role of human capital within the triple bottom line. Human capital can mean the actual employees that work for a company, the surrounding community, or even society at large. Accounting for human capital in the triple bottom line means acting in a fair and responsible manner in an attempt to provide a positive impact on society. Taking voluntary measures to positively impact society may come at an economic cost for a business. But the cost of ignoring the social context of a business can be greater. For example, in the United States, when a local community does not have faith in a company's ability to protect the environment, operating permits and applications and/or renewals may be challenged and held up in the courts. In this scenario, the company has lost its human (or social) capital necessary to operate.

Increasingly, companies are forming focus groups or panels of local representatives to guide their efforts to manage their impacts on local communities. For global businesses operating in the developing world, investors are increasingly demanding social responsibility with regard to human rights. A common sustainable business practice for global companies is to implement policies on human rights and child labor, guaranteeing that they will meet minimum ethical standards for the treatment of workers at all their operations in the globe, even in countries that do not require these standards by law. Some companies are even requiring all suppliers and contractors to operate under these standards as well. In this way, sustainability concerns drive businesses to act in an ethical manner regarding social impacts.

Similar to managing their social impacts, sustainable businesses must also manage their environmental impacts. In the context of sustainable business, the environment is considered natural capital. Businesses with a commitment to the triple bottom line make a commitment to conduct their business in the most environmentally sustainable manner.

One of the most prevalent ways to minimize a product's impact on the environment is to perform a complete life cycle assessment (LCA) on a product. This cradle-to-grave approach considers the environmental impacts of all aspects of a product's manufacture, use, and ultimate disposal. It often includes voluntarily redesigning products to account for ease of disassembly and recycling throughout a product's life cycle. Some of these components may cost more but ultimately produce less impact on the environment.

In the past, profit was the only measure of success of a business. With the evolution of environmental protection laws, businesses were forced to make a profit within the limits society placed on their impacts to the environment. For example, it may cost a business additional funds to control its air pollution through the use of bag houses or some other pollution-control technology, but when they are required by law, these technologies simply become another cost of doing business. The concept of triple bottom line implies that all three elements of society, environment, and economy are balanced, and the company no longer measures success only in economic terms.

For a business to be able to manage and balance its triple bottom line, there must be some metric by which the impacts on each of the three pillars can be measured. Economic impacts can be quantified using currency, such as U.S. dollars. But social and environmental impacts do not have a similar common unit by which to measure all impacts. A business may affect both the environment and society in a myriad of ways.

Indicators are measurable effects that can be tracked over time. Common indicators have been developed for each of the three pillar areas, which are usually selected and tailored to serve as meaningful metrics for a unique business. For example, common environmental indicators may include annual emissions of hazardous air pollutants and annual freshwater withdrawn from surface water; social indicators may include impacts to native cultures and lifestyles. For an urban dry cleaning operation, the annual emissions of hazardous air pollutants is a more meaningful metric than the annual freshwater withdrawn from surface water, or the impacts to native cultures and lifestyles. But for a rural trout farm, the more meaningful metrics may be the annual freshwater withdrawn from surface water and the impacts to native culture and lifestyles. This example demonstrates the importance of selecting meaningful indicators for a specific business. However, it is also important that indicators be comparable across businesses.

The Global Reporting Initiative (GRI) is an international organization with the goal of standardizing sustainability indicators and reporting internationally. GRI has developed a sustainability reporting framework, including general sustainability reporting guidelines and special guidance for different industry sectors, called sector supplements. Sector supplements are continually being developed for new sectors as needed and include the following: airports, apparel and footwear, automotive, construction and real estate, electric utilities, events, financial services, food processing, logistics and transportation, media, mining and metals, nongovernmental organizations (NGOs), oil and gas, public agency, and telecommunications.

Both the sector supplements and the general sustainability reporting guidelines have been created through consensus seeking processes involving representatives from business, academia, governments, labor, professional societies, and others worldwide. The sustainability reporting guidelines include guidance on appropriate indicators, and are meant to be reviewed and updated as needed. The third version of the guidance, commonly referred to as "G3 Guidelines," was published in 2003, and is available for free for use globally. Guidelines include suggested sustainability indicators for use in reporting. Sector supplements include additional indicators that may be unique to a given sector of business. Protocols for each indicator provide its definition, reporting units, and the methodology for data gathering and calculation.

See Also: Business Ethics, Shades of Green; Consumption, Business Ethics and; Greenwashing; Precautionary Principle.

Further Readings

Elkington, J. *Cannibals With Forks: The Triple Bottom Line of 21st Century Business.* Gabriola Island, British Columbia, Canada: New Society Publishers, 1998.

Elkington, J. "Toward the Sustainable Corporation: Win-Win-Win Business Strategies for Sustainable Development." *California Management Review,* 36/2 (1994).

Global Reporting Initiative. "Sustainability Reporting Guidelines Version 3.0" (2006). http://www.globalreporting.org (Accessed April 2010).

Michelle E. Jarvie
Independent Scholar

Sustainability, Consumer Ethics and

Responsibility of consumers to consider sustainability in their consumption habits is a common topic of discussion in the "greening" of society. Moral concerns were outlined by foundational free-market economist Adam Smith in the 1700s. Rogene Buchholz suggests that the Protestant ethic provided moral limits on consumption by individuals during the early stages of industrialization in Western Europe and the United States, and that this ethic eventually weakened during the development of a consumer society.

Academic studies on consumption did not flourish until the 1950s, when consumer society greatly magnified. Yet Dan Miller's anthology *Acknowledging Consumption* cites studies as early as the 1890s by William James, founder of pragmatism, linking self-identify of common man with possessions. Not long after James, Thorstein Veblen coined the term *conspicuous consumption* in 1902 to illustrate the rising importance of material goods in societal status, one that grows today in emerging economies such as China's.

Aldo Leopold, perhaps best known for his seminal 1948 essay "The Land Ethic," wrote a 1928 piece called "The Homebuilder Conserves," illustrating this concern over a growing consumer society: "A public which lives in wooden houses should be careful about throwing stones at lumbermen, even wasteful ones, until it has learned how its own demand [for] lumber helps cause the waste it decries."

Harvard professor John Kenneth Galbraith in his 1958 book *The Affluent Society* highlights (as did earlier Marxist critiques of capitalism) the strong influence of production, marketing, and advertising in determining consumption preferences and behavior. At the other end of the 1960s' environmental movement, the Club of Rome and its "Limits to Growth" effort in 1972 offered a prominent evaluation of capitalist society's thresholds regarding consumption.

Laura Westra and Patricia Werhane edited *The Business of Consumption,* a 1998 anthology that mixes business, consumer, and environmental ethics. In this volume, biologist John Lemons analyzes north–south inequity in per capita consumption rates and draws from the ethics of risk analysis. Donald Mayer illustrates how corporate marketing and domestic and international law have institutionalized overconsumption. Rogene Buchholz and Sandra Rosenthal examine the roots of consumption behavior, calling for "qualitative" economic growth.

Bill McKibben's book *Deep Economy* echoes this call for a revision of neoclassical economics, citing global happiness survey findings that once incomes rise above a base level of approximately $10,000 per year, happiness does not increase in sync with income and consumption levels.

Despite these critiques that tend to place responsibility on the capitalist system, corporate marketing, or deficient government policies, consumers themselves are still a focus of moral discussions about overconsumption. Some reviewers such as John Peloza call for refocusing responsibility on the consumer. He notes that a recent move to small cars has not been for environmental reasons but primarily due to high gas prices.

Douglas MacCleery's article (2000) suggests Leopold's land ethic is incomplete without a corresponding consumption ethic—otherwise we simply export our environmental impacts to other nations. MacCleery points out that since the first Earth Day in 1970, the average family size in the United States dropped by 16 percent, while the size of the average single-family house being built increased by 48 percent. Today the United States holds about 5 percent of the world population, yet consumes about 30 percent of its resources. A corresponding formulaic acronym is IPAT (Impact − Population times Affluence times Technology).

Ramachandra Guha's 2006 book *How Much Should a Person Consume?* continues Lemons's and MacCleery's concern about exporting environmental impacts, while anecdotally highlighting the large gap between elitist and nonelitist environmentalism in the respective developed and developing worlds. Deep ecologist Bill Devall proposes that the consumer arena is so complex that it is inappropriate to expect the consumer to hold primary responsibility for sustainability.

Other, more-nuanced critiques suggest multiple responsibilities. Planning professor William Rees highlights in *The Business of Consumption* the concept of an "ecological footprint," a method to compare environmental impacts of citizens in different nations. Cofounder Mathis Wackernagel and Rees estimate a nation's per person impact in acres.

Rees argues that a solution cannot be sought through our growing information-based rather than product-based economy as our new, increased buying power maintains consumption levels. He sees solutions not only in gradual implementation of taxes and quotas on consumption levels, but investing in nonmonetary networks of mutual support or social capital.

Following this research track of mutual support and social capital, studies in China and conservation psychology, respectively, suggest that self-identity formed by peer pressure and early life experiences with nature are primary motivators in sustainable behavior. Thus a status symbol of conspicuously lower consumption may evolve, along with green technology and green marketing, as field-based environmental education expands and improves.

The anthology most tied to individual consumer ethics and environment is David Crocker's 1997 *Ethics of Consumption: The Good Life, Justice, and Global Stewardship*. Robert Goodland's entry follows Rees's taxation solution, suggesting taxes on food items that are higher on the food chain (and thus less environmentally benign).

Martha Nussbaum outlines the Aristotelian capability approach published by herself and Nobel Prize–winner Amartya Sen, which suggests basing most policy decisions on commonly agreed norms for human flourishing (not simply basic needs but also not opulent consumption for all—capability for a certain life expectancy, health status, educational opportunities, etc.). These norms can often be agreed on by surprisingly diverse cultures, sometimes found within a single nation-state.

One of the more recent, large investigations into consumerism is the Cultures of Consumption research program in the United Kingdom (UK) that involved 26 projects over a five-year span. Key results include that (1) despite globalization, there remains diversity among consumers—although U.S. citizens are reading less, those who do read are reading more; elite UK citizens consume both elite (e.g., opera) and nonelite (e.g., cinema) culture;

French citizens still have dinner at home 96 minutes a day (roughly twice the time of UK and U.S. residents); and (2) consumption is not as much individual choice as habit formed by historically evolving interaction of culture and technology.

Results investigating ethical consumption conclude that ethical consumerism can function as a pathway to broader political engagement; however, such campaigns are more effective at the collective level, such as creation of Fair Trade towns. As individuals (echoing Devall's claim), "consumers feel easily overwhelmed by appeals to change their own lifestyle to save the planet." It may be more effective to emphasize "sensual and spiritual pleasures of a different lifestyle, say, one less dependent on cars, noise, and traffic jams." Just one example of the complexity offered by too many choices is the suggestion of "natural" products as automatically sustainable: to the contrary, chicle harvest from its natural state deforests the Yucatan peninsula, while use of its synthetic substitute does not. Approximately half of citizens surveyed have vague notions of the definition of sustainability.

One of the projects more specific to ethical consumption was conducted by geographers, led by Clive Barnett, and titled "Governing the Subjects and Spaces of Ethical Consumption." A working paper produced by this project, in taking a look at the complexity of ethical consumption, brings up positive aspects of a topic typically seen as highly problematic due to spatial separation of consumer and producer. Barnett et al. suggest that ethical consumption opens up new ways to look at consumption, a relationship typically seen as linear and information based—falsely assuming that if an individual consumer has the right information, good decisions and policy can follow to adjust a linear commodity chain.

The authors point out that ethical consumption, following actor-network theory, involves transnational networks of individual consumers, producers, corporations, governments, and activist campaigns that operate quite differently than traditional ethical frameworks. Thus research and practice in ethical consumption by collectives, which define place for both consumer and producer, provide a new, spatial forum for addressing consumption ethics. By collectives, the authors mean members of communities of practice, for example, "members of faith groups, schoolchildren, or residents of distinctive localities."

The most philosophically oriented recent book on consumers and sustainability also comes out of this project, titled *The Ethical Consumer.* One author uses existential-phenomenological interview techniques to determine life experiences of ethical consumers. The introductory chapter addresses consequential and deontological aspects of ethical consumption and finds that virtue ethics may be more applicable for participants. The authors also differentiate between ethical consumption, a mode to display personal values, and the ethics of consumption, more of a critique of modern capitalism. They call for more public participation in debates over the meaning and purpose of ethical consumption.

See Also: Consumption, Business Ethics and; Ecological Footprint; Sustainability, Business Ethics and.

Further Readings

Barnett, Clive, Phil Cafaro, and Terry Newholm. "Philosophy and Ethical Consumption." In *The Ethical Consumer,* Rob Harrison, Terry Newholm, and Deirdre Shaw, eds. London: Sage, 2005.

Barnett, Clive, Paul Cloke, Nick Clarke, and Alice Malpass. "Articulating Ethics and Consumption." Working Paper Series: Cultures of Consumption. School of Geographical Sciences, University of Bristol, UK, 2004.
Clayton, Susan and Gene Myers. *Conservation Psychology: Understanding and Promoting Human Care for Nature.* Oxford, UK: Wiley-Blackwell, 2009.
Crocker, David A. and Toby Linden, eds. *Ethics of Consumption: The Good Life, Justice, and Global Stewardship.* Philosophy and the Global Context Series. Lanham, MD: Rowman & Littlefield, 1997.
Devall, Bill. *Simple in Means, Rich in Ends: Practicing Deep Ecology.* Layton, UT: Gibbs Smith, 1988.
Galbraith, John Kenneth. *The Affluent Society.* Cambridge, MA: Riverside Press, 1958.
Guha, Ramachandra. *How Much Should a Person Consume? Environmentalism in India and the United States.* Berkeley: University of California Press, 2006.
Leopold, Aldo. "The Homebuilder Conserves." *American Forests and Forest Life,* 34/413 (1928).
MacCleery, Douglas W. "Aldo Leopold's Land Ethic: Is It Only Half a Loaf Unless a Consumption Ethic Accompanies It?" *Journal of Forestry,* 98/10 (2000).
McKibben, Bill. *Deep Economy: The Wealth of Communities and a Durable Future.* New York: Times Books, 2007.
Miller, Dan, ed. *Acknowledging Consumption: A Review of New Studies.* New York: Routledge, 1995.
National Science Foundation. "Peer Pressure Plays Major Role in Environmental Behavior." Press Release 09-132. http://www.nsf.gov/news/news_summ.jsp?cntn_id=115049 (Accessed March 2010).
Nicholls, Alex and Charlotte Opal. *Fair Trade: Market-Driven Ethical Consumption.* London: Sage, 2005.
Peloza, John. "What About Consumer Responsibility?" Commentary. Canadian Imperial Bank of Commerce Centre for Corporate Governance and Risk Management, 2008. http://business.sfu.ca/cibc-centre/comments (Accessed March 2010).
Trentmann, Frank. "4-1/2 Lessons About Consumption: A Short Overview of the Cultures of Consumption Research Programme." Birkbeck College, University of London: http://www.consume.bbk.ac.uk/researchfindings/overview.pdf (Accessed March 2010).
Westra, Laura and Patricia H. Werhane, eds. *The Business of Consumption: Environmental Ethics and the Global Economy.* Lanham, MD: Rowman & Littlefield, 1998.

William Forbes
Stephen F. Austin State University

Sustainability, Seventh Generation

The term *seventh generation* most frequently refers to an ecological sustainability idea stemming from traditional Iroquois law that stated, "In our every deliberation we must consider the impact of our decisions on the next seven generations." This principle links concerns for cultural norms and current environmental practices with future generations.

The law has become a popular maxim for some environmentalists. The phrase *seventh generation* has also been co-opted by business and corporate entities who use it to position their practices as green and moral as well as by many businesses and marketing entities that use it in direct marketing to sell products that may or may not be environmentally friendly.

The Iroquois Confederacy, or Haudenosaunee, also known as People of the Longhouse or the League of Peace and Power or the Five Nations, was a group of Native Americans that consisted of five Indian tribes: the Mohawk, the Seneca, the Cayuga, the Onondaga, and the Oneida. The epicenter of the Iroquois Confederacy was located in Upstate New York when the Europeans first arrived.

The Iroquois had a mix of skills and were hunters, gatherers, fishers, farmers, warriors, and statesmen. Their ability to strategically grow corn, beans, and squash, and to then store food, enabled the confederation to attain a large population. They were most numerous and powerful in the beginning of the 17th century with a population of around 12,000. Before the Iroquois Confederacy was established, each of the tribes intermittently fought against each other. Two prophets are believed to have brought a message of peace that united the tribes. After the tribal infighting stopped, the confederacy emerged as the strongest (Native American) force in eastern North America during the 17th and 18th centuries.

As Europeans settled North America, the Iroquois became involved in trading and in wars with the Dutch, French, English, and Americans. The peaceful form of government within the Iroquois Confederacy is believed to have begun in 1142 and coincided with an eclipse of the sun. In the confederacy, each Indian nation remained independent. The confederacy created a Great Council with representatives called Sachems and they were all equal in authority and rank. Unanimity was the goal of the council. Iroquois treaties were not binding unless ratified by three-quarters of the male voters and three-quarters of the mothers in the nation. The confederacy had equality between the sexes. In particular, women had the power to veto war declarations and treaties.

The Iroquois nation certainly had a type of democratic government and political union between different tribes. Some historians have convincingly argued the Articles of Confederation and the United States Constitution were inspired by the Iroquois system of government; however, there is much debate about the directness of that influence. Benjamin Franklin and Thomas Jefferson were, at times, directly involved with the Iroquois, but the nature of the relationship and its effect on their subsequent political ideas will probably never be determined with perfect clarity.

Native American Perspective

Many Native American groups, including those tribes associated with the Iroquois Confederacy, had an ecological perspective that was holistic and Earth centered. They considered humans to be a part of the natural environment, and because of this belief, they sought a harmonious relationship between themselves and nature. The Iroquois law demanded a consideration for future generations in current decision making and provided a built-in context for securing sustainability while avoiding other actions, like excessive consumption, that might lead to a despoiled Earth and environmental problems.

Traditional Iroquois culture understood the idea of renewable resources and that near-term consumption should not be done at a level that compromises long-term use that can be sustained. They saw forests, fish, and animals as renewable resources that should be protected for future generations. Precontact North American indigenous peoples did have extensive trade and economic systems, and sometimes those systems led to environmental decay. However, those systems were usually characterized by a sense of equilibrium and stewardship and a longer sense of time and process than that of settling Europeans. Many writers argue that the Native American sense of the environment was more ecological and focused on process and connection, while the European sense of the environment was frequently more objectifying and reductionistic.

In the European mind, characterized by dualistic objectivism, science, reductionism, consumerism, private property, and massive trade, the rights of future people are not likely to enter into consideration. For the Iroquois, however, the rights of future people provided a central context for daily decisions. The European context and languages stressed a more utilitarian- and "progress"-oriented approach toward the environment that often viewed nature as something to be vanquished and brought under dominion. The Iroquois language and sense of nature are more apt to describe a sense of nurturing, relating, and communion. Some writers argue that the Iroquois understanding accents not only a conscious connection between people and nature but a basic difference in perception as well. This long-term, holistic perception of the Iroquois, then, naturally leads to a different sense of appropriate public policy. The European view, driven by objectivism, distance, and data, created a distorted view of humans and their relationship to nature. The Iroquois view encompasses an ecologically based perception that sees man as part of nature and nature as possessing an inherent value apart from people.

Western European Perspective

Many environmental writers further suggest that while the Western European perspectives are/were scientific and data driven, they frequently failed in understanding or communicating fundamental values. The capitalistic system within European perspectives failed at valuing nature, natural things, animals, ecological processes, and whole ecosystems. The European valuing system was instead utilitarian and driven by short-term profit. Western economic value systems have a schema to describe the cost of interference with the development and exploitation of natural resources, but they have no such schema to value natural resources in their undeveloped state. That natural systems themselves have an intrinsic value and contribute things such as clean drinking water and fresh oxygen is beyond the interpretive schema of the current Western economic system.

When there is a conflict, for instance, between a watershed or threatened species and the development of forest products, in the current system, the value of the forest products prevails.

Some writers go so far as to describe the current Western European, capitalistic valuing system as a kind of psychosis; they propose that science in its current state is incapable of being ecological. These writers hypothesize that the current psychosis is a kind of addiction and that the real environmental crisis is but a crisis of perception. These writers say that a wholesale change to our ideas and our experience of the natural world is required for one to understand the ecological issues of modernity. It is possible that insightful ecologists will not emerge from the sciences and biology but from philosophers, ethicists, poets, children, and ministers. It is these groups that are closer to the interpretive framework of the seventh generation cultures and may have the ability to show the ecological connections between things in ways that the scientists have thus far failed to communicate.

Individuals who espouse the ethic of the seventh generation have a palpable sense of people in the future that appears different from that of most Westerners. The Iroquois and others who speak with concern for the seventh generation do so in a way that communicates they are real people. These future people are in waiting, but held in the current sidelines of time. They are incapable of acting for their own benefit. They are waiting for their turn to live and for our time and current future to become their present. This sort of intergenerational subjectivity is one thing that distinguishes the Iroquois type of thinking from Western European thinking. Critics of this style of thinking write that seventh generation mentality is unrealistic and could bring an end to modern industrial society because it might challenge current materialism and economic growth as an end goal.

There is, however, an emerging recognition that modern industrial society must strive to create sustainable economic systems and that the current systems are not sustainable. At the level of self-interest, some corporations acknowledge that if they engage in unsustainable practices, they will eventually go out of business. One of the recognized leaders in the United States on issues of corporate responsibility owns a corporation called Seventh Generation, and he has written extensively on the issue of big business, moral practices, and sustainable business plans. He argues that companies must not only focus on profits; they should be responsible for what they create and do to other people, other business, communities, and their employees. This business leader argues that long-term growth can be secured when a company's business relationships and culture establish a socially responsible path. He also argues that environmentally conscious consumers of the future are going to insist on eco-friendly products and that customer loyalty can be secured by providing those products.

The field of business ethics is beginning to ask larger questions about social responsibility and environmental sustainability. Corporate social responsibility is now a heavily researched topic. Critics argue that corporate structure itself is about short-term profiteering, cannot be designed to have a conscience, and that most corporate discourse about the environment is greenwashing, merely designed to sell more products. Other environmentalists and socially responsible corporations argue that government alone cannot prescribe a positive and sustainable environment. However, demanding consumers and partnering businesses together can create sustainable systems well into the seventh generation.

See Also: Biocentrism; Climate Ethics; Deep Ecology; Human Values and Sustainability; Precautionary Principle; Sustainability and Spiritual Values; Tragedy of the Commons.

Further Readings

Armstrong, Virginia. *I Have Spoken: American History Through the Voices of the Indians.* Chicago, IL: Swallow Press, 1971.

Jennings, Francis. *The Ambiguous Iroquois Empire: The Covenant Chain Confederation of Indian Tribes With English Colonies.* New York: W. W. Norton, 1984.

Morgan, Lewis H. *Ancient Society.* Chicago, IL: Charles H. Kerr & Co., 1907.

Nollman, Jim. *For the Seventh Generation: The Interspecies Newsletter* (2000). http://www.interspecies.com/pages/onlinenews.html (Accessed April 2009).

Parkman, Francis. *A Half-Century of Conflict: France and England in North America.* Boston: Little, Brown, 1897.

Smith, Dean Howard. *Modern Tribal Development.* Walnut Creek, CA: AltaMira Press, 2000.

John O'Sullivan
Gainesville State College

Sustainability and Distributive Justice

Sustainability can be understood as an ability to endure in all circumstances. Sustainability in the natural world, for example, refers to ecosystems and how chemical processes, visible or not to the naked eye, create and maintain natural habitats like forests and plains. Sustainability in human terms is the ability to survive and to produce basic elements that help support human life, which generally means using natural resources. Combining human and natural aspects, sustainability can be defined as the healthy continuity of natural habitats and the responsible use of natural resources. The most commonly used definition of sustainability is from the 1987 report by the World Commission on Environment and Development, commonly known as the Brundtland Report. The term is defined as "development that meets the needs of the present without compromising the ability of future generations to meet their own needs." This definition provides at least three indispensable goals of sustainability: economic growth, environmental protection, and social justice.

While "distributive justice" is an inherent part and a goal of sustainability, "sustainability and distributive justice" refers to a harmonious relationship between growth and profit, the environment, and human welfare. Distributive justice refers to the allocation of resources to every social stratum in a just and egalitarian manner. Broadly speaking, the term can be understood as equal distribution of both goods (e.g., resources) and bads (e.g., pollution, wastes) across society regardless of race, class, gender, location, and sexual orientation.

Environmental discourses in the 1970s and 1980s focused largely on the growing tension between the economy and the environment because industrialization as a form of economic enterprise had negative impacts on the environment. For example, *The Limits to Growth* (1972), published by the Club of Rome, predicted ecological collapse if current economic growth trends continued in population, industry, and resource use. There were also discussions and debates on the use of appropriate and intermediate technologies, soft-energy paths, ecodevelopment, and north–south economic disparity. While the limits to growth debate asked whether environmental protection and continued economic growth are compatible, the notion of sustainability assumes that the two are complementary. Currently, *sustainability* is a very lucrative term and is used by almost all development institutions and organizations as their goal. However, there remains a lack of clarity in the meaning of sustainability and distributive justice. For example, deciding what should be "sustained," such as the environment, the economy, or human welfare, and what should be developed, such as the gross national product, environmental resources, or human welfare. Questions also arise around deciding whose needs sustainability serves, and to what degree socioeconomic inequality is sustainable.

Due to the lack of clarity and heavy use and abuse of the phrase *sustainability and distributive justice,* different perspectives emerged. At least three perspectives on sustainability and distributive justice can be identified. Free market environmentalism does not question the existing economic and political arrangements, but rather opines that free market is the best way to address social and environmental problems through voluntary actions by corporations and consumers. It advocates change through green consumerism and green technology as well as boycotting harmful products. The policy/reformist sustainability and distributive justice approach also does not question the existing political and economic structures; rather, it looks at how policies can be reformed to integrate sustainable development. Al Gore, for example, advocates greening within the bounds of economic growth. It invites political actors at all levels—local, state, national, and international—to take a reformist approach by adapting to the goals of environment, economy, and social equality. The critical structural approach strongly contests the existing economic practices and consumption patterns. According to this approach, sustainability and distributive justice actually serves as a legitimate guise for the dominant groups to perpetuate their interest, which eventually causes further inequality. Sustainability is used to hide the exploitative relations of production. Economic logic, according to this approach, will always win over ecological and social logic as long as free markets dominate.

Inequality and Sustainability

Sustainability and distributive justice is very crucial to addressing all major problems facing the planet, including global warming. It becomes very obvious that we cannot deal with environmental issues without addressing inequality and establishing distributive justice. Inequality matters in addressing climate change because, first, there is an apparent inequality in suffering from the odds of climate change. The poor are largely paying the price for the crimes of the rich, and the poor are less capable of dealing with disasters largely resulting from climate change. While the effects and the ability to handle climate change are unequally distributed, responsibility for the problem is even more unequally distributed. For instance, with only 4 percent of the world's population, the United States is responsible for 24 percent of global emissions. Second, poor regions have less bargaining power than wealthy nations, which can attend international meetings with legal experts, scientists, economists, skilled diplomats, and observers, allowing them to read every document, attend every committee meeting, and can weigh the pros and cons of many proposals. It is difficult to overstate the importance of this impact of global inequality on how the climate crisis is being negotiated. Poor nations feel they have been repeatedly overlooked and underrepresented in initiatives like the World Trade Organization (WTO), the TRIPS treaty on intellectual property rights, the International Monetary Fund (IMF), or allocation of loans from the World Bank. Third, a painful lesson was learned from the failure of the Kyoto Protocol: unless inequality is addressed, an effective climate deal cannot be reached. It was found that inequality drives the deadlock over Kyoto in various ways: (1) inequality within and between nations drives desperation and vulnerability in the global south, (2) it drives anger at the injustice of the distribution of goods (wealth) and bads (emissions), since "waste flows downhill," (3) it drives inability and unwillingness of poor countries to participate effectively in international efforts to address climate change (e.g., participation in Kyoto and other environmental treaties), and (4) in some key nations like the United States and Australia, extreme wealth has allowed the continuing delay in seriously addressing the issue. Each of these impacts of inequality is complex and must be addressed to move climate policy forward.

Solutions to Attaining Distributive Justice

On a global level, considering ecological and economic divides among nations, sustainability and distributive justice calls for at minimum a three-phase implementation of justice to ensure proper responsibility allocation. The first phase would be supplying "material justice" to poorer nations to cope with disasters they have encountered from the vicious cycle, and to address different varieties of inequality they might face from more developed and wealthier nations. The second phase is "procedural justice," where fairness in decision making between nations, developed or less developed, is to be made equal to work toward a better environment. The last phase is "compensatory justice," where carbon emitters should pay the price via, for example, carbon taxes, and technologically advanced nations should rigorously introduce the development of carbon-free environments to lessen or end carbon emissions entirely. International aid must be encouraged for nations responsible for heavy carbon emissions to apply their commitment worldwide.

The political climate and decision-making processes, from global issues like climate change and rainforest destruction to state and local issues, are thought to currently be skewed toward destruction of the ecological, bio-geo-physical, and other processes that support life on the planet. For sustainability and distributive justice to be effective, an Earth ethic is needed. Scholars have provided an effective model of sustainability and distributive justice: evaluating the consequences and acting upon three realms of the geosphere (including soil, air, water, and climate); the biosphere (our bodies, those of other species, and plant life); and the sociosphere (human concerns, such as worker conditions).

See Also: Development, Ethical Sustainability and; Environmental Justice; Human Values and Sustainability.

Further Readings

Conca, K. and G. D. Debalko, eds. *Green Planet Blues: Environmental Politics From Stockholm to Johannesburg.* Boulder, CO: Westview Press, 2004.

Frey, R. Scott, ed. *The Environment and Society Reader.* Boston: Allyn & Bacon, 2001.

Goleman, Daniel. *Ecological Intelligence: How Knowing the Hidden Impacts of What We Buy Can Change Everything.* New York: Broadway Books, 2009.

Gould, K. A. and T. L. Lewis, eds. *Twenty Lessons in Environmental Sociology.* New York: Oxford University Press, 2009.

Md Saidul Islam
Nanyang Technological University

Sustainability and Spiritual Values

While spirituality and spiritual values have been an enduring human interest, the concept of sustainability achieved an enhanced profile with the publication of the Brundtland Commission's report in 1987. The various interpretations and nuances of sustainability and spiritual values have consequences for practice. If sustainability means living in a manner that preserves the interests of subsequent generations, sustainability and spirituality may both have common ground in an individual's concern for that which is beyond oneself. More broadly, sustainability and spiritual values involve an acknowledgment and active attention to that which is the "other." In the spiritual realm, according to one's point of view, the other may be human, other life forms, other matter, or a personal or other divinity, all of which may be included in a reference to "spirit." Where there is an interest in spiritual values, there is also provision for an additional dimension to anthropocentric and biocentric worldviews or an alternative to them.

Providing for future generations could not only lead to a sustainable life but also include spiritual values. Indeed, it may not be completely inaccurate to consider sustainability itself to be a spiritual value. Sustainability and spiritual values may be contrary to each other where one operates with a dualistic worldview. For example, in Manichaeism, where the physical and the "flesh" is seen as evil and spiritual life is a renunciation of materiality, an interest in sustainability would be an interest in the physical and may be seen as antithetical to spiritual enlightenment.

The anthropocentric, biocentric, and divinity-centered perspectives in combination with varying meanings given to sustainability have implications for its relation to spiritual values.

In an anthropocentric view, the human is the central concern. Sustainability then prioritizes human interests, with the rest of existence only relevant to sustaining values of significance to humans. In this approach and, more narrowly, where sustainability is seen solely as an economic matter, the goal of maximizing present profits is modified to include improved prospects for subsequent human generations. Even a broader view of sustainability that includes the social and the environmental aspects in addition to the economic retains the human-centric valuations and is sometimes driven by a concern for survival of humanity as a species.

Spiritual values and interests as they may be aligned with human interests retain the fundamental priorities inherent in the anthropocentric perspective. Recognizing the social, environmental, and economic dimensions of sustainability permits a discussion of the "other" in greater complexity. Such discussions remain centered on the human interest even as they allow for a greater attention to the nonhuman. Whether one adopts a narrow or expansive meaning of sustainability, it is the humans who would determine the horizon of sustainability brought to bear in decision making.

In a biocentric view, the interests of the present generation of humans are not the central consideration. For instance, in a deep ecology perspective, humans are part of the larger world and are dependent on a balanced environment for existence. It is the health of the environment that is the predominant concern, and the task of sustainability is to determine what humans can contribute to the rest of existence. It is a necessary and spiritual value to appreciate the Earth as the source of life and sustainability is primarily that of the Earth, with humans as an incidental species. Damage to the Earth is an offense and repairing the damage is a virtue and an exercise that improves the biosphere and hence spiritual life.

Where one subscribes to a view of a divine reality in addition to the human and physical earthly experience, spiritual values are located in the service of the divine. For some, sustainability may be either a central or an incidental interest in spiritual life. Where divine direction includes the imperative to care for the economic, social, and natural environments, spiritual values could operate to enhance sustainability.

Traditionally, monastic communities and lifestyles globally have typically included features that correspond to spiritual values as well as sustainability. For such communities, the spiritual is inseparable from the religious and may even be one and the same. Such views have gained prominence through the work of Thomas Berry and others who have asserted that sustainability can only be achieved as a function of spirituality. In their view, the depth of change necessary to achieve sustainability has been referred to in terms of a conversion experience. The fundamental change involved is seen as located in a value system. Spiritual values then are essential to achieving the kind of sustainability allegedly needed.

Spiritual values would not be dissimilar to ethical values in treating the social, natural, and economic aspects of the human environment. Where the spiritual may have additional interest is perhaps inherent in the term *spirit* itself. From the Greek *pneuma,* the spirit may be seen as entity as well as an intimate part of all life. Related to the terms *respire, inspire,* and *expire,* spiration or breathing is a critical feature of life. In this latter sense, spirituality as the breath of living beings may often be appreciated more acutely in its unavailability than its routine presence. The association between spirituality as breathing and sustainability may perhaps be more obvious as an immediately available experience. It should nevertheless be noted that while some religious views include sustainability, others such as those similar to the Manichaean example noted above may, in fact, militate against it. In this sense, not all religiosity or spiritual values are either consistent with or conducive to sustainability.

The Will to Power and Sustainability

The interest in power or the will to power operating as a spiritual value does provide a critical nuance to sustainability. Power to do either beneficial or harmful actions when deployed with intent can be directed by an interest in spirituality. A more subtle and delicate consideration is the ability to distinguish between a harmful action and one of purification. In this critical endeavor, it is imperative to act with attention to values of significance. Where sustainability is seen to include a social dimension, for instance, the individual's will to power is attenuated by the appreciation of the societal or aggregate will to power. In a biocentric view, the aggregate would be extended to all of the Earth or cosmos.

To the extent that spiritual values are about ultimate truth, they could exceed the scope of an interest in a sustainability that is concerned only about the subsequent generation of humanity. In this context, sustainability may be a contributing interest or a stepping-stone on the path to ultimate truth. In this vein, some religious traditions have embraced and supported the advancement of sustainability as a value consistent with their teachings. For some, such as deep ecologists, sustainability could be a founding principle and a principal spiritual value.

An association between the concepts of sustainability and spiritual values being those of our age are not without similarities with other ages and times. An appreciation for sustainability has led to a renewed interest in some indigenous peoples as well as some Eastern cultures that have been cited to illustrate the directions possible when spiritual values underlie lifestyles generally or, more particularly, sustainable lifestyles. As with others, Greek philosophers have expended considerable resources in considering the complexity of the unity and the difference. A philosophy demonstrating an appreciation of a unity in difference has been seen in the views of Thales of Miletus, an Ionian from about the sixth century B.C.E., and is referred to in the work of Aristotle. Thales subscribed to the general view of the time and context that the material world was without beginning or end. In his view, as described by Aristotle, matter in all its diversity originates or arises out of water. Anaximander, an associate of Thales, differed and believed the material cause for all is a substance without limits and neither water nor other material entity. For him, motion provides a means of causing matter to differentiate and permits him to recognize a plurality of coexistent worlds.

It was only later in Greek thinking that a philosophical distinction between spirit and matter was developed and accorded varying values. Where matter and spirit are seen as distinguishable but inseparable, Thales's view of the role of water might find a resonance in current ecological concerns for the material unity in all of the diversity and viewing water as an inseparable spiritual value. In this context, treating water in a sustainable manner is then both a spiritual value and necessary to treating all matter, including humans, sustainably. Anaximander's view of the role of motion may find echoes in the more dynamic views and processes of sustainability. In Anaximander's type of approach, spiritual values may be more closely associated with valuing motion, while motion and the unlimited originating substance interact to materially cause differentiated matter.

Many years later in the era of the Roman Empire and subsequently, the Greek terms *kataphatic* and *apophatic* have been used to describe alternate paths of spirituality. The *kataphatic* path emphasizes the way of light, using images and other features of material reality to achieve spiritual development. The contrasting *apophatic* path relies on negation of images, emptiness, and darkness for the same goal. These paths may be further described and variously coordinated with speculative and affective approaches in practice. For instance, sustainability in the *kataphatic* speculative path of spiritual growth would appear to be a desirable, constructive imagined future to provide direction. The *apophatic* appreciation of the affective features associated with the losses resulting from the discontinuation of unsustainable practices as well as with processes to address the emptiness of living without the previously valued goal of economic growth. Thus, alternatives are recognized spiritual paths to sustainability.

Other Approaches to Sustainability

Other approaches to sustainability as a matter of spiritual growth are developing in contexts where spirituality has not always been considered a distinct category in the manner of Judaism, Christianity, or Islam. For instance, in the *advaita* (Hindi) or nondualistic approach of Indian thought, the self, or *atman,* is the Eternal, and does not provide for religion as a category—there is no word for "religion" in Hindi. In practice, the usage in Hindi is the word *dharma,* which is also translated as "duty." Doing one's duty is a spiritual practice necessary to progress spiritually. Sustainability in this sense is a duty and an obligation to society necessary to progress spiritually.

Another related concept is encountered in the process of achieving the realization of ultimate enlightenment through the values and practices of yoga. Etymologically related to the English "yoke," yoga essentially involves linking or achieving union with ultimate "other," or Truth. In this process, it would not be inaccurate to recognize the achievement of union with subsequent human generations and with all of reality. Through yoga, spirituality then can be seen as inseparable from sustainability.

Dualistic approaches also operate in the context of Indian thought. Here materiality and the self and ultimate reality are eternally distinct. Dharma and yoga in this view involves completing a transition from materiality to ultimate reality. Sustainability in this context is of value to the extent it contributes to the ultimate goal and is otherwise an impediment to final spiritual achievement.

As noted above, indigenous peoples, the nature-centric Celtic, Wiccan, and deep ecology views as well as the Eastern approaches seen in Buddhism, Confucianism, and other systems provide values and practices that can be understood as necessary for spiritual development. Even if these views do not explicitly or inherently define spirituality or sustainability, they do approximate these values on their own terms. In Confucianism, for instance, respect for elders is a central value that can only operate if the elders have subsequent generations to show them respect. In this sense, sustainability as an intergenerational consideration is included and necessary. Intergenerational respect as a value is not opposed to and may even be consistent with what is considered a spiritual value in Western thought. Perhaps the domestic and societal intergenerational theme may be associated with an ecological view. Etymologically, the word *ecology* is from the Greek *oikos,* meaning home, which seems to be consistent with fostering spiritual values and an attentiveness to intergenerational or sustainable practices in social contexts.

Sustainability may be seen as a consequence of spiritual values, a corollary to spiritual life, or in opposition to spiritual enlightenment.

See Also: Anthropocentrism; Berry, Reverend Fr. Thomas; Biocentrism; Brundtland Report; Deep Ecology; Religious Ethics and the Environment; Sustainability, Seventh Generation.

Further Readings

Carroll, J. E. *Sustainability and Spirituality.* Albany: State University of New York, 2004.
Copleston, F. *A History of Philosophy.* New York: Image Books, 1962.
Healey, C. J. *Christian Spirituality.* New York: Society of St. Paul, 1999.
Holthaus, G. *Learning Native Wisdom: What Traditional Cultures Teach Us About Subsistence, Sustainability, and Spirituality.* Lexington: University Press of Kentucky, 2008.

Lester de Souza
Independent Scholar

Taylor, Paul W.

One of the most important books during the emergence of environmental philosophy as an academic discipline was *Respect for Nature: A Theory of Environmental Ethics* by Paul W. Taylor. Published in 1986, the book is considered the primary reference supporting a biocentric viewpoint in environmental ethics. Biocentrism in general prioritizes individuals in nature, including humans, but does not assign humans higher priority. This approach is opposed to anthropocentrism, which focuses on human benefit, and ecocentrism, which enlarges concern to include the nonliving elements of ecosystems and focuses on species preservation over the welfare of the individual members of those species. Taylor's book came at a time when environmental philosophy was a new subdiscipline heavily scrutinized by mainstream philosophers. Thus the philosophical rigor Taylor used to outline his theory of biocentrism in *Respect for Nature* helped build credibility for the field.

Paul Warren Taylor is a professor emeritus at Brooklyn College, City University of New York. He received his Ph.D. in philosophy from Princeton University in 1950. His two other books, both preceding *Respect for Nature,* covered more traditional philosophy: *Normative Discourse* (1961) and *Principles of Ethics: An Introduction* (1975). Taylor also published 15 peer-reviewed articles in standard philosophy journals, mostly on ethics and moral philosophy, before entering the realm of environmental philosophy in 1981 with "The Ethics of Respect for Nature" in the journal *Environmental Ethics.* He followed this with "Frankena on Environmental Ethics" in the *Monist,* also in 1981, "In Defense of Biocentrism" in *Environmental Ethics* in 1983, and "Are Humans Superior to Animals and Plants?" in *Environmental Ethics* in 1984. Thus, by 1986, Taylor was well qualified to outline a book-length theory of environmental ethics within the realm of traditional philosophy.

A key strength of Professor Taylor's theory is its use of commonly understood norms in human ethics to build a foundation for environmental ethics. Taylor details the following three areas that make respect for nature similar to respect for humans:

1. A belief system that humans are members of Earth's community of life; that Earth's ecosystems are a complex web of interconnected elements; and that individual organisms, like humans, are teleological (goal-oriented) centers of autonomous choice, that each organism is an end-in-itself.

2. An ultimate moral attitude that respect for individuals in nature is like respect for "persons as persons," with inherent worth or "human dignity." Taylor cites philosopher Immanuel Kant in this respect. It is a rational basis of "respect" for each organism as inherently valuable and worthy of equal moral consideration; this is different from an emotion or "love" of nature.
3. Rules of conduct and an ethical system of duties that are acknowledged to be owed by everyone (Taylor cites philosopher John Rawls's duties of human morality). Taylor suggests using the rules of maleficence, noninterference, fidelity, and restitutive justice.

Taylor sees as groundless the claim that humans, by their very nature, are superior to other beings. For him, this is the most important point that justifies his theory. Taylor suggests that the roots of our claim to human superiority are found in classical Greek humanism, where rationality is the basis for superiority (Taylor suggests this is only one of many valuable capacities); in Cartesian dualism of a separate soul and body, which also prioritizes rationality (Taylor asks, why is thinking better if it is not needed by an organism to meet its goals?); and in Judeo-Christianity's Great Chain of Being, a hierarchy of God, angels, humans, and beasts, with humans "made in God's image."

An example Professor Taylor gives (in *Respect for Nature*) to support his egalitarianism is the value of humans and monkeys. Humans may be considered more valuable due to their mathematical ability, however he stated the following:

> It is true that a human being may be a better mathematician than a monkey, but the monkey may be a better tree climber than a human being. If we humans value mathematics more than tree climbing, that is because our conception of civilized life makes development of mathematical ability more desirable than the ability to climb trees. But is it not unreasonable to judge nonhumans by the values of human civilization, rather than by values connected with what it is for a member of *that* species to live a good life? If all living things have a good of their own, it at least makes sense to judge the merit of nonhumans by standards derived from *their* good.

Taylor refers to philosopher David Hume's is/ought dichotomy, or separation of fact and value, to reject holism or "balance of nature" as a moral foundation for consideration of the environment. He claims that science and ecology (facts) alone do not give us clear direction on morals (values). However, science does inform our understanding of individual organisms as "teleological centers of life," helping us to judge what is good for them. Taylor acknowledges a need to address competing claims on a case-specific basis, through five priority principles: self-defense, proportionality, minimum wrong, distributive justice, and restitutive justice.

Taylor's biocentric theory is one of several nonanthropocentric positions. The central feature of his theory is inherent worth, rather than the more-usual intrinsic value, which he bases on what he calls "recognition respect" rather than appraisal respect. Although Holmes Rolston III's theory of objective nonanthropocentric intrinsic value in nature is similar to Taylor's view in some respects, Rolston holds that more complex organisms have more intrinsic value, with humans at the top of the pyramid, a rejection of Taylor's egalitarianism. J. Baird Callicott, the leading proponent of Aldo Leopold's land ethic within philosophy, advocates a more subjective nonanthropocentric intrinsic value theory, in which value does not exist in nature but instead is created by a valuing agent.

Although Leopold's land ethic was once considered a biocentric approach, Taylor's writings have so altered our understanding of the term *biocentrism* that Leopold's position is now viewed as a more holistic, ecocentric approach, primarily because of its extension of concern to nonliving elements of ecosystems. Although Taylor also places important value in ecosystems and species, his concern for individual organisms separates his theory from the more recent term of *ecocentrism.* Taylor's theory has been defended and sympathetically extended by other philosophers, especially by James Sterba. Taylor's biocentric theory of environmental ethics remains an important standard in the field.

See Also: Biocentric Egalitarianism; Biocentrism; Rolston, III, Holmes.

Further Readings

Callicott, J. Baird. "Rolston on Intrinsic Value: A Deconstruction." *Environmental Ethics,* 14 (1992).
Norton, Bryan G. "Environmental Ethics and Weak Anthropocentrism." *Environmental Ethics,* 6 (1984).
Rolston, Holmes, III. "Are Values in Nature Subjective or Objective?" *Environmental Ethics,* 4 (1982).
Sterba, James P. "Biocentrism and Human Health." *Ethics and the Environment,* 5 (2000).
Sterba, James P. "A Biocentrist Strikes Back." *Environmental Ethics,* 20 (1998).
Sterba, James P. "From Biocentric Individualism to Biocentric Pluralism." *Environmental Ethics,* 17 (1995).
Taylor, Paul W. "Are Humans Superior to Animals and Plants?" *Environmental Ethics,* 6/2 (1984).
Taylor, Paul W. "The Ethics of Respect for Nature." *Environmental Ethics,* 3/3 (1981).
Taylor, Paul W. "Frankena on Environmental Ethics." *The Monist,* 64/3 (1981).
Taylor, Paul W. "In Defense of Biocentrism." *Environmental Ethics,* 5/3 (1983).
Taylor, Paul W. *Normative Discourse.* Upper Saddle River, NJ: Prentice-Hall, 1961.
Taylor, Paul W. *Principles of Ethics: An Introduction.* Encino, CA: Dickenson, 1975.
Taylor, Paul W. *Respect for Nature: A Theory of Environmental Ethics.* Princeton, NJ: Princeton University Press, 1986.

William Forbes
Stephen F. Austin State University

Technology

If our technology becomes so bold that we become reliant on it to provide for our food, water, clean air, and regulation of the climate itself, critics raise concerns that what we are really passing along to future generations is a dependence on technology. Optimists are unfazed by such concerns and point out how reliant we already are on technology for our food production systems, and so on. However, skeptics fear that such a dependence on technology threatens the ability of future generations to meet their basic needs should the technologies ever fail. This gets to the heart of how we understand

A crew watches as the damaged blowout preventer from the Deepwater Horizon oil drill is lifted out of the Gulf of Mexico on September 4, 2010. The many failed attempts by experts to stop the oil spill raised public awareness of the limitations of technological solutions to environmental problems.

Source: U.S. Coast Guard/Thomas Blue

the role of technology: does technology allow us to use resources more efficiently and thus live better within certain natural limits, or does technology give us the ability to live in defiance of natural limits and dream up creative new ways to meet the needs of humans that had previously been provided for by the natural world?

Lynn White, Jr.'s 1967 article in *Science,* "The Historical Roots of Our Ecological Crisis," is an often-cited landmark essay in the field of environmental ethics. When the typical person thinks about or refers to technology, it is often high technology that comes to mind. However, even mundane objects such as shoes and reading glasses are technological devices that help us as humans to improve our lives in the world. Setting the technological bar at this level often raises heated debates as to whether humans are the only technological animals or whether others in the animal kingdom also use technology to their own benefit.

While not contributing to this debate directly, White does maintain that all creatures are reactive and actively modify their surroundings to their own benefit. This, of course, includes human beings. Humans have always modified our environment in order to survive. However, a decisive moment, according to White, was when humans combined science and technology, giving us the ability to radically transform the environment around us in ways that no other species, including our own, had ever been able to do before. With further development and innovation, the transformative power of technology has only increased.

As an example, consider the effects of warfare on the environment. The destruction that two hunter-gatherer groups at war could have caused the environment would have been minimal and local. With the development of gunpowder and canons, White says we began to plunder resources from the Earth to fuel our wars, and the destructive effects of our wars increased. However, with the current state of weapons of mass destruction, an all-out war has the capacity to eliminate the majority of life on this planet. Never before have we had the power to effect such radical change on the global environment.

White famously claimed that science and technology are distinctively Western, and more specifically, are distinctively Christian in their conceptual origins. He makes the philosopher's assertion that in order to ameliorate inevitable environmental crises, we must shift our conceptual frameworks in regard to the natural world and our place in it. That is, the power of our human technology has facilitated environmental degradation and destruction, and the solution to environmental problems, contrary to how we often perceive it, will not be a technological solution, but a conceptual or philosophical one. For this reason, this article is understandably a centerpiece in philosophical and religious approaches to environmental ethics. It is not technology that will solve environmental problems, but rather new conceptual frameworks. Indeed, White claims the further application of science and technology will only exacerbate environmental problems. Technology is thus seen as the culprit, not the savior of environmental crises. At least, not until we can reenvision our place in the natural world.

And so, in the early stages of the literature in environmental ethics, as represented by authors like Lynn White, Jr., environmental problems were not to be conceived of as technological problems in need of a techno-fix. In fact, technology was largely seen as part of the problem. This can be further emphasized by the significant impact Rachel Carson's book *Silent Spring* had on society. Chronicling the negative impacts of DDT, some argue that this book first alerted the American public to the dangers of technological solutions to environmental problems and raised the question of whether technological fixes are the most desirable ones. *Silent Spring* was a springboard for the environmental movement, and it tells a striking cautionary tale of the environmental dangers of science and technology.

Early approaches to technology by environmental ethicists were largely negative. Or rather, environmental ethicists challenged the tendency of policy makers, government officials, and the media to turn to scientific and technological experts for solutions to environmental problems. At the time of this writing a similar drama is sadly playing out in the Gulf of Mexico. In order to fuel our societies, we are seeking ever-newer and cleaner forms of energy. However, as we are still largely reliant on fossil fuels and supplies are harder to locate, we are forced to drill in riskier areas. When the 2010 British Petroleum oil spill first occurred, our leaders turned to scientific and technological experts for advice and solutions. The general public watched, however, as numerous technological attempts by experts to stop the spill failed. As with Rachel Carson's call to alarm about the effects of DDT, commentators are currently pointing out the dangers inherent in environmental technologies and again raising questions about technological solutions to environmental problems.

Such situations open the door for philosophers and ethicists to perhaps be overly critical of technology while seizing the opportunity to draw attention to the philosophical and ethical dimensions that need to be addressed in environmental problems. And while there are those who are skeptical of the role technology ought to play in environmental problems, there are just as many people, if not more, that see such problems as largely the provenance of science and technology. Many place a great deal of confidence in our technological ability to devise a solution to whatever environmental problems we may encounter, both locally and globally. Indeed, while philosophers typically seek ways to shift our conceptual understanding of our role in the natural world to combat enormous crises like global climate change, geoengineers are working on devising technological methods to manipulate the climate itself and bring it under our control. Debates about how to address environmental issues can thus quickly become polarized, with one side arguing that the application of further technology will only aggravate environmental problems while the other argues that massive technological intervention is the only way to address such problems.

Technological Optimism and Pessimism

In a book titled *Infinite Nature*, professor of natural resources R. Bruce Hull captures the tension between the different attitudes toward technology nicely in a chapter titled "(In) finite Nature." Depending on our attitudes toward technology, nature and its resources can alternatively appear to us as either finite or infinite. If technology is ultimately incapable of overcoming natural limits and gaining control of ecosystemic processes and systems, then the natural world is finite, and we must alter our actions and lifestyles accordingly to be able to inhabit this planet well and within certain limits. We must learn to live within the limits of the finite world to produce sufficient food and clean water, and to filter our wastes, and so forth. For years critics have been warning that we have already surpassed the carrying capacity of the Earth. This raises ethical concerns regarding whether we are doing a grave injustice to future generations by potentially depriving them of the same access to a relatively healthy environment that our predecessors and we have enjoyed. And so, if we are unable to geoengineer the climate and future carbon emissions will drive the planet toward inhospitable temperatures, then we must reorganize large parts of how our societies currently operate. If, however, one has faith that technology can overcome most problems, then this ostensibly places no limits on how much carbon we put into the atmosphere, for example, because in the future we will be able to remove it technologically, and thus no real change in the course and behaviors of our global societies is required. According to technological optimistic positions, the most prized of all natural resources are not water, energy, or food, but rather human ingenuity, creativity, and innovation. These resources enable us to put other natural resources to more efficient use and potentially even clean and improve upon the current state of the environment. To limit technology, according to optimists, would be to inhibit progress and stifle opportunities for a better tomorrow and doom us to an unnecessarily diminished quality of life in the future.

Two radically different visions of the future appear in these competing stances toward technology. This is why discussion of the role of technology and of what kind of future we want as humans on this planet is so important. However, this is not so easily accomplished because there is not one thing called *technology,* but rather many different *technologies.* Deep-sea drilling and geoengineering technologies are technologies, and so are wind turbines and solar panels. Debates about the role of technology can sometimes obscure the reality that we will not cease to live in a world with technologies; however, different technologies can reflect different visions of the future and different ways of life. There can be so-called green technologies that seek to recognize natural limits and live within those—solar and wind power technologies arguably among them—and there can be technologies that seek to manipulate or alter the Earth's natural process and deny any limits placed upon our human endeavors, such as some geoengineering proposals to technologically alter the Earth's climate. Thus the debate should not be construed as a pro- versus anti-technology issue, but rather, which technologies are desirable and facilitate the lifestyles and communities we want.

Dependence on fossil fuels may reflect a cultural leaning toward techno-optimism. We are repeatedly told that sometime in the near future we will exhaust the Earth's fossil fuels; the deepwater oil spill has already revealed how far we must go to tap the last remaining deposits. However, this has arguably done little to change human behavior. For a time, when oil prices rise, many attempt to cut back on driving where possible or buy more fuel-efficient vehicles. However, there does not seem to be general alarm at the depletion of fossil fuels and the prospect that we might not drive automobiles and fly in airplanes in the future, but rather a general feeling that we'll find solutions to enable us to continue these activities we've grown accustomed to, be it with natural gas–powered cars or 100 percent electric vehicles, or some as-of-yet unknown technology. Thus techno-optimist solutions do not necessarily entail behavioral changes but rather technological improvements to maintain and improve upon the current standards of living.

This relates to a concern of many philosophers about the popularity of sustainable development discourses. Many critics have reiterated that when we talk about sustainability or sustainable development, we must question what it is that is being sustained.

Techno-pessimists fear it is our consumptive lifestyles that are being sustained, and new technologies will simply allow us to continue our lifestyles without requiring any difficult transformations. This is in opposition to those who wish to interpret sustainable development as sustaining the capacity of the Earth to flourish and maintain itself for its own sake and to provide for our basic needs.

The key definition from the famed Brundtland Report on sustainable development is "sustainable development is development that meets the needs of the present without compromising the ability of future generations to meet their own needs." Critics worry that this definition can be and has been construed to mean, as mentioned above, that it is current Western consumptive lifestyles that ought to be sustained and not nature itself, so long as we can provide similar lifestyles and opportunities to future generations. Many hold out hope that new developments in technology will allow us to do just that. The concern is just how prominent a role technology will play in the ability of future generations to provide for their needs.

Technology and Society

A further significant feature philosophers of technology discuss is the extent to which our technologies shape not only our ways of life but also our political structures and ways of living within a society. For example, R. Bruce Hull and others point out that certain technologies such as wind and solar are compatible with decentralized and local forms of political organization. Individuals can operate their own solar panels and thus control their own energy production to a certain extent. Solar and wind technologies can also be operated on a neighborhood or village scale, in addition to large scale. Thus such technologies are compatible with a range of different forms of political organization. Technologies such as nuclear power, on the other hand, require more centralized control due to the risks posed by nuclear waste and the threats of day-to-day operation. Nuclear technologies then will likely require more centralized power and political control. Therefore, to the extent that we find certain forms of social organization more or less desirable, we should also take into account which technologies will be most compatible with those visions.

So while early environmental ethicists and philosophers have warned of the environmental dangers posed by technology and the misguided attempts to find techno-fixes to our various environmental problems, today it rather seems that solutions will be techno-fixes; however, we have a richer understanding of technology and what a techno-fix can entail. Our solutions to environmental problems will involve technology, but we must collectively make decisions about what kinds of technology to employ and understand that our choices in technologies are reflective of our cultural understandings of our place within nature and will shape how we will be able to live with the environment in the future and how our societies will be organized. For this reason it is important that we are aware, as much as possible, of the various futures offered by different technologies and make conscious decisions about which technologies we want and which risks are acceptable. Philosophy of technology and environmental ethics are beginning to come together to address these important issues. There will likely be even more convergence in the future as we come to understand the philosophical and social dimensions of various technologies and grasp the centrality of technology in addressing environmental problems.

See Also: Brundtland Report; Carson, Rachel; "Ecological Crisis, The Historical Roots of Our"; Ethics and Science; Future Generations; Human Values and Sustainability; *Silent Spring;* Strong and Weak Sustainability.

Further Readings

Carson, Rachel. *Silent Spring*. Boston: Houghton Mifflin, 1962.
DesJardins, Joseph R. *Environmental Ethics: An Introduction to Environmental Philosophy*. Boston: Wadsworth, Cengage Learning, 2006.
Hull, R. Bruce. *Infinite Nature*. Chicago, IL: University of Chicago Press, 2006.
White, Lynn. "The Historical Roots of Our Ecological Crisis." *Science,* 155 (1967).

Jonathan Parker
University of North Texas

Ten Key Values of the Greens

The Ten Key Values of the Greens are guiding principles that provide a framework for green party activities including organizing, education, outreach, activism, and the development of political platforms. The current version of the Ten Key Values adopted by the Green Party of the United States was ratified in 2000 at the Green Party Convention in Denver, Colorado. It consists of the following principles:

1. Grassroots democracy
2. Social justice and equal opportunity
3. Ecological wisdom
4. Nonviolence
5. Decentralization
6. Community-based economics and economic justice
7. Feminism and gender equity
8. Respect for diversity
9. Personal and global responsibility
10. Future focus and sustainability

The content of the Ten Key Values can be attributed to and is derived from green activism and politics outside the United States. Although environmentalism and green politics have a rich and active history in the United States, formal green parties appeared relatively late on the global scene. Tasmania, New Zealand, and the United Kingdom are credited with the earliest incarnations of green political parties, which emerged in the early 1970s. Founded in 1980, the German green party Die Grünen is generally regarded as the touchstone of the green movement because of its significant electoral and political successes, particularly in the middle and late 1980s. As the movement developed and grew in numbers, German greens organized around four main principles, also referred to as the "four pillars" of the green party: ecology, social responsibility, grassroots democracy, and nonviolence. Taken together, the four pillars express a broad, holistic, social program that can incorporate a range of political issues coming from the Left. The four pillars have proven to be foundational, and they continue to underpin green political thought and practice today.

In 1984, 62 activists met at Macalester College in St. Paul, Minnesota, to discuss further organizing of a national green movement. Naming themselves the Committees of Correspondence—after grassroots political organizations within the 13 colonies during the American Revolution—they drew up and approved a set of principles that adopted the four pillars and added six more. These principles became the initial version of the Ten Key Values of the Greens.

The Ten Key Values serve as guiding principles; they are not to be understood as rigid laws or party doctrine. The initial version of the Ten Key Values, which remains substantively intact, was intended as an invitation for thinking about and debating green ideas. The Ten Key Values today continue to serve as a vehicle for discussion. Moreover, in keeping with two of the professed values—decentralization and grassroots democracy—the Ten Key Values are adapted, interpreted, and expressed in different ways by different local and state-level chapters. Indeed, the Website of the Green Party of the United States invites readers to compare different versions of the Ten Key Values across dozens of green party organizations, as well as to explore how the Ten Key Values help shape different green political platforms.

While the Ten Key Values are adaptable to different contexts and are useful as discussion points, the unsettled nature and variant interpretations of green principles are reflected in the fractious history of the green movement, which perhaps should not be surprising given the broad, albeit left-leaning, political spectrum of green groups and individuals. In *Green Politics,* Charlene Spretnak and Fritjof Capra describe how the four pillars of the greens, as well as other integral green principles, are contested and ascribed different meanings by groups from variant political traditions within the German green movement in the early 1980s. For example, "radical-left greens" coming from Marxist traditions interpreted the principle of social responsibility as attainable only through socialism, whereas the green center wanted to keep open a range of possibilities. Similarly, Stephen Rainbow depicts the German greens as being "racked with party splits" between fundamentalists who wanted structural transformation and realists who were focused on reforming existing institutions.

In *Against All Odds: The Green Transformation of American Politics,* John Rensenbrink suggests that the green movement in the United States can also, in part, be characterized by ideological fissures and political fracturing. Rensenbrink shows how political divides emerged between green party members who adhered to ideas around deep ecology, which positions human beings within the context of the natural environment, and those who prioritized the importance of addressing social inequalities. This split was followed by ongoing conflicts that eventually positioned green party members who were in favor of more involvement in electoral politics against a more radical, anticapitalist left that was wary of being co-opted by capitalist institutions.

The indeterminacy of the Ten Key Values allows for open debate and evolving ideas, which on the one hand might encourage balkanization and ideological conflict. On the other hand, many greens believe that debate and openness is an important part of the political process. Moreover, the Ten Key Values of the Greens have helped to keep divergent green ideas in dialogue and created a sense of unity. They have remained a constant presence throughout the history of the green movement, its conflicts, and political divides. This dynamic can be seen at the national level where ruptures within the movement have led to the existence of two national green parties: the more prominent Green Party of the United States, which focuses on electoral politics, and the Greens/Green Party USA, which is more activist and movement oriented. While these national parties have different visions and statures, they are both guided by the political and philosophical contours of the Ten Key Values of the Greens.

See Also: Bright Green Environmentalism; Ecopolitics; Green Party, German.

Further Readings

Global Greens. http://www.globalgreens.org/index.php (Accessed March 2010).
Green Party of the United States. http://www.gp.org/index.php (Accessed March 2010).
Rainbow, Stephen. *Green Politics.* Oxford, UK: Oxford University Press, 1993.
Rensenbrink, John. *Against All Odds: The Green Transformation of American Politics.* Raymond, ME: Leopolog Press, 1999.
Spretnak, Charlene and Fritjof Capra. *Green Politics.* Santa Fe, NM: Bear and Co., 1986.

Boone Shear
University of Massachusetts Amherst

Thoreau, Henry David

Henry David Thoreau (1817–62) was an American naturalist and essayist and an important figure in the New England transcendentalist movement. He was born in Concord, Massachusetts, graduated from Harvard in 1837, and returned to live in Concord for most of his life. His best-known works are *Walden,* "Civil Disobedience," and "Walking." Although he was not widely appreciated during his lifetime, he has become one of the United States' most celebrated authors and has had a lasting impact on environmental thought and American nature writing.

Walden recounts Thoreau's two-year stay in a rustic cabin he built on the edge of Walden Pond in a wooded area near Concord. This was not, as many people think, a wilderness experience; Walden lies only a mile and a half from Concord. But Thoreau was not seeking wilderness. He moved to Walden for personal self-development—to live deliberately and "front only the essential facts of life." He wanted to "live deep and suck all the marrow out of life" and "put to rout all that was not life." For Thoreau, this could only be achieved by living in close contact with nature, away from neighbors, and with minimal possessions.

Thoreau was deeply influenced by his friend Ralph Waldo Emerson. In Emerson's view, nature is continuous with humanity and the Divine. Divinity pervades nature, and a true understanding of nature points to an understanding of higher spiritual laws. Since humanity is continuous with nature, and the divinity in nature is also within every human individual, understanding nature reveals knowledge of ourselves. Thus, at Walden Thoreau sought to understand both nature and himself. The luxuries and complexities of society are, in his view, hindrances to both kinds of understanding; they obscure our vision of nature and they enslave people to needless toil, resulting in "lives of quiet desperation." He was very critical of what today is called consumerism. His solution: "Simplify, simplify."

Thoreau's values of simplicity and living in harmony with nature are basic green values. If everyone were to follow Thoreau's example of simple living, there would likely be no environmental crisis. And if Thoreau is right about our relationship to nature, our lives would be enriched in the process. Thoreau presents a radical utopian challenge to the status quo and invites the reader to take up the challenge. This utopian element, coupled with Thoreau's compelling writing style, did much to inspire the back-to-the-land movement of the 1960s and 1970s, a movement that provided an important experiential base for green ethics.

Thoreau was rarely concerned with the fate of nature. Presumably, nature could take care of itself. When he does comment on the degradation of nature, he is often concerned with the effect on human aesthetic experience. To modern critics this might seem superficial. But for a transcendentalist, a developed aesthetic appreciation of nature can reveal nature's unity and deeper truths. Thoreau is also concerned about the effects of the degradation of nature on our creative imagination. In "Walking" he remarks that most so-called improvements "deform the landscape," noting that "fewer pigeons visit us every year. Our forests furnish no mast for them." Similarly, "fewer thoughts visit each growing man from year to year, for the grove in our minds is laid waste. . . . Our winged thoughts are turned to poultry." Thoreau's writings express a love of nature. Though he rarely advocated the preservation of natural landscapes, in today's circumstances it is a short step to a preservationist stand from the love, appreciation, and knowledge of nature that we find eloquently expressed in Thoreau.

The essay "Walking" celebrates wildness. Thoreau announces that he wishes "to speak a word for Nature, for absolute freedom and wildness . . . to regard man as an inhabitant, as part and parcel of nature." He proclaims that "In wildness is the preservation of the world." Wildness is a source of nourishment and regeneration for the individual and for society and is even equated with life itself. Wildness is not necessarily wilderness; it can be found in all sorts of places. Thoreau was particularly attracted to swamps and the "impermeable and unfathomable bog." Wildness is also found in myth and music. When society turns away from the wildness of nature, it becomes inbred and weak.

Thoreau's transcendentalism stimulated him to look beyond the details of nature and seek its wider patterns. As he became more scientific in his investigations, this appreciation of pattern led to his understanding of mechanisms of seed dispersal and his discovery of forest succession, scientific achievements that are still respected today. He pioneered what later became the discipline of ecology.

Thoreau's most important political essay is "Civil Disobedience," originally published as "Resistance to Civil Government." It influenced the thought and tactics of Mohandas Gandhi and Martin Luther King, Jr., and the movements they led. Following the success of these movements, civil disobedience has been used by green activists around the world as a method of protest and attracting media attention, and as a way of effecting social change.

See Also: Emerson, Ralph Waldo; Gandhi, Mohandas; Philosophy and Environmental Crisis.

Further Readings

Buell, Lawrence. *The Environmental Imagination: Thoreau, Nature Writing, and the Formation of American Culture.* Cambridge, MA: Belknap Press, 1995.

Nash, Roderick Frazier. *Wilderness and the American Mind,* 4th ed. New Haven, CT: Yale University Press, 2001.

Oelschlaeger, Max. *The Idea of Wilderness.* New Haven, CT: Yale University Press, 1991.
Thoreau, Henry D. "Civil Disobedience." In *The Higher Law: Thoreau on Civil Disobedience and Reform,* Wendell Glick, ed. Introduction by Howard Zinn. Princeton NJ: Princeton University Press, 2004.
Thoreau, Henry D. *Walden: An Annotated Edition.* New York: Houghton Mifflin Harcourt, 1995.
Thoreau, Henry D. "Walking." In *Excursions,* Joseph Moldenhauer, ed. Princeton NJ: Princeton University Press, 2007.
Worster, Donald. *Nature's Economy: A History of Ecological Ideas,* 2nd ed. Cambridge, UK: Cambridge University Press, 1994.

Kelvin J. Booth
Thompson Rivers University

Tragedy of the Commons

First coined by Garrett Hardin in 1968, "tragedy of the commons" refers to a dilemma in which multiple individuals acting out of personal interest, freedom, and rationality in the pursuit of their self-interests will eventually ruin shared limited resources, even when it is clear that it is in nobody's interest for this to happen. Hardin's original paper gives the example of a common land where everyone has the right to graze their cattle and where, even if the land becomes overgrazed, people will continue to put their animals on the damaged fields and even add to their herd. This is especially true when some see the need to limit one's actions, while others may continue to utilize the common land with no restraint. To put it simply, even if everyone is aware of the risk of abuse, the mix of selfishness, competitiveness, and unregulated exploitation eventually makes the land unusable for all.

The tragedy of the commons is often illustrated with the example of overgrazed public land. The cattle on this hillside in the Chase Lake Wetland Management District in North Dakota have overgrazed the pasture behind the fence; overgrazing has been prevented on the land in the foreground.

Source: U.S. Fish & Wildlife Service/Rick Bohn

While the case in question centers on the "common land," the "common" may be taken to refer to any resource in the environment that is shared by people, such as air or water. This concept can therefore be extrapolated to aid in the understanding and explanation of the current environmental problems the world is facing: for example, humankind's use of the Earth's natural resources or negative atmospheric changes like ozone depletion and global warming. The actions of individual states, corporations, and other stakeholders working to serve their own interests have resulted in environmental catastrophes that affect the planet at large.

It would be useful to employ the practices of withdrawals and additions to better comprehend Hardin's concept. The withdrawing of resources from a common land leads to an accumulation of wealth by an individual or a group, but with a cost borne by all. Some are able to accumulate wealth more rapidly than others, and this forms the basis of the capitalist system where there is social inequality, and a certain privileged group will inevitably have greater access to natural resources compared to the less-privileged group. Everyone, however, shares the same environment, and due to the rapid increase in population, the environment is straining to sustain a growing and resource-dependent population. Resource withdrawals that have environmental impacts thus affect the entire community.

On the other hand, it costs less to discharge waste without treatment compared to purifying wastes before release. Since this is an adopted behavior deemed rational by most, the world is locked into a system of "fouling our own nest," so long as we behave only as independent, rational, free enterprisers. The accumulation of waste, coupled with a growing population, can have devastating impacts on the environment—impacts that are shared by all. And because the burden is shared by all, people believe they can continue in such activities, being under the impression that they have more to gain and less to lose. This fuels the accumulation of wealth through more withdrawals of resources and additions of wastes that eventually lead to exponentially devastating impacts on the environment, and, in so doing, on the human race. The "tragedy of the commons" then becomes a much greater and more pressurizing issue when such selfish interests are pursued within and on a state level. In the contemporary global capitalist system, most, if not all, nation-states claim to act in accordance with the interests of their own country and people, evidenced in the actual actions of many nation-states around the world.

The Kyoto Protocol serves as an excellent example of how one is moved more by one's own narrowly conceived self-interests than by being concerned about the implications of one's actions. In 1997, the heads of several countries met to sign the Kyoto Protocol. Bound by a physical contract, the protocol required the countries to reduce their greenhouse gas emissions. However, not all the countries of the world signed the Kyoto Protocol. The countries that did sign it account for 55 percent of the world's greenhouse gas emissions. Two groups hesitated to endorse it—developed countries like the United States and Australia. On the one hand, the United States claimed that it would work to reduce emissions unilaterally. On the other hand, some developing countries saw no reason to participate in saving and protecting the environment and reducing their gas emissions because they held the global north responsible for the Earth's current predicament, and therefore deemed it the global north's responsibility to rectify the problem.

Hardin's theory of the tragedy of the commons is even more important in today's world, and has an immense explanatory power to account for many of the environmental catastrophes humankind currently faces. International agreements currently serve as one of the significant ways in which the tragedy of the commons can be avoided, and perhaps rectified. However, in a world fueled by capitalistic profit-seeking interests, it is much more complicated and difficult to address such issues. The grim truth is that resources are limited, and the consequences of depleting natural resources are already making themselves apparent. The hole in the ozone layer and global warming are real. Environmental problems such as these do not discriminate, and, if they are not already, will soon be a common tragedy.

See Also: Climate Ethics; "Ecological Crisis, The Historical Roots of Our"; Future Generations; Globalization.

Further Readings

Agence France Presse. "Global Warming: 'Tragedy of the Commons'" (2005). http://www.commondreams.org/headlines05/0213-06.htm (Accessed February 2010).

Baden, J. and D. Noonan. *Managing the Commons.* Bloomington: Indiana University Press, 1998.

Gardiner, L. "The Kyoto Protocol Is in Effect! Around the World, 141 Countries Are Taking the First Step to Decrease Greenhouse Gases!" (2005). http://www.windows.ucar.edu/tour/link=/headline_universe/earth_science/stories_2004/kyoto_news.html (Accessed February 2010).

Hardin, G. "The Commons" (1997). http://oto2.wustl.edu/bbears/trajcom/commons.htm (Accessed February 2010).

Kerns, B. "Climate Change: The Tragedy of the Commons?" (2005). http://www.met.utah.edu/reichler/6030/Sample_paper.pdf (Accessed February 2010).

Md Saidul Islam
Nanyang Technological University

United Nations Conference on the Human Environment

The United Nations (UN) Conference on the Human Environment, also known as the Stockholm Conference, was the first UN conference focused on international environmental issues. Held in Stockholm, Sweden, from June 5 to 16, 1972, the conference reflected a growing interest in environmental issues worldwide and laid the foundation for global environmental governance. The final Declaration of the Stockholm Conference was an environmental manifesto that remains a forceful statement of the finite nature of Earth's resources and the necessity for man to safeguard them. A further result of the Stockholm Conference was creation of the United Nations Environment Programme (UNEP) in December 1972 to provide leadership and to coordinate global efforts to promote sustainability and safeguard the natural environment.

The roots of the Stockholm Conference lie in a 1968 proposal from Sweden that the UN hold an international conference to examine environmental problems and identify those that required international cooperation to solve. The 1972 conference was attended by delegations from 114 governments: it was boycotted by Eastern bloc countries due to the exclusion of East Germany. Documents created during the conference have proven influential in international environmental law: in particular, the Final Declaration, which elucidates 26 principles concerning the environment. The conference also produced an Action Plan containing 109 recommendations in six categories: human settlements, natural resource management, pollution, educational and social aspects of the environment, development, and international organizations.

The final declaration was a statement of human rights, as well as of the need for environmental protection. The first principle began by stating that "Man has the fundamental right to freedom, equality and adequate conditions of life, in an environment of a quality that permits a life of dignity and well-being," a statement comparable in scope to the statement in the preamble of the constitution of the World Health Organization, which declared that all human beings have a fundamental right to enjoy the highest possible state of health. The need to preserve the environment was not placed in opposition with economic development: in fact, their interdependence was explicitly stated in principle 8: "Economic and social development is essential for ensuring a favourable living and working environment for man," and principle 9, which stated that wealthier countries should provide financial and technological assistance to poorer countries to aid development, as well as to improve environmental conditions.

Other topics treated by the declaration included the necessity of conservation, including the preservation of wildlife habitat (principle 4), avoidance of polluting the seas (principle 7), wide use of nonrenewable resources (principle 5), the importance of developing coordinated planning (principles 13–17), the importance of environmental education (principle 19), scientific research and the free flow of information (principle 20), development of international law regarding environmental pollution and damage (principle 22), and the elimination and destruction of nuclear weapons (principle 26).

The "Framework for Environmental Action" plan from the Stockholm Conference consists of 109 recommendations for specific actions to be taken to protect and support the environment. The first 18 referred to organizational or bureaucratic actions, some of which were noncontroversial (establishing various intergovernmental bodies, arranging for the exchange of scientific expertise across national boundaries, increasing available training within regions), while others were more controversial recommendations, such as increasing assistance for family planning (birth control). A second set of recommendations involved natural resources management and included recommendations to evaluate the use of biological controls as a substitute for pesticides, and research into ways to reduce the use of chemical fertilizers; to assist activities related to recycling and control of wastes, establishing environmental monitoring systems, including wildlife resources; and calls for the establishment of protected areas such as national parks, the preservation of genetic resources through seed banks as well as wildlife conservation programs, creation of an international inventory of endangered species, and international monitoring of world fishery resources. In the category "Identification and control of pollutants of broad international significance," the action plan called for consideration at an international level of the effects of pollution on the climate and called on countries to avoid discharging toxic and dangerous substances in general, including persistent substances such as heavy metals.

See Also: Earth Day 1970; Earth Summit; United Nations Environment Programme.

Further Readings

Long, B. L. *International Environmental Issues and the OECD 1950–2000: An Historical Perspective*. Paris: Organisation for Economic Co-operation and Development, 2000.

Meadows, D., et al. *The Limits to Growth: A Report for the Club of Rome's Project on the Predicament of Mankind*. New York: Universe Books, 1972.

Sachs, I. *L'Ecodéveloppement*. Paris: Syros, 1993.

United Nations. Stockholm 1972. "Report of the United Nations Conference on the Human Environment." http://www.unep.org/Documents/Default.asp?documentID=97 (Accessed February 2010).

United Nations Environment Programme (UNEP). "Declaration of the United Nations Conference on the Human Environment" (June 1972). http://www.unep.org/Documents.Multilingual/Default.asp?DocumentID=97&ArticleID=1503&l=en (Accessed August 2010).

Wards, B. and R. Dubos. *Only One Earth: The Care and Maintenance of a Small Planet.* Preface by M. Strong. London: A. Deutsh, 1972.

Philippe Boudes
Paris 7 University, LADYSS-CNRS

United Nations Environment Programme

The United Nations Environment Programme (UNEP) is an organization set up by the United Nations (UN) at the 1972 UN Conference on the Human Environment, held in Stockholm, Sweden. Its secretariat is based in Nairobi, Kenya. The UNEP budget is provided by voluntary contributions from member states, and the UNEP employs around 900 people. It is not a funding agency, but rather seeks to promote policies and programs that will be funded by other agencies and provides scientific advice to organizations such as the World Bank and the UN Development Programme (UNDP).

The UNEP Website states it "is the voice for the environment in the UN system," and that its mission is "to provide leadership and encourage partnership in caring for the environment by inspiring, informing, and enabling nations and peoples to improve their quality of life without compromising that of future generations."

Like many other international environmental agencies, the UNEP's mandate and objectives are largely determined by Agenda 21, adopted at the UN Conference on Environment and Development (UNCED, the "Earth Summit") in 1992. In 1999, the UN General Assembly endorsed a proposal to institute an annual ministerial-level global environmental forum, in which participants can review important and emerging policy issues in the field of the environment. The Global Ministerial Environment Forum (GMEF), which meets annually as part of the UNEP Governing Council's regular and special sessions, has greatly enhanced the UNEP's capacity to identify and evolve consensus on current and emerging environmental challenges.

The first GMEF, which met in May 2000 in Malmö, Sweden, issued the Malmö Declaration, bringing the major environmental challenges of the 21st century to the attention of the 55th session of the UN General Assembly. At the Millennium Summit, which marked the commencement of the 55th UN General Assembly, world leaders adopted the Millennium Declaration that includes a set of time-bound objectives, collectively known as the Millennium Development Goals, which include the goal of ensuring environmental sustainability.

Whereas much "environmentalist" literature is preservationist and ignores problems of underdevelopment, and much "development" literature ignores environmental issues, the UNEP has always focused on the essential relationship between environment and development, a key outcome of the 1972 conference.

Various international meetings and documents have emphasized the UNEP's central role in fostering sustainable development, from the 1986 Brundtland Report to the 2002

World Summit on Sustainable Development Plan, the 2005 Bali Strategic Plan emphasizing the UNEP's role in capacity building and technology support programs for sustainable development, and the 2005 World Summit, which sought to encourage "institutional frameworks for international environmental governance"—especially in the key area of climate change, which is its particular current focus.

In its literature, the UNEP emphasizes its role in raising global awareness about the importance of environment for development. It provides advice, primarily to governments, on environmental policy and law, as well as capacity building and technology support. Specific programs include the Great Apes Survival Project and the International Coral Reef Action Network. It promotes events such as World Environment Day, a program on sport and sustainability (including the Olympics), and numerous youth programs.

Effectiveness

The UNEP's most important achievement is generally held to be the 1987 Montreal Protocol to protect the ozone layer. In addition, along with the World Meteorological Organization, it established the Intergovernmental Panel on Climate Change (1988). It administers the Basel Convention on the Control of Transboundary Movements of Hazardous Wastes and Their Disposal (1992), the main goal of which is to discourage industrialized countries from disposing of their hazardous wastes in developing countries. The UNEP also hosts the Convention on International Trade in Endangered Species of Wild Fauna and Flora in Geneva, the Convention on Biological Diversity in Montreal, the Convention on Migratory Species in Bonn, and the Convention on Persistent Organic Pollutants in Geneva.

On the positive side, the UNEP has fulfilled its mission as an advocacy organization by helping to create an environmental constituency within some governments, and by getting environmental issues like desertification, ozone depletion, hazardous wastes, and toxic chemicals on the international agenda. The *Global Environmental Outlook* (GEO), published by the UNEP, is respected as a source of information about emerging environmental issues and for placing national issues in broader perspective and for developing a methodology since adopted by regional and national governments.

However, the UNEP has also been criticized as a "weak and ineffectual" organization, in the words of Maria Ivanova of the Yale Center for Environmental Law and Policy. Evaluating the performance of the UNEP poses a challenge because, unlike many other international agencies, it has no specific and quantifiable goals, such as reduction in cases of infectious diseases, or improvement in the status of rare species. The UNEP's power and influence is also limited by its status as a program rather than an agency (it is a subsidiary of the UN General Assembly rather than an autonomous organization), and it has a relatively small budget ($215 million in 2005), compared to $3.2 billion for the UN Development Programme. In addition, the UNEP has no guaranteed source of financing, but depends on voluntary contributions from member nations (around 12 have been regular contributors).

The UNEP has been criticized for poor internal organization that results in frequent duplication of efforts and inconsistent quality control in published information, damaging the organization's credibility. Poor organization and large amounts of missing data in the UNEP's databases and publications limit the usefulness of information disseminated as well. Inefficiency and duplication of effort have also hampered the UNEP's ability to become a "go-to" source for environmental information, as the World Health Organization has for global health information.

The UNEP has also been criticized in the area of setting goals and priorities. The UNEP's ability to set ambitious goals is limited because the agency has no publicly stated long-term vision or goals nor any plan for achieving them. The fact that the UNEP is dependent on voluntary contributions for its financing means that contributing states have been able to manipulate the agenda according to their own priorities, while the lack of clear long-term goals has made it difficult to attract new funding. Its lack of formal authority and its remote location (in Nairobi, Kenya) have made it difficult for the UNEP to coordinate among or exert influence on other international environmental organizations.

See Also: Brundtland Report; Earth Summit; United Nations Millennium Development Goals.

Further Readings

Anderson, S. E. and K. G. Rosendal. "The Potential to Increase the Effectiveness of the UN in Global Environmental Governance." Paper presented at the annual meeting of the International Studies Association 48th Annual Convention, Chicago, Illinois, February 28, 2007. http://www.allacademic.com/meta/p179097_index.html (Accessed January 2010).

Ivanova, Maria. "Assessing UNEP as Anchor Institution for the Global Environment—Lessons for the UNEO Debate." Yale Center for Environmental Law and Policy Working Papers Series 05/01. http://www.yale.edu/gegdialogue/uneo-wp.pdf (Accessed January 2010).

United Nations. "New UNEP Guidebook Launched on Producing Effective Environmental Campaigns." http://www.un.org/News/Press/docs/2005/unep310.doc.htm (Accessed January 2010).

World Commission for Environment and Development. *Our Common Future*. New York: Oxford University Press, 1987.

Alastair S. Gunn
University of Waikato

United Nations Millennium Development Goals

The eight Millennium Development Goals (MDGs) are global policy targets for 2015, adopted by acclamation at the 2000 UN Millennium Summit by world leaders from 191 countries in response to urging by the United States, the European Union, and Japan. The overarching goal was to "end extreme poverty worldwide" by 2015, which was defined entirely in economic terms. Despite all good intentions and moderate progress, by the 2005 "Millennium+5" summit it became evident that the targets would not be achieved as planned, and the 2009 report of the MDG Gap Task Force confirmed this. The shortfall is probably caused by inadequate models and indicators, overly vague definitions, and a lack of international commitment—and not, as the United Nations claims, by the current global economic crisis.

The eight Millennium Development Goals are as follows:

1. Eradicate extreme poverty and hunger
2. Achieve universal primary education
3. Promote gender equality and empower women

4. Reduce child mortality
5. Improve maternal health
6. Combat human immunodeficiency virus and acquired immune deficiency syndrome (HIV/AIDS), malaria, and other diseases
7. Ensure environmental sustainability
8. Develop a Global Partnership for Development

According to the program, chronic hunger in developing countries is to be addressed through economic development and through technological advances. The first MDG calls for a 50 percent reduction of the proportion of people living on less than $1 a day and of people who suffer from hunger. These targets seem now quite elusive, especially as they do not include targets for population size. At the turn of the millennium, the global human population already consumed in excess of 40 percent of the Earth's net photosynthetic production and about 15 percent more renewable resources than can be sustainably delivered with existing technologies. Further increases in food production would exacerbate ecological overshoot, stressing and destabilizing already overtaxed ecosystems, which renders the goal unrealistic in theory as well.

Goals 2 and 3 recognize the importance of education in empowering global citizens to make decisions toward their own security and peaceful coexistence. All girls and boys are to be ensured a full course of primary schooling, and gender equity is to be promoted. Gender disparity at all levels is to be eliminated by 2015, and in primary and secondary education by 2005, which has not yet eventuated. The signatories recognized the fact that the empowerment of women is an essential requirement for the reduction of poverty and for the achievement of all the other MDGs. However, it seems unlikely that education reforms by themselves will result in sufficient female empowerment within the given time frame without appropriate reforms in legislation, monitoring, and enforcement. Nevertheless, some progress in the political representation of women has been made, even though the other targets remain elusive.

Goals 4, 5, and 6 are directed at population health. The mortality among children under 5 is to be reduced by two-thirds, which still appears achievable everywhere except in sub-Saharan Africa. The maternal mortality ratio is to be reduced by three-quarters and universal access to reproductive health is to be achieved; both targets now seem quite out of reach. The spread of HIV/AIDS is to be halted and a reduction initiated; while the pandemic has undeniably peaked, it may have done so for intrinsic biological reasons. Malaria and tuberculosis have been reduced as major causes of infant mortality, but to lesser extents than was hoped. These goals, should they be achieved, would stimulate population growth significantly in the short term. In light of the severe constraints on food production and other essential life support, it seems unlikely that life expectancies would increase to the envisioned extent. Rather, a shift in the causes of mortality is to be expected.

Environmental sustainability is to be achieved by "integrating the principles of sustainable development into national policies" and "reversing the loss of environmental resources" by reducing by half the proportion of people without sustainable access to safe drinking water, and by achieving significant improvement in the lives of at least 100 million slum dwellers by 2020. Environmental sustainability itself is actually left undefined and unaddressed, while the plan merely calls for more effective extraction. The origins of resources and safe drinking water, namely ecosystems, are not mentioned, nor has much progress been made in measures by which they could be strengthened, such as reforestation. Surprisingly, the present shortfall only amounts to about a third below target.

"Global partnership" is to be achieved through a more-open trading and financial system that addresses the special needs of individual countries, making the debts of developing countries sustainable, developing decent and productive work, making essential drugs affordable, and the benefits of new technologies available. Aid from developed countries has fallen precipitously, rendering those goals, poorly defined as they are, unattainable.

The MDG Gap Task Force has proposed no conclusive explanations for the shortfall, other than the global economic crisis. However, some intrinsic shortcomings of the program are likely to have contributed. First, some MDGs, such as global partnership, are not defined in terms of quantitative end states; others are defined in excessively vague terms. This makes it difficult to assess progress and to determine when the goal might be achieved. Second, the MDGs are based on some highly questionable assumptions: that sovereign countries can agree on how to preserve the global commons and emancipate women; and that population growth will halt without the help of specific population targets before humanity encounters the natural mechanisms that normally regulate populations (malnutrition, aggression, and epidemics), which the projected 10 billion by 2050 are likely to face.

However, the most influential reason for the shortfall may lie in a fundamental flaw in modeling. Hunger and poverty are generally caused by shortages of food and of economic capital. Both shortages occur when ecosystems lose their productivity in the course of the global environmental crisis, which manifests in ever-faster resource depletion; in the effects of pollution on climate, habitat quality, and public health; and in the spiraling extinction of species as ecosystems are lost. Five self-reinforcing causes have been identified for the crisis: economic growth, population growth, technological expansion, arms races, and growing income inequality. It seems more than coincidental that the two main remedies prescribed by the MDG program—economic development and technological expansion—are also two of the causes for the global environmental deterioration, which in turn creates hunger and poverty. The MDG gap should have come as no surprise to anyone.

See Also: Development, Ethical Sustainability and; Ecological Footprint; United Nations Environment Programme.

Further Readings

Amin, Samir. "The Millennium Development Goals: A Critique From the South." *Monthly Review* (March 2006). http://www.monthlyreview.org/0306amin.php (Accessed February 2010).

Lautensach, A. "Human Security: A Comprehensive Perspective." In *Encyclopedia of Earth*, Cutler J. Cleveland, ed. Washington, DC: National Council for Science and the Environment, 2007. http://www.eoearth.org/article/Human_security:_a_comprehensive_perspective (Accessed February 2010).

Meadows, Donella, Jørgen Randers, and Dennis Meadows. *Limits to Growth: The 30-Year Update*. White River Junction, VT: Chelsea Green Publishing, 2004.

Sachs, Jeffrey D. and J. W. McArthur. "The Millennium Project: A Plan for Meeting the Millennium Development Goals." *The Lancet*, 365 (2005).

United Nations. "End Poverty 2015: Millennium Development Goals" (2009). http://www.un.org/millenniumgoals (Accessed February 2010).
Vitousek, P. M., H. A. Mooney, J. Lubchenco, and J. M. Melillo. "Human Domination of Earth's Ecosystems." Science, 277 (1997).

Alexander K. Lautensach
University of Northern British Columbia

United Tasmania Group

The contested Lake Pedder as seen from Mount Eliza in the Tasmanian Wilderness World Heritage Area in 2009. The United Tasmania Group was unable to prevent the construction of a dam on this lake but did inspire many other groups and political parties.

Source: Wikipedia

The United Tasmania Group (UTG) was the world's first green political party and its formation marked a major turning point for the global environmental movement. The UTG was created in 1972 from protest groups opposed to the construction of a dam in the southwest of the Australian state of Tasmania. Although the group was relatively short-lived, its legacy is twofold. First, its manifesto identified principles that have recurred in later international initiatives. Second, it catalyzed the creation of new environmental groups and paved the way for the success of later green political parties around the globe.

The history of the UTG began with the 1967 announcement by Tasmania's Hydro-Electric Commission of its intention to dam an area that included Lake Pedder. The region had great natural beauty and a distinctive environment. Several community groups opposed the dam, including the Lake Pedder Action Committee. A meeting of these groups on March 23, 1972, created the UTG to contest the state election. It received 3.9 percent of the vote, which was not enough to win a seat in parliament despite the more accessible proportional electoral system. The best performance was in the seat of Franklin where it received 8 percent (one of the group's candidates, Bob Brown, won 2,077 of their 3,078 votes). The UTG ran again in the 1976 state election and won 2.2 percent of the vote. The group officially disbanded in 1979, but reformed briefly for the 1990 federal election when they ran four candidates for the senate without success.

The philosophy and ethics of the UTG were summarized in their 1972 manifesto *The New Ethic*. Human dignity, cultural heritage, natural ecosystems, and the unique natural beauty of Tasmania were all valued. Divisive ideologies, the misuse of power, and inappropriate displays of wealth were condemned. Members undertook to live in a way that preserved Tasmania's environment for future generations, created "aesthetic harmony" between human settlements and nature, reduced resource use, and minimized environmental damage. The UTG also committed itself to creating institutions that promoted more open and participatory decision making, justice and equal opportunity, cooperation instead of alienation or competition, communal life, and fairer parliamentary and legal systems.

The principles of *The New Ethic* have reappeared persistently in subsequent international initiatives. Intergenerational equity, for example, became central to the idea of sustainable development in the 1987 World Commission on Environment and Development report and the United Nations 1992 Agenda 21. Reducing resource use and environmental damage became common goals of environmental management systems in the 1990s. Similar ethics and principles were enshrined in the Earth Charter, which was launched in 2000 by a global network of community organizations.

While the UTG did not achieve its political ambition of gaining seats in parliament or preventing the Lake Pedder dam, it did start a chain of events that led to the rise of subsequent environmental groups and political parties. In 1976, some members of the UTG helped to form the Tasmanian Wilderness Society (later renamed The Wilderness Society [TWS]) and Bob Brown became its director in 1978. TWS was one of the key organizations that successfully campaigned to prevent the Franklin dam in 1983 and it remains a significant force in Australian environmental politics. Brown entered state parliament in 1983 and, together with four other green independents, forged the Green-Labor Accord in 1989 that supported a minority Labor government in return for major environmental initiatives. The Tasmanian and Australian Greens were formed in 1992, offering an alternative to the major political parties (Labor and Liberal). Brown won a senate seat in 1996, and by 2007 there were five Green senators.

Influence of the United Tasmania Group

Internationally, the UTG's lead was soon followed by the formation of the Values Party in New Zealand (May 1972) and the PEOPLE party in the UK (1973), both of which developed into green parties. In 1979, a group to coordinate green and radical parties in Europe was established that eventually led to the formation of the European Greens. Die Grünen (the Greens) formed in West Germany in 1980, merged with the East German group Bündnis 90 after reunification, and formed a national coalition government with the Social Democratic Party between 1998 and 2005. In the United States, the Green Party first contested elections in 1985 and had some success in local councils and state legislatures. In 2001, the Global Greens was formed by representatives of green parties from 72 countries. Their charter and the principles of green parties around the world harken back to ideas expressed in the UTG's manifesto.

The legacy of the UTG has been substantial. Its manifesto espoused philosophical and ethical principles that have recurred in subsequent international environmental initiatives. It blazed the trail for The Wilderness Society and the Australian Greens that forced significant changes to state and national environmental policies. Finally, it led the way for green parties around the world.

See Also: Bright Green Environmentalism; Earth Charter; Ecopolitics; Global Greens Charter; Green Party, German.

Further Readings

Archives Office of Tasmania. http://www.archives.tas.gov.au (Accessed January 2010).
Earth Charter. http://www.earthcharter.org.au (Accessed January 2010).

Hutton, D. and L. Connors. *A History of the Australian Environmental Movement.* Cambridge, UK: Cambridge University Press, 1999.
Pybus, C. and R. Flanagan. *The Rest of the World Is Watching: Tasmania and the Greens.* Sydney: Pan Macmillan, 1990.
South East Queensland Regional Plan 2009–31. http://www.dip.qld.gov.au/regional-planning/regional-plan-s.html (Accessed January 2010).
Tasmanian Greens. http://www.tas.greens.org.au (Accessed January 2010).
Tasmanian Parliamentary Library. http://www.parliament.tas.gov.au/tpl/tplmain.htm (Accessed January 2010).
United Nations. Agenda 21. Division for Sustainable Development, 1992. http://www.un.org/esa/dsd/agenda21/index.shtml (Accessed January 2010).
The Wilderness Society. http://www.wilderness.org.au (Accessed January 2010).
World Commission on Environment and Development. *Our Common Future.* Oxford, UK: Oxford University Press, 1987.

Michael Howes
Griffith University

Urbanization

This article explores the alternative meanings of the concept of urbanization, describes the processes that facilitate it and the different forms it takes, the problems that are associated with it, and concludes with a brief overview of recent philosophies of urbanization.

Urbanization is a process of population concentration by which large numbers of people live in relatively small areas called cities, towns, or urban areas. It can most generally be defined as the increase in the urban area's density, extent, or population through migration and/or natural growth. Defining a city is not as simple as it looks. Commonly used quantitative thresholds are rarely consistent from one place to another. Alternatively or sometimes additionally, functional, administrative, or political criteria are used as well. As with the question of what makes a city, defining what the urbanization process entails is no easy matter. Definitions of urbanization range from the demographic to the structural to the behavioral.

As a demographic process, urbanization may refer to the absolute growth of populations living in urban areas as well as the growth of urban population relative to the whole due to the redistribution of population between rural and urban areas, or it may even refer to the rate at which the proportion of urban populations is increasing over time. Concentration of population in the larger urban settlements of a given territory or increasing density of population within urban settlements may also be termed urbanization.

There are arguments, however, that residing in an urban area is not sufficient for being "urban." It is not possible to separate the urbanization discussion from that of economic shifts and technological and societal developments. "Structural urbanization" refers to the shift from a nomadic, hunter-gatherer, or agrarian society to an urban one. This involves a shift in employment and brings an economic transition from primary production to industrial, commercial, and service economy. With this shift, household self-sufficiency is lost and specialization in production begins. These changes follow from industrialization

A dense slum in Rio de Janeiro, Brazil. Essential infrastructure and services such as housing, clean water, waste removal, and law enforcement have not been able to keep up with the high rates of growth in urbanization in less-developed countries.

Source: iStockphoto.com

and capitalization processes. Urbanization and structural economic change have a complex relationship.

"Behavioral urbanization" refers to the sociocultural effects of an increasing part of the population living in cities. This is intermingled with the related concept of urbanism, which is sometimes used interchangeably with urbanization and refers to a way of life associated with living in an urban area. Urban areas are centers of social change: values, attitudes, and behavioral patterns are modified in urban life and new forms are spread through the urban system by diffusion processes. Ideas such as democracy are products of urban life. On the other hand, more frequent and diverse but less personal human interactions mark urban life, diluting urban dwellers' ties with the natural environment. Viewed in this light, some see urbanization as a social pathology of modern society that results in alienation. The argument is that the three criteria for distinguishing urban places—their size, density, and heterogeneity—result in social disorganization.

Studies of urbanization focus on the structure and form of urban growth including trends of suburbanization, centralization and decentralization, or emergence of edge cities; distribution of urban areas; their relative size and hierarchy; and their relationship with each other.

While the origin of cities remains uncertain, urbanization has been accelerating, first as a result of developments in agriculture and transportation and then by industrialization. Increase in agricultural productivity created agricultural surplus and freed large numbers of people from rural lands. Evolving technology and transportation systems made carrying the surplus and other necessities to urban areas possible. These early cities were limited in size by availability of water and transportation technology. Development of these technologies enabled urbanization to continue. The Industrial Revolution was another important driver of urbanization, both with its insatiable need for workers and with developments it brought technologically.

Current trends in worldwide urbanization indicate that scale and rate of urbanization are escalating. More and more people are living in cities. Today half the world's population is urbanized. Cities are increasing in number as well as in size. The very high rate of urbanization in some areas of the world and the sheer size of some cities are unprecedented. The world's fastest-growing and largest cities are in developing countries. With these extraordinary sizes, rates, and densities, a number of problems appear.

"Overurbanization" is the case where urban population growth is significantly in excess of the growth of jobs and housing in the area. This is closely related to the condition of "pseudo-urbanization" in which a large city has formed in an area without a functional infrastructure to support it. This is the dominant form of urbanization in the third world. The rate of urbanization is generally inversely correlated with the level of economic development and the rates of growth observed in less-developed countries result in a lag in the provision of essential infrastructure and services such as housing, education, transportation, clean water, and waste removal, law enforcement, and so on. A widespread belief argues that the optimum city size over which costs of growth are higher than the benefits has been exceeded in many cases. The urban entities that transcend the traditional concept of a city are labeled as "metropolis," "megalopolis," "conurbation," "urban region," or "megacity."

Along with multiple definitions of urbanization, there are diverging and sometimes contrasting trends in worldwide urbanization. "Counterurbanization" is the trend of decrease or slower levels of increase of population of the central metropolitan areas relative to the small and medium-sized urban areas and the countryside. This trend that started in the last quarter of the 20th century is yet again a reflection of structural changes in the economy that go hand in hand with technological developments. This phenomenon is observed more commonly in developed countries.

For a long time, the urbanization process took place without guidance or planning, although some interventions such as aqueducts to carry water supplies were provided to facilitate it. With the problems observed in cities during the Industrial Revolution, urban planning efforts gained pace to protect people from hazards caused by urbanization and to manage urban growth in an orderly fashion. The current scale and rates of urbanization and the size of the cities offer new challenges on this front and require more aggressive planning efforts. In response, novel philosophies of urbanization such as "new urbanism/neotraditional development/transit-oriented development," "growth management/smart growth," "sustainable urbanization/green urbanism" appeared. Some of these ideas target parts of cities, whereas others bring principles to be adhered to throughout the whole urban area. The purpose is to provide sustainable human settlements in an urbanizing world by balancing environment and economy. This involves providing adequate shelter for all as well as infrastructure and services without depleting or destroying the world's limited resources. Strategies have to be comprehensive to include provision of food and other supplies to an increasing population and sustainable land management, water use, waste management, energy, transportation, and housing. Strategies also have to be fair and just, decided democratically, and implemented locally with stakeholder participation.

See Also: Democracy; Mumford, Lewis; Technology; United Nations Conference on the Human Environment.

Further Readings

Allen, Adriana and Nicholas You. *Sustainable Urbanisation: Bridging the Green and Brown Agendas*. University College, London: Development Planning Unit, 2002.

Beatley, Timothy. *Green Urbanism: Learning From European Cities*. Washington, DC: Island Press, 2000.

Mumford, Lewis. *The City in History: Its Origins, Its Transformations, and Its Prospects*. New York: MFJ Books, 1961.

Aysin Dedekorkut
Griffith University

Urban Tree Management Ordinances

A soil conservation expert from the U.S. Department of Agriculture, Natural Resources Conservation Service (left), discusses urban tree management with a member of a local urban greening project on a Los Angeles, California, street.

Source: U.S. Department of Agriculture Natural Resources Conservation Service/Bob Nichols

Urban planning has an ethical duty to consider trees, bushes, and other vegetation within its broad policy framework. The urban forest reduces air and noise pollution, changes air temperature and wind speed, reduces energy consumption, slows rainfall runoff, offers habitat for wildlife, increases urban biodiversity, improves urban aesthetics and community image, and has manifest human and environmental health benefits among other positive impacts on city welfare. For that reason, trees need to be maintained (e.g., pruned, fertilized, and watered), to be preserved, and to be protected, for example, from construction works, both public and private.

Urban tree management includes, in numerous cities, an action plan (e.g., a tree plan, or an urban forest program) and norms or administrative rules (e.g., a tree ordinance, a section in a zoning regulation, or a section in the municipal development code), whose complexity varies, in both cases, according to city size. They include, among other aspects, the definition of rules and standards about the planting, maintenance, protection, and removal of trees, both in public spaces and on private properties near streets, sidewalks, public parks, and other public spaces, as well as standards to prevent the removal of native plants and the destruction of soil before (re) development takes place.

A municipal tree plan is a local government document, often part of a wider urban environmental plan, whose main goal is to define a long-range view for the preservation and improvements of the urban forest. It has, in general, the following objectives: to achieve a certain amount and proportion of tree cover in the total urban area; to improve the health of the city's trees; to achieve and to maintain a certain level of urban biodiversity; to allow efficient municipal management of the urban forest; to reduce air and noise pollution within the urban area; to control rainfall runoff, air temperature, and wind speed in certain parts of the urban area, contributing indirectly to the reduction of energy consumption in both public and private buildings; and to promote citizen participation in municipal environmental policy.

A tree plan starts with a basic inventory of resources (e.g., tree species, number, and location of trees), problems affecting the trees in the city, and problems that may be solved or reduced by tree planting (e.g., noise and air pollution or excessive wind speed), as well as an assessment of how the urban forest has been managed in the past. Based on this inventory, the plan identifies what needs to be done. Among other aspects, the plan addresses biological needs, such as the need to increase urban biodiversity (e.g., number of tree species), and the diversity of tree age in order to ensure long-term sustainability of the urban forest, as well as other urban environment needs (e.g., to reduce air and noise pollution). It deals also with the need to provide trees for new and for redeveloped urban areas (e.g., tree bank), as well as the need to guarantee compatibility between species, size, and number of trees in each urban sector in the city. Management needs, namely those related to staff, financial resources, and interdepartmental coordination, as well as the conditions for effective citizen participation should also be addressed by the tree plan.

The tree plan includes technical specifications for planting and maintenance of trees in public urban spaces, a list of species allowed to be planted, the classification of exceptional trees, and also the creation of protected areas within large urban parks. It can also include environmental education programs and the establishment of botanical gardens. In addition, the tree plan can enforce land use planning solutions that take into consideration the need to use trees in order to reduce air and noise pollution, can advise private developers on the sort of tree species that they shall plant in new or in redeveloped urban areas, can set standards for tree maintenance, and can impose the obligation to replace trees previously removed.

A tree ordinance, or an equivalent urban management tool, focuses on the following aspects: the objectives of the tree ordinance and its institutional and geographical context; technical aspects covered and regulated by the ordinance (e.g., standards for tree care in urban areas, prevention of damage to trees in public spaces, etc.); the institutional framework and administrative procedures; and different types of penalties for unlawful actions against trees in public spaces (e.g., fines, forced replacement of trees removed without consent). It may include norms that regulate how to protect trees in the urban space (e.g., with a fence or any other appropriate device) against potential negative impacts from excavations, tunnels, drainage works, buildings and urban infrastructures built in its vicinity, and from the waste produced by such works, or from objects that may affect water drainage or sunlight, with the concomitant negative effects on the trees' overall development. In addition, a tree ordinance includes norms intended to ensure safety for citizens and to safeguard public health (e.g., how to treat disease and to disinfest trees), and norms to be considered in land use planning and management, namely, norms aimed to enhance air quality and the attainment of acceptable levels of noise in residential areas within the city and to preserve the aesthetics of public urban spaces.

Besides these actions, municipalities can enhance the urban forest through other means, such as individual programs of tree planting in urban parks, in school yards, and in other public urban spaces. It may also apply taxes to encourage or to avoid certain tree species being planted or removed, and incentive programs for planting new trees or certain species, namely, native species.

In brief, the long-term management of trees in urban areas requires a comprehensive vision of the urban forest, protecting, for example, both old and young trees, defining rules for the removal of trees as well as for the planting of new trees in new, consolidated, or redeveloped brownfield urban areas, in floodplains and along stream banks, to address the urban forest as a whole rather than to approach trees on an individual basis, to record and to analyze existing trees prior to site cleaning, to transplant trees whenever possible, to recommend certain species and to forbid others, to control land alteration in order to preserve natural habitats (e.g., native plants, wetlands). In some cases, the possibility of extending municipal control over trees on privately owned land in urban areas should be considered. The management of trees in urban areas should also make one of its main goals the improvement of urban air quality, the attainment of noise levels in urban areas that are acceptable for human beings, in particular in residential areas, the reduction of wind speed, and indirectly, the reduction of energy consumption. This formal regulatory framework, be it a tree ordinance or any equivalent urban management tool, standing alone or integrated into a comprehensive urban planning instrument, needs to be flexible and articulated with other components of the urban planning strategic framework, and, above all, needs to have the support of local citizens and other urban stakeholders—a critical factor for success in this policy field.

See Also: Environmental Law; Environmental Policy; Forest Preservation Laws; Green Laws and Incentives; Preservation; Urbanization.

Further Readings

Johnston, G. R. *Protecting Trees From Construction Damage: A Homeowner's Guide.* Minneapolis: University of Minnesota Press, 2010.

Simpson, J. R. and E. G. McPherson. "Simulation of Tree Shade Impacts on Residential Energy Use for Space Conditioning in Sacramento." *Atmospheric Environment,* 32/1 (1998).

Starbuck, C. J. *Preventing Construction Damage to Trees.* Columbia: University of Missouri Press, 1994.

Swiecki, T. J. and E. A. Bernhardt. *Guidelines for Developing and Evaluating Tree Ordinances.* Sacramento: Urban Forestry Program, California Department of Forestry and Fire Protection, 2001.

Carlos Nunes Silva
University of Lisbon

Utilitarianism

Utilitarianism is an ethical doctrine that holds that right conduct consists in doing whatever will maximize the well-being of everyone affected by an action. It is often summed up by British philosopher Jeremy Bentham's slogan "the greatest good for the greatest number." While relatively few philosophers today defend a pure version of utilitarianism, it has had substantial impact on the practice of conventional environmental management and is an important reference point for other ethical theories.

There is controversy among utilitarians about precisely what constitutes "well-being"—often termed "utility"—and how it is to be maximized. Utilitarians agree that what constitutes an individual's well-being should be defined by the individual him- or herself, not by any external or objective standard of the proper way to live. The earliest explicit formulation of the modern version of utilitarianism by Bentham defined subjective well-being as pleasure or happiness, and the absence of pain. Bentham's follower John Stuart Mill disputed whether all happiness counts equally or whether some forms of happiness are "higher" and worth more. More contemporary utilitarians such as R. M. Hare and Peter

Singer have defined subjective well-being as preference satisfaction, arguing that people often want things other than the psychological state of happiness and that utilitarianism should maximize what people actually want for themselves. Philosophers also disagree about whether what matters is the actual consequences of a course of action or the consequences that the actor would have reasonably predicted at the time of choosing to act.

Utilitarian views on how to maximize well-being are conventionally divided into "act utilitarian" and "rule utilitarian" camps. Act utilitarianism holds that each act ought to be judged according to whether the actor chose the available option that would lead to the greatest well-being. This does not mean that each act must be preceded by a conscious calculation of the consequences of the various options but rather that such calculation is the standard that would determine if an action was right. Rule utilitarians, by contrast, hold that the criterion of maximizing well-being should be applied to rules for conduct, which should then be followed in practice without consideration of the effect on well-being of each application. Thus, a rule utilitarian might judge that "never tell a lie" is the rule that would produce the most subjective well-being out of all the possible rules for when to lie, and would then refuse to lie under any circumstances, even if one particular lie would, in fact, maximize well-being. An act utilitarian, on the other hand, would decide for each individual lie whether telling it would maximize subjective well-being.

Utilitarianism is opposed to a number of components of other common ethical theories. Utilitarians recognize no foundational role for rights (though utilitarians may grant them instrumental usefulness), because it may be necessary to violate someone's alleged rights for the greater good. Utilitarians do not give credence to the act-omission distinction—which says it is worse to do something bad than to fail to stop an equivalent bad thing—because the rightness or wrongness of an action is simply a function of the well-being it produces. And utilitarians recognize no direct ethical value for any entity such as a nation or an ecosystem that is not a well-being-experiencing individual.

Environmental Applications

Utilitarianism famously entered the environmental domain through Gifford Pinchot, the first chief of the U.S. Forest Service. Pinchot saw utilitarianism as providing a justification for natural resource conservation and expanded Bentham's slogan to "the greatest good for the greatest number for the longest time" to emphasize the implications of utilitarianism for future sustainability. Pinchot believed that unregulated exploitation of natural resources was benefiting contemporary business at the expense of the overall long-term well-being of society. However, the well-being or rights of nature itself did not factor into his thinking.

Utilitarianism is one of the foundations of welfare economics and cost-benefit analysis. These approaches are commonly used to evaluate projects and policies that affect the environment by adding up the benefits and costs of each option, then selecting the one with the greatest balance of benefits over costs. There is some question about whether the practice of measuring costs and benefits in terms of monetary prices accurately reflects well-being in the sense intended by utilitarian philosophers, but both defenders and critics typically see economic cost-benefit analysis as rooted in utilitarianism.

Utilitarianism has also had a significant influence on the field of animal rights. Peter Singer argues that if what is right is maximizing well-being, then it is an irrational prejudice—"speciesism"—to be concerned with only human well-being. Instead, we should be concerned for the well-being of all creatures that are able to consciously experience well-being.

This move expands utilitarianism's concern and potentially its ability to justify "green" policies that may improve the well-being of wild and domestic animals even if they do not so benefit humans.

Criticisms

Utilitarianism has come under a variety of critiques by philosophers both in and out of environmental philosophy. One of the more common critiques is that utilitarianism fails to respect the distinctness of persons. Instead, critics say, utilitarians view persons as mere containers for happiness, which could as easily be held in another person. Related to this is the concern that utilitarianism sanctions sacrificing individuals for the greater good. For example, critics point out that there is a plausible utilitarian case that it would be right for a surgeon to kill an innocent person off the street to obtain organs that, when transplanted, would save the lives of five people.

Other critics of utilitarianism focus on whether something that is good for people is really the same as what they want or feel happy about. These critics may hold that some preferences should not count because they are bad for you no matter how much you want them (e.g., a smoker's preference for having a cigarette), or because they are the result of indoctrination or false consciousness (e.g., a woman in an intensely patriarchal society who has been socialized to be happy with her subordinate place).

Green critiques of utilitarianism also focus on the limits it sets on what entities count as ethically considerable. Rocks, trees, species, and ecosystems are not (at least, according to Western science) capable of experiencing happiness, preferences, or any other ground for self-defined well-being. According to utilitarians, it is therefore incoherent to claim an ethical duty toward those entities since there is no way to say what is good for them. They can be preserved only insofar as their preservation benefits humans and animals—and they may be destroyed if doing so would produce greater benefit in terms of human and animal well-being. This position is anathema to biocentrists, ecocentrists, and other deep green philosophers, who see utilitarianism as thereby unable to respond to the ecological crisis.

See Also: Animal Ethics; Cost-Benefit Analysis; Instrumental Value; Pinchot, Gifford; Singer, Peter.

Further Readings

Bentham, Jeremy. *Introduction to the Principles of Morals and Legislation.* Oxford, UK: Clarendon Press, 1996.

Mill, John Stuart. *Utilitarianism.* Indianapolis, IN: Hackett, 2001.

Singer, Peter and Jim Mason. *The Way We Eat: Why Our Food Choices Matter.* New York: Random House, 2006.

Stentor Danielson
Slippery Rock University

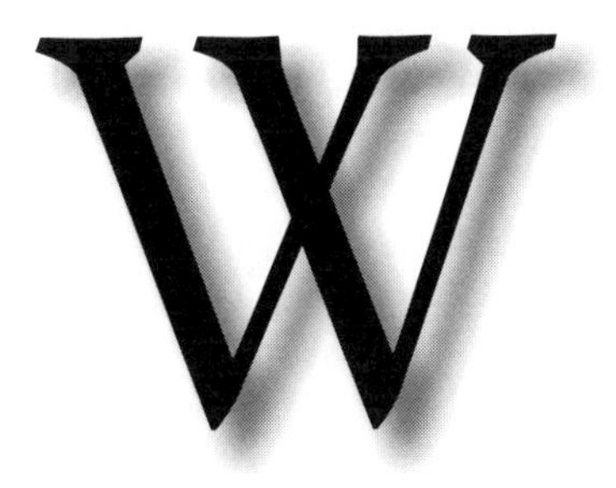

WARREN, KAREN

Karen J. Warren was a professor of philosophy at Macalester College in St. Paul, Minnesota, until her retirement in 2010. She is best known for her contributions to feminist philosophy and environmental ethics, especially ecofeminism. She has taught philosophy in a variety of organizations outside the university, notably in schools.

Warren's teaching interests include critical thinking, ethics, social and political philosophy, peace studies, the history of women philosophers, and philosophy for children. Warren has described herself as a "street philosopher" and a "public philosopher" because of her commitment to bringing philosophy to diverse populations, especially groups to whom philosophy is not normally available. She has taught philosophy in schools from K–12 level and initiated one of the first high school philosophy programs in the United States in 1972. She has also taught a course in a correctional facility and has worked with many environmental organizations and civic groups. Thus, as well as being a distinguished academic philosopher, Warren has also spent much of her career promoting philosophical ideas in the larger community, with a view to making a difference in people's lives.

Karen Warren's Philosophy

In her pioneering and most influential work, "The Power and the Promise of Ecological Feminism," Warren noted that what she described as the four leading versions of feminism—"liberal feminism, traditional Marxist feminism, radical feminism, and socialist feminism"—were all inadequate because they failed to incorporate environmental concerns. She therefore set out to develop a feminist philosophy that would include all forms of oppression, including oppression of natural systems.

The central idea in Warren's environmental philosophy is that our society, and indeed most societies, are characterized by oppression of one group by another, and that all forms of oppression are both conceptually and historically linked. The historical link, she believes, is clear enough: societies such as the United States have always been racist, sexist, and environmentally exploitative. Moreover, environmental degradation often worsens the lot of women, especially in developing countries. For example, deforestation causes erosion and loss of soil fertility and also reduces the supply of freshwater. Since obtaining freshwater for the household is traditionally "women's work" in many parts of the world, deforestation has increased women's already heavy workload because they are forced to travel ever-greater distances to locate water. Similarly, the replacement of indigenous forests in parts of India by commercial eucalypt cultivation has meant the loss of traditional forest products, including foods and medicines by which women formerly sustained their households.

The conceptual connection between various forms of oppression arises from the pervasiveness of what she calls oppressive conceptual frameworks. A conceptual framework is a set of beliefs, values, attitudes, and assumptions that determine how one views the world and one's place in it, a "socially constructed lens." Western thought is characterized by what she refers to as "value-hierarchical dualisms" and "the logic of domination." Dualism is the division of phenomena and ideas into pairs, such as animate-inanimate, mind-body, male-female, reason-emotion. Hierarchical thinking is characterized by distinguishing one member of a pair as superior to the other. In itself, she believes, hierarchical dualism is not necessarily wrong in a particular context. Human beings are able to engage in environmental transformation in ways that other animals, plants, and inanimate objects are not, so in terms of their capacities, humans are "better" than other species at what we might today call environmental management. However, this does not provide a basis for claiming that humans are morally superior to other species and are therefore entitled to exploit and oppress the natural world. This assumption of moral superiority on the basis of difference is the basis for the logic of domination that is at the heart of the oppressive conceptual framework that has been at the heart of most, if not all, societies. The role of feminist philosophy—and indeed of the promotion of philosophical ideas in general—is to make explicit and undermine the logic of domination and thus help to end all forms of exploitation.

See Also: Anthropocentrism; Biocentric Egalitarianism; Biocentrism; Ecofeminism/Ecological Feminism; Sustainability and Spiritual Values.

Further Readings

Karen Warren. http://www.macalester.edu/~warren (Accessed July 2010).

Warren, Karen J. *Eco-Feminist Philosophy: A Western Perspective on What It Is and Why It Matters.* St. Louis, MO: San Val, 2000.

Warren, Karen J. "The Power and the Promise of Ecological Feminism." *Environmental Ethics,* 12 (1990).

Alastair S. Gunn
University of Waikato

WESTERN "WAY OF LIFE"

When considering the Western "way of life," one should recognize that a psychological and emotional disconnection with nature has been shaped and nurtured throughout time by the influence of underlying assumptions about our relationship with the natural environment.

These assumptions are deeply embedded within the Western way of life, but may often go unnoticed. Perhaps if we grow in awareness of how we have become disconnected from nature, then we can work with greater focus to overcome the challenges associated with the consequences of this disconnect, which is in some ways responsible for the current condition of rapidly deteriorating health of the natural world. How can this disconnect be explained and understood? How did this disconnect become so profound that humans, instead of living like true members of ecological systems, now live as if they are not a part of the natural world? This article presents a brief historical account of how underlying assumptions about the relationship of humanity and the natural environment have become ensconced in the Western way of life.

We can begin answering the questions posed above by looking into the lives of prehistoric humans. As hunters and gatherers, Paleolithic people lived in harmony with their environment. Evidence suggests that they viewed nature as sacred, accepted it, revered it, and lived in harmony with it. Living as a part of their natural surroundings, they flourished and functioned as part of the living ecosystem. This is not meant to create a romantic, unrealistic picture of a life in Eden but rather to recognize that nature was not apart from the moment-by-moment existence of Paleolithic humanity and, in that state, humans thrived.

Neolithic people ushered in the dawn of agricultural society and introduced new boundaries between "wild" nature and civilization. As crops were cultivated and animals were domesticated, "wild" creatures such as insects and predators became a threat. Home became a place where nature was subdued, overcome, cultivated, and domesticated. There is little evidence to suggest that such a separation from wild nature was the goal of this period or its people; rather, it appears to have been an outcome. This transition from living as hunter-gatherers to agriculturalists marks the beginning of a now long-standing separation between humanity and wild nature.

The idea that civilization meant developing and maintaining boundaries between humans and wild nature was well established by the time of the Egyptian and Sumerian cultures. Evidence of this can be found in a prevailing philosophy that nature was valueless without being humanized. By the time of the Greek philosophers, civilization afforded unprecedented leisure, and the movement to understand the world on the basis of reason was under way. This time period marks the beginning of the next major step in broadening the human–environment disconnect. Perhaps the most influential transition in thinking occurred when Aristotle began to question the notions of reality previously held by Plato.

Plato supposed that human reason dictated reality and that ideas were the highest form of reason. Therefore, human ideas, achieved through the exercise of reason, constituted the source of reality. According to such ideals, the very essence of anything in the natural world lies not in the thing itself or how it functions, but within whatever ideas and knowledge humans ascribe to it through their ability to reason.

In contrast, Aristotle suggested that reality is outside our ideas, that we are a part of nature, and that by observing the natural world, we can understand it. Aristotle maintained that it is not ideas but nature that holds the source of reality. The goal of reason, then, was not to contemplate one's way into understanding but to observe the natural world and thereby learn about reality from nature through careful observation. This was the birth of empiricism and the embryo that ultimately grew into the scientific revolution. One guiding principle that emerged from empirical thinking was that if nature can be understood, then it can be mastered.

The scientific revolution was a maturation of empiricism. It represents the entrenchment of positivistic, reductionist science in which the observer's position had now irreversibly moved from participant in a complex, organismic, natural world to that of outside observer. Hence, this era solidified the Western compartmentalized, mechanized view of nature. While we praise the progress of the scientific revolution and appreciate its many contributions to modern Western life, we should also recognize that it promoted an agenda beyond that of intellectual curiosity or the exercise of human reason through empiricism. The leaders of this movement believed that the woes of the uncivilized world could be overcome through science and they worked to achieve mastery over nature. From physics to the circulatory system, nature became understood as a machine.

Following the scientific revolution, the Industrial Revolution brought humans even further from living in response to principles of ecology. With the Industrial Revolution firmly established, it became evident that Western capitalism, and its view of wild nature as a collection of resources and commodities, had essentially disregarded ecological principles in favor of a growth and development economy. This is further recognized within the work of authors such as Mathis Wackernagel and William Rees, who have noted the impacts of capitalism on environmental health and, in response, have devised a way to quantify and calculate an individual's impact on the natural environment.

The Western way of life is plagued by its own history. The separation between humans and the natural environment has grown throughout time, with notable occurrences around the era of the Greek philosophers and followed by the scientific and industrial revolutions. It is the work of the environmental movement to develop a new way of Western life that overcomes the artificially constructed psychological borders that have been reinforced since Neolithic times. This requires a renewed awareness that we are nature and are therefore deeply connected to natural systems. Although science has been implicated in inadvertently fostering the disconnect, it is now the work of Western science and education to restore rightful understandings of our true position as members of complex ecological systems and to promote Western ways of living that are consistent with that awareness.

See Also: Anthropocentrism; Ecological Footprint; Ecology; Human Values and Sustainability.

Further Readings

Dare, B., G. Welton, and W. Coe. *Concepts of Leisure in Western Thought: A Critical and Historical Analysis,* 2nd ed. Dubuque, IA: Kendall Hunt, 1998.

Nash, R. *Wilderness and the American Mind,* 4th ed. New Haven, CT: Yale University Press, 2001.

Oelschlaeger, M. *The Idea of Wilderness.* Binghamton, NY: Vail-Ballou, 1991.

Oliver, B. "Nature, Capitalism, and the Future of Humankind." *South African Journal of Philosophy,* 24/2 (2005).

Orr, D. W. *Earth in Mind: On Education, Environment, and the Human Prospect.* Washington, DC: Island Press, 2004.

Wackernagel, M. and W. E. Rees. *Our Ecological Footprint: Reducing Human Impact on the Earth.* Gabriola Island, British Columbia, Canada: New Society Publishers, 1996.

Wynn Shooter
Monash University

Wilderness Act of 1964

A 3,360-acre section of Glacier National Park in Montana, shown here in early October 2008, was among the first lands to be recommended for wilderness designation after the Wilderness Act of 1964.

Source: U.S. Fish & Wildlife Service/Ryan Hagerty

The United States Wilderness Act of 1964 created 54 wilderness areas (9.1 million acres) in national forests, and established a National Wilderness Preservation System (NWPS). Philosophers J. Baird Callicott and Michael P. Nelson (1998) claim that the definition of wilderness in this act is the classic expression of the "received wilderness idea," an idea that is fraught with problems and has been under attack for decades.

Wilderness preservation in the United States is often defined as the creation and management of national parks, but this is not correct. The "wilderness" of early U.S. national parks largely consisted of lands the National Park Service had not yet developed into campgrounds, scenic roads, and access areas for motorized travel. National park management was focused to a considerable extent on showcasing and even altering natural landscapes for presentation to the modern, industrial tourist. Early wilderness proponents such as Aldo Leopold and Robert Marshall formed the Wilderness Society in the mid-1930s in large part to champion roadless, undeveloped lands to serve as a foil against the industrial tourist wilderness of national parks.

The U.S. Forest Service (USFS) administratively designated wilderness areas in 1924, 1929, and 1939, in part because of recommendations made by people such as Leopold and Marshall—both of whom worked for the USFS. By the 1950s, environmental groups spearheaded by the Wilderness Society and the Sierra Club began clamoring for more permanent wilderness protection beyond USFS administrative whims. The first wilderness bill was introduced into Congress in 1956. After 65 rewrites, the Wilderness Act of 1964 was passed.

Section 2c of the act famously defines wilderness as "an area where the earth and its community of life are untrammeled by man, where man himself is a visitor who does not remain," as contrasted with areas inhabited and/or dominated by people. Untrammeled is further qualified in a legal sense as "undeveloped Federal land retaining its primeval character and influence, without permanent improvements or human habitation, which is protected and managed so as to preserve its natural conditions." These natural conditions are qualified in a subjective sense in that wilderness "generally appears to have been affected primarily by the forces of nature, with the imprint of man's work substantially unnoticeable." Debate over how "pure" an area had to be of prior human impacts in order to qualify for wilderness designation ensued after 1964, and eventually led to a legislative, judicial, and philosophical interpretation of wilderness in terms of diachronic naturalness that extended into the future as human impacts disappeared across time and naturalness washed back into an area.

Section 2c of the act also stipulates that wilderness "has outstanding opportunities for solitude or a primitive and unconfined type of recreation," is of a sufficient size so as to "make practicable its preservation and use in an unimpaired condition," and may be valuable for education, history, scenery, and science. This encapsulates the idea that wilderness is valuable for recreation and aesthetics, two historical sources of the received wilderness idea.

Two rationales support legal wilderness preservation: (1) concerned that expanding settlement, growing mechanization, and increasing population can "modify all areas within the United States and its possessions, leaving no lands designated for preservation and protection in their natural conditions," Section 2a expresses a rationale that wilderness be preserved in its natural condition; and (2) preserving wilderness for solitude and nonmotorized recreation expresses a rationale that wilderness be preserved for human use and enjoyment. Because "wilderness" solitude and recreation are possible only in an actual wilderness setting, the naturalness preservation rationale seems to have lexical priority over the use and enjoyment rationale. These two rationales correspond to defining wilderness in Section 2c as natural areas and as areas for solitude and recreation.

In spite of a requirement for naturalness, the act permits a number of human activities that run counter to wilderness preservation. Section 4d permits the continued use of aircraft and motorboats, when such use had been well established, grazing, mineral leasing, mining, water developments, and access road construction for these activities. These exceptions to full preservation express a compromise rationale that resulted from political negotiations between wilderness proponents and various commercial interests opposed to wilderness preservation during the legislative history of wilderness bills between 1956 and 1964.

Since 1964, the U.S. Congress has passed over 100 additional wilderness bills that have expanded the NWPS into other lands, including national parks. As of 2010, there were 756 federal wilderness areas in 44 states that collectively totaled 109,494,500 acres, or about 4.7 percent of the landmass of the United States.

See Also: Callicott, J. Baird; Environmental Law; Forest Preservation Laws; Leopold, Aldo; Preservation.

Further Readings

Allin, Craig W. *The Politics of Wilderness Preservation.* Westport, CT: Greenwood Press, 1982.

Callicott, J. Baird and Michael P. Nelson, eds. *The Great New Wilderness Debate.* Athens: University of Georgia Press, 1998.

Nelson, Michael P. and J. Baird Callicott, eds. *The Wilderness Debate Rages On: Continuing the Great New Wilderness Debate.* Athens: University of Georgia Press, 2008.

Wilderness Act of 1964, Public Law 88-577, 78 Stat. 890 (codified as amended at 16 U.S.C. §§ 1131-1136 [2000]).

Woods, Mark. "Wilderness." In *Companion to Environmental Philosophy,* Dale Jamieson, ed. Hoboken, NJ: Basil Blackwell, 2001.

Mark Woods
University of San Diego

Green Ethics and Philosophy Glossary

A

Adaptation: An evolutionary process by which organisms become better suited to their habitat over many generations, through gradual change. The process by which heritable traits that benefit the organism become more common throughout the population over many generations is called adaptive selection or natural selection.

Agriculture: The production of goods, including food, through the cultivation of crops and the herding of livestock, often also encompassing the breeding and hybridization of plants and the selective breeding of animals; one of the fundamental human enterprises. The combination of agriculture and cooking (which maximized the nutrients available to humans) enabled the creation of the earliest human civilizations, as food surpluses made possible the dense populations and non-nomadic lifestyles of ancient cities.

Alternative Fuels: Various possible substitutes for the fossil-fuel-based gasoline and diesel used to power most motor vehicles; includes biofuels (see below), alcohol-based fuels, and mixtures of fossil fuels with other fuels.

Anthropocentric: Literally, "human centered." Anthropocentrism is the belief, taken for granted in most cultures through most of human history and argued more explicitly by some schools of philosophy today, that humans are the figurative "center of the universe" and that ethical systems should thus be principally concerned with human benefit.

Anthropogenic: Man-made; used especially to underscore the human origins of a substance or phenomenon, as in "anthropogenic climate change" or "anthropogenic toxic compounds."

B

Biodiversity: The total variety of life on Earth. Modern science considers biodiversity to be an inherently good thing for the ecosystem and the loss of species and of species diversity to be an alarming consequence of environmental damage. From an evolutionary standpoint, genetic diversity—the diversity of genes within a species—is also especially important.

Bioethics: The philosophical study of ethical issues arising in the medical field and the biological sciences. Bioethics, concerned with issues that arise as a result of scientific advances, is sometimes contrasted with medical ethics, the questions of which may be centuries old or older. Other usages may emphasize medical ethics as applied ethics for a particular profession, while bioethics can encompass more abstract questions. The distinction is more a matter of the speaker's preference than a formal one. Both studies include the political and legal dimensions of their issues.

Biofuel: Fuel derived from biological material (biomass), including alcohol, hydrogen gas, excreta, and plant materials (which are usually processed into some sort of burnable fuel).

By-Product: Any material other than the product that is generated as a consequence of a process or as a breakdown product.

C

Carbon Offsets: Financial instruments, expressed in metric tons of carbon dioxide equivalent, which represent the reduction of carbon dioxide or an equivalent greenhouse gas. Carbon offsets allow corporations and other entities to comply with caps on their emissions by purchasing offsets to bring their totals down to acceptable levels. The smaller voluntary market for carbon offsets exists for individuals and companies that purchase offsets in order to mitigate their emissions by choice. There is a great deal of controversy over the efficacy and truthfulness of the offsets market, which is new enough that, in a best-case scenario, the kinks have not yet been worked out, while in the worst, it will turn out to be a dead end in the history of environmental reform.

Chlorofluorocarbons (CFCs): A family of inert nontoxic chemicals used in a variety of applications including refrigeration and air conditioning, solvents, and aerosol propellants. Because CFCs are not destroyed in the lower atmosphere, they drift into the upper atmosphere, where their chlorine components destroy ozone. Also see fluorocarbons, of which CFCs are a subtype.

Climate Change: Increasingly, a term preferred over "global warming," which refers to only one phenomenon. Climate change refers to all changes in the properties of the climate system over a long period of time (an unseasonably warm summer or an El Niño period do not constitute climate change) and may be used in reference to a smaller area than the whole Earth.

Consequentialism: An ethical framework in which the consequences of an action determine its moral value: things are right or wrong according to the effects they have. Utilitarianism is the best-known articulation of consequentialism, but altruism is essentially a consequentialist position, and not all consequentialisms need to follow utilitarianism's goal of maximized pleasure and minimized pain. Naturally, consequentialism's greatest weakness is that it is not always easy, or possible, to predict the consequences of an action—but note that this weakness cannot invalidate consequentialism's moral stance; it simply implies a morally difficult universe.

Conservation: The protection and management of a thing; in green contexts, usually refers to the conservation movement, which predates the environmentalist movement and is concerned chiefly with the sustainable use of natural resources by humans rather than the protection of the ecosystem for its own sake.

Cornucopian: The position that environmental crises—including climate change, finite nonrenewable resources, water/air/soil pollution, and problems caused by population growth or finite carrying capacity—either do not exist or will be solved by the combination of technological advances and free market forces.

Cost-Benefit Analysis: A process that weighs the expected costs of an action against the expected benefits in order to determine the best possible action.

D

Deep Ecology: The belief, first formulated by Norwegian philosopher Arne Naess, that human identity should be conceived of within a larger framework called the ecological self; deep ecology was conceived as an ecological ideology that would emphasize the common ground among people of different cultural and spiritual backgrounds and has led to a social movement of the same name.

Distributive Justice: An allocation of goods and resources within a society that is socially just; distributive justice is often, but not always, an explicit goal of consequentialist approaches and is a concept under much consideration (if not always named) in any discussion of globalization.

E

Ecocentrism: Any nature-centered belief system; contrast with anthropocentrism. Ecocentrism emphasizes the agency of the human species in the damage to its environment and holds sustainability and the health of the ecosystem as goods unto themselves.

Ecofeminism: A sociopolitical philosophy that argues that ecological destruction and global inequality are the inevitable outcomes of the unequal power distribution between the global north and global south (see north–south), between humans and nature, and between men and women.

Egalitarianism: The belief that all people should be treated as equals, or that all economic inequalities should be remedied. Egalitarianism can be articulated in many different ways, as the other political beliefs of the speaker impact his views on the cause and nature of these inequalities.

Emission: The discharge into the atmosphere from exhausts, vents, smokestacks, chimneys, surface areas, and so forth, typically used when such discharge constitutes pollution.

Environmental Justice: The equitable distribution of the impact of environmental activity among the populations of the Earth. Proponents of environmental justice point to the frequency with which the owners of wealthy corporations guilty of environmental damage live far from the places where that damage occurs, or can afford the remedies to that damage, thus insulating themselves from the consequences of their actions.

Ethics: The study of moral questions. Ethics can refer to specific types of ethics (such as applied ethics or medical ethics) or to specific systems of ethics (such as Catholic ethics or Marxist ethics). Though the religions of the world always include an ethical dimension to their belief systems, ethics and religion are not coequal, and the term *secular ethics* is sometimes used to describe systems of ethics that derive their conclusions from logic or moral intuition rather than from religious teachings or revealed truths. Secular ethics and religious ethics can and often do reach the same conclusions and may do so by the same means; there are both secular and religious articulations of utilitarianism, for instance. Major types of ethics include descriptive ethics (which describes the values people live by in practice), moral psychology (the study of how moral thinking develops in the human species), and applied ethics (which addresses the ethical concerns of specific real-life situations and putting ethics into practice).

Etiquette: The rules of proper social behavior, either received (as "unwritten rules") or prescribed by popular authorities such as Miss Manners and other guides to polite behavior. While ethics govern moral behavior, etiquette is simply the avoidance of appearing rude, crude, or offensive; while this may seem trivial in comparison to weighty moral issues, matters of etiquette govern a great deal of day-to-day behavior and often have greater implications. For instance, American etiquette encourages the shaking of hands and other casual physical contact between acquaintances but is uncomfortable with the idea of wearing face masks (which conceal the expression); the combination of these points of etiquette has a significant impact on the spread of contagion in the United States.

F

Fallibilism: The doctrine that all claims of knowledge could be wrong (that is, fallible), that things we believe we know could be proved wrong by later evidence, and in its most robust articulations, that even truths are impossible to prove with absolute certainty.

Fluorocarbons (FCs): Organic compounds analogous to hydrocarbons in which one or more hydrogen atoms are replaced by fluorine. Primarily used in coolants and industrial processes.

Footprint: A figurative term for the impact of an activity or entity: The ecological footprint of a thing is its total environmental impact, while the carbon footprint is the total carbon emissions released by its activity.

Free Market Fundamentalism: The belief that free market forces are the best remedy to economic and social problems. Though the "fundamentalism" of the term is usually figurative, in the wake of the global financial crisis of the early 21st century, religious free market fundamentalists who combine strict laissez-faire economics with strict literalist interpretations of Christian scripture have become more prominent, interpreting Adam Smith's invisible hand (a metaphor for the market's self-regulation) rather more literally.

G

Gaia Hypothesis: A view of the Earth in its entirety as a single living superorganism.

M

Moral Absolutism: In contrast with consequentialism or robust expressions of moral relativism (which see), holds that an action can be inherently good or bad and that neither context nor consequence affects this.

Moral Nihilism: The view that nothing is either moral or immoral. Though some schools of thought are legitimately nihilist, it is very often used as an exaggerated criticism: for instance, moral absolutists may accuse a philosopher of nihilism who says that an action is not inherently right or wrong but must be considered in terms of its consequences and context.

Moral Relativism: The acknowledgment that different cultures have different moral standards. There are various levels of moral relativism, from the weak descriptivist articulation that simply acknowledges and describes those differences to the normative position that says that there is no universal moral standard, only culturally derived morals. Moderate positions often propose that there are certain key moral standards that form a universal ethical core such as taboos on murder, incest, or parental neglect. The question of which moral standards are universal becomes important when cultures deal with one another and when international bodies mediate between them. Most differences are not about matters as obvious or seemingly clear-cut as murder but may instead bear on matters of justice or on the distribution of responsibility: the questions of who has the responsibility to do something about climate change or of the ethical importance of avoiding polluting behaviors vary widely around the world. The opposite of relativism is universalism.

Moral Skepticism: A framework of belief that claims that no one has moral knowledge—that no one confidently knows right from wrong, in some ways the most robust case of normative moral relativism. The strongest articulations of moral skepticism say that not only does no one possess moral knowledge but such knowledge is impossible.

Moral Syncretism: The attempt to reconcile differing moral beliefs. Though moral syncretism descends in some sense from the efforts of early Christian missionaries to emphasize the similarities among the cultures they encountered and to find the ways that Christianity would suit those cultures, it is predicated on the certainty that religion cannot be the sole arbiter of moral truths.

N

Normative: Describes how a thing ought to be. A normative statement describes what should be done, regardless of what is done. Normative ethics are concerned with how people ought to behave and what actions they ought to take. "Alcohol intake impairs judgment" is a descriptive statement; "drunk driving is wrong" is a normative statement.

North–South: A model of the world that contrasts the industrialized, developed, wealthy countries of the global north with the developing, poorer countries of the global south. Geography here is partially figurative, with Australia and New Zealand included in the global north, and a number of African, Middle Eastern, and Asian nations in the Northern Hemisphere included in the global south. The term became popular in the wake of the Cold War, when a new way of distinguishing between the developed (First and Second Worlds) and developing (Third World) was desired. However, while there are many political and cultural ties between the nations of the global north, the global south—much like the Third World—is varied enough to invite criticism of the model's accuracy and usefulness.

P

Polychlorinated Biphenyls (PCBs): A group of toxic chemicals used in electrical transformers and capacitors as insulators and in gas pipeline systems as lubricants. Because of their toxicity, the sale and new use of PCBs is banned by law in the United States and many other countries.

S

Sick Building Syndrome: Various ailments associated with a workplace or residence, usually due to poor air quality as a result of allergenic or toxic molds, flaws in HVAC systems, insufficient fresh air (and too much recirculation of the same air throughout the building), and chemicals used in workplace activities and their resulting emissions.

Spandrel: In biology, an observable characteristic that is a by-product of the evolution of a different characteristic, not a product of adaptive selection. The term was coined by Stephen Jay Gould and Richard Lewontin in their landmark paper "The Spandrels of San Marco and the Panglossian Paradigm: A Critique of the Adaptationist Programme," which faulted adaptationists in evolutionary biology for attributing seemingly every feature of any organism to adaptive selection, when many such features must have resulted indirectly. In recent years, both religion and the faculty for language have been suggested as possible spandrels in the human species: characteristics of the organism that resulted as a by-product of some other needed characteristic's development.

Strip Mining: A machine process that scrapes soil and rock away from mineral deposits just beneath the surface of the Earth, most commonly used to mine tar sand or coal. Similar surface mining methods include open-pit mining, dredging, and mountaintop removal. All such methods are criticized for their impact on the environment, on water resources, on the health of mine workers, and on aesthetics.

Sustainability: The capacity to continue; for instance, the capacity for a process to continue without using up the resources it consumes or for a species or an ecosystem to endure.

U

Utilitarianism: A form of consequentialism (which see) according to which the moral value of an action is determined by the total pleasure (or happiness) that results from it. The term *utilitarian,* meaning a pragmatic approach focused on "the bottom line," does not necessarily connote a stance of utilitarianism, which is defined by its very specific goal of "greatest happiness" and its denial of moral absolutism.

W

Waste: Unwanted materials left over from a process; see also by-products.

Bill Kte'pi
Independent Scholar

Green Ethics and Philosophy Resource Guide

Books

Allen, Adriana and Nicholas You. *Sustainable Urbanisation: Bridging the Green and Brown Agendas*. London: University College, Development Planning Unit, 2002.

Allen, Catherine and George Henry Stankey, eds. *Adaptive Environmental Management: A Practitioner's Guide*. New York: Springer, 2009.

Allin, Craig W. *The Politics of Wilderness Preservation*. Westport, CT: Greenwood Press, 1982.

Attfield, Robin. *Environmental Philosophy: Principles and Prospects*. Aldershot, UK: Avebury, 1994.

Attfield, Robin. *The Ethics of Environmental Concern*. New York: Columbia University Press, 1983.

Beatley, Timothy. *Green Urbanism: Learning From European Cities*. Washington, DC: Island Press, 2000.

Berry, Wendell. *The Unsettling of America: Culture and Agriculture*. New York: Avon, 1977.

Brinkley, Douglas. *The Wilderness Warrior: Theodore Roosevelt and the Crusade for America*. New York: HarperCollins, 2009.

Buell, Lawrence. *The Environmental Imagination: Thoreau, Nature Writing, and the Formation of American Culture*. Cambridge, MA: Belknap Press, 1995.

Callicott, J. Baird and Michael P. Nelson, eds. *The Great New Wilderness Debate*. Athens: University of Georgia Press, 1998.

Carson, R. *Silent Spring*. Greenwich, CT: Fawcett, 1962.

Chappell, T. D. J. *The Philosophy of the Environment*. Edinburgh, UK: Edinburgh University Press, 1997.

Clayton, Susan and Gene Myers. *Conservation Psychology: Understanding and Promoting Human Care for Nature*. Oxford, UK: Wiley-Blackwell, 2009.

Cohen, M. P. and S. Club. *The History of the Sierra Club, 1892–1970*. New York: Random House, 1988.

Crocker, David A. and Toby Linden, eds. *Ethics of Consumption: The Good Life, Justice, and Global Stewardship*. Lanham, MD: Rowman & Littlefield, 1997.

Croly, Herbert. *The Promise of American Life*. Cambridge, MA: Belknap Press of Harvard University Press, 1965.

Daily, Gretchen C. and Katherine Ellison. *The New Economy of Nature: The Quest to Make Conservation Profitable*. Washington, DC: Island Press, 2002.

De-Shalit, Avner. *Why Posterity Matters*. London: Routledge, 1995.

Devall, Bill. *Simple in Means, Rich in Ends: Practicing Deep Ecology*. Layton, UT: Gibbs Smith, 1988.

Evans, L. T. *Feeding the Ten Billion: Plants and Population Growth*. Cambridge, UK: Cambridge University Press, 1998.

Ferriss, Susan and Ricardo Sandoval. *The Fight in the Fields: Cesar Chavez and the Farmworkers Movement*. New York: Harcourt Brace, 1997.

Freeman, R. Edward, Jessica Pierce, and Richard H. Dodd. *Environmentalism and the New Logic of Business: How Firms Can Be Profitable and Leave Our Children a Living Planet*. New York: Oxford University Press, 2000.

Freeman, R. Edward, Jessica Pierce, and Richard H. Dodd. *Shades of Green: Business Ethics and the Environment*. New York: Oxford University Press, 1995.

Galbraith, John Kenneth. *The Affluent Society*. Cambridge, MA: Riverside Press, 1958.

Gaston, Kevin J. and John I. Spicer. *Biodiversity: An Introduction*, 2nd ed. Malden, MA: Blackwell, 2004.

Graham, F. *Since Silent Spring*. Boston: Houghton Mifflin, 1970.

Guha, Ramachandra. *How Much Should a Person Consume? Environmentalism in India and the United States*. Berkeley: University of California Press, 2006.

Gunningham, Neil A., Robert Allen Kagan, and Dorothy Thornton. *Shades of Green: Business, Regulation, and Environment*. Palo Alto, CA: Stanford University Press, 2003.

Hawken, Paul, Amory Lovins, and L. H. Lovins. *Natural Capitalism: Creating the Next Industrial Revolution*. Boston: Little Brown, 1999.

Hayward, Tim. *Ecological Thought*. Cambridge, UK: Polity Press, 1995.

Hill, David, Matthew Fasham, Graham Tucker, Michael Shewry, and Philip Shaw, eds. *Handbook of Biodiversity Methods: Survey, Evaluation and Monitoring*. Cambridge, UK: Cambridge University Press, 2005.

Holling, C. S., ed. *Adaptive Environmental Assessment and Management*. Toronto, Canada: Wiley Interscience, 1978.

Janick, Jules. *Classic Papers in Horticultural Science*. Englewood Cliffs, NJ: Prentice Hall, 1989.

Jefferson, Thomas. *Notes of the State of Virginia*. New York: Penguin, 1999.

Jennings, Francis. *The Ambiguous Iroquois Empire: The Covenant Chain Confederation of Indian Tribes With English Colonies*. New York: W. W. Norton, 1984.

Levin, Simon A., ed. *Encyclopedia of Biodiversity*. 5 vols., 2nd ed. Amsterdam: Elsevier, 2007.

Levine, E. *Rachel Carson: A Twentieth-Century Life*. New York: Viking, 2007.

Long, B. L. *International Environmental Issues and the OECD 1950–2000: An Historical Perspective*. Paris: Organisation for Economic Co-operation and Development (OECD), 2000.

Makower, Joel. *Strategies for the Green Economy: Opportunities and Challenges in the New World of Business*. New York: McGraw-Hill, 2009.

McKibben, Bill. *Deep Economy: The Wealth of Communities and a Durable Future*. New York: Times Books, 2007.

McManis, Charles, ed. *Biodiversity and the Law: Intellectual Property, Biotechnology, and Traditional Knowledge*. London: Earthscan, 2007.
Meffe, Gary K., et al., eds. *Ecosystem Management: Adaptive, Community-Based Conservation*. Washington, DC: Island Press, 2002.
Miller, Dan, ed. *Acknowledging Consumption: A Review of New Studies*. New York: Routledge, 1995.
Mumford, Lewis. *The City in History: Its Origins, Its Transformations, and Its Prospects*. New York: MFJ Books, 1961.
Nash, Roderick Frazier. *Wilderness and the American Mind*, 4th ed. New Haven, CT: Yale University Press, 2001.
Nelson, Michael P. and J. Baird Callicott, eds. *The Wilderness Debate Rages On: Continuing the Great New Wilderness Debate*. Athens: University of Georgia Press, 2008.
Nichols, Alex and Charlotte Opal. *Fair Trade: Market-Driven Ethical Consumption*. London: Sage, 2005.
Oelschlaeger, M. *The Idea of Wilderness*. Binghamton, NY: Vail-Ballou, 1991.
Orr, D. W. *Earth in Mind: On Education, Environment, and the Human Prospect*. Washington, DC: Island Press, 2004.
Pybus, C. and R. Flanagan. *The Rest of the World Is Watching: Tasmania and the Greens*. Sydney: Pan Macmillan, 1990.
Rainbow, Stephen. *Green Politics*. Oxford, UK: Oxford University Press. 1993.
Regan, Tom. *The Case for Animal Rights*. Berkeley: University of California Press, 1983.
Rensenbrink, John. *Against All Odds: The Green Transformation of American Politics*. Raymond, ME: Leopolog Press, 1999.
Roe, Dilys and Joanna Elliott, eds. *The Earthscan Reader in Poverty and Biodiversity Conservation*. London: Earthscan, 2010.
Rolston, H. *Environmental Ethics: Duties to and Values in the Natural World*. Philadelphia: Temple University Press, 1988.
Roosevelt, Theodore. *Theodore Roosevelt: An Autobiography*. New York: Macmillan, 1913.
Schweitzer, Albert. *Civilization and Ethics*. London: Adam and Charles Black, 1946.
Scott, James. *Seeing Like a State: How Certain Schemes to Improve the Human Condition Have Failed*. New Haven, CT: Yale University Press, 1998.
Shiva, Vandana. *Biopiracy: The Plunder of Nature and Knowledge*. Cambridge, MA: South End Press, 1997.
Singer, Peter. *Animal Liberation: A New Ethics for Our Treatment of Animals*. New York: New York Review of Books, 1976.
Spretnak, Charlene and Fritjof Capra. *Green Politics*. Santa Fe, NM: Bear and Co., 1986.
Taylor, Paul W. *Normative Discourse*. Upper Saddle River, NJ: Prentice Hall, 1961.
Taylor, Paul W. *Principles of Ethics: An Introduction*. Encino, CA: Dickenson, 1975.
Taylor, Paul W. *Respect for Nature: A Theory of Environmental Ethics*. Princeton, NJ: Princeton University Press, 1986.
Ten Have, Henk A. M. J. *Environmental Ethics and International Policy*. Paris: United Nations Educational, Scientific and Cultural Organization (UNESCO) Publishing, 2006.
Wackernagel, M. and W. E. Rees. *Our Ecological Footprint: Reducing Human Impact on the Earth*. Gabriola Island, British Columbia, Canada: New Society, 1996.

Walters, Carl J. *Adaptive Management of Renewable Resources*. New York: Macmillan, 1986.

Westra, Laura and Patricia H. Werhane, eds. *The Business of Consumption: Environmental Ethics and the Global Economy*. Lanham, MD: Rowman & Littlefield, 1998.

Williams, Byron K., et al. *Adaptive Management: The U.S. Department of the Interior Technical Guide*. Washington, DC: U.S. Department of the Interior, 2007.

Wilson, Edward O., ed. *Biodiversity*. Washington, DC: National Academy Press, 1988.

World Commission on Environment and Development. *Our Common Future*. New York: Oxford University Press, 1987.

Zimmerman, M. E. *Environmental Philosophy: From Animal Rights to Radical Ecology*. Englewood Cliffs, NJ: Prentice-Hall, 1993.

Journals

Forest Products Journal

Journal for Nature Conservation
Journal of Applied Philosophy
Journal of Forestry
Journal of Heritage Tourism
Journal of International Wildlife Law and Policy
Journal of Materials Processing Technology
Journal of Solar Energy Engineering
Journal of the History of Ideas
Journal of the Institute of Conservation
Journal of Transdisciplinary Environmental Studies
Journal of Water Resources Planning and Management

South African Journal of Philosophy

Websites

Center for Environmental Philosophy
www.cep.unt.edu

Earth Charter
www.earthcharter.org.au

Environmental Movement Timeline: A History of the American Environmental Movement
www.ecotopia.org/ehof/timeline.html

Environmental Protection Agency
www.epa.gov

Global Greens
www.globalgreens.org

Green Part of the United States
www.gp.org

International Association for Environmental Philosophy
www.environmentalphilosophy.org

International Society for Environmental Ethics
www.cep.unt.edu/ISEE.html

Leave No Trace
www.lnt.org

Pew Center on Global Climate Change
www.pewclimate.org

Sierra Club
www.sierraclub.org

Slow Food Movement
www.slowfoodusa.org

United Nations Environment Programme
www.unep.org

The Wilderness Society
www.wilderness.org

Green Ethics and Philosophy Appendix

Caring for God's Creation

http://www.nccbuscc.org/sdwp/ejp

This is the Website of the Environmental Justice Program (EJP) of the United States Conference of Catholic Bishops. The program, founded in 1993 and located within the Department of Social Development and World Peace, aims to "educate and motivate Catholics to deeper reverence and respect for God's creation, and to engage parishes and dioceses in activities aimed at dealing with environmental problems, particularly as they affect the poor." The EJP also acts as a resource for Catholic dioceses and conferences and maintains close ties with the National Religious Partnership for the Environment. The EJP focuses its activities on four main areas: scholarship, leadership development, public policy and advocacy, and special projects. Resources available from the Website include an overview of EJP policies and positions and the ethical reasoning that ties their values and activities to Catholicism, public policy, an archive of relevant documents, current campaigns, news about the EJP and about domestic and international issues, and key articles including "An Ecological Spirituality" by the Reverend Joseph A. Tetlow, S.J., and "The Good Life From a Catholic Perspective: The Problem of Consumption" by Monsignor Charles Murphy.

The Center for Environmental Philosophy

http://www.cep.unt.edu

This Website, maintained by the University of North Texas, is an excellent source of basic information about environmental philosophy and ethics (including the history of the subject as an academic field) and provides many links to other sources of information. It includes a cumulative index to the journal *Environmental Ethics* and other information including instructions to authors and ordering information. The Website also includes an annotated bibliography of books relevant to environmental philosophy, photo and video presentations, links to relevant graduate programs and associations, funding opportunities, a searchable, annotated database of books and articles relevant to environmental ethics (also available as a PDF file), a collection of syllabi for courses in environmental philosophy and related subjects (viewable by course title, instructor, region, school, and textbook), and links to other sources of information available on the Internet.

Coalition on the Environment and Jewish Life: Protecting Creation, Generation to Generation

http://www.coejl.org/index.php

This is the official Website of the Coalition on the Environment and Jewish Life (COEJL), founded in 1993 to catalyze "a distinctively Jewish programmatic and policy response to the environmental crisis." The COEJL's mission includes partnering with Jewish organizations to integrate environmental stewardship into Jewish life, bringing environmental education and opportunities for environmental action to Jewish people, bringing a Jewish vision and voice to issues of sustainability and environmental justice, and participating in civic coalitions and interreligious efforts to protect the environment. The Website includes information about the COEJL programs including the Jewish Environmental Leadership Institute, the COEJL Environmental Policy Platform, the Green Synagogues program, the COEJL Global Climate Change and Energy Campaigns, the Jewish Global Environmental Network, and links to other Jewish organizations working on environmental issues. The Website also includes resources for those who are interested in starting a Jewish environmental organization or who are interested in the topic as well as information about ordering COEJL publications.

Environmental Justice

http://www.epa.gov/environmentaljustice

This Web page, part of the U.S. Environmental Protection Agency (EPA), defines environmental justice as "the fair treatment and meaningful involvement of all people regardless of race, color, national origin or income with respect to the development, implementation, and enforcement of environmental laws, regulations and policies." The Web page includes an FAQ page, documents, and publications about environmental justice; a searchable bibliography about environmental justice; a database that can be searched by geographic location for facilities that have been inspected for air, water, and hazardous waste compliance; an interface to search for information specific to different regions of the United States; and the Environmental Justice Geographic Assessment Tool that allows users to identify geographic areas that may suffer increased exposure to environmental harm. The Website also offers information about the National Environmental Justice Advisory Council, the Federal Interagency Working Group on Environmental Justice, grants and programs of the EPA related to environmental justice, the National Achievements in Environmental Justice Awards Program, and links to multimedia resources including podcasts, photos, and video.

Greenpeace International

http://www.greenpeace.org/international

This is the Web page of the global nongovernmental organization, founded in 1971, which focuses on research, lobbying, and direct action to further its goal to "ensure the ability of the earth to nurture life in all its diversity." The Website includes information about Greenpeace's many activities organized by chief area of concern including global warming, energy, oceans, forests, nuclear, toxics, and sustainable agriculture and genetic engineering. The Website also contains a media center including official Greenpeace statements on various issues; downloadable reports, factsheets, toolkits, and other information relevant to Greenpeace campaigns; a catalog of images and videos; information for those who wish to become involved with Greenpeace; links to national Greenpeace Websites; and a link to a downloadable copy of the Greenpeace Code of Ethics (http://www.greenpeace.org/usa/press-center/reports4/greenpeace-code-of-ethics). The Website also offers access to several Greenpeace blogs (focused on campaign, grassroots, and community actions) and the opportunity to sign up for e-mail alerts of Greenpeace news and action alerts.

Peter Singer

http://www.princeton.edu/~psinger

This is the official Princeton University Web page of Peter Singer, Ira W. DeCamp Professor of Bioethics at Princeton and professor laureate at the Centre for Applied Philosophy and Public Ethics at the University of Melbourne. Singer is perhaps the best known of modern philosophers and is recognized for taking controversial (but well-reasoned and defended) positions on many questions related to bioethics. Singer laid the foundations of the animal liberation movement with his 1975 book *Animal Liberation,* which introduced the term *speciesism* to refer to the common practice of valuing humans more than other animals. In his 2009 book *The Life You Can Save: Acting Now to End World Poverty,* Singer argues that there is a moral imperative for people living comfortably in affluent nations to donate some of their income to help raise people in less fortunate countries out of poverty. Singer has also taken well-known positions on other controversial subjects such as abortion, euthanasia, bestiality, and vegetarianism. The Princeton Website includes an FAQ page on his philosophical positions, his curricula vitae (CV), descriptions of current research, his speaking schedule, and links to resources including many articles written by Singer that are available online at the official Websites of organizations that he belongs to or supports.

Sarah Boslaugh
Washington University in St. Louis

Index

Article titles and their page numbers are in **bold**.